Springer Series in Molecular Biology

Series Editor
Alexander Rich

Springer Series in Molecular Biology

Series Editor: ALEXANDER RICH

Kensal E. van Holde

Chromatin

With 138 Illustrations

Springer-Verlag
New York Berlin Heidelberg
London Paris Tokyo

Kensal E. van Holde
Department of Biochemistry and Biophysics
Oregon State University
Corvallis, OR 97331-6503
U.S.A.

Series Editor:
Alexander Rich
Department of Biology
Massachusetts Institute of Technology
Cambridge, MA 02139
U.S.A.

Library of Congress Cataloging-in-Publication Data
Van Holde, K.E. (Kensal Edward), 1928–
  Chromatin.
  (Springer series in molecular biology)
  Bibliography: p.
  Includes index.
  1. Chromatin. I. Title. II. Series.
QH599.V36 1988   574.87'32   87-36922

Typeset by David E. Seham Associates, Inc., Metuchen, New Jersey.
Printed and bound by Arcata Graphics/Halliday, West Hanover, Massachusetts.
Printed in the United States of America.

9 8 7 6 5 4 3 2 1

ISBN 0-387-96694-3 Springer-Verlag New York Berlin Heidelberg
ISBN 3-540-96694-3 Springer-Verlag Berlin Heidelberg New York

This book is dedicated
with respect and affection
to the memory of
Irvin and Cynthia Isenberg

Series Preface

During the past few decades we have witnessed an era of remarkable growth in the field of molecular biology. In 1950 very little was known of the chemical constitution of biological systems, the manner in which information was transmitted from one organism to another, or the extent to which the chemical basis of life is unified. The picture today is dramatically different. We have an almost bewildering variety of information detailing many different aspects of life at the molecular level. These great advances have brought with them some breath-taking insights into the molecular mechanisms used by nature for replicating, distributing, and modifying biological information. We have learned a great deal about the chemical and physical nature of the macromolecular nucleic acids and proteins, and the manner in which carbohydrates, lipids, and smaller molecules work together to provide the molecular setting of living systems. It might be said that these few decades have replaced a near vacuum of information with a very large surplus.

It is in the context of this flood of information that this series of monographs on molecular biology has been organized. The idea is to bring together in one place, between the covers of one book, a concise assessment of the state of the subject in a well-defined field. This will enable the reader to get a sense of historical perspective—what is known about the field today—and a description of the frontiers of research where our knowledge is increasing steadily. These monographs are designed to educate, perhaps to entertain, certainly to provide perspective on the growth and development of a field of science which has now come to occupy a central place in all biological phenomena.

The information in this series has value in several perspectives. It provides for a growth in our fundamental understanding of nature and the manner in which living processes utilize chemical materials to carry out a variety of activities. This information is also used in more applied areas. It promises to have a significant impact in the biomedical field where an understanding of disease processes at the molecular level may be the capstone which ultimately holds together the arch of clinical research and medical therapy. More recently in the field of biotechnology, there is another type of growth in which this science can be used with immense practical consequences and benefit in a variety of fields ranging from agriculture and chemical manufacture to the production of scarce biological compounds for a variety of applications.

This field of science is young in years, but it has already become a mature science. These monographs are meant to clarify segments of this field for the readers.

Cambridge, Massachusetts   Alexander Rich
                                                                Series Editor

Preface

A Greek legend recounts how Sisyphus was condemned to the eternal trial of pushing a heavy boulder to the top of a hill. Each time, before he could bring it to the summit, it would roll back down, and he had to begin anew. There have been many moments during the preparation of this monograph when I was reminded of his story. To attempt to write on a subject as broad as "chromatin," during a period of intense and revealing research, is to place oneself in the role of Sisyphus.

The project was begun in 1980, with the sanguine expectation that it might be completed in two years. But as new results appeared, it became necessary to revise nearly every chapter, not once but several times. The two years have now grown into eight.

It would be naive to assume that the pace of chromatin research has finally slowed, or that startling new developments may not soon appear. It does seem, however, that we have reached a point where the picture of chromatin *structure,* at least in its more elementary aspects, has clarified to the point where a coherent presentation is possible. Unfortunately, the same cannot be said of chromatin *function,* in particular with respect to the problems of transcription. Here, the field seems to be at the edge of a major breakthrough, and I fear that much of what I have written will soon be obsolete.

However, not being immortal, I have decided to present the book at this point. I suffer no delusions as to its permanence; indeed, even as this Preface is being written, new and important results are being published. I realize full well that some questions I present herein as unanswered will be resolved by the time the book comes to print and that some of my most cherished conclusions will be dashed to bits by clever experiments.

Given these lamentations, can *any* justification be found for the immense labor involved in such an enterprise? I believe that there is justification and that it goes beyond self-delusion.

First, such a book brings together a mass of *factual* information that has been scattered helter-skelter through the literature and allows comparisons that are not self-evident. Second, it permits a summing-up of our understanding of the field, at one point in time. Although in some cases the summation of current knowledge is only confusion, in others it is possible to make a reasonable stand on issues. I have not hesitated to do so. Last, but not least, a book like this should provide a *source* to give workers in the field ready access to much of the earlier work.

This latter point requires clarification. I have not attempted to include *all* references to each topic, even up to some arbitrary date. Indeed, my card files for the project were only about half-exhausted in compiling the bibliography to the text. I have tried to emphasize what I feel to be the seminal papers in each area and have made frequent references to reviews. One of the sobering lessons of such an enterprise is an appreciation of the fact that much of the scientific literature (including one's own contributions) is either trivial or questionable.

Much of this book was written in the library of the Marine Biological Laboratory, Woods Hole, Massachusetts. I wish to express my appreciation to the MBL and its benefactors for providing and sustaining such a remarkable resource.

In putting the book together, I have received wonderful cooperation from many scientists, both in providing illustrations and in thoughtful answers to queries. Their names are far too many to list here, but I thank them one and all. I would also like to express my appreciation to the publishers and to Dr. Alexander Rich, the Series Editor, for their patience during this long effort. Special thanks are due Dr. Don Olins for a careful reading and many helpful comments. And finally, my deepest appreciation to Barbara Hanson and Barbara van Holde, who typed the innumerable drafts with skill and patience.

Contents

1

The First Hundred Years

No area of scientific endeavor progresses along a smooth path from inception to resolution. All science is episodic, and most fields are characterized by what are seen, only in retrospect, to have been a series of advances and retreats. Such is the history of attempts to understand the structure and biological function of chromatin. In this brief first chapter, I shall attempt to summarize this tangled story, from the appearance of Miescher's first paper in 1871 to the moment, almost exactly one hundred years later, when a detailed picture of the molecular structure of chromatin finally began to emerge. Since the principal focus of this book will be on *current* knowledge of chromatin structure, I do not aim for an exhaustive history. For details, the historically minded reader is referred to a number of excellent books and reviews. I have found Olby's *The Path to the Double Helix* (1974) especially useful; but I would also recommend the concise summary by Mirsky (1968) in *Scientific American*. Views of the state of the field at various times can be gained by perusing the Cold Spring Harbor Symposium volumes for the 1941, 1947, and 1973 conferences, Volume 1 of the Symposia of the Society for Experimental Biology (1947), and the collection of papers, *The Nucleohistones*, edited by J. Bonner and P. Ts'o in 1964. However, the best source is, as always, the original literature.

Friedrich Miescher began his work in 1868, in the laboratory of F. Hoppe-Seyler at Tübingen. Isolating white cells from the pus adhering to bandages obtained from hospital patients, he carried out what appears to have been the first purification of nuclei. Such cells, which are easily isolated and lysed, were an excellent choice given the technical limitations of the time. The nuclei were extracted with slightly alkaline solutions, and the extract was then acid-precipitated. The resulting material, which Miescher found to contain a relatively high phosphorus content (~2.5%), he named *nuclein*. We would recognize it today as a crude chromatin preparation; the phosphorus content was well below that of pure DNA (~10%) or even purified chromatin (~4%). Very likely, considerable degradation of both nucleic acid and proteins had occurred, both from endogenous and exogenous enzymes; the isolation process was slow, and Miescher used a crude extract from pig stomach to help clean the nuclei of cytoplasmic debris. But the fact that the preparation was imperfect by modern standards is of little consequence; what is important is that Miescher recognized it as containing a unique substance (Miescher, 1871). The concept of "nuclein" as a definable and biochemically important material became well accepted in the following decades.

More significant for the work to follow, however, were Miescher's subsequent investigations of salmon sperm (Miescher, 1874). In this case, he extracted the sperm first with hydrochloric acid, removing a basic substance that he termed *protamin*. The acid-extracted sperm were then treated with dilute alkali to yield an acid-precipitable material that Miescher considered to be identical to the "nuclein" obtained earlier. The salmon sperm "nuclein," however, must have more nearly represented a true DNA preparation, since most of the basic protamine had been removed. This is supported by the reported phosphorus content—an average of 9.6%, which compares remarkably well with the presently accepted value of 9.5% for salmon sperm DNA. But perhaps the most significant point is that Miescher had now shown that the nuclear material consisted of a combination of acidic and basic substances.

The next major advance came in 1884, when Albrecht Kossel, also a student of Hoppe-Seyler, extended Miescher's studies to the nucleated erythrocytes of geese. Kossel isolated the nuclei and extracted them with dilute acid. This process yielded a preparation of basic proteins, but these were clearly different from Miescher's salmon sperm protamin. Kossel named these proteins *histon* (Kossel, 1884). Thus, by this early date both of the major components of chromatin had been isolated in reasonably pure form. There is a temptation, I believe, to make a bit too much of this precocity. It must be remembered that aside from elementary composition, virtually nothing was known about these substances. The concepts of proteins and nucleic acids as complex macromolecules lay far in the future. Furthermore, little more was even knowable, considering the limitations of the biochemical techniques of the time.

Although further advances in the analysis of chromatin *structure* were impossible at this date, major new discoveries in cell biology had opened the door to a remarkably clear intuition of the general features of chromatin *function*. Flemming's studies of mitosis and the work of Oskar Hertwig and others on fertilization in the sea urchin led to the understanding that it is the nucleus and in particular the chromosomes that are the repositories of genetic information. See Darlington (1966) for a discussion of these early hypotheses. A major reason for the connection, in the biologists' minds, between chromosomes (or chromatin) and nuclein arose from the development, in this same period, of cytological staining techniques. Flemming (1880) had given the name *chromatin* to the readily stainable material in nuclei. The circle was closed when Zacharias (1881, 1893a, 1893b) showed that the substance of chromosomes was probably identical to nuclein. Miescher's "nuclein" then became equivalent, or very similar to, Flemming's "chromatin." Therefore, even as early as 1885, Hertwig felt able to state:

I believe at the least to have made it very probable that the nuclein is the substance which not only fertilizes but also transmits the hereditary characteristics (Hertwig, 1885; for the translation see Voeller, 1968).

Thus, well before the turn of the century it was perceived that a major substance in the nucleus consisted of a complex composed of an acidic, phosphorus-rich component and a basic, proteinaceous component. Furthermore, it seems to have been widely held at that time that this complex was, or contained, the genetic material.

In the next decades, progress was rapid. Lilienfeld (1894) isolated calf thymus nuclei and obtained from them the histones and DNA; he coined the word *nucleohistone* for their complex. During the preceding twenty years, there had been some confusion over terminology. For example, Miescher and his contemporaries tended to use the term *nuclein* for what must have been both chromatin and DNA preparations. Although Kossel (1892) seems to have been the first to use the specific word *nucleinsäure*, the criteria for nucleic acid purity had already been clearly stated by Altman (1889), who prepared these substances from a variety of cell types. Clarification of these distinctions cleared the way for meaningful studies of the components of chromatin. In a remarkable series of papers published in the early 1900s, Huiskamp (1901a, 1901b, 1903) described many of the properties we now associate with chromatin: (1) solubility in low (<0.02 M) and high (1 M) salt, (2) precipitability by both isotonic salt solutions and by very low levels of divalent cations, and (3) the salt-like (we would say electrostatic) nature of the interaction between histones and nucleic acid.

Perhaps Huiskamp's most remarkable experiments (at least to modern eyes) were his studies of the "electrolysis" of nucleoprotein and histones (Huiskamp, 1901b). These were, in fact, primitive electrophoresis experiments in which he demonstrated that nucleoprotein migrated toward the anode, and thus carried a net negative charge, whereas isolated histones moved toward the cathode, and therefore were positive. From such experiments he was able to conclude:

Die Bindung des Histons am nuclein muss wahrscheinlich als eine saltzartige ausgesact werden, und zwar so, dass die Säurekomponente durch das Nuclein, der basisch Besandteil durch das Histon vertreten ist (Huiskamp, 1903).

The statement can stand with little modification at the present time.

Contemporary with Huiskamp's studies of the nucleic acid–protein complexes, rapid advances were also being made in the chemistry of the nucleic acids themselves. The major figure in this effort was P.A. Levene, working at the Rockefeller Institute. Between 1909, when he determined the structures of inosine and inosinic acid (Levene and Jacobs, 1909), and 1929, when he demonstrated that the sugar in DNA was 2-deoxyribose (Levene and London, 1929), Levene and his collaborators dominated the field. Unfortunately, this dominance also played a major role in the entrenchment of one of the erroneous *ideés fixes* that were to become so prevalent in this field—the "tetranucleotide hypothesis."

While the work of Levene and others had clearly established the presence of four bases in both DNA ("thymonucleic acid") and RNA ("yeast nucleic acid"), as well as the nature of the phosphodiester linkages between residues, the concept of a macromolecule still lay in the future. This was the golden age of colloid chemistry, and slowly diffusing or gelatinous substances were generally held to be colloidal aggregates of small molecules. Even as late as 1925, we have the account of H. Staudinger shouting to an audience of authorities derisive of his concept of long-chain molecules: "Hier stehe ich, ich kann nicht anders" (Olby, 1974). Thus, it is not surprising that the existence of four bases, in what seemed from incomplete data to be present in roughly equal quantities, suggested the idea of a cyclic or linear tetranucleotide as the unitary structure of the nucleic acids. The concept seems to have originated with Osborne and Harris (1902), but Levene's espousal of the idea gave it enormous weight. Nor was the concept seriously challenged by the attempts at physicochemical studies made in the first decades of the century. For example, Hammarsten (1924) carried out osmotic pressure measurements with DNA and obtained apparent molecular weights of about 800, only a bit *smaller* than anticipated for a tetranucleotide. There seem to be two likely explanations for results of this kind. First, many of the early nucleic acid preparations were highly degraded, and workers preferred to use the "soluble" component. Second, the behavior of polyelectrolytes, particularly with respect to the Donnan effect, was perhaps not fully appreciated.

The remarkable durability of the tetranucleotide concept is illustrated by the following anecdote: In 1936, Dorothy Wrinch published a chromosome model in which cyclic tetranucleotides were arrayed at right angles to the chromosome axis (Wrinch, 1936). The model was objected to, not because of the postulate of tetranucleotides, but because their orientation was inconsistent with the observed birefringence of chromosomes (see Olby, 1974, p. 116). It was not until the late 1930s that the application of such techniques as viscometry, flow birefringence, and the ultracentrifuge began to convince some biochemists that nucleic acids (and, by inference, chromatin) were indeed macromolecular substances (see, for example, Signer et al., 1938; Schmidt and Levine, 1938). But even then, the idea retained its grip. If DNA was not a tetranucleotide, then perhaps it was constructed of tetranucleotide repeating units. As late as 1947, one finds no less a luminary than W. Astbury arguing that X-ray diffraction data indicate "a sequence of nucleotides that is a multiple of four" (Astbury, 1947).

However, I think it can be argued that the tetranucleotide hypothesis was not nearly such an impediment as it may seem at first hindsight. After all, what was most needed in nucleic acid research in the early 1900s was the elucidation of the structures of the bases, nucleosides, and nucleotides, and how these were linked together. This was a task for organic chemists, and it would seem that the tetranucleotide model provided a convenient, finite framework. Had it been realized that nucleic acids were enormous molecules of complex sequence, the organic chemists of the day would likely have turned from them in horror. Even an erroneous model can be useful for a time. It was only in the later years that the tetranucleotide hypothesis became a straightjacket.

But even as the tetranucleotide hypothesis was slowly dying, another shift in belief was developing that would prove much more inhibiting to serious research on chromatin. Recall that as early as 1885 Hertwig had proclaimed "nuclein" to be the genetic material. This idea seems to have been accepted by virtually all researchers in the latter decades of the nineteenth century and the first decade or so of the twentieth. But with the continued development of cytological staining techniques, the impression grew that the nucleic acid content of the nucleus varied strangely during the cell cycle. Strasburger (1909) stated the view most specifically: "Chromatin cannot itself be the hereditary substance, [as] the amount of it is subject to considerable variation in the nucleus, according to its stage of development." Such observations, together with the increasing appreciation of the complexities of protein structure, led to the "protein theory" of the gene: Genes were proteins, and the nucleic acid served at most as an auxiliary substance, perhaps associated in some undefined manner with protein replication. The idea gained strength with the establishment of the tetranucleotide hypothesis. How could so simple and regular a structure as a tetranucleotide play an informational role? Even when physical studies began to convince many that DNA was a macromolecule, the old ideas carried on; the tetranucleotide was now thought of (without justification) as the building block of a structural polymer. The transition to the protein theory can be seen very clearly by comparing successive editions of E.B. Wilson's influential book *The Cell in Development and Heredity*. In the first edition (1896), Wilson strongly defends the idea that nucleic acids are the genetic material. The same statements stand in the second edition (1911), but by the third edition (1925), they have been removed. The physiological role of the components of chromatin is now seen to be ambiguous.

Remarkably, the protein theory received its strongest support from the most sophisticated techniques yet devised to investigate the contents of nuclei in situ: ultraviolet microspectrophotometry. Caspersson (1936) observed marked changes in the ultraviolet absorption of nuclei during the cell cycle; in some circumstances, the UV absorption actually decreased. It now seems likely that the experiments were confusing because the method did not distinguish between DNA and RNA, but the different roles of these substances were unknown at the time. In any event, Caspersson's experiments had enormous impact. Indeed it required nothing less than the bacterial transformation experiments of the 1940s to finally dislodge the protein–gene model.

Even after Avery et al. (1944) and a number of other workers had clearly demonstrated that the transforming principle was DNA and not protein, the concept of "protein genes" died hard. The difficulty of the demise can be observed in the Cold Spring Harbor Symposium of 1947. Two quotations from papers at that meeting characterize the debate. First, Stedman and Stedman:

There is probably general agreement that the essential component of the chromosome must belong to the proteins, for no other known class of naturally occurring chemical compounds would be capable of possessing the properties or existing in the variety necessary to account for their genetic function.

The conventional view could not be expressed more succinctly. Then, Boivin states very cautiously the new heresy:

In bacteria—and in all likelihood in higher organisms as well—each gene has as its specific constituent, not a protein but a particular desoxyribonucleic acid, which, at least under certain conditions, . . . is capable of functioning *alone* as the carrier of hereditary character.

In a sense, we are finally back to Hertwig and 1885! It should be added that the recorded discussion shows that Dr. Boivin's paper was greeted with evident skepticism by a number of authorities.

I believe that unlike the tetranucleotide hypothesis per se, the protein–gene hypothesis had a strongly inhibiting effect on chromatin research. Essentially, it subsumed the questions of chromatin structure and function under the seemingly hopeless problem of protein structure. It reduced the nucleic acids to the role of uninteresting support substances, with the consequence that relatively little attention was paid them over a period of several decades.

A telling blow against the tetranucleotide hypothesis came in 1950, when Chargaff (1950) showed that the four bases were *not* present in equal quantities in most DNA. Furthermore, differences in composition in different organisms argued for an informational role.

It might be thought that Watson and Crick's (1953) striking and elegant structure for DNA, with its boundless implications for nucleic acid function, would lay the protein–gene hypothesis to rest once and for all. But no; even in 1954 we find:

It will be recalled that in Furberg's zig-zag arrangement of the phosphate groups in DNA there are two suitably placed rows of phosphate groups, and it would be quite possible for one peptide chain to be 'anchored' on one row, while its copy was being formed on the other (Davison et al., 1954).

The concept is clearly that of a protein copying itself, with the nucleic acid acting as a passive framework on which the event could occur. In fairness to Davison et al., it should be added that they come out in the main for DNA as the genetic material. I cite the above passage only to illustrate how strong is the grip of established ideas.

While the 1940s and 1950s saw the gradual emergence of the nucleic acid theory of the gene, the accompanying studies of *chromatin* structure in this period were quite confused. Some of this confusion resulted from the development of a technique that employed 1 M NaCl for the extraction of "nucleohistone" from lysed cells (Mirsky and Pollister, 1942). This method avoided some of the technical difficulties of the older techniques of aqueous extraction and allowed easier studies of "chromatin" properties (see, for example Mirsky and Ris, 1947; Mirsky, 1947). Unfortunately, it had the disastrous effect of stripping many of the histones and other proteins from the solubilized DNA. Even though many of these proteins would rebind to the DNA when the salt concentration was subsequently diluted to 0.14 M to reprecipitate the nucleohistone, it is very likely that whatever "reconstitution" occurred was largely aberrant. Even more unfortunate was the practice of some of conducting physical studies on the supposed complex in concentrated salt. For example, J.A.V. Butler and L. Gilbert carried out sedimentation studies of "thymus nucleoprotein" in 10% NaCl, which are illustrated in Davison et al. (1954). Examination of these patterns shows what is almost surely a hyper-sharp DNA boundary followed by a rapidly diffusing protein boundary.

A number of workers (Cohen, 1945; Mirsky and Pollister, 1946, 1948; Peterman and Lamb, 1948; Stern, 1949) began to perceive the dangers of salt extraction. They were also, by this time, cognizant of the hazards arising from degradation by proteases and deoxyribonucleases when studying chromatin at low ionic strengths (see especially Cohen, 1945). To avert this, attempts were made to inhibit the enzymes. In 1946, the first purification of a deoxyribonuclease (from beef pancrease) was accomplished by McCarty (1946). He found the enzyme required magnesium and could be inhibited by complexing the magnesium ions with citrate. Accordingly, Peterman and Lamb utilized citrate in their studies, and Stern (1949) extracted chromatin in 0.002 M sodium azide or citrate solutions.

In the overall confusion of the period, a generally neglected paper by Steiner (1952) stands out as quite remarkable. Steiner prepared thymus chromatin (which he called "genoprotein") in low salt, with arsenate present as an enzyme inhibitor. He carefully centrifuged out large aggregates, obtaining a soluble product with a molecular weight of approximately 6×10^6 g/mol. This material was studied by sedimentation, light scattering, viscometry, and flow birefringence, as was the DNA prepared from it. He concluded that the chromatin particles were much less extended than the DNA and could best be described as rods about 2000 Å long by 80 Å in diameter. This is, in fact, very close to what would be expected for a polynucleosome of this size (see Chapter 7). His product appeared to have a protein/DNA ratio of about 1:1, a value in good agreement with modern estimates. In 1 *M* salt, the particles became more extended, and the weight average molecular weight dropped to about that of the nucleic acid component; this would be expected if proteins dissociated, since these low molecular weight molecules would contribute little to the weight average molecular weight. In short, Steiner appears to have had an excellent chromatin preparation and to have characterized it accurately. Unfortunately, the paper was enough ahead of its time that little could come from it. The main problem was that so little was then known about histones and other chromosomal proteins that a model for chromatin structure could not be constructed from Steiner's work. There had in fact been virtually no significant advances in the study of histones between the time of Kossel and Huiskamp and 1950. Indeed, in a volume summarizing Kossel's research (Kossel and Schenck, 1928) the view is expressed that histones probably exist only in the nuclei of certain types of tissue. This view seems to have been widely accepted and must have inhibited research on these proteins.

In 1951, the Stedmans published a paper that had considerable impact in renewing interest in the histones (Stedman and Stedman, 1951). In the first place, it seems to represent the first attempt to isolate and compare histones from a variety of tissues. Second, it describes the first attempts at histone fractionation. And finally, the Stedmans put forth a hypothesis of histone function that was both to spark renewed interest in these proteins and to serve as a working model for nearly two decades. The Stedmans found that histones which had been acid-extracted from different types of nuclei could in each case be fractionated into two components. These they referred to as the "main" and "subsidiary" histones, respectively. The latter appear to have approximately corresponded to what we now refer to as the "very lysine rich" histones, the former to a mixture of the remaining histones. Most important, the Stedmans realized that these two components were themselves heterogeneous and suggested that the specific distribution of histone types might be both organism- and tissue-specific. On this basis, they postulated that the histones could play a role in gene regulation; that is, different local distributions of histones might allow the expression of different genes, which in turn could differentiate cell types. The proposal is remarkable in two respects. First, it marks a final break with the protein–gene theory, which the Stedmans had been defending only four years earlier (see above). Implicit is the idea that DNA is the genetic material; the proteins are now considered to be regulators. Second, it was wrong; the data in fact provided a flimsy basis for such a sweeping hypothesis. Most of the experimental results consisted of analyses for total nitrogen, amide nitrogen, arginine, and tyrosine. The difference between organisms or cell types was in most cases far from spectacular.

It seems probable to me that the Stedman hypothesis gained such wide acceptance because subsequent studies in many laboratories soon confirmed and extended the concept of histone heterogeneity. The next two decades saw intensive activity in the fractionation and amino acid analysis of histones from a number of sources. The situation about ten years after the Stedmans' paper is best appreciated by perusal of the first few chapters of the book *The Nucleohistones* (Bonner and Ts'o, 1964). The histone field was, in a word, confused, partly because of the diversity of isolation and fractionation methods employed and partly because of the inadequacy of characterization and identification techniques. Unlike enzymes, histones possess no catalytic activity to aid in their identification. In 1964, it was by no means clear how many histones there were, or how extensive interspecies differences might be. Numerous statements in *The Nucleohistones* imply that most workers at this time believed there to be *many* histones.

Real progress in histone research came only after the development of chromatographic fractionation methods and gel electrophoresis techniques in the late 1950s (see Crampton et al., 1957, and Luck et al., 1958, for early applications). The subsequent use of these procedures to develop systematic methods for the fractionation and identification of histones can be largely credited to E.W. Johns and co-workers. In a long series of papers (see, for examples, Johns et al., 1960; Johns and Butler, 1962; Johns, 1964; Philips and Johns, 1965), they refined the techniques to the point where the presently recognized classes of histones were revealed. The progress of this analysis, starting from the "main" and "subsidiary" histones of Stedman and Stedman (1951) is schematized in Table 1-1. Also listed here are the numerous changes in nomenclature that have at times confused the field. For details concerning the properties of these histone classes, the reader is referred to Chapter 4.

It is not always easy to identify the "fractions" obtained by the earlier workers with the currently recognized histone classes. For example: While it seems possible from their aggregation properties that the "β histones" of Cruft et al. (1954) may have been an H3/H4 mixture and the "γ histone" an H2A/H2B mixture, insufficient data are available for certainty. However, it is clear that the distinct properties of the "very lysine rich" histones were recognized at an early date. In their 1962 paper, Johns and Butler showed that this component could be isolated separately by perchloric acid extraction.

In any event, by 1965 all of the major categories of histones recognized today had been identified. The fact that only a few classes were found cast some doubt upon the Stedmans' hypothesis.

At about this time, there began direct attempts to assess the role of

Table 1-1. Resolution of the Histone Classes: Histone Nomenclature

Kossel (1884)	Stedman & Stedman (1951)	Cruft et al. (1954)	Johns et al. (1960)	Johns & Butler (1962)	Philips & Johns (1965)	Current nomenclature	Other nomenclatures
					F2c →	H5	V .KAP
	Subsidiary histones →	α →	f1 →	f1 →	F1 →	H1	I
				f2(b) →	F2b →	H2B	IIb2 .KAS
					F2a2 →	H2A	IIb1 .LAK
Histon	Main histones	β, γ	f2, f3 →	f2(a) →	F2a1 →	H4	IV .GRK
				f3 →	F3 →	H3	III .ARE

histones as transcriptional regulators. Such experiments monitored in vitro RNA synthesis, using isolated chromatin as a template. Bonner and Huang (1964) showed that pea nucleohistone was a much less efficient template for *E. coli* polymerase than the corresponding free DNA. Allfrey and Mirsky (1964) examined transcription in calf thymus nuclei and demonstrated that addition of histones inhibited RNA synthesis, whereas removal of histones was stimulatory. At first glance, such results might be thought to lend support to the idea that histones acted as regulators. However, Allfrey and Mirsky seem not to have accepted the Stedmans' hypothesis of *specific* regulation by histones, for they concluded with the statement:

The experimental findings . . . are consistent with the view that histones, by combining with DNA, cause it to assume a condensed state, no longer active as a primer in the synthesis of ribonucleic acid.

Thus, histones were envisioned by Allfrey and Mirsky not to be a set of specific regulators for particular genes, but rather to be *nonspecific* agents in a general mechanism for chromatin inactivation. As will be seen, this viewpoint subsequently gained favor and with some elaboration is widely held to the present day.

The final decline in the concept of histones as genetic regulators can be traced, however, to the further studies on purification and sequencing of the histones that took place in the late 1960s. Methods were continually improved, and by the time the landmark papers of Johns (1967) and of Panyim and Chalkley (1969a, 1969b) appeared it was clear that there were only a *few* basic types of histones, although numerous variants could exist. (See Fig. 1-1.) Perhaps the most stunning blow was the discovery that not only did cows and peas each have a single, homogeneous type of H4 molecule, but that the sequence of these proteins from such distantly related sources differed in only two residues (DeLange et al., 1969a, 1969b)! Such conservatism was the very antithesis of the features expected for a regulatory molecule. Consequently, the histones came to be viewed primarily as structural components of the nucleus. Perhaps this explains the relative lack of interest (apart from the sequencers and a few dedicated laboratories) in histone chemistry during the late 1960s and early 1970s. In this context, it is noteworthy that the report of the 1973 Cold Spring Harbor Symposium on "Chromosome Structure and Function" contains only one paper devoted primarily to histones, out of a total of 95. The circle had come full: Nucleic acids, once disregarded as uninteresting structural or metabolic materials, were now the center of attention, and the nuclear proteins, once considered to be themselves the loci of genetic information, had very much passed out of vogue.

The emphasis now turned to the study of chromatin structure. If the histones were not specific gene regulators, could the conformation of the whole DNA–protein complex somehow fulfill that function? The spectacular success of X-ray diffraction and scattering techniques in revealing

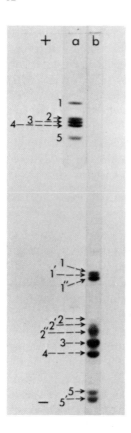

Fig. 1-1. Gel electrophoresis of calf thymus histones, according to Panyim and Chalkley (1969b). The acid-urea gels used here (particularly the long gel, to the right) allowed a much better resolution of histones than previously possible. The numbers used in this figure correspond to modern nomenclature as follows: 1 = H1, 2 = H3, 3 = H2B, 4 = H2A, 5 = H4. In the longer gel, several variants of H1 and modified forms of H3, H2A, and H4 can be resolved. The fact that a limited number of bands were observed, with very similar patterns from a number of tissues, tended to discredit the idea that histones might serve as specific gene regulators. (Reprinted with permission from Biochemistry 8, 3972, Panyim and Chalkley. Copyright 1969 American Chemical Society.)

the detailed structure of nucleic acids and proteins during the 1950s led a number of workers to attempt to use the same methods for the study of chromatin. The pioneer in this field was M.H.F. Wilkins. At the 1956 Cold Spring Harbor Symposium, he had presented data on fibers prepared from a number of kinds of sperm and somatic cell chromatins (Wilkins, 1956; see also Feughelman et al., 1955). While diffraction patterns were obtained for both types, the protamine-containing sperm exhibited both better orientation in fibers and more detail in the diffraction patterns.

In 1956, there were still serious problems in the isolation of chromatin with even a moderate degree of integrity. Therefore, it is not surprising that the next major advances in the diffraction studies were made only after Zubay and Doty (1959) had critically reexamined the conditions for isolation. Like Steiner (1952), they eschewed high salt extraction and extracted at low ionic strength. They also included EDTA to chelate nuclease-activating divalent ions and used relatively vigorous shearing to solubilize the gel-like material. In this fashion, soluble material with a molecular weight of about 19 million was obtained. Since the DNA content was

~47% and the DNA molecular weight was found to be 8 million, Zubay and Doty concluded that the particles contained individual DNA molecules. From hydrodynamic and electron microscopy studies, they judged that the DNA must somehow be coiled in the individual particles, presumably through the agency of the protein. The Zubay–Doty procedure became the standard method for preparation of soluble chromatin for the next 15 years.

The first X-ray diffraction studies of oriented fibers of such chromatin preparations were relatively unrevealing with respect to structure. Only a diffuse diffraction ring corresponding to a 35–38 Å reflection could be definitely ascribed to the nucleoprotein structure (see, for example, Wilkins et al., 1959). However, the major progress in this area came through the efforts of John Pardon (Pardon et al., 1967; Richards and Pardon, 1970; Pardon and Wilkins, 1972). With much better diffraction patterns, reflections were observed at 105 Å, 55 Å, 35 Å, 27 Å, and 22 Å, all of which could definitely be ascribed to the chromatin (as opposed to DNA) structure. (See Fig. 1-2.) Stretching the fibers or removing the histones by salt extraction destroyed the pattern; relaxation or reconstitution of the chromatin by salt gradient dialysis restored most features. From these data, Pardon and his co-workers suggested the existence of a superhelical coiling of the DNA, stabilized by histone interactions. Calculations showed that the X-ray data were consistent with a uniform superhelix of pitch 120 Å and diameter 100 Å.

The Pardon–Wilkins model (as it came to be known) was never very specific concerning the disposition of the histones; it was basically a description of the DNA coiling. But it seems clear from many discussions of it that the authors, and most other scientists as well, assumed a more or less uniform "coating" of the DNA superhelix with protein (see, for example, Pardon and Wilkins, 1972). The model was persuasive and widely accepted for a number of good reasons. First, it was consistent with the observations by electron microscopists of "100 Å fibers" in chromatin preparations (see, for example, Ris and Kubai, 1970; Bram and Ris, 1971). It explained at least qualitatively the poor orientation of the DNA strand in unstretched chromatin fibers, as indicated by X-ray diffraction, and the change in orientation upon stretching such fibers. Finally, it was a simple structure that could account for at least a part of the compaction of DNA in the nucleus.

As we shall see, the model was basically an incorrect one. However, the conviction it carried can be felt in reading the careful review by Ris and Kubai (1970). These authors present electron micrographs that depict knobby, irregular fibers. Ris and Kubai were clearly disturbed by the fact that they could not observe a smooth, helical structure of the kind expected from the Pardon–Wilkins model. They compromised with the model by invoking the possibility of preparation artifacts in the electron microscopy.

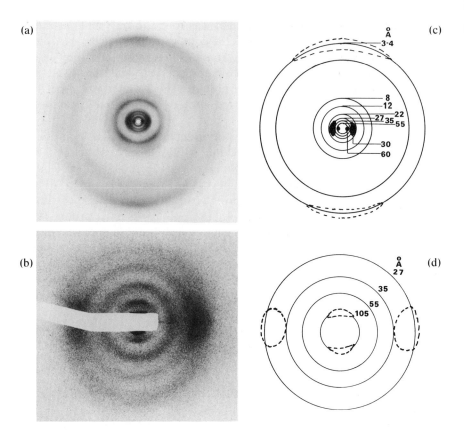

Fig. 1-2. X-ray diffraction patterns from calf thymus chromatin. (a,b) High-angle pattern showing the 3.4 Å DNA spacing and 8 Å and 12 Å spacings from the second and third DNA layer lines. (c,d) Low-angle pattern emphasizing the 22 Å, 27 Å, 35 Å, 55 Å, and 105 Å reflections arising from the higher order nucleo-protein structure. Courtesy of J. Pardon.

The idea of a superhelical chromatin structure received support from other kinds of experiments as well. While Pardon and his co-workers were studying X-ray scattering from fibers, others were carrying out similar measurements on unoriented solutions and gels of chromatin. Both Luzatti and Nicolaieff (1963) and Bram and Ris (1971) interpreted data in terms of rod-like particles, with a cross-sectional radius of gyration of about 30 Å, and a mass per unit length of 1100 to 1500 daltons/Å. While Luzatti and Nicolaieff postulated a rather complicated series of multistrand structures, Bram and Ris argued for an irregular DNA superhelix, of a diameter of about 60 Å and an average pitch of 45 Å. While the dimensions, particularly the pitch, were at variance with the Pardon–Wilkins model, the basic concept was very similar. It seemed that the principal problems would henceforth lie in reconciling the dimensional discrepancies.

Thus, exactly one hundred years after Miescher's first paper, a structural model for chromatin had finally emerged. The model did not survive for long; the next few years were to see dramatic advances culminating in an entirely different model and an enormous resurgence of interest in chromatin structure and biological function.

2

Development of the Nucleosome Model for Chromatin Structure

A eukaryotic chromosome made out of self-assembling 70Å units, which could perhaps be made to crystallize, would necessitate rewriting our basic textbooks on cytology and genetics! I have never read such a naive paper purporting to be of such fundamental significance. Definitely it should not be published anywhere!
—Anonymous review of paper submitted by D.F.L. Woodcock, 1973.

By the early 1970s, the uniform supercoil model of chromatin structure had become quite firmly established as a reasonable hypothesis, a concept consistent with the facts as they were then known. But a few dissonant experimental results had begun to appear, and some researchers were beginning to express doubts. Indeed, even Pardon et al., in their paper in the 1973 Cold Spring Harbor Symposium, considered the possibility that the supercoil regions might be very short—"one turn or less." In fact, a new and radically different model was to emerge within less than a year. This chapter will attempt to recount, in a somewhat journalistic fashion, the events that led to this dramatic change in view. The aims are twofold: to describe that exciting period in research, between 1971 and 1975, and to provide a basic description of the elements of chromatin structure as a background for the following chapters.

It is almost impossible to give a strict chronological account of the earlier events of this period. The clues that led to the dissolution of the supercoil model were diverse and at first seemingly unrelated. The crucial experiments were performed in a number of laboratories, and the participants were, in many cases, unaware of the progress, or even the existence, of other groups. To reconstruct the *thinking* of the researchers is even more difficult, and I shall not, except for the case of my own laboratory, even attempt it. The reader can find a description of the conceptual developments in the MRC laboratory at Cambridge in a *Scientific American* article by two of the principals, R. Kornberg and A. Klug (1981).

With these caveats, I begin what is admittedly a personal view of the events, emphasizing what were, in my opinion, the critical experiments and conceptual turns.

It has always seemed to me that the first *experiments* to really challenge the supercoil model were the nuclease digestion studies of Clark and Felsenfeld (1971). They were not the first to employ these enzymes in the study of chromatin (see, for example, Cohen, 1945; Murray, 1969), but Clark and Felsenfeld examined both the kinetics of the reactions and the nature of the products with exceptional care. Their studies revealed two quite remarkable features: First, whereas micrococcal nuclease digested free DNA rapidly and completely into small oligonucleotides, the digestion of *chromatin* seemed to come to a halt when only about 50% of the DNA had been rendered acid-soluble. Second, the resistant portion was shown by sedimentation experiments to contain not random-sized pieces of DNA, but reasonably homogeneous fragments on the order of 100 to 200 base pairs (bp). Clark and Felsenfeld speculated that this resistance must be due to periodic protein protection and suggested that the readily digested regions were less thoroughly covered by protein. Support for this concept came from experiments in which polylysine was used to titrate the putative open regions. It was found that this treatment increased protection to nuclease.

On the basis of this work, they stated: "Our results show that the histones are not evenly distributed along the DNA backbone in a manner that results in complete covering." It is possible to interpret such a statement as a precocious hint at a "subunit" model; certainly it is inconsistent with uniform supercoils. But a close reading of the paper suggests that Clark and Felsenfeld were thinking more in terms of *extended* open regions.

While this paper aroused interest in a number of laboratories, it did not lead immediately to further research of this kind. There was in fact some argument to the effect that the limit to digestion was artifactual, a consequence of the precipitation of the chromatin in the latter stages of the reaction; see, for example, Itzahki (1971). But by and large, the implications of this paper seemed so much at variance with accepted concepts that they were merely ignored. Indeed, papers presented in the Cold Spring Harbor Symposium in 1973 carry few references to the substance of this work, even though further evidence was given at this meeting (Axel et al., 1973).

In the two or three years following Clark and Felsenfeld's paper, several groups, working independently, developed separate lines of research that completely undermined uniform supercoil theories of chromatin structure and converged toward a quite different model. At Flinder's University of

South Australia, Hewish and Burgoyne (1973) began to systematically utilize DNA gel electrophoresis to study the degradation of chromatin by
endonucleases. Some years earlier Williamson (1970), examining what was
termed "cytoplastic DNA" in necrotic liver cells, had reported a strange
result. This DNA, when analyzed on acrylamide gels, gave a series of
distinct bands. Furthermore, Williamson extracted the DNA from these
bands, estimated the sizes from the sedimentation velocities, and found
the molecules to be multiples of about 200 bp. However, the relationship
of this material to nuclear DNA was uncertain at the time, and the possible
significance of this bizarre result was largely unappreciated.

Hewish and Burgoyne, on the other hand, were clearly working with
nuclear DNA, for they carried out their first digestion experiments in isolated rat liver nuclei, making use of the endogenous Ca–Mg-dependent
nuclease contained therein. Like Williamson, they observed a series of
DNA bands on polyacrylamide gels (see Fig. 2-1). These, too, seemed to
occur in sizes that were multiples of a minimum size. Even more important
was the observation that the nuclear DNA itself was being gradually degraded to produce these fragments. Hewish and Burgoyne prepared a crude
isolate of the nuclease and showed that it did *not* behave in this way toward
naked DNA; instead, the expected heterogeneous population of fragments
was obtained. They therefore concluded that the relatively precise cutting
of the DNA in chromatin must reflect some features of its in vivo structure.
They stated: "It is proposed that chromatin has some simple, basic, repeating sub-structure with a repetitive spacing of sites that are potentially
accessible to the Ca–Mg endonuclease."

Others were beginning to question the uniform supercoil model on quite
different grounds. Dusenbery and Uretz (1972) examined the linear dichroism of oriented fibers in squashed chromatin preparations. To their
surprise, they observed a dichroism quite different from that expected
from the supercoil models, which would have required the DNA base
planes to be tilted in directions roughly parallel to the fiber axis. The
dichroism observed by Dusenbery and Uretz was in the opposite sense,
as if portions of the DNA chains were stretched parallel to the axis, with
their base planes *perpendicular*. Suggestive though this result was, it was
difficult to interpret unambiguously because of artifacts that might have
been produced by stretching the fibers. After all, Pardon et al. (1967)
had already shown pronounced changes in the X-ray fiber pattern upon
stretching.

Our own initial doubts concerning the supercoil models arose from similar experiments. We had been carrying out linear dichroism studies with
molecules oriented in solution by an electric field. Dr. R. Rill in my laboratory began such experiments with calf thymus chromatin in the early
1970s. At the beginning, we anticipated adding only a few refinements to
the uniform superhelix model. We were wholly ignorant at that time of
the researches of Hewish and Burgoyne and did not know what to make

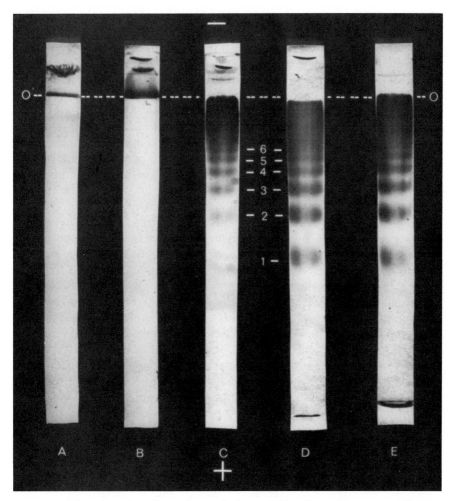

Fig. 2-1. Digestion of rat liver chromatin in nuclei by the endogenous nuclease. From left to right, nuclei were incubated for successively longer times in $1 mM$ Ca^{2+}, 10 mM Mg^{2+}. The band pattern observed is the first clear indication for a subunit structure in chromatin. (Reprinted with permission from Biochem. Biophys. Res. Comm. *52*, 504, Hewish and Burgoyne. Copyright 1973 Academic Press.)

of the Clark and Felsenfeld experiments. But it soon became evident that uniform supercoils were quite inconsistent with our results. The experiment measures the relative dichroism $\Delta\varepsilon/\varepsilon$, where ε is the extinction coefficient for an unoriented sample and $\Delta\varepsilon$ is the difference in extinction coefficient for light polarized parallel and perpendicular to the applied field. The relative dichroism is then extrapolated to infinite field, at which complete orientation of the particles should be approached. The Pardon and Wilkins model predicted a value of $(\Delta\varepsilon/\varepsilon)_\infty = +0.42$; the Bram and

Ris (1971) model a value of $+0.55$. Our observed result was about -0.70; thus, even the sign was wrong! The only way in which the data could be reconciled with such supercoils was if the structure were *heterogeneous*, containing both supercoiled regions and appreciable segments of structure like free DNA, for which $(\Delta\varepsilon/\varepsilon)_x \cong -1.4$. Although these data did not finally see print until about two years later (Rill and van Holde, 1974), by which time their significance had become obvious, they had enormous influence on our own early thinking about chromatin. They led us to consider more carefully the Clark and Felsenfeld experiments, which also pointed to a heterogeneous structure. Most important, they led Rill to use nuclease digestion to attempt to separate the hypothesized "supercoiled" and "linear" regions of the structure.

The first particles so produced (Rill and van Holde, 1973) were clearly heterogeneous, but they had quite unusual properties. The protein/DNA ratio seemed high; the circular dichroism spectrum was not only startlingly different from DNA, but also unlike anything that had been seen before in chromatin; and the electric dichroism very low—all properties that we might expect from histone–DNA complexes with highly coiled DNA. A graduate student, C. Sahasrabuddhe, continued the research and by improving the preparation methods was able to obtain about 50% of the chromatin DNA in homogeneous nucleoprotein particles that had the properties listed in Table 2-1 (Sahasrabuddhe and van Holde, 1974).

The most striking thing about these particles was their compactness. They appeared to be nearly globular objects ($f/f_o \cong 1.1$) with a Stokes diameter of about 80 Å. Yet sedimentation analysis of the DNA contained therein indicated a size of about 110 bp, which would correspond to an extended length of about 375 Å. Obviously, the DNA had to be coiled or folded in the particle. Furthermore, it appeared that protein was respon-

Table 2-1. Properties of Chromatin Particles[a]

	Native particle	Trypsinized[b] particle	DNA[c]
Sedimentation coefficient ($S_{20,w}$)	11.5S	~7S	4.8S
Molecular weight (M)	176,000[d]	158,000	72000[d]
Frictional ratio (f/f_o)	1.1	2.0	2.4

[a]Data from Sahasrabuddhe and van Holde, 1974.
[b]Particles in which the histones had been partially proteolyzed by trypsin.
[c]The DNA that was extracted from either native or trypsinized particles.
[d]These values are somewhat lower than the currently accepted values for "core" particles (about 200,000 and 90,000, respectively—see Chapter 5). The difference lies in the DNA weight, indicating that the DNA had been degraded somewhat beyond the core particle limit.

sible for this compaction, for limited trypsinolysis led to a drastic reduction in the sedimentation coefficient with little loss in particle mass (Table 2-1). It seemed obvious to us at this point that much of the DNA in chromatin was bound in compact histone-containing *particles* of some kind.

Meanwhile, other researchers had been working independently along very different lines to arrive at almost the same conclusions. Two groups, one headed by C.F.L. Woodcock at the University of Massachusetts, and the other by Ada and Donald Olins at the Oak Ridge National Laboratory, had been utilizing the Miller spreading techniques (Miller and Beatty, 1969) to visualize chromatin fibers in the electron microscope. The Olinses describe their initial experiments in a "Citation Classic" (Olins and Olins, 1983):

Returning to Oak Ridge in September 1971, we set about to visualize the "naked" stretches of DNA in chromatin, fully expecting to see very long regions between structures resembling supercoils. We initially tried the method of critical point drying but the results were hopeless; chromatin became a tangled, twisted mess under these conditions. Oscar L. Miller, Jr., was in Oak Ridge at that time, and we decided to try his method of centrifuging swollen nuclear contents onto a carbon film. For several months during the winter of 1972–1973, we accumulated micrographs of well-spread erythrocyte nuclei, never suspecting what could be revealed under a simple magnifying lens. One evening in February 1973, we happened to look closely at some negatively stained nuclei. To our excitement and surprise we saw that everywhere the chromatin looked like "beads-on-a-string." We called these particles v (nu) bodies because they were new in the nucleohistone field.

In November 1973, both groups reported their results at the annual Cell Biology Meeting (Olins and Olins, 1973; Woodcock, 1973). The Olinses soon thereafter published a more complete account of their work in *Science* (Olins and Olins, 1974). Woodcock's parallel paper was sent to another journal and, unfortunately, rejected. (See the quote that heads this chapter.) In brief, both groups made the same observation: Extended chromatin fibers had a "beaded string" appearance (see Fig. 2-2). The beads were reported to be 60 to 80 Å in diameter (Olins and Olins, 1973, 1974) or about 100 Å in diameter (Woodcock, 1973). It seemed to us almost certain that these were the particles which we had been isolating. This was later supported by electron microscope examination of the isolated particles (van Holde et al., 1974a). Furthermore, it now seemed evident that these particles could be the elements of the "simple, basic, repeating structure" that Hewish and Burgoyne had postulated.

But why should histones bind to DNA in such a way as to produce particulate structures of this kind? Clues to this had already begun to emerge from discoveries of specific histone–histone interactions. It had been known for some time that individual, purified histones tended to self-aggregation at even moderately low salt concentration (see, for example, Edwards and Shooter, 1969). But now evidence began to appear

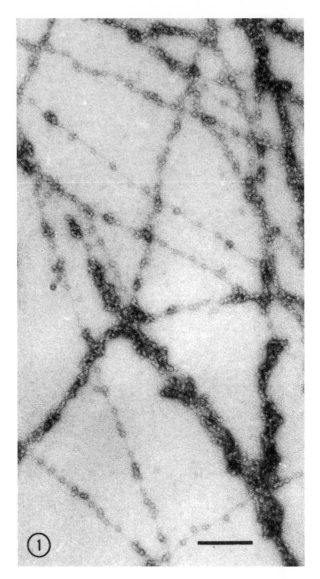

Fig. 2-2. An early electron micrograph showing the "beaded string" structure of chromatin fibers (courtesy of C.L.F. Woodcock).

that histone *mixtures* behaved quite differently. For example, Skandrani et al. (1972) reported chromatographic studies showing the formation of a specific H2A·H2B complex. This was confirmed, and the complex shown conclusively to be a dimer, by Kelley (1973). But the generality of the phenomenon only became apparent when D'Anna and Isenberg (1973, 1974a, 1974b, 1974c) published a series of detailed and meticulous studies of the interactions between these and other pairs of histones. They demonstrated first the association of H2B with H4, and then that H2A formed both a strong 1:1 complex with H2B (in agreement with the earlier reports) and a weak 1:1 complex with H4. The details of these and other interactions are given in Chapter 4; at this point, it is sufficient to note that the mere existence of specific histone complexes provided a *raison d'etre* for the formation of some kind of particulate structure in chromatin.

The crucial discovery was the existence of the H3/H4 tetramer. This was reported by two groups almost simultaneously. At Hershey, Pennsylvania, Dennis Roark and his collaborators were carrying out sedimentation equilibrium studies of salt-extracted histones. They reported, in an abstract for the 1974 ASBC meetings (Geoghegan et al., 1974; see also Roark et al., 1974), the existence of an H3/H4 dimer that exhibited reversible association to a tetramer. (See also D'Anna and Isenberg, 1974d.)

During the same period, J. Thomas and R. Kornberg were carrying out extensive studies of the salt-extracted histones at the Medical Research Council Laboratory, Cambridge, England. Since 1971, researchers at the MRC had been seeking alternatives to the supercoil model. In 1972, Roger Kornberg, a postdoctoral researcher in the laboratory of F. Crick, began studies of histones isolated by the salt-extraction technique of van der Westhuyzen and von Holt (1971). In June 1973, the first evidence for a specific complex was obtained by gel chromatography. Kornberg was aware of the work of Hewish and Burgoyne but had the initial impression that the DNA repeat size they had observed was about 500 bp. The rest of the story is best told in Kornberg's words:

I made no further progress until the beginning of August. The next few weeks were very quiet in the lab with Crick, Klug, and just about everyone else away on holiday, and the absence of distractions must partly explain why the problem was solved during that time. I pursued the notion of an H3-H4 complex by molecular weight determination in the analytical ultracentrifuge and, at Jean Thomas's suggestion (and with her help) by chemical crosslinking. The results were unequivocal, pointing to an $(H3)_2(H4)_2$ tetramer, and I drew the obvious conclusions: the stoichiometries of H2A, H2B, and DNA corresponded to a set of two each of all four histones and 200 base pairs of DNA; the globular nature of the H3-H4 tetramer called for wrapping the DNA around a histone core. I had no sense of denouement at this point. I could not even be sure the H3-H4 tetramer had physiological relevance. The moment of discovery came some days later, after turning the ideas over in mind, when I thought again of the digestion pattern of Hewish and Burgoyne. What if the unit length of DNA had been measured in a neutral rather than alkaline sucrose gradient and corresponded to 250 rather than 500 base pairs? And what if the result was a bit off, the true value being nearer 200 base pairs? The possible agreement with the value I had conjectured from the stoichiometry of the tetramer was electrifying. I remember it was about 6 o'clock on a weekday evening and I rushed to the library. Hewish and Burgoyne referred to a paper by Williamson on a similar pattern of DNA fragments from the cytoplasm of necrotic liver cells. Williamson reported a sedimentation coefficient consistent with a unit size of about 200 base pairs. I was euphoric. (D. Kornberg, in a letter to the author, April 16, 1984.)

The first public presentation of the model was at a colloquium at the MRC laboratories in November 1973, followed by a lecture at Oxford in March 1974. An insight into the state of the field in early 1974 can be gained from a perusal of proceedings of the Ciba Foundation Symposium held in London in April of that year (see Fitzimmons and Wolstenholme, 1975). The recorded discussions are especially illuminating. Although the electron microscope results of the Olinses and Woodcock were known, they appear to have been dismissed by some as artifactual. Kornberg's model was presented during the discussion and seems to have been the object of considerable interest.

A more detailed exposition of the model appeared shortly thereafter, in the May 24 issue of *Science* (Kornberg, 1974), accompanying a paper on histone association (Kornberg and Thomas, 1974). Kornberg's proposal was built on three principal observations: (1) the existence of a 200 bp "repeat" in chromatin, as evidenced from nuclease digestion, (2) the $(H3)_2(H4)_2$ tetramer, and (3) the approximate molar equivalence of the histones in chromatin. There is a curious parallel here to the role played by a similar equivalence (A = T, G = C) in the discovery of the DNA structure.

The model postulated that each 200 bp of DNA was coiled about an $(H3)_2(H4)_2$ tetramer. The addition of two molecules each of H2A and H2B made up the protein octamer, but the precise structural role of these latter histones was not clearly defined. It was suggested that they might play some part in connecting tetramers to one another. As we shall see, the model had certain deficiencies. However, these are quite irrelevant in assessing its importance to the field. It was a novel, specific model that provided a new synthesis of the data. After the Kornberg paper, chromatin structure would never be viewed in the same way again.

The Kornberg model postulated that a histone octamer interacting with the entire 200 bp of DNA constituted the universal subunit of chromatin. This idea was supported by the first determinations of chromatin repeat size from micrococcal nuclease digestion experiments (Burgoyne et al., 1974; Noll, 1974a). However, those of us who were working with the nucleoprotein particles themselves were obtaining considerably smaller values for the number of DNA base pairs associated with the histone octamer -values in the range of 110 to 175 bp (Shaw et al., 1974). Furthermore, electron micrographs of both chromatin (Olins and Olins, 1974; Woodcock et al., 1974) and of isolated oligomers of the particles (van Holde et al., 1974a) indicated the existence of significant lengths of "spacer" DNA between the particles.

Consequently, the chromatin model that we proposed postulated a smaller particle, with about 120 bp of DNA wrapped about a histone octamer (van Holde et al., 1974b). This model, too, had its deficiencies; the DNA size is too small for even a core particle, and the possible interaction of the amino-terminal histone "tails" with the DNA was probably over-stressed. (See Chapter 6.)

Basically, however, the differences between the models were less important than their similarities. Both emphasized the role of a histone octamer in coiling and therefore compacting the DNA. Both assumed the DNA to be coiled on the outside of a histone core. With such models in hand, the way was now opened for more directed investigations of chromatin structure.

These came swiftly. In an incisive study, Noll (1974a) showed that micrococcal nuclease digestion could be carried out on whole nuclei, with results identical to those observed when this enzyme digested isolated chromatin, or when nuclei were digested by the endogenous activity. At the same time, he showed that at least 87% of the DNA in chromatin could be accounted for as part of the repeating structure. These two observations were particularly important at this stage. The first proved that the subunit structure was neither an artifact of chromatin isolation nor a consequence of some peculiar property of endogenous nucleases; cleavage to the series of DNA oligomers could be performed at will, within intact nuclei. The second observation demonstrated once and for all that these subunits represented a major feature of chromatin structure; they were not just some peculiarity of a minor fraction. In a sense, it is this paper by Noll that put the model on a solid experimental foundation. It is appropriate that Noll received the 1978 Friedrich Miescher Award for this work.

It should be clear from the above that the chromatin field was revolutionized in the relatively brief period between November 1973 and November 1974. A short chronology of these events is given in Table 2-2. By autumn 1974, the situation could be summarized as follows: The existence of a "subunit" of chromatin was now firmly established. There was evidence that this structure involved between 100 and 200 bp of DNA, coiled in some fashion by an octamer of the histones H2A, H2B, H3, and H4. Two major issues remained ambiguous: Exactly *how much* DNA was involved in each subunit, and *where* was it located relative to the protein—inside, outside, or intermixed?

A tentative answer to the second question was provided by Noll (1974b). He showed that if nuclei were digested with pancreatic DNase 1 (which makes single-strand nicks in double-strand DNA) and the digested DNA were electrophoresed under denaturing conditions, a dramatic "ladder" of DNA fragments was obtained (see Fig. 2-3a). The spacing between bands turned out to be very nearly 10 nucleotides. Two hypotheses were put forward to explain these results. Crick and Klug (1975) suggested that the DNA helix, in winding about a histone core, was periodically "kinked." Similar proposals with different "kink" types have since been

Table 2-2. Chronology of Important Events in the Development of the
Nucleosome Model for Chromatin Structure[a]

Date	Event
Nov. 14–17, 1973	Cell Biology Meeting, Miami Beach: Olins and Olins and Woodcock present first EM pictures of "beaded" structures
Nov. 1973	Colloquium at MRC: first presentation of Kornberg model
Jan. 10, 1974	Sahasrabuddhe and van Holde paper on particle structure
Jan. 25, 1974	Olins and Olins paper on EM studies
April 3–5, 1974	Ciba Foundation Meeting, London: Kornberg model presented
May 7, 1974	D'Anna and Isenberg paper on H2B·H4 and H2B·H2A interactions
May 1974	Abstract by Geoghegan et al. describing dimer–tetramer association of H3 and H4
May 24, 1974	Paper by Kornberg and Thomas on $(H3)_2(H4)_2$ tetramer; first publication of model in accompanying paper by Kornberg
Sept. 20, 1974	Noll paper on digestion of nuclei by micrococcal nuclease
Oct. 23, 1974	Paper by van Holde et al. showing EM pictures of individual particles
Nov. 1974[b]	Noll paper on DNase 1 digestion
Nov. 1974[b]	van Holde et al. paper on model
Nov. 19, 1974	D'Anna and Isenberg paper on histone cross-complexing pattern

[a]The period covered is one year: November 1973–November 1974. I have included only those published papers or symposia that I feel to have been of maximal importance.
[b]These two papers appeared together in the November issue of *Nucleic Acid Research*. An exact date of publication is not given on the journal issue.

advanced by other workers (see Chapter 6). However, the explanation originally proposed by Noll is simpler and more probably correct: If the DNA is wrapped smoothly about the outside of the histone core, a given chain should "surface" (e.g., become most accessible) about every 10 nucleotides and hence should be cut with this periodicity (see Fig. 2-3b). This can be taken as the first, albeit indirect, evidence that the DNA lies on the outside. More compelling evidence was soon to come from X-ray and neutron scattering and diffraction experiments, but these will be described in Chapter 6.

The question as to the *amount* of DNA in each subunit proved to be complicated. The Hewish and Burgoyne experiments, as well as the early nuclease digestion experiments of Noll (1974a), all pointed to a repeating unit of about 200 bp. Thus, Kornberg (1974) had postulated a universal chromatin subunit with DNA of this size, associated with the histone core. In contrast, our experiments were yielding nucleoprotein particles with somewhat less DNA (110 to 175 bp) (Shaw et al., 1974). We believed that the electron microscope studies clearly indicated *two* kinds of DNA regions within each subunit; one coiled and associated with the histones (except H1) and the other more extended and perhaps carrying H1. Support for this concept came from studies of the time-course of digestion, which

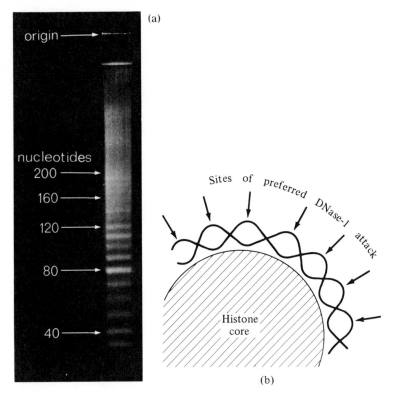

Fig. 2-3. (a) The "ladder" of single-strand DNA fragments produced by DNase 1 digestion of nuclear chromatin (Noll, 1974b). (Reprinted with permission from Nucleic Acids Research, vol. 1, issue 11, p. 1574AV Noll. Copyright 1974 IRL Press.) (b) The simplest explanation for this pattern: The nuclease cuts preferentially where the DNA is maximally exposed on the nucleosome surface.

showed that micrococcal nuclease cleavage produced first particles with repeat-sized DNA that were then further digested to yield stable "core" particles (Sollner-Webb and Felsenfeld, 1975; van Holde et al., 1975). The core particles were shown to contain about 140 bp of DNA and an octamer of the histones, excluding the very lysine rich histones. These latter were retained in the oligomer fractions, hence we suggested that they were associated with the "spacer" regions (see Fig. 2-4).

Comparative studies of chromatin structure in a variety of nuclei shed further light on the problem. Dr. D. Lohr carried out the first studies of yeast chromatin (Lohr and van Holde, 1975). The value found for the repeat (135 bp), although subsequently shown to be about 20% low, was sufficiently different from values reported for vertebrate chromatins to indicate that the "200 bp repeat" was not ubiquitous. Compton et al.

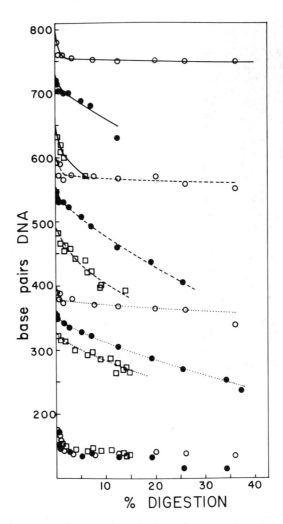

Fig. 2-4. Nucleosome oligomer sizes during micrococcal nuclease digestion of chromatin. The decrease in DNA fragment size during digestion results from ex-onucleitic attack on the DNA. Data are given for chicken erythrocyte (○), HeLa (●), and yeast (□). The sizes shown here are about 10% low, due to incorrect size calibration. From Lohr et al. (1977).

(1976) studied a number of vertebrate and invertebrate cell types, with results demonstrating unequivocably that the repeat length varied over a wide range of values. Remarkably, the core particle DNA size appeared to be essentially the same in all species studied. The situation can perhaps be best appreciated by examination of Table 7-1, which summarizes data accumulated to the present time. Repeat length, as defined by the common size factor for oligomers generated early in the digestion, varies from ~160 bp (some fungi) to 260 bp (some invertebrate sperm). On the other hand, the core particle DNA seems very rigidly fixed at 146 ± 2 bp (see Table 6-4, Chapter 6).

The apparent controversy concerning linker DNA size turned out to be largely illusory. As is often the case, further experience has shown that both of these seemingly disparate views of chromatin substructure have elements of validity and are in fact reconcilable. As we shall see in Chapter 7, spreading of chromatin by the Miller technique, as employed by the Olinses and Woodcock in early studies, tends to extend the structure. Electron micrograph studies of chromatin under more nearly physiological conditions show a closer packing of particles; the linker DNA can evidently be tightly coiled. In this sense, Kornberg's (1975) description of chromatin as a string of closely packed beads is both correct and appropriate. On the other hand, the linker DNA remains operationally distinguishable, as a segment of variable length and greater susceptibility to nucleases, even under conditions where the chromatin structure is compacted.

The model of chromatin structure that had evolved by late 1975 is as schematicized in Figure 2-5. At about this time the nomenclature, which had hitherto been rather chaotic, began to stabilize. A short glossary of current usage may help at this point:

Nucleosome (Oudet et al., 1975): the entire quasi-repeating structural unit of a particular chromatin, containing the histone octamer, the lysine-rich histones, and often nonhistone proteins as well, together with one repeat length of DNA.

Core particle (van Holde et al., 1975): the histone octamer together with 146 bp of DNA.

Linker (Noll and Kornberg, 1977): earlier called "spacer" by van Holde et al. (1974). The remaining DNA of the nucleosome; in some usages it also connotes any attached proteins.

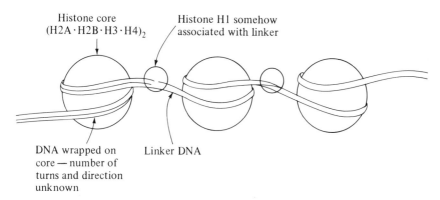

Fig. 2-5. A model of chromatin structure, based on evidence available in 1975. The distinction between "core" and "linker" is clear, but fine details are lacking.

Earlier terms, such as "PS particle" (Rill and van Holde, 1973) and "nu body" (Olins and Olins, 1974) are less precise. Both would seem to have corresponded roughly to what are now called core particles. The terms listed above will be utilized henceforth throughout this book.

The foregoing has been intended only to provide a brief summary of the development of a concept and to provide a basis from which the remainder of this book can more easily be read. We must turn now to more recent and detailed information, beginning with some background descriptions of the components of chromatin: the DNA and the proteins.

3

The Structures of DNA

The preceding chapters have briefly reviewed the history of chromatin research and provided a glimpse of modern ideas about the structure of this substance. In later chapters, that structure will be explored in depth, and current concepts of chromatin function will be presented. Before entering into these matters, it seems wise to describe in some detail the major constituent parts of chromatin: the eukaryotic DNA and the chromosomal proteins. This and the following chapters are dedicated to that purpose.

In a fundamental sense, the DNA is the more important component. It is not unreasonable to take the position that chromatin structure, and indeed the eukaryotic nucleus itself, exists because of the peculiar requirements of eukaryotic DNA. The genomes of higher organisms are very large, often comprising more than 10^{10} bp. This very long DNA, which would extend for roughly 3 m if unraveled, must be compacted into a nucleus about 10^{-5} m in diameter. So one function of the proteins is to accomplish the necessary compaction. More interesting are the constraints on DNA *processes* in eukaryotic cells. The DNA is not replicated continually, but only in a certain portion of the cell cycle. Most of the DNA in a eukaryotic cell is never transcribed, and different cell types select different portions of the genome for expression. It is presumed, with good evidence, that much of the proteinaceous structure exists to facilitate and regulate these DNA functions.

In considering DNA structure, I shall focus on those features that may be of importance in understanding how DNA and proteins can interact to form chromatin. The emphasis will be primarily on current views. A reader interested in the fascinating chain of events leading to the Watson–Crick model should consult *The Path to the Double Helix* (Olby, 1974).

Composition and Modification of Eukaryotic DNA

Eukaryotic DNA is composed almost entirely of the four "canonical" nucleotides, with strict adherence to the rule for equivalence of A and T, G and C. Although $(A+T)/(G+C)$ ratios vary considerably, the DNA of most higher eukaryotes is significantly A-T rich (see *The Handbook of Biochemistry and Molecular Biology*, 3rd ed., Vol. II, G. Fasman, ed., 1976, for extensive data).

The only chemical modification that has been observed in eukaryotic DNA is methylation. Doerfler (1983) has provided an excellent review. There are also three recent books dedicated to the subject (Taylor, 1984; Trautner, 1984; Razin et al., 1984). Most methylation occurs on position 5 of cytosine, although a number of lower eukaryotes (and some plants and higher animals) contain very small amounts of 6-methyl adenine as well (see Fig. 3-1). The fraction of 5-methyl cytosine varies enormously among the eukaryotic taxa. In most animal cells, about 5% of cytosines are modified in this way, whereas values in excess of 50% have been observed in some plants. On the other hand, certain particular organisms appear to have exceedingly low levels of DNA methylation. For example, it has not been possible to detect any methylation in yeast DNA (Profitt

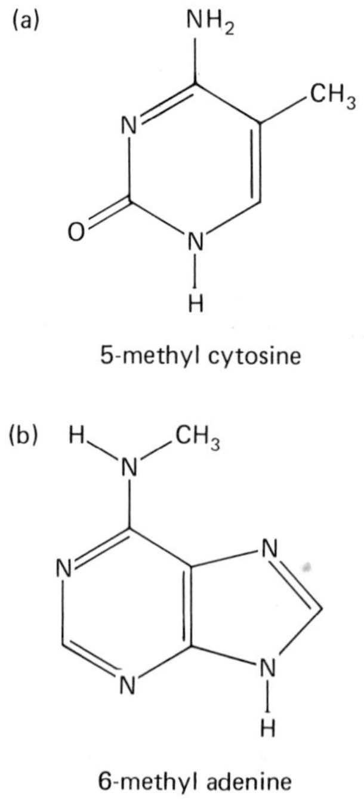

Fig. 3-1. Methylation sites on eukaryotic DNA bases: (a) 5-methyl cytosine, (b) 6-methyl adenine.

et al., 1984). While one might expect yeast, as an untypical eukaryote, to be anomalous, the studies on *Drosophila* have been much more disconcerting. Despite many attempts, over a long period, no evidence for methylation could be found. A very careful study by Urieli-Shoval et al. (1982) led to the conclusion that methylation in *Drosophila melanogaster* could not occur on more than 0.1% of cytosine residues. Neither could adenine methylation be detected. However, the development of ultrasensitive methods enabled Achwal et al. (1984) to show that 0.008% of all residues in *Drosophila* DNA were 5-methyl cytosine—about 0.03% of all cytosines. This corresponds to only one methylated site in every 12,500 nucleotides. Possible implications of this will be considered later.

In those eukaryotes in which methylation is more abundant, it tends to be found in quite specific locations. Gruenbaum et al. (1982) assert that >95% of all methylation in animal cells occurs on (5'. . .CG. . .3') sites. Approximately 70% of such sites are methylated in these cells, and most are modified on both strands. There exists a remarkable system for the preservation of such a methylation pattern through DNA replication (see Fig. 3-2). Stein et al. (1982) have carried out experiments in which methylated, double-strand ϕX-174 DNA was inserted into mouse L-cells by DNA-mediated gene transfer. The pattern of modification at CG sites was found to be maintained for 100 generations! A "maintenance" methyl transferase that catalyzes this reaction in vitro has been isolated by this same group (Gruenbaum et al., 1982). The enzyme shows strong specificity for hemi-methylated CG sites and presumably maintains the methylation pattern in the manner depicted in Figure 3-2. Such a mechanism was first hypothesized by Holliday and Pugh (1975) and by Riggs (1975). The special nature of CG sites is further emphasized by their rarity in eukaryotes. Since there are 16 possible dinucleotides, we might expect about 6% would be CG, if nearest neighbor frequencies were random. This is, in fact, what is found in most viruses and bacteria. But in higher eukaryotes, the values range about 1 to 2%. This suggests that there may be something special about CG pairs. However, Bird (1980) has pointed out that the low level of CG may only be a consequence of the tendency of 5meC to mutate to T.

Not all methylation is at CG sites, nor is all necessarily conserved. Gjerset and Martin (1982) have discovered a demethylating enzyme in murine erythroleukemia cells that removes methyl groups from the internal C in CCGG. Thus, the earlier suggestion of Razin and Riggs (1980) that demethylation occurs via inhibition of the maintenance system (and hence only following replication) is at least an oversimplification.

Considering the specificity of methylation and the existence of mechanisms for its maintenance and control in eukaryotes, we must presume that it performs some significant function. In subsequent sections, we will encounter strong hints as to what that function might be. Later in this chapter, evidence will be presented that methylation in (CG)$_n$ runs can

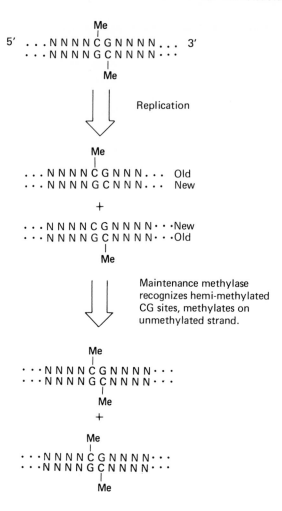

Fig. 3-2. Maintenance of specific DNA methylation sites.

facilitate the B→Z structural transformation. In Chapter 8, data correlating methylation levels with gene activation will be summarized. The idea that methylation can maintain specific patterns of gene expression in differentiated cells is an attractive one. However, as we shall see, the data are not as yet compelling. Furthermore, the extremely low level of DNA methylation in *Drosophila* casts a shadow of doubt over the whole concept. Either methylation is a regulatory mechanism used by some organisms and not by others, or the sites so involved must represent a very small subset of all the modified sites in most species. As will be shown later, there is evidence to support the latter view. If this is the case, the question as to what all the rest of the methylation is for remains a quandary.

Conformations of DNA in the Solid State

The early X-ray diffraction studies of DNA all followed a more or less uniform protocol, and to understand the information they provided—and the limitations thereof—it is necessary to consider the methods used.

Beginning with the earliest studies of Asbury in the 1930s, continuing through the crucial experiments of Wilkins and colleagues in the 1950s, and up until the late 1970s, all X-ray diffraction analysis of polynucleotides was based upon fiber patterns (see Olby, 1974). These experiments were done in the following way: A concentrated solution of the particular DNA to be investigated, in a salt solution of defined composition, was pulled into a fiber, with additional stretching as the fiber dried, to orient the molecules as much as possible. This fiber would then be mounted in an atmosphere of controlled humidity and the diffraction pattern recorded. For a number of reasons, "fiber patterns" obtained in this manner can provide only limited information. First, axial orientation of the molecules, even in stretched fibers, is never perfect. Second, there is nearly random rotational orientation of the molecules about their axes. Third, most of the molecules studied (with the exception of some synthetic polynucleotides) will present a random selection of sequences at any point along the fiber axis. Since we now know that the helix parameters themselves vary locally with nucleotide sequence, it is clear that such studies could at best describe an *average* conformation for the given DNA.

Most important, it must be realized that fiber diffraction patterns are not "solved" in the sense that crystal diffraction patterns are solved for molecular structures. The pattern itself can provide only limited direct information. Most of the structural details are obtained by a trial and error procedure, in which a reasonable model is proposed, the expected fiber pattern is calculated, and the intensities of the predicted spots are compared with those observed. The model is then systematically modified, always within the bounds of steric restrictions and acceptable bond lengths and angles, so as to improve the correspondence as much as possible.

In such a procedure, there are elements of subjectivity, and preconceived notions can easily be incorporated into the proposed structure. A particularly apt example concerns the conformation of the pentose ring in the high-humidity form of DNA (B-DNA). Crick and Watson (1954) assumed the C3'–*endo* conformation (see Fig. 3-3) since the only experimental evidence available to them (from Furburg's 1950 diffraction study of cytidine crystals) showed this conformer. Subsequent refinements of the Crick–Watson structure led to the proposal of first C3'–*exo* (Arnott and Hukins, 1972) and later C2'–*endo* (Arnott and Chandrasekaran, 1981) ring puckers. Very recent studies of single crystals of oligonucleotides indicate that B-DNA probably cannot be described in terms of any single sugar conformation. Rather, a range of structures is observed, depending on local sequence effects (Wing et al., 1980; Drew et al., 1981; Fratini et al., 1982; Dickerson, 1983a). We shall consider this in greater detail later, but it is

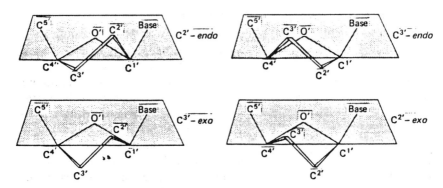

Fig. 3-3. Some frequently observed conformations of the deoxyribose ring. (From Biophysical Chemistry, Part I by Cantor & Schimmel. W.H. Freeman and Company. Copyright 1980.)

mentioned here because it illustrates two levels at which preconceptions can be introduced into the interpretation of fiber patterns. First, a *particular* sugar conformation was originally assumed, but even when the first choice was abandoned, it was still necessary to assume some *uniform* conformation in order to make usable models to test against the fiber patterns. The example also sets forth a central theme of this chapter—the gradual realization of greater complexity and nonuniformity in DNA structure.

In order to describe DNA conformations in detail, a set of conventions concerning rotational isomers about bonds in the sugar–phosphate backbone is necessary. The purine–pyrimidine rings themselves are relatively rigid, but as Figure 3-4 shows, there are six bonds in the backbone about which freedom of rotation is possible. We shall utilize the IUB-IUPAC nomenclature for these rotational angles (see Fig. 3-4). All of the angles α, β, γ, δ, ε, ζ are defined with their zero values corresponding to the conformation of the bonds as drawn in Figure 3-4, with positive rotation as indicated in the figure. The reader is cautioned that other symbols and conventions for these angles have been used in older publications; Arnott and Chandrasekaran (1981) provide a translation. Specific models for DNA structures require particular values for these angles, and published descriptions of such structures will usually include a table thereof. However, it should be remembered that, at least for fiber diffraction studies, the precise values given are simply those for the best structure found. The exact values stipulated should not be taken too literally in such cases; other conformations with slightly different values would probably fit the data equally well. It is perhaps more appropriate to describe the structures in terms of the notations introduced by Arnott and co-workers, in which a particular orientation symbol is used for a range of angles, thus:

gauche minus (g⁻) −60° ± 60°
trans (t) ±180° ± 60°
gauche plus (g⁺) +60° ± 60°

Clearly, there will be cases (such as the range around 120°) where the
designation may become ambiguous, but the convention usually provides
a useful shorthand for a qualitative description of the structure. In addition
to the six backbone orientation angles listed above, one more angle is
required to specify the structure of a nucleotide unit; this is the rotation
(χ) about the sugar-base bond (the "glycosyl" rotation). The two favored
orientations are defined as syn (s) (χ = +45° ± 90°) and anti (a) (χ =
−135° ± 90°), respectively (see Fig. 3-4). With this set of definitions, the
conformation of a particular residue can be specified by a 7-letter code,
in the order $\alpha\beta\gamma\chi\delta\epsilon\zeta$; for example, g⁻g⁺g⁺attg⁻. The angle δ is of particular importance, for it largely dictates the sugar ring conformation, as
is illustrated in Figure 3-4.

The set of angles listed in Figure 3-4 will describe the backbone conformation of a single residue in a polynucleotide. But for a description of
the overall structure of the helix, other parameters are more useful since

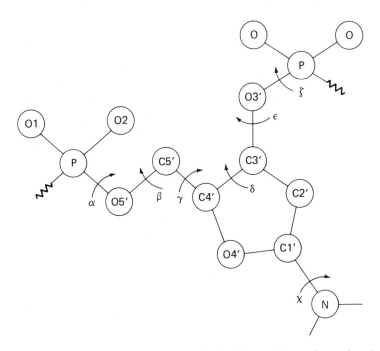

Fig. 3-4. Rotational angles in the nucleic acid backbone. The conformation shown
corresponds to zero values for all of the backbone angles. Positive rotation is as
indicated by the arrows. (From Biophysical Chemistry, Part I by Cantor & Schimmel. W.H. Freeman and Company. Copyright 1980.)

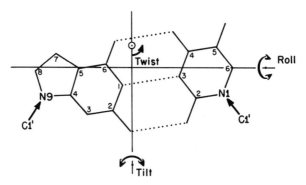

Fig. 3-5. Angular parameters of twist *(t)*, tilt (θ_T), and roll (θ_R) for a 2-strand polynucleotide helix. From Fratini et al. (1982), Journal of Biological Chemistry.

they allow a clearer visualization of the structure. One is the axial translation per residue (the "rise," *h*) defined as the distance between equivalent points on successive residues projected onto the helix axis. The other parameters are the angles *t*, Θ_T, and Θ_R shown in Figure 3-5. The rotation per residue about the helix axis, *t*, is defined as the "twist." In a uniform helix, this will be related to the number of residues per turn, *n* (which may be integral or nonintegral), by $n = 360°/t$. The pitch of the helix, *P*, is obviously $P = nh$. For each base pair there is a local pseudo dyad axis, which passes perpendicularly through the helix axis (see Fig. 3-5). Rotation by 180° about this axis will superimpose the sugar–phosphate moieties of the two antiparallel chains. In general, a plane containing this pseudo dyad axis and any corresponding pair of atoms in the two chains will not lie perpendicular to the helix axis. The deviation from perpendicularity defines the "tilt," Θ_T. Finally, the bases may be rotated about a third axis perpendicular to the pseudo dyad; this defines the "roll" angle, Θ_R. If the roll angle for two bases in a pair is not the same, the bases are said to exhibit a "propellor twist." Even if a base pair exhibits a propellor twist, the angle between the normal to a median base plane and the helix axis can be used to define an average roll.

Armed with these conventions, we can describe, as in Table 3-1, a number of structures that have been proposed for DNAs of various compositions, under a variety of conditions. All of these models are derived from fiber diffraction studies. More detailed descriptions, including stereo drawings, are provided for most of these by Arnott and Chandrasekaran (1981); references to the original work are also given therein. Several points of clarification are necessary:

1. Arnott and Chandrasekaran classify all nucleotide residues as being in either A or B forms, depending upon whether the sugar conformation is C3′–*endo* (A) or C2′–*endo* (B). As we shall see later, this is probably an oversimplification of a very complex situation, but for the moment

Table 3-1. Some DNA Structures from Fiber Diffraction

Designation	Conformational description	Rise, Å	Rotation angle	$\frac{bp}{turn}$	Tilt[a]	Reference
A family						
A'	g⁻tg⁺ag⁺tg⁻	3.0	30°	12	10°	Arnott et al. (1973)
A	g⁻tg⁺ag⁺tg⁻	2.6	32.7°	11	16–19°	Arnott and Hukins (1972) Arnott and Chandrasekaran (1981)
unnamed[b]	tttag⁺tg⁻	3.1	36°	10	~0°	Chandrasekaran et al. (1980)
B family						
B	g⁻tg⁺attg⁻[c] g⁻tg⁺attt	3.4	36°	10	6°	Arnott and Hukins (1972) Arnott and Chandrasekaran (1981)
B'	g⁻tg⁺attg⁻	3.3	36°	10	8°	Arnott and Selsing (1974)
C	g⁻tg⁺attg⁻	3.3	38.6°	9.33	8°	Arnott and Selsing (1975)
D	g⁻tg⁺attt	3.0	45°	8.0	16°	Arnott et al. (1974)
E	tttattt	3.3	48.0	7.5	15°	Chandrasekaran et al. (1980) Leslie et al. (1980)
(A+B) family						
Z	(g⁺ttsg⁺tg⁺) + (ttg⁺atg⁻g⁺)	(7.3/2)[d]	(−60°/2)[d]	12	7°	Arnott et al. (1980)

[a]Defined as angle between normal to median plane of base pairs and helix axis.
[b]This conformation has only been observed with the DNA-RNA hybrid poly(dI)poly(C). It is included here to show the conformational range of A forms. This corresponds to the structure first proposed by Crick and Watson (1954).
[c]The conformation is still a matter of dispute. See text.
[d]The repeating unit is a dinucleotide. Half-values are only averages.

it is convenient. In terms of the orientation angles, this distinction depends primarily on δ; for C3'–*endo* δ is g^+, for C2'–*endo* δ is t (see Fig. 3-3). An amusing consequence is that the structure originally assigned to B-DNA by Crick and Watson (1954), which may be written as (tttag$^+$tg$^-$) is now classified as an "A" form, and is quite different, at least in backbone conformation, from current models for B-DNA structure (g$^-$tg$^+$attt) or (g$^-$tg$^+$attg$^-$) (Table 3-1).

2. Within the A and B families, enormous variation in the helix parameters h, t, and n is possible. There is overlap between the two families in all three quantities. Arnott and Chandrasekaran argue that it is therefore no longer possible to classify DNA structures on the basis of such parameters; since it is the backbone rotations that determine the conformation, classification should be through these. However, recent studies of crystalline DNA oligomers (see below) indicate that even these distinctions are blurred. In a thoughtful analysis of the problem, Conner et al. (1984) conclude that the principal difference between A and B helices resides in the location of the bases with respect to the helix axis. In B-DNAs, the axis passes between the paired bases; in A-DNAs, the bases are pushed outward from the axis (compare Figs. 3-6 and 3-7).

3. It will be noted that Table 3-1 contains two different descriptions of "B-DNA." This reflects the confusion that has continued to plague this structure. In particular, there has been disagreement over the value for the angle ζ. In their first revision of the Crick–Watson structure, Arnott and Hukins (1972) gave as their best model one containing a 3'–*exo* conformation with $\zeta = -96°$, which would be designated (g$^-$tg$^+$attg$^-$). Subsequently, this was revised to a 2'–*endo* conformation, with $\zeta \cong -157°$, (g$^-$tg$^+$attt) (Arnott et al., 1980). However, on the basis of energy minimization calculations Levitt (1978) arrived at a value of $\zeta \cong -90°$, again giving (g$^-$tg$^+$attg$^-$). This latter seems to be supported (albeit ambiguously—see below) by studies on single crystals of DNA oligomers. Even today, the structure of B-DNA in fibers remains a matter for controversy.

4. Certain structures contain a dinucleotide as the repeating unit, with a regular alternation of conformation along each chain. As an example, consider the left-handed "Z-DNA" form listed in Table 3-1. This is one of a number of possible (A + B) structures in which alternating nucleotides are in A and B conformation. In such cases, it is necessary to specify the conformation of both nucleotides in the asymmetric unit, and h and t now refer to the dinucleotide. For comparison with other DNA structures, we can think of the individual nucleotides in Z-DNA as having an average rise of $7.25/2 = 3.63$ Å, and an average rotation of $-60°/2 = -30°$.

As we shall see in further discussions, the forms described above are probably idealizations, especially for DNA molecules in a complex mo-

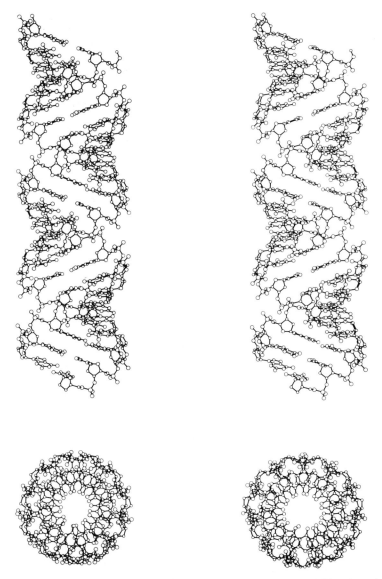

Fig. 3-6. The classical A structure for DNA. (From Arnott and Chandrasekaran, 1981. Reprinted with permission.)

lecular environment. Nevertheless, it is useful to further describe some of the classical forms in qualitative terms, in order to point out some salient features. Note that the features described pertain only to the specific forms indicated; within any family, wide variation is possible.

A-DNA: The classical A form (Fig. 3-6) is broad (about 23 Å diameter). The bases are strongly tilted and lie well off the helix axis. The minor

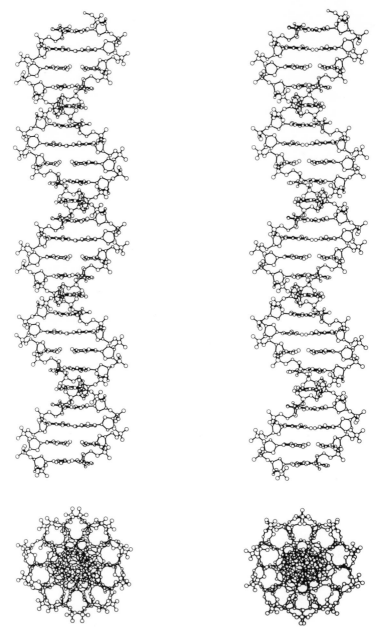

Fig. 3-7. The B-DNA structure, according to recent fiber diffraction studies. (From Arnott and Chandrasekaran, 1981. Reprinted with permission.)

groove is exceedingly deep and thin, the major groove shallow. Calculations (Levitt, 1978) indicate A to be relatively inflexible. It is observed in fibers at low relative humidity and in certain dehydrating conditions in solution.

B-DNA: The B helix (Fig. 3-7) is much narrower ($\sim$19 Å), with relatively open but shallow grooves. A major and minor groove can be distinguished; the former, in particular, allows easy access to the bases. The bases are nearly perpendicular to the helix axis. The molecule is relatively easily bent (Levitt, 1978). It is found in fibers at high humidity and, with some structural modification, in solutions of low to moderate ionic strength (see *DNA Molecules in Solution,* below).

Z-DNA: This left-hand helix (Fig. 3-8) appears to be restricted to certain molecules containing alternate purine and pyrimidine bases. The conformation of these bases differs, even in the glycosyl angle χ, which has the usual anti orientation for the pyrimidines, but is syn for the purines. As a consequence, the helix has a somewhat irregular outline, and the backbone follows a zigzag path. Unlike the A and B forms, which have evenly spaced phosphate groups, the phosphates in Z-DNA are alternately close and far apart. Depending upon the bases involved, moderate to high ionic strength or alcohol concentration is required for observation of the Z form (see *Transformations from the B Form to Other DNA Conformations in Solution,* below).

It should be clear from even this brief introduction that DNA structure is exceedingly plastic. Double-strand DNA can, in response to differences in base composition, ionic milieu, or even degree of hydration, assume a startling variety of conformations. But just how locally irregular the structure can be has only become evident in recent years, mainly as a result of X-ray diffraction studies of single crystals of DNA oligomers. The most detailed and comprehensive study of this kind has utilized crystals of the double-strand form of the dodecamer CGCGAATTCGCG. This sequence contains the EcoRI restriction site GAATTC, flanked by alternating CG segments. The structures of this molecule and a derivative brominated at the ninth residue have been examined in great detail by R. Dickerson and his co-workers, who have carried the analysis to high resolution (see, for example, Wing et al., 1980; Drew et al., 1981; Dickerson and Drew, 1981; Fratini et al., 1982; Kopka et al., 1983; Dickerson, 1983a; Dickerson et al., 1983).

The results of these studies are enlightening and in some ways unsettling to our preconceptions of DNA structure. A drawing of the molecule is shown in Figure 3-9. The first thing one notices is that the helix is *bent;* there is a curvature of about 19° over the length of the molecule, corresponding to a radius of curvature of about 112 Å. At first, it seemed likely that this bending was simply a consequence of the crystal packing, in particular of the ways in which the ends of each molecule interact with molecules in adjacent layers. The energy required to produce such bending

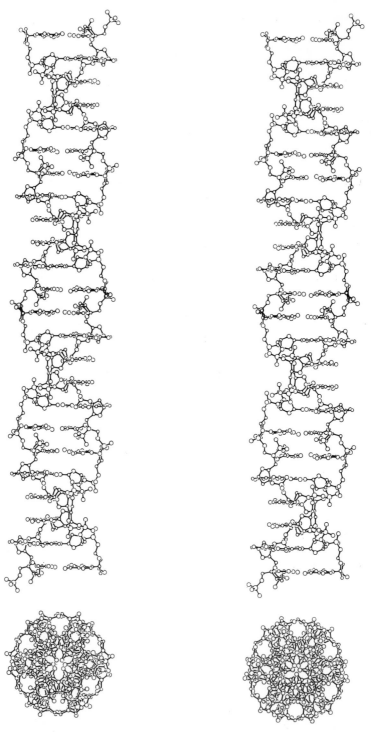

Fig. 3-8. The helical structure of Z-DNA. (From Arnott and Chandrasekaran, 1981. Reprinted with permission.)

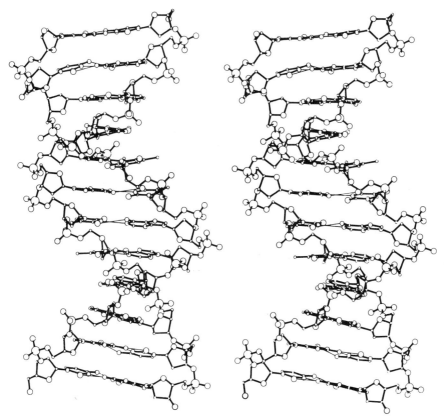

Fig. 3-9. The structure of the double-strand dodecamer with sequence CGCGAATTCGCG from single-crystal X-ray diffraction studies. (Reprinted by permission from *Nature*, Vol. 287, p. 755. Copyright 1980 Macmillan Journals Limited.)

has been calculated to be only about 0.5 kcal/mol, according to the methods of Levitt (1978) (see below). Furthermore, it is now recognized that DNA molecules can be bent simply as a consequence of sequence (see below). However, more recent studies indicate that the situation in the dodecamer is complex (Fratini et al., 1982; Dickerson et al., 1983). The degree of bending differs in the brominated derivative and depends upon the medium from which the molecules have been crystallized. In particular, the presence of a spermine molecule bound to the DNA fragment seems to have a major effect on the bending. Evidently, even minor perturbations can markedly change B-DNA conformation.

The same conclusion is supported by examination of the local variations in conformation. Table 3-2 lists the glycosyl and main-chain torsion angles, as well as the sugar conformations and the propellor twist angles. It is clear from these data that *no single structure or class of structures can*

Table 3-2. Conformation of the Dodecamer CGCGAATTCGCG[a]

Residue	Glycosyl angle χ	Main chain torsion angles						Adjacent phosphorus atom separation (Å)	Observed sugar conformation	Base pair propellor twist ψ
		α	β	γ	δ	ε	ζ			
C1	−105°	—	—	174°	157°	−141°	−144°	—	C2'-endo	13.2°
G2	−111	−66°	170°	40	128	−186	−98	6.64	C1'-exo	11.7
C3	−135	−63	172	59	98	−177	−88	6.47	O1'-endo	7.2
G4	−93	−63	180	57	156	−155	−153	6.83	C2'-endo	13.2
A5	−126	−43	143	52	120	−180	−92	6.88	C1'-exo	17.1
A6	−122	−73	180	66	121	−186	−89	6.90	C1'-exo	17.8
T7	−127	−57	181	52	99	−186	−86	6.29	O1'-endo	17.1
T8	−126	−59	173	64	109	−189	−89	6.87	C1'-exo	17.1
C9	−120	−58	180	60	129	−157	−94	6.70	C1'-exo	18.6
G10	−90	−67	169	47	143	−103	−210	6.55	C2'-endo	4.9
C11	−125	−74	139	56	136	−162	−90	7.05	C2'-endo	17.2
G12	−112	−82	176	57	111	—	—	—	C1'-exo	6.2
C13	−128	—	—	56	137	−159	—	—	C2'-endo	6.2
G14	−116	−51	164	49	122	−182	−93	6.62	C1'-exo	17.2
C15	−134	−63	169	60	86	−185	−86	6.45	O1'-endo	4.9
G16	−115	−69	171	73	136	−186	−98	7.12	C2'-endo	18.6
A17	−106	−57	190	54	147	−183	−97	6.77	C2'-endo	17.1
A18	−108	−57	186	48	130	−186	−101	6.71	C2'-endo	17.1
T19	−131	−58	174	60	190	−181	−88	6.70	C1'-exo	17.8
T20	−120	−59	179	55	122	−181	−94	6.70	C1'-exo	17.1
C21	−114	−59	185	45	110	−177	−86	6.17	C1'-exo	13.2
G22	−88	−67	179	50	150	−100	−188	6.60	C2'-endo	7.2
C23	−125	−72	139	45	113	−174	−97	6.68	C1'-exo	11.7
G24	−135	−65	171	47	79	—	—	6.68	C3'-endo	13.2
Avg.	−177	−63	171	54	123	−169	−108	6.68		
Std. Dev.	(14)	(8)	(14)	(8)	(21)	(25)	(34)	(0.23)		

[a]From Dickerson and Drew (1981).

define this molecule. Consider, for example, the sugar conformations: A wide range of types are observed, with the angle δ ranging from about 80° to 160°. There seems to be a strong correlation between the values of δ and the glycosyl angle χ. Figure 3-10 from Fratini et al. (1982) shows that as the δ increases, χ becomes less negative. In addition, there is an anti-correlation between the χ, δ loci for base-paired residues. That is, points in Figure 3-10 corresponding to such pairs (1/24, 2/23, etc.) seem to lie at approximately equal distances from the center of the diagonal line that could be passed through the points in the figure.

In addition to δ and χ, the angle ζ also shows considerable variation. The most common value is around −90°, corresponding to a (g⁻tg⁺attg⁻) conformation, rather than the (g⁻tg⁺attt) form recently suggested for B-DNA by Arnott and Chandrasekaran (1981).

Finally, even such parameters as the base twist and rise are nonuniform. The local base twist, defined with respect to the local helix axis, ranges from 32.2° to 44.7°, and the rise between adjacent base pairs from 3.14 Å to 3.56 Å! Similar variances are found in the bromo derivative. Although the central region of the molecule (steps 4–8) is somewhat more regular, with a mean local twist of 36.9° corresponding to 9.8 bp/turn, the ends of the molecule contain some base pairs whose helical conformations resemble much more closely A- or D-DNA (see Dickerson and Drew, 1981). The same kind of single-crystal X-ray analysis has been applied to A and Z structures as well. As Table 3-3 illustrates, the structures resemble, on the average, the idealized forms listed in Table 3-1. However, the standard deviation values show how large is local variance.

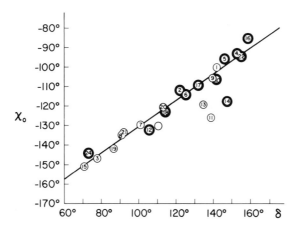

Fig. 3-10. Correlation of the rotational angles χ and δ in the dodecamer structure. Numbering of the residues proceeds from the 5′ end of one chain to its 3′ end, and then from 5′ to 3′ on the other chain. From Fratini et al. (1982), Journal of Biological Chemistry.

Table 3-3. Helix Parameters from Single-Crystal X-Ray Diffraction[a]

Parameter	A-DNA	B-DNA	Z-DNA	
Twist (°)	33.1 ± 5.9	35.9 ± 4.3	G-C: −51.3 ± 1.6	
			C-G: − 8.5 ± 1.1	
Rise (Å)	2.92 ± .39	3.36 ± .42	G-C: 3.52 ± .22	
			C-G: 4.13 ± .18	
Tilt (°)	13.0 ± 1.9	−2.0 ± 4.6	8.8 ± .7	
Propellor twist (°)	15.4 ± 6.2	11.7 ± 4.8	4.4 ± 2.8	
Roll (°)	5.9 ± 4.7	−1.0 ± 5.5	3.4 ± 2.1	

[a]Adapted from Dickerson (1983b). Values are given ± standard deviation.

Why is the local conformation of DNA so sensitive to sequence? While many factors may be involved, a simple mechanical analysis by Calladine (1982) provides considerable insight. Calladine points out that while the propellor twisting of bases increases stacking overlap (and thus should be energetically favorable), it can lead to steric interference between purines on opposite sides of the axis. There are several kinds of adjustments the helix can make to relieve this interference. These have been analyzed in detail by Dickerson (1983a), who has shown how such principles can be used to predict local structure.

The implications of this work are clear. DNA conformation is much more sensitive to perturbation from sequence differences and/or molecular environment than most scientists had heretofor expected. The significance to those who are interested in DNA–protein interactions is profound. If what appear to be small perturbing forces can easily bend and locally reconfigure a DNA molecule, how drastic may be the effects of winding DNA about a histone matrix? It now seems very naive to visualize a nucleosome in terms of an idealized "B-DNA" wrapped about a histone core. Indeed, the very concepts of "B-DNA" and "A-DNA" now become abstractions; it may well be that simple DNA conformations are only possible for simple-sequence polynucleotides under idealized conditions.

Hydration of DNA

Although it has been long believed, from hydrodynamic experiments, that DNA was highly hydrated, it is only with the advent of single-crystal X-ray diffraction studies that we have learned anything specific concerning this hydration. Kopka et al. (1983) have examined the B-form dodecamer at very high resolution and find water atoms associated with virtually every accessible hydrogen bond donor or acceptor. Of particular interest is a "spine" of regularly ordered water molecules that occupies the minor groove in the A-T–rich portion of the molecule. It is disrupted in G-C regions, probably because of the projection of guanine-NH_2 groups into the groove. Kopka et al. speculate that it is this hydration feature in particular that stabilizes B-DNA (as opposed to A-DNA) under conditions of high humidity. (See also Lee et al. 1981.) A corollary of this hypothesis would be that this hydration spine should not be observed in A-DNA crystals. This is indeed the case, but the situation is clouded by the fact that the minor groove in the only A-form crystal so far studied ([1]C-C-G-G) is blocked by intermolecular packing (Conner et al., 1984). Furthermore, the composition of this molecule would mitigate against a water spine. The A-form molecule is, in fact, highly hydrated. Further studies with other A-form molecules may be required to confirm or refute this speculation concerning the water spine.

DNA Molecules in Solution

Until recently, very little was known concerning the conformations of DNA in solution. The random orientation of the molecules and the effects of thermal fluctuations tend to blur structural details. The applicable physical techniques, such as low-angle X-ray scattering, circular dichroism, linear dichroism, and nuclear magnetic resonance, are much more difficult to interpret in terms of absolute conformation than are crystal or even fiber diffraction patterns.

Intimations that DNA in dilute aqueous solution might differ in conformation from the fiber B form appeared as early as 1971. Bram (1971) compared X-ray scattering curves from such solutions with the predicted scattering for B-DNA. The data did not fit the predictions exactly, so Bram tried structures with varying twist angles. The best model, which gave an excellent fit to the data, had a twist of 33.2°, corresponding to 10.8 bp/turn. However, Bram's contention was not generally accepted. Most researchers in the field seem to have been more strongly impressed by the work of Tunis-Schneider and Maestre (1970), who had compared circular dichroism spectra of hydrated films of DNA (which they believed should have the same B structure as hydrated fibers) with the spectra of DNA in solution. The curves were virtually identical, leading to the reassuring but erroneous conclusion that the fiber structure persisted unchanged in dilute salt solutions.

The problem was reopened by Levitt's (1978) energy-minimization calculations. Starting with the fiber structure for B-DNA, atoms in a hypothetical "random sequence" 20-mer were allowed to relax to a minimum energy conformation, holding constant a particular value of n, the number of base pairs/turn. The value of n was then systematically varied, and the minimization procedure was repeated. The results are shown in Figure 3-11, curve L. The minimum energy was found for a conformation with 10.6 bp/turn. There are a number of other interesting differences between this minimum energy conformation and the fiber B form. The angle between the vector normal to the bases and the helix axis has increased from about 6° to 18°, caused largely by a propellor-like twisting of the base pairs.

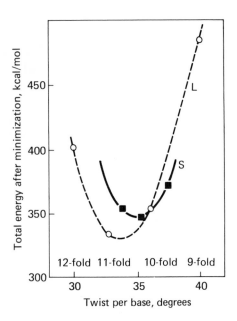

Fig. 3-11. Energy of a DNA helix as a function of the number of base pairs/turn, according to Levitt (1978). Two curves are shown: (L) linear DNA; (S) superhelical DNA, with radius 45 Å, pitch 55 Å. Note the difference in the positions of energy minima.

There is evidence from linear dichroism measurements to confirm this increased base twisting. For example, Hogan et al. (1978) report electric dichroism measurements that indicate a 17° twist; the same value has been obtained by Dougherty et al. (1983) from flow dichroism experiments.

Levitt's calculations aroused renewed interest in the problem of DNA conformation in solution for two reasons. First, many studies of the solution properties and reactivity of DNA had pointed to structural flexibility; the idea of conformational changes under mild conditions had become respectable. Second, Levitt's work had major implications for nucleosome structure, as we shall see later. There followed, therefore, a multitude of attempts to obtain experimental evidence for the structure in solution. A summary of these is given in Table 3-4. As can be seen from a glance at the table, the story is a complicated one, both because of seemingly conflicting results and because different experiments involve different assumptions and measure different quantities.

The first attempts utilized electron microscopy of phage DNAs. Vollenweider et al. (1978), using λ-DNA freeze-dried on EM grids, obtained a value for the rise consistent with the classical B form (3.43 Å). At the time Vollenweider et al. carried out their measurements, the molecular weight of λ-DNA was not known exactly, but the estimate they used turns out to have been only 0.8% low, so a corrected value from their data would be 3.40 Å. In contrast, Griffith (1978) employed φX174 DNA and found $h = 2.9$ Å, indicating a quite different structure in solution! It is not easy to compare these experiments, for procedural details differ considerably. Since Griffith also used a DNA of precisely known size, the contradiction is difficult to resolve. Rhodes and Klug (1980) have suggested that Griffith's results may have been compromised by structural changes during drying. If we accept the value of Vollenweider et al., it would appear that at least the *rise* in solution is very close to that found in fiber studies.

A very different approach to the problem of DNA conformation in solution is based on studies of superhelical DNA molecules. Wang (1979) examined a series of closed circular DNAs that differed in length by numbers of base pairs that were not integral multiples of the helix repeat. The electrophoretic mobilities of topoisomers of these molecules, containing different numbers of superhelical turns, were shifted in a manner that allowed calculation of the number of base pairs/turn. In essence, the experiment asks how much the closed circular DNA molecule is twisted for each additional base pair inserted. The value obtained for n was 10.4 bp/turn, in quite good agreement with the prediction of Levitt (10.6 bp/turn). A further analysis of the data by Strauss et al. (1981) leads to a value of 10.6 bp/turn. These authors have also investigated poly(dA)·poly(dT) and poly(dA-dT)·poly(dA-dT) by the same technique, obtaining 10.1 bp/turn for the former, and 10.7 bp/turn for the latter. As we shall see in Chapter 6, this difference may have important implications for nucleosome structure.

Another approach to the problem has utilized the periodicity of DNase I cutting of DNA. It will be recalled (Chapter 2) that Noll (1974b) had explained the periodic DNase I nicking of nucleosomal DNA by the idea that each strand becomes maximally accessible to the enzyme once each turn. Thus, if DNA molecules were absorbed from solution onto a flat surface, without distortion, their cutting frequency should reveal their periodicity. Rhodes and Klug (1980) employed calcium phosphate, magnesium phosphate, and mica as absorbants and digested short DNA molecules on these surfaces. High-resolution gel electrophoresis allowed a precise band counting. The results, for all three surfaces, were 10.6 ± 0.1 bp/turn. Similar experiments by Behe et al. (1981) gave an almost identical result, $n = 10.4 \pm 0.4$ bp/turn. Probably the best current estimate, from all data, is 10.5 bp/turn.

All of these experiments would seem to be strongly supportive of Levitt's predictions. However, there is one set of contradictory results that cannot be lightly dismissed. Zimmerman and Pheiffer (1979a) have carried out X-ray diffraction experiments with DNA fibers highly swollen in glass capillaries. The swollen fibers contained up to 80% water, yet typical "B" patterns were obtained, with $h = 3.35$ Å and $n = 9.91$ bp/turn. Equatorial

Table 3-4. Evidence Concerning DNA Structures in Solution

Reference	Technique[a]	DNA source	Solution conditions	Rise (Å)	bp/turn	Other features, comments
1. B-like structures						
Bram (1971)	XS	calf	0.05–0.15 M NaCl	—	10.8	
Griffith (1978)	EM	φX174	0.12 M NaCl 1.0 mM MgCl₂ 2.0 mM spermidine	2.9	—	
Liu and Wang (1978) Vollenweider et al. (1978)	EM	λ-phage	See ref.	3.43	—	Value should be reduced to 3.40 Å. using recent sequence data for λ
Wang (1979)	SC	*E. coli* plasmids	0.2 M NaCl 10 mM Tris 0.1 mM EDTA	—	10.4	
Zimmerman & Pheiffer (1979a)	XD	salmon sperm	0.02–1 M NaCl	3.34	9.91	Highly swollen fibers; salt conc. somewhat uncertain
Rhodes and Klug (1980)	DNase	nucleosomal cores	Varied, see ref.	—	10.6	DNA absorbed on calcium or magnesium phosphate or mica surfaces and digested thereon
Strauss et al. (1981)	SC	*E. coli* plasmid with DNA insert	40 mM Tris-acetate 20 mM Na acetate 2 mM EDTA	—	10.6	
		Same, with poly (dA) · poly (dT), insert	Same	—	10.1	Value pertains to inserted DNA
		Same, with poly (dAdT) · poly(dAdT) insert	Same	—	10.7	As above
H.M. Wu et al. (1981)	ED	Various, see ref.	0.385 mM Tris/HCl 0.0385 mM Na₂EDTA	3.4	—	From rotational relaxation time, assuming 26 Å diameter; evidence that bases are propellor twisted

Table 3-4. *Continued*

Reference	Technique[a]	DNA source	Solution conditions	Rise (Å)	bp/turn	Other features, comments
II. A-like structures						
Griffith (1978)	EM	φX174 DNA · RNA	0.12 M NaCl. 1 mM MgCl$_2$, 2mM spermidine	2.55	—	
Vollenweider et al. (1978)	EM	λ-phage	90% ethanol	2.58	—	This is the smallest rise; molecules with h = 3.17 Å, 2.90 Å were also found; all should be reduced by 0.8% from recent sequence data
H.M. Wu et al. (1981)	ED	fragment of *E. coli* plasmid	0.385 mM Tris/HCl 0.0385 mM Na$_2$EDTA. 70–80% ethanol	2.8	—	From rotational relaxation time, with assumed values for diameter; evidence that bases are not propellor twisted
III. Z-like structures						
H.M. Wu et al. (1981)	ED	poly(dGdC) · poly (dGdC)	0.385 mM Tris/HCl 0.0385 mM Na$_2$EDTA 55–60% ethanol	3.7	—	From rotational relaxation time, with assumed values for diameter; evidence that bases are not propellor twisted
Behe et al., 1981	DNase	poly(dG-M^{5}dC) · poly (dG-M^{5}dC) poly(dG-dC) · poly(dG-dC)	On Ca phosphate or Ca oxalate	— —	13.6 ± 0.4 14.0 ± 0.4 13 ± 1	
Peck and Wang (1983)	SC	Plasmids with (dCdG)$_n$ inserts, superhelically wound	90 mM Tris-boric acid, pH 8.3 2.5 mM EDTA	—	11.6	

[a]Code for techniques: DNAse, cleavage by DNAse of absorbed molecules; ED, electric dichroism; EM, electron microscopy; SC, supercoil shift method; XD, X-ray diffraction; XS, X-ray scattering from solution.

reflections proved that the DNA molecules had indeed separated during the swelling, to give center-to-center spacings as large as 40 Å. Similar experiments with comparable results by J.T. Finch and L. Lutter are mentioned by Rhodes and Klug (1980). Rhodes and Klug argue that the fiber structure may be retained in such gels by interaction between DNA molecules through several structured layers of water molecules. The explanation seems at first glance to be labored, yet there is now evidence that this may indeed be the case.

Mandelkern et al. (1981) have discovered that at moderately high concentrations DNA forms swollen "bundles" of parallel fibers. When these bundles were investigated by the electric dichroism technique, the base-plane transition moments were found to be nearly perpendicular to the fiber axis. This is the result expected from the DNA fiber patterns and indicates that the propellor twisting observed in *dilute* DNA solutions has disappeared. Thus, it would seem that loose aggregation of DNA molecules in concentrated solution can give rise to significant conformational changes. In the same context, it should be noted that Rill et al. (1983) finds NMR evidence for a condensed phase in concentrated DNA solutions, which exhibits quite different molecular dynamics from diluted DNA (see below).

It now seems that the preponderance of evidence is in favor of a DNA structure in dilute solution that is slightly underwound (by about one-half bp/turn) as compared to the structure found in fibers or concentrated solutions. However, it also seems to me that the question has been given more attention than it deserves. If, as DNA *crystal* structures indicate, the *local* conformation of DNA can vary in ways more extreme than the one-half bp/turn that has caused so much controversy, the whole problem of "the" average conformation becomes almost meaningless.

There is, in fact, abundant evidence of a quite different kind to indicate that DNA molecules in dilute solution are capable of considerable conformational flexibility. A series of NMR studies, utilizing ^1H, ^{31}P, and ^{13}C spectra all point to the same conclusion: Atoms in the DNA molecule undergo fast (nanosecond) motions that are too rapid to be explained in terms of axial rotation of a rigid helix (see, for example, Klevan et al., 1979; Early and Kearns, 1979; Bolton and James, 1980; Hogan and Jardetzky, 1980; Levy et al., 1981; Bendel et al., 1982; the subject is reviewed by Kearns, 1984). Insight into the possible nature of these motions has been provided by theoretical calculations of local DNA flexibility by Keepers et al. (1982). Individual backbone torsion angles in a dodecamer were "forced" into alternative conformation (i.e., g⁻ to t) and an energy minimization calculation then performed to find the best values for other angles. The remarkable result is that virtually any of the backbone angles could be deformed to any of the three allowed orientations at very low energy cost; the remaining angles were able to "compensate" for the strain. Deformation was largely local.

In a sense, this study provides a companion picture to Dickerson's analysis of the crystal structure of the same dodecamer. The small barriers to local deformation predicted by Keepers et al. are reflected in the local nonuniformity of structure observed in the crystal. Once again, the image that emerges is one of considerable localized flexibility in structure.

The NMR studies by Rill et al. (1983) provide additional perspective on these motions and may help to reconcile the apparent contradictions between DNA structure in dilute solution and the fiber diffraction results. Rill et al. find that at high concentrations of DNA a cooperative phase transition occurs. The critical concentration decreases with increasing DNA molecular weight. Most important, they observe that many of the nanosecond motions seen at low concentration are "frozen" above this phase transition. As Rill et al. point out, the possibility that *different* selections of possible conformations are averaged above and below the transition might account for the seeming contradiction between the results obtained with swollen fibers and those found in dilute solution (see above).

In summary, the secondary structure of DNA appears to be both dynamic and easily mutable. In viewing DNA as a major component of chromatin, we must take care not to regard it as having a fixed structure, to which the proteins must adapt. Rather, *mutual* adaptation of DNA and protein conformations would seem more likely. The first high-resolution diffraction studies of a DNA–protein complex support this point quite dramatically. Fredrick et al. (1984) have determined the tertiary structure of the EcoRI-DNA complex. They find that the DNA is kinked in the region of contact with the enzyme. In Chapter 5, we shall see that similar deformations appear in the DNA of the nucleosomal core particle. The nucleosome is undoubtedly more complex than a simple sum of its component parts.

Bent or Curved DNA

Quite aside from the conformational *flexibility* of DNA, there now exists evidence that some DNA sequences are inherently *curved* or *bent* in solution. Several years ago, E. Trifonov and his colleagues suggested that periodicities observed in the sequences of many eukaryotic DNA molecules might lead to a regular bending or curving (Trifonov and Sussman, 1980; Trifonov, 1980). In particular, it was proposed that such dinucleotides as AA/TT might be "wedged," so that repetition of these at ~10-bp intervals would lead to a regular curvature, allowing the DNA to fold more easily about a nucleosome (but also see Zhurkin, 1981, for an alternative explanation).

This concept gained strong support from the subsequent discovery that certain naturally occurring DNAs, which contained such sequences, gave anomalous behavior in electric dichroism and gel electrophoresis, consistent with the idea that they were bent or curved (Marini et al., 1982; Lee and Charney, 1982; Ross and Landy, 1982; Hagerman, 1984).

The precise reasons for such bending or curvature are still a matter of discussion. In an excellent review, Trifonov (1985) points out that A/T runs are distributed with an approximate 10-bp periodicity in a number of curved DNAs. Furthermore, Ulanovsky et al. (1986) have been able to prepare such DNAs, which showed a strong tendency to form small circles. An alternative explanation for bending has been presented by Prunell et al. (1984), who suggest that poly(dA)·poly(dT) may adopt a unique conformation as a consequence of wedging (see also Jolles et al., 1985). It is argued that the *junctions* between A/T stretches and normal B-DNA are the sites of bending.

Transformations from the B Form to Other DNA Conformations in Solution

In the preceding section, we have seen that structures such as the B form, but with significant differences from the fiber B, appear to characterize DNA in dilute aqueous solution at low to moderate ionic strengths. We have also observed that DNA in the solid state is capable of adopting entirely different conformations in response both to environment and base sequence. It should not be surprising, therefore, to find that dissolved DNAs can, under special conditions, change from the B family of structures to quite different conformers. We shall be most interested, of course, in those conformational transitions that can occur under physiological, or quasi-physiological, conditions of ionic strength, pH, and temperature.

The B→A Transition

Although no physiologically relevant conditions are presently known that cause this conformational change, it has been observed in concentrated (>65%) ethanol solutions. The phenomenon is of historical interest, since it was the first example of such a major conformational transition to be observed in solution (Brahms and Mommaerts, 1964). Although Brahms and Mommaerts did not identify the conformation in concentrated ethanol as an A structure, they noted that the CD spectrum resembled that of double-strand RNA, which exists in this form in solution. Proof that the conformation was A-like came from the studies of Gray et al. (1979), who obtained powder patterns from DNA precipitated from ethanolic solution, and Zimmerman and Pheiffer (1979b), who exposed fibers of DNA to various alcohol concentrations, observing an A pattern above 70% ethanol. More recently, the solution structure has been examined (in 70 to 80% ethanol) using electric dichroism measurements (H.M. Wu et al., 1981). Measurement of the length of homogeneous, short DNA molecules from the rotational relaxation time yields a value for the rise of 2.8 Å, in good agreement with the value for the A form in the fiber. Furthermore, H.M. Wu, et al. find that the orientation of DNA transition moments with the helix axis is consistent with an A structure. The overall conclusion is that A-DNA in such solutions is very similar to the form observed in fibers at low humidity. Presumably, alcohol promotes this change by lowering the water activity. Whether there are other, more physiologically relevant solution conditions that can cause the B→A transition is not known. It should be recalled, of course, that there is long-established evidence that double-strand RNA and RNA-DNA hybrids adopt an A-form structure in aqueous solutions (see, for example, Spencer et al., 1962).

The B→Z Transition

The first experimental studies to provide any suggestion of a left-hand polynucleotide structure were presented by Mitsui et al. (1970). Studying fibers of the synthetic polymer poly(dI-dC)·poly(dI-dC), they noted that the diffraction data could be explained by either left-hand or right-hand helices with 8 residues/turn. The concept of a left-hand helix was heterodox at the time, so Mitsui et al. turned to circular dichroism to provide additional information. The polymer exhibited a very unusual CD spectrum; the bands were of the opposite sign to those they observed for all other polymers containing dI, dC, rI, or rC. In the early 1970s, however, left-hand helices were not scientifically popular, and the structure proposed by Mitsui et al. was considered "bizarre" (Arnott et al., 1973). In the same period, Pohl and Jovin (1972) observed and studied a conformational change of poly(dG-dC)·poly(dG-dC) in concentrated salt solutions. Even though the transition produced an "inverted" CD spectrum like that of poly(dI-dC)·poly(dI-dC) (see Fig. 3-12), Pohl and Jovin no more than hinted at the possibility of a left-hand helix. In retrospect, this caution seems wise, for recent experiments have shown that an inverted CD spectrum does not *always* signify left-handed DNA (Tomasz et al., 1983).

Since the concept of right-hand DNA was so firmly fixed, the first report from the laboratory of A. Rich (see Wang et al., 1979) that a crystallized alternating (dG-dC) hexamer exhibited a left-hand helical structure was met with considerable surprise. Evidence for this new form (Z-DNA) in fibers of poly(dG-dC)·poly(dG-dC) was provided shortly thereafter by Arnott et al. (1980). A similar structure was found for alternating (dG-dC) tetramers crystallized from high salt by Drew et al. (1980). The structure of Z-DNA has been described in the preceding section. It now seems very probable that the high-salt form of poly(dG-dC)·poly(dG-dC) observed in solution by Pohl and Jovin is very like this structure. This contention is supported by the electric dichroism studies of H.M. Wu et al. (1981). These experiments were carried out in alcoholic solution (>50% ethanol), a condition under which the characteristic inversion of the CD spectrum is observed (Pohl, 1976). The length of the molecules, as judged from the rotational relaxation times, corresponds to a rise of 3.7 Å, and the limiting dichroism at high field can be explained by assuming that the bases exhibit very little propellor twist. Both results are compatible with the crystal structure for Z-DNA. A slightly different result has been obtained from the DNase 1 digestion experiments of Behe et al. (1981); the periodicity

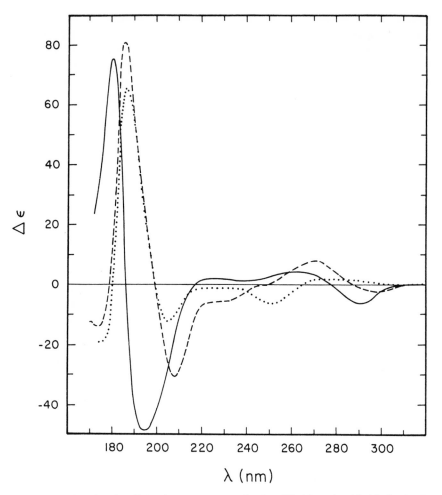

Fig. 3-12. The circular dichroism spectrum of poly (dG-dC)·poly(dG-dC) in the B form, A form, and in the Z conformation. Legends: (·····) B-DNA in 10 mM phosphate; (---) A-DNA in 0.67 mM phosphate, 80% trifluoroethanol; (—) Z-DNA in 10 mM phosphate, 2M NaClO$_4$. (From Nucleic Acids Research, vol. 13, issue 13, p. 4986. Riazance et al. Copyright 1985 IRL Press.)

of cutting indicates 13–14 bp/turn, rather than the 12 bp/turn expected for the Z helix (see Table 3-3). More recent experiments by Peck and Wang (1983) yield a value of 11.6 bp/turn, slightly *less* than the crystal value. Thus, as in the case of B-type DNAs, there is some uncertainty concerning details of the structure in solution.

Despite these ambiguities, it seems clear that the transition observed by Pohl and Jovin involved a change from a B-like conformation to a very different Z-like form, and their careful studies of the equilibrium and kinetics are of interest as an example of a radical conformational change in solution. The enthalpy change for the process was found to be nearly zero, indicating that the B→Z transformation must be driven by a higher entropy of the Z form in high salt. The source of this entropy difference is unknown. The kinetics of the transition could be explained by a model in which a transition zone, about 3 bp in length, passed through the molecule. In the model of Pohl and Jovin, this discontinuity between the two states of DNA was assumed to involve base unstacking and opening of the helix. Other models have been proposed in which a B-Z junction could be made by a few (2 to 4) bases in an unusual but stacked and H-bonded conformation (see Wang et al., 1979; Arnott and Chandrasekaran, 1981). The only direct evidence we have concerning the nature of the junction comes from digestion studies. Singleton et al. (1984) examined short (13 to 16 bp) blocks of (dC-dG) in circular, superhelical DNA. They observed that S1 nuclease, which has a strong preference for single-strand regions, cuts almost exactly at the edges of these blocks when they have been forced into the Z form by supercoiling (see below). This argues for at least a limited open region at the B-Z interface. In accord with this concept, experiments by Peck and Wang (1983) show that the energy requirement for initiating a stretch of Z-DNA is much greater than for its extension.

The existence of the Z-DNA structure, first observed at high salt or high ethanol concentrations, would be of little interest to the molecular biologist, but for two observations. The first is that negative supercoiling can apparently induce (dG-dC) blocks to adopt the Z conformation *even at physiological ionic strength* (200 mM NaCl) (see Peck et al., 1982; Brahms et al., 1982; Peck and Wang, 1983; Singleton et al. 1984). A more detailed discussion must await further description of supercoiled DNA; at this point, it suffices to note that the underwinding of a whole, closed DNA corresponding to negative supercoiling can obviously be relieved in part by unwinding or reverse-winding a small region of that molecule. The transition has been examined mathematically by Benham (1980, 1981), whose calculations indicate that a critical superhelix density must be achieved before the transition will occur.

The other observation that has deep implications concerning possible physiological roles of Z-DNA comes from the work of Felsenfeld and his collaborators (Behe and Felsenfeld, 1981; Behe et al., 1981). They have found that 5-methylation of the C-residues in poly(dG-dC)·poly(dG-dC) greatly enhances the tendency of this polymer to convert to the Z form. Some data are shown in Table 3-5. Particularly striking is the enormous reduction in the critical Mg^{2+} concentration resulting from even partial methylation. The mechanism by which cytosine methylation stabilizes Z-DNA has been clarified by an X-ray diffraction study of the oligomer $(m^5CG)_3$ (Fujii et al., 1982). Apparently, the methyl group rests in a hydrophobic pocket in the Z structure, in which it is better protected from surrounding water than in the B form. Behe and Felsenfeld also observed

Table 3-5. The Effect of
Methylation on the Ionic
Requirements for the B → Z
Transition in poly(dC-dG) ·
poly(dC-dG)[a]

% of C residues methylated	Midpoint (mM) for B → Z conversion	
	NaCl	MgCl$_2$
0	2500	700
30	—	20
70	—	1.0
100	700	0.6

[a]Data of Behe and Felsenfeld (1981).

that polyamines frequently found associated with DNA, such as spermine and spermidine, had nearly as strong an influence in stabilizing the Z form.

As a consequence of these results, we are now obliged to consider the likelihood that left-hand DNA conformations may play a significant role in physiological processes. Evidence for the existence of such structures in chromatin will be presented and discussed in later chapters. The reader is also referred to the comprehensive review by Jovin et al. (1983).

Topology of DNA: Supercoils and Cruciforms

The first intimation that DNA topology might be of major significance in molecular biology came from studies of polyoma virus DNA by Vinograd et al. (1965). They had observed that a single nick in one strand was sufficient to produce a substantial change in the sedimentation coefficient of the closed circular DNA molecules. With remarkable acuity, Vinograd realized that the most likely explanation was that the DNA was *supercoiled*, and that the nicking reaction allowed the compact supercoils to relax. Within a year, Vinograd and Lebowitz (1966) had worked out the topological rules that should govern such supercoiling, opening a whole new area of molecular biology. (See also Vinograd et al., 1968; Pulleyblank et al., 1975.)

The rules are basically simple. They have been analyzed in a rather sophisticated fashion by Fuller (1971), and Crick (1976) has explained this analysis in a way more comprehensible to biologists and biochemists. The topic has also been reviewed in some detail by Bauer (1978).

We may consider the two sugar–phosphate backbones of a duplex DNA molecule as the edges of a ribbon, which will usually be a twisted ribbon. The *twist, Tw,* of the molecule is the integral of rotation about the helix axis of the twisted ribbon, right-hand helical rotation being taken as positive. The twist is expressed in numbers of turns of DNA; it may be an integral or nonintegral number. For a relaxed DNA molecule in dilute aqueous solution, there will be about 10.5 bp/turn, or per unit of twist. Thus, a relaxed DNA molecule of 1000 bp will have $Tw = 95.24$.

For a closed circular DNA, we can define another number, the *linking number, Lk*. The linking number is defined as the number of times one chain (or ribbon edge) crosses the other. Again, a right-hand crossing can be taken as positive. The absolute value of Lk is equal to the number of times the two chains are interlinked (see Fig. 3-13). Obviously, Lk must be an integer. Furthermore, it is a topological invariant of a closed circular DNA molecule; it can be changed by cutting the molecule and rejoining, but not by *any* deformation. There is a simple way to visualize the linking number. Imagine a closed circular ribbon, as in Figure 3-13a. Since the edges do not cross each other, the linking number is zero. Now imagine cutting the ribbon and, while holding one end, rotating the other through some integral number of turns before refastening. Since DNA chains are directional, we must rotate through *whole* turns—half turns would involve reconnecting 3'-3' and 5'-5' ends. The number of rotational turns defines the new value of the linking number.

Now it might seem at first glance that the twist and linking number of a closed circular molecule should be identical. That is indeed the case if the axis of the molecule is required to lie in a plane. But if the *axis* of the molecule is allowed to wrap about itself ("writhe"), then the invariant Lk can be apportioned in an infinite number of ways between twist *(Tw)* and writhe *(Wr)*. That is:

$$Lk = Tw + Wr \qquad (1)$$

This is the result first deduced by Vinograd and Lebowitz in 1966. It is more useful to subtract from equation 1 the values for the DNA in its relaxed state. In this case $Wr = 0$, and $Lk^0 = Tw^0$, so

$$Lk - Lk^0 = Tw - Tw^0 + Wr \qquad (2)$$
$$\Delta Lk = \Delta Tw + Wr$$

The closed circular DNA molecules found in nature always exhibit negative supercoiling; that is, $Lk < n/10.5$ where n is the number of base pairs, or $\Delta Lk < 0$. Therefore, ΔTw and/or Wr must differ from their equilibrium values of zero. The DNA can be underwound ($\Delta Tw < 0$), or there can be negative supercoiling (see Fig. 3-13b,c). Either will require energy, which must be provided when the negatively supercoiled molecule is formed and can be considered to be stored therein. In prokaryotic organisms, this energy is provided from ATP hydrolysis, using topoisomerase II, or DNA "gyrase." In eukaryotes, the roles of such enzymes are less clear, but negative superhelical turns are induced by wrapping the DNA about nucleosome cores. The energy involved here comes from the DNA–histone interactions. A convenient measure of supercoiling is the "su-

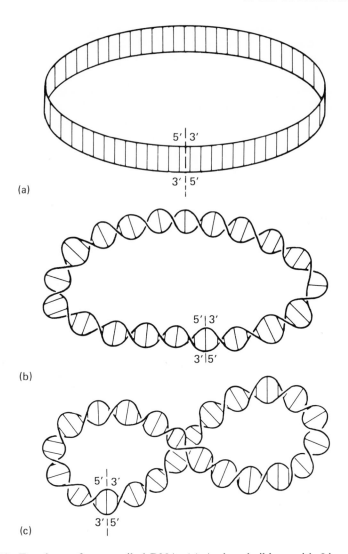

Fig. 3-13. Topology of supercoiled DNA. (a) A closed ribbon with $Lk = 0$. (b) The ribbon has been cut and twisted 10 times to the right. In this case, the ribbon is just long enough (contains just enough base pairs) to make 10 turns with the axis lying in a plane. (c) Base pairs have been added, but not corresponding to an integral number of turns. Now the ribbon must either be twisted differently or supercoiled, keeping the original twist.

percoil density," σ, defined as $\Delta Lk/Lk^0$. Values observed for native DNAs are often in the range $\sigma = -0.06$.

The energy stored in DNA supercoiling was first estimated by Depew and Wang (1975). It can reach quite impressive values. According to Seidl and Hinz (1984), the free energy is given by

$$\Delta G \cong 10\, RTN\sigma^2 \tag{3}$$

where N is the number of base pairs. For a circular DNA of 10,000 bp, superhelix density -0.06, and $T = 37°C$, we obtain

$$\Delta G = 222 \text{ kcal/mol} \qquad (4)$$

It might be expected that this free energy is predominantly stored in torsion and bending of the DNA and therefore is predominantly an enthalpy term. This has been confirmed by calorimetric measurements of the heat of unwinding of a supercoiled plasmid (Seidl and Hinz, 1984).

Most of the studies of supercoiling have utilized small, circular DNA molecules, such as bacterial plasmids or polyoma DNA. The topological constraints on the very long DNA molecules of the eukaryotic genome are not nearly so well understood. But it should be pointed out that a DNA circle can, in principle, be closed as effectively by a tightly bound protein molecule as by covalent closure of the DNA chain itself. Such a "domain" must obey the same general topological rules as a closed DNA circle, although it is not necessary that the quantity corresponding to linking in such a domain be integral. As we shall see later, there is evidence for such structures in chromatin. Since such domains are very large (often of the order of 50 kbp, it might be thought that a small number of twists would have little effect on the supercoil density. However, it seems likely that the DNA on the nucleosomes is fixed in twist, so that changes of coiling of the whole domain would be concentrated in the linker regions.

With this perspective, we are now in a position to better understand the relationship between superhelicity and the B→Z transition, and its possible significance in chromatin function. If, in a negatively supercoiled DNA there exists a G-C–rich block capable of the B→Z transition, the tension of negative supercoiling can be relieved by converting such a region to the Z form. If about one turn is so transformed, there is a change of twist of -2 (from $+1$ to -1), thereby relieving two negative superhelical turns.

The problem has been analyzed in some detail by Benham (1980, 1981), who predicts that at superhelix densities observed in some plasmids, G-C–rich stretches of DNA will be converted to the Z form. With increasing winding, this process will continue until such regions have been exhausted; only then can $|\sigma|$ again increase. While Benham's analysis is not difficult, the same principle can be seen in an even simpler way. Equation 3 can be used to show that at a supercoil density of -0.06, unwinding an additional turn costs about 8 kcal in free energy. On the other hand, conversion of B-DNA to the Z form requires about 0.8 kcal/bp (Pohl and Jovin, 1972). Therefore, the two processes become energetically equivalent at about this density, the only requirement being that Z-susceptible tracts exist.

The B→Z transition is not the only way in which superhelicity might be relaxed. One negative superhelical twist could be removed by wholly unfolding one turn of DNA. But the unpairing of so many bases would

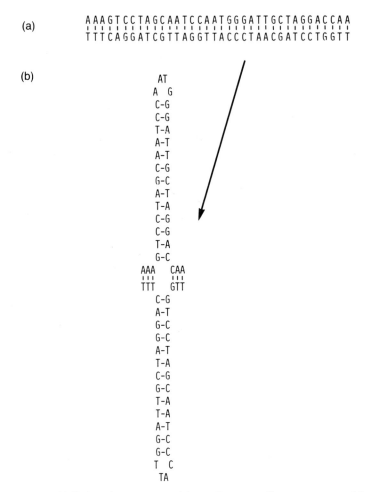

(a)   AAAGTCCTAGCAATCCAATGGGATTGCTAGGACCAA
      | | | | | | | | | | | | | | | | | | | | | | | | | | | | | | | | | | | |
      TTTCAGGATCGTTAGGTTACCCTAACGATCCTGGTT

(b)

Fig. 3-14. Palindromic sequences (a) can form cruciform structures (b). (Reprinted by permission from Biochemical Society Transactions, vol. 12, p. 127. Copyright 1984 The Biochemical Society, London.)

require a much larger energy expenditure and is not a likely process (see Benham, 1981).

However, there does exist one way in which certain regions of DNA can be "denatured" at a very low energy cost. Such regions are like that shown in Figure 3-14a, which contain an "inverted repeat." As early as 1966, Gierer (1966) proposed that sequences like this could exist in an alternate "cruciform" conformation (Fig. 3-14b). Wang (1974) first pointed out that supercoiling might stabilize cruciforms. The point is that upon formation of such a structure, all of the "arms" are effectively removed from contributing to the topology of the circular DNA molecule (see Fig.

3-14b). In effect, it is as if this whole region had been melted. The subject has been extensively studied by D. Lilley and is succinctly described in a recent review (Lilley, 1984). As Lilley points out, if the inverted repeat contains $2n + m$ base pairs (Fig. 3-14b), the change in twist will be

$$\Delta Tw = \frac{2n + m}{10.4} \tag{5}$$

Substantial relaxation of superhelices can be accomplished in this way. Since only a few bases need be unpaired, the energy cost is low. Vologodskii and Frank-Kamenetskii (1982) have carried out theoretical calculations of the probability of cruciform formation at known inverted repeats in various circular DNA molecules. The results are shown in Figure 3-15. Two aspects are particularly striking: (1) There is a rather abrupt onset of cruciform formation at a critical superhelix density, which lies in the range observed for such molecules in vivo, and (2) the cruciform probability quickly rises to quite large values.

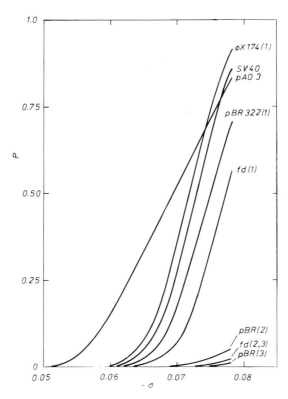

Fig. 3-15. Prediction of formation of cruciform structure as a consequence of DNA supercoiling, according to Vologodskii and Frank-Kaminetskii (1982). (Reprinted with permission.)

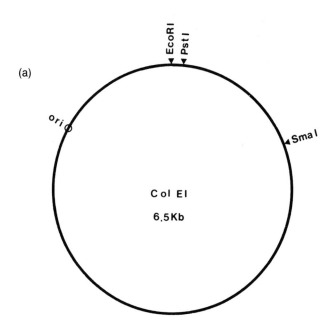

Fig. 3-16a. Site-specific cleavage of the plasmid Col E1 by S1 nuclease. The S1 site lies in a palindrome about 100 bp counterclockwise from the EcoRI site.

There is now a considerable body of experimental data to demonstrate cruciform structures in solution, much of it obtained by D. Lilley (see Lilley, 1980, 1981a, 1981b). Most of the evidence is based on S1 nuclease cutting. Lilley has shown that a number of circular DNA molecules are cleaved at a few specific loci by S1, *but only if the molecule is sufficiently supercoiled*. Furthermore, the cutting loci always appear at the center of the inverted repeat, which corresponds to the loops at the ends of the cruciform arms. Where comparisons are possible, the favored inverted repeats correspond well to those predicted by Vologodskii and Frank-Kamenetskii. Figure 3-16 shows an electrophoretic analysis of products of such an S1 digestion followed by cleavage with restriction nucleases to map the location. The extreme sharpness of the bands and the lack of background points up the high selectivity of S1 under these conditions. Similar experiments with wholly palindromic circular DNAs have been conducted by Mizuuchi et al. (1982).

Thus, it appears that the torsional energy introduced into DNA by negative supercoiling can be relaxed in at least two ways: by the B→Z transition if appropriate GC blocks are present, and by cruciform formation, if sufficiently large inverted repeats exist in a circle or domain. Which transitions actually occur will depend upon the frequency and

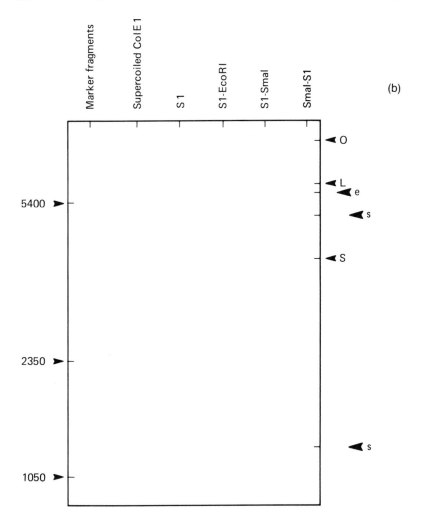

Fig. 3-16b. The supercoiled plasmid was cleaved with S1 nuclease. The left lane shows the supercoiled (S) form of the plasmid. In further lanes, cleavage was either by S1 alone, or by S1 plus the designated restriction nuclease. L represents the linear form of the plasmid. The bands s and e demonstrate that the S1 cleavage is at the site of the inserted palindrome. (Reprinted by permission from Biochemical Society Transactions, vol. 12, p. 127. Copyright 1984 The Biochemical Society, London.)

length of palindromic and potential Z-DNA sequences. Both kinds of sequences do in fact exist in eukaryotic DNA (see Cavalier-Smith, 1976; and Jovin et al., 1983). In a sense, both of these transitions represent mechanisms for "action at a distance" along a DNA molecule. Superhelical winding strain, produced at one locus either by topoisomerase activity or nucleosome formation, can be transmitted through the molecule to produce specific conformational changes at other loci. The possible functional roles of these phenomena in chromatin will be discussed in later chapters.

4

The Proteins of Chromatin. I. Histones

The histones are commonly regarded as unpleasant proteins for rigorous studies (Luck et al., 1956).

All of the proteins associated with chromatin may be divided into two categories: histones and nonhistone chromosomal proteins. The distinction is by no means an arbitrary one, for the histones represent a well-characterized group of proteins with quite unique compositions, sequence characteristics, and functions. Their relative amounts and stoichiometry with respect to DNA are nearly constant throughout the eukaryotic kingdom. The nonhistone chromosomal proteins (NHCPs, henceforth) might be operationally defined as all of the other proteins isolated with chromatin. The definition here is far less precise, for different methods of releasing chromatin from the nucleus will surely yield different subsets of the total nuclear proteins (see Comings, 1978, for an excellent review, and Chapter 5 for further discussion). The remaining proteins of the eukaryotic nucleus could be defined as *nonchromosomal nuclear proteins* (NCNPs). These will include proteins associated with the nuclear envelope and nuclear matrix (see Chapters 5, 7) as well as proteins soluble in the nuclear sap. For an overview of these proteins and the associated structures, the reviews by Agutter and Richardson (1980) and Hancock (1982) are recommended. The enumeration of the NCNPs will be especially difficult, for the class must overlap (depending on chromatin isolation methods) with the NHCPs, and there are surely proteins that may be found in either the cytoplasm or the nucleus, or both, depending on methods of nuclear isolation. Therefore, at the present time, only operational definitions for these latter two classes can be given. In this book, we shall be concerned primarily with those proteins associated with chromatin isolated by standard techniques: histones and nonhistone chromosomal proteins.

Table 4-1. Composition of Chromatins: Mass Per Unit Mass of DNA

Organism	Tissue	Histones	Nonhistone chromosomal proteins	RNA	Ref.
Human	HeLa cells	1.08	0.70	0.05	1
Rat	Liver	0.93	0.80	0.10	2
		0.96	0.32	0.06	3
		1.06	0.65	0.05	4
		1.12	0.62	0.04	5
	Prostate	1.07	0.45	0.05	3
	Kidney	0.91	0.38	0.03	2
	Uterus	1.04	0.47	0.05	3
	Spleen	0.88	0.27	0.03	2
	Thymus	0.88	0.20	0.03	3
		0.89	0.24	0.10	2
	Brain	0.95	0.87	0.10	2
	Rhodamine sarcoma	0.96	1.22	0.15	2
Mouse	Liver	0.95			6
	Kidney	0.93			6
	Brain (adult)	1.00			6
	Brain (neonatal)	1.05			6
	Brain (fetal)	1.1			6
	Teratomona	0.97			6
Chicken	Erythrocytes	1.02	0.36	<0.01	7
	Thrombocytes	0.96	0.45	0.04	7
Drosophila	Embryos	0.79	1.2	0.06	8
Strongylocentrotus purpuratus	Spermatozoa	1.02	0.13	0.04	9
	Averages:	0.99	0.58[a]	0.06	

[a]Not including *S. purpuratus* sperm.

References:
1. Bhorjee and Pederson (1973).
2. Miyazaki et al. (1978).
3. Hamana and Iwai (1979).
4. Gottesfeld et al. (1975).
5. Garrard and Bonner (1974).
6. Choie et al. (1977).
7. Krajewska and Klyszerjbo-Stefanowicz (1980).
8. Elgin and Hood (1973).
9. Ozaki (1971).

Table 4-1 lists some representative data resulting from attempts at quantitation of the amounts of histone and nonhistone proteins in chromatin. While values given for the NHCPs are highly variable, the histone/DNA ratios cluster closely around a value of 1 g/g.

Occurrence and Distribution

The histones enjoy the distinction of being one of the first groups of proteins to be recognized as a distinct class, with unique properties. As is mentioned in Chapter 1, Kossel (1884) described a method for the isolation of histones and recognized them as distinct from protamines and the few other kinds of proteins known at that time. By 1928, when Kossel's summary monograph was published, considerable progress had been made in elucidating their general properties. Despite this auspicious beginning, interest in the histones then languished. They presented formidable difficulties to the kind of physico-chemical characterization that was popular during the period from about 1930 to 1960. The main obstacles to further study lay in their heterogeneity and in the tendency of purified histones to aggregate in dilute buffer solutions. In addition, the lack of enzymatic activity made histone isolation difficult and uncertain, given the techniques of the period. The conventional protocol for enzyme purification, which involved following specific activity through successive stages, could not be applied to the histones.

Indeed, it was not until the 1960s, when chromatographic and gel electrophoretic methods became widely available, that real progress in the study of individual histones became possible. By this time, there was serious interest in these proteins as potential genetic regulators (see Chapter 1), and a flurry of activity led quickly to the isolation, characterization, and sequencing of a number of histones. Today, they are certainly the best characterized of chromosomal proteins.

The histones are found in the somatic cells of *all* eukaryotic organisms, with the exception of some dinoflagellates (see *Noneukaryotic Histone-Like Proteins and the Evolution of the Histones,* below). In addition, the sperm of some (but by no means all) eukaryotes contain histones, as do a number of the animal viruses. The reason for this ubiquity of histones among eukaryotes is now clear. The common element of chromatin structure in eukaryotic somatic cells is the nucleosome, and this requires a set of histones that will interact so as to form a core about which DNA can be wrapped. This kind of organization seems to be synonymous with the existence of recognizable histones. But it should be noted that not *all* cells, even in the higher organisms, exhibit such chromatin organization. For example: In many types of sperm, histones are replaced by protamines as the spermatid matures, and nucleosomal organization can no longer be detected. Such seems to be the general pattern among the vertebrates. However, there is recent evidence to indicate that at least a small fraction of the DNA in some mammalian sperm may remain complexed with tightly bound histones (Avramova et al., 1980; Uschewa et al., 1982).

It is still not clear whether or not all chromatins that exhibit nucleosomal structure contain *all* of the common major classes of histones. For example, lysine-rich histones such as H1 have still not been convincingly demonstrated in yeast. Several reports of a fifth acid-soluble protein in yeast chromatin have indeed appeared (see Suchilene and Gineitis, 1978a, 1978b; Sommer, 1978; and Pastink et al., 1979). None of these, however, have been proved to be H1 proteins, and at least one now appears to be a mitochondrial protein contaminant (Certa et al., 1984). The failure to find lysine-rich histones in yeast is surprising, since such are observed in both *Neurospora* (Goff, 1976) and *Aspergillus* (Felden et al., 1976). Negative results in the search for a particular histone must always be viewed with some suspicion, for histones can often be lost through proteolysis by endogeneous proteases. Thus, for some time it was believed that yeast did not possess H3. Now, the protein has not only been found, but sequenced (Brandt and von Holt, 1982). It is also quite possible that chromatin organization in primitive eukaryotes may involve proteins that will be difficult to recognize as histones by the usual criteria. This may be a particular problem with the lysine-rich histones, which appear to exhibit by far the greatest evolutionary variation in their primary structures. Indeed, anomalous, small "H1s" have even been found in annelids (Kmiecik et al., 1985). Thus, a yeast protein that plays the functional role of H1 might have a size, an amino acid composition, and solubility characteristics quite different from the H1s of higher eukaryotes. The lysine-rich protein of *Dictyostelium* may be an example (Parish and Schmidlin, 1985).

The participation of the "inner" histones (H2A, H2B, H3, and H4) in formation of the nucleosome octamer implies that they should be present in equimolar amounts in the chromatin of all somatic cells. Indeed, an early observation of approximate equivalence (Johns, 1967) was crucial in the first formulations of nucleosome models (see Chapter 2). There have been many subsequent analyses of the relative concentrations of the histones, but a large fraction of these studies are flawed by the use of gel stains that do not stain the histones with equal intensity. To obtain reliable results, one must either select a stain carefully and calibrate the staining procedure or, better, use radioactive labeling of the histones. Table 4-2 lists some of the data that satisfy one or the other of these criteria. Although there is some scatter in the results, the overall conclusion is that molar equivalence is indeed observed for the inner histones. However, a *caveat* of sorts may be required for some protista. In certain of these organisms, a single histone may serve the role of both H2A and H2B (see, for example, Rizzo, 1985, and Rizzo et al., 1985). The amount of this histone is twice the amount of either H2A or H2B found in higher organisms.

The situation with respect to the very lysine rich histones (H1 and H5) is more complex. As Table 4-2 shows, these proteins appear to be present in rather different amounts in different cell types. This variability, which may be related to chromatin higher order structure, will be discussed in Chapter 7.

Isolation and Separation of Histones

Histones are bound to the DNA by noncovalent forces, among which the ionic interactions between positively charged residues on the histones and DNA phosphate are probably the most important. There may be some hydrophobic and/or van der Waals interactions as well, but evidence for such is much less definite. Most methods of separating histones from DNA depend upon diminishing the ionic interactions, either by partially neu-

Table 4-2. Relative Amounts of Histones in Chromatin

Organism	Tissue or cell line	H5	H1	H2A	H2B	H3	Ref.
				Molar ratio to H4			
Chicken	Primitive erythroid cells	0.30	0.76	— (2.03) —		1.07	1
	Erythroblasts	ND	ND	0.92	0.98	0.99	2
	Adult erythrocytes	0.83	0.33	— (1.84) —		0.97	1
	Adult erythrocytes	0.92	0.45	0.94	1.15	1.01	3
	Adult erythrocytes	0.44	0.30	1.04	1.03	0.93	4
Mouse	Mastocytoma cells		0.30	0.93	0.99	1.08	5
Rat	Ehrlich ascites cells		0.50	0.98	1.02	1.00	6
	Testes		0.39	0.93	1.09	0.97	6
	Spermatagonia		0.32	0.97	0.98	1.05	7

References:
1. Urban et al. (1980).
2. Joffe et al. (1977). The H3 value is an average from two methods; the H2A value is by difference from (H2A + H2B) and H2B.
3. Olins et al. (1977a).
4. Bates and Thomas (1981).
5. Albright et al. (1979). H2A value includes uH2A.
6. Chiu (1982).
7. Chiu and Irvin (1983).

tralizing the DNA phosphates (acid extraction) or by the use of high ionic strength conditions (salt extraction). That forces other than ionic may be involved is suggested by the fact that it is also possible to dissociate histone complexes by the use of moderate concentrations of reagents such as guanidine hydrochloride or urea. This cannot be taken as *prima facie* evidence for hydrophobic DNA–histone interactions, however, since such substances may exert their effect through the stabilization of denatured forms of the histones, which bind less strongly to DNA. Methods for the isolation, fractionation, and purification of histones are described in considerable detail in a number of chapters in Vol. 16 of *Methods in Cell Biology* (Johns, 1977; von Holt and Brandt, 1977; Spring and Cole, 1977). Therefore, they will be only briefly summarized here.

1. *Acid extraction:* This is the primoidal technique, dating in its conception to Miescher and Kossel in the late nineteenth century. Multitudinous variants of the acid-extraction procedures have since been developed and utilized. In most cases, the process is carried out on isolated chromatin or lysed nuclei, although Lawson and Cole (1979) have shown that H1 can be selectively removed from intact nuclei at low pH. Some early experiments utilized whole cell homogenates, but this increases the danger of both contamination and proteolysis. The history of the development of acid extraction methods has been recounted in part in Chapter 1. The procedures developed by E.W. Johns and co-workers employed extraction with HCl/ethanol mixtures, followed by either

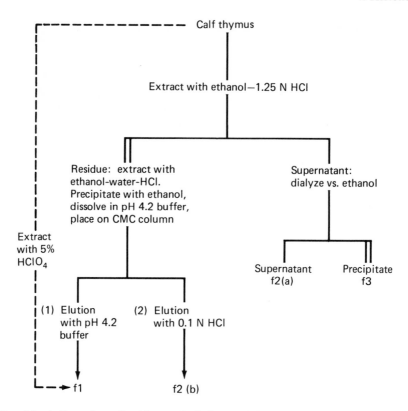

Fig. 4-1. A flow sheet for histone isolation and fractionation according to the original methods of Johns (see text for references). In modern nomenclature, f1 = H1, f3 = H3, f2 (b) = H2B, f2 (a) = H2A + H4.

fractional precipitation with acetone or ethanol or ion exchange chromatography.

A flow sheet for such a fractionation is shown in Figure 4-1. The closely related technique of Bonner et al. (1968) substitutes extraction with 0.25 N H_2SO_4 and precipitation with ethanol. Both procedures, and their many recent variants, are effective in isolating histones largely free of other nuclear proteins. The histones are denatured in the process, but the resulting powders or lyophylizates are freely soluble in distilled water or solutions of low ionic strength.

There has been some concern that histones so prepared might be irreversibly denatured since it is clear that the molecules exist in an unfolded conformation in acidic solution; however, Isenberg and co-workers have convincingly demonstrated that acid-extracted histones are capable of the specific associations involved in nucleosome formation. (For review and discussion, see Isenberg, 1979.) Both Lewis (1976) and Beaudette et al. (1981) have shown that the $(H3 \cdot H4)_2$ tetramer can be dissociated (with accompanying histone denaturation) in urea or at low pH, and then quantitatively reassociated. Last, it should be noted that a number of workers (i.e., Laskey et al., 1978) have employed acid-extracted histones successfully in chromatin reconstitution experiments. It seems likely that the reported cases of "irreversible" histone denaturation in acid may be explained either by competing histone aggregation or by H3 sulfhydryl oxidation under the conditions employed (see Lewis, 1976; Lindsey et al., 1982).

Nevertheless, there may still be subtle irreversible changes produced in a few histones by acid extraction. Smith and collaborators (Smith et al., 1973, 1974; Chen et al., 1974) have demonstrated the existence in some nuclei of a protein kinase that phosphorylates histidyl and lysyl residues (see *Phosphorylation*, below). Such phosphate groups are acid-labile and may be lost under most acid-extraction conditions. It seems unlikely that such modification can greatly affect histone behavior, since a number of studies have shown that acid-extracted histones are fully capable of normal interactions (see above). The possibility of other, more subtle effects does not seem to have been investigated.

2. *Salt extraction:* Concentrated salt solutions can be used either to remove all histones from chromatin, or only certain histones in a selective extraction. Figure 4-2 shows data of Burton et al. (1978) describing the removal of proteins as the salt concentration is increased. Histone H1 is removed first, followed by H2A and H2B at about 0.8 *M*, with H3

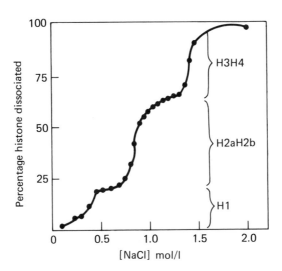

Fig. 4-2. The dissociation of histone types from chromatin as a function of salt concentration. (Reprinted from Nucleic Acids Research, vol. 5, issue 10, p. 3646, Burton et al. Copyright 1978 IRL Press.)

and H4 not entirely removed until about 1.5 M salt. Such partial extraction by 1 M salt must have occurred in some of the early techniques for chromatin isolation (see Chapter 1). However, this phenomenon seems to have been first employed in a systematic manner for histone isolation and fractionation by Ohlenbusch et al. (1967) and Dick and Johns (1969). As Figure 4-2 shows, NaCl concentrations of 2 M or greater should remove all histones from chromatin. This is the basis for the procedure of van der Westhuyzen and von Holt (1971). These authors added protamine to the 2 M salt solution to ensure displacement of the DNA and to allow its precipitation upon dialysis to low salt. Such salt extraction procedures have an important advantage over extraction by acids or other denaturants; the histones are not denatured and can even retain some of the quaternary interactions that existed in the native chromatin. This fact was of the greatest importance in the work of Kornberg and Thomas (1974), for it allowed identification of the $(H3 \cdot H4)_2$ tetramer in extracts of native chromatin (see Chapter 2).

3. *Other isolation methods:* A number of other techniques have been employed for the isolation of histones from chromatin. For example, Bhorjee and Pederson (1976) extracted chromatin with a mixture of 0.4 M guanidine hydrochloride and 6.0 M urea. Faulhaber and Bernardi (1967), Bloom and Anderson (1978), Rhodes (1979), and Simon and Felsenfeld (1979) have all made use of the affinity of DNA for hydroxyapatite to effect separation. In Rhodes' technique, a 2 M NaCl solution of the chromatin is applied to the hydroxyapatite column; the DNA is retained and the proteins elute in a single peak. By applying the chromatin at a lower salt concentration and then eluting with a salt gradient, Simon and Felsenfeld were able to obtain partial histone fractionation as well as separation from the DNA. In the method of Bloom and Anderson, elution is with various mixtures of salt, urea, and guanidine hydrochloride. Fractionation of both histones and nonhistone proteins is possible.

4. *Fractionation of histones:* For some purposes, separation from the DNA of the whole histones, or even the complete complement of chromosomal protein, is adequate. However, most studies require some fractionation of the components. This may be such as to simply isolate the major histone classes (H1, H2A, etc.), or it may be desirable to purify individual histone variants and/or modified forms (see *Histone Variants* and *Histone Modifications*, below). As with histone isolation, the methods that have been employed are legion. The older techniques relied upon combinations of acetone and ethanol precipitation from the acid extract (see Fig. 4-1) and served mainly to isolate the major classes. As has been noted above, the final products were usually denatured lyophilizates. However, as early as 1957 ion exchange chromatography was employed for some separations (Crampton et al., 1957). In recent

years, techniques have become gentler, more discriminating, and more rapid. For example, van der Westhuyzen and von Holt (1971) subjected their salt-extracted histones to gel filtration, obtaining an immediate separation of $(H3 \cdot H4)_2$ from $H2B \cdot H2A$. This technique, often combined with ion exchange chromatography for further resolution, has formed the basis for many methods; see Spring and Cole (1977) for a review. Partial histone fractionation during preparation can also be obtained by the hydroxyapatite techniques of Bloom and Anderson and Simon and Felsenfeld (see above).

Recently, a number of rapid techniques of high discrimination or resolution have been developed. For example, individual histones can be isolated in high purity by immuno absorption (Absolom and van Regenmortel, 1977). Preparative gel electrophoresis has been employed by many workers. Figure 4-3 shows the kind of resolution of histones obtained by a two-stage electrophoretic method. A recent development of great promise is the use of reversed-phase high-performance liquid chromatography (Certa and von Ehrenstein, 1981; Gurley et al., 1983a, 1983b, 1983c, 1984; Jackson and Gurley, 1985; see Fig. 4-4). As is evident from Figures 4-3 and 4-4, these newer methods are capable of preparative isolation of specific variants and modified forms of individual histones.

The Primary Structures of Histones

The development of reliable methods for the isolation and fractionation of histones was soon followed by the first determinations of primary sequences. In 1969, Delange et al. (1969a) and Ogawa et al. (1969) independently sequenced calf thymus H4. Since then, approximately one hundred other sequences have been determined, covering all of the major

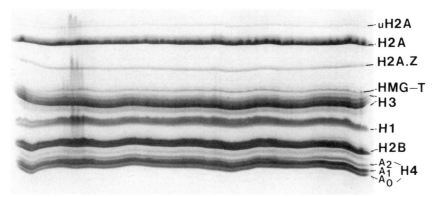

Fig. 4-3. Preparative gel electrophoresis for the separation of histones. Five hundred mg of trout testis histones on an acetic acid—urea, 15% acrylamide gel. Courtesy of Dr. James Davie.

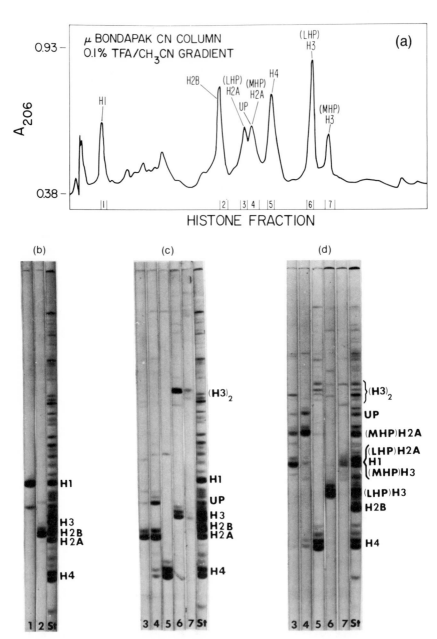

Fig. 4-4. Separation of histone fractions by high pressure liquid chromatography (a) and identification of fractions on gels (b,c,d). Gel types: (b,c) acid-urea polyacrylamide, (d) Triton DF-16-polyacrylamide. Fractions are indicated by numbers; St is the whole mixture as standard. (From Journal of Chromatography, vol. 266, p. 609, Gurley et al. Copyright 1983 Elsevier Science Pubishers.)

Table 4-3A. Complete Sequences of Core Histones

Organism	H4	H3	H2B	H2A
1. Plants				
Cycad				
Pea	DeLange et al. (1969b)	Brandt and von Holt (1986)		
Wheat	Tabata et al. (1983) Tabata and Iwabuchi (1984)	Patthy et al. (1973) Tabata et al. (1984)		Rodrigues et al. (1985)
2. Fungi				
Neurospora crassa	Woudt et al. (1983)	Woudt et al. (1983)		
Physarum polycephalum	Wilhelm and Wilhelm (1984)			
Saccharomyces cerevisiae	Smith and Andresson (1983)	Brandt and von Holt (1982) Smith and Andresson (1983)	Wallis et al. (1980)	Choe et al. (1982)
3. Protists				
Tetrahymena pyriformis	H. Hayashi et al. (1984)	T. Hayashi et al. (1984)	Nomoto et al. (1982)	Fusauchi and Iwai (1983)
Tetrahymena thermophila	Bannon et al. (1984)			
4. Sipunculids				
Sipunculus nudus				Kmiecik et al. (1983)
5. Molluscs				
Patella granatina			Van Helden et al. (1979)	
Sepia officinalis				Wouters-Tynou et al. (1982)
6. Insects				
Drosophila melanogaster			Elgin et al. (1979)	

Table 4-3A. *Continued*

Organism	H4	H3	H2B	H2A
7. Echinoderms				
Starfish				
Asterias rubens				Martinage et al. (1983)
Sea Urchins				
Lytechinus pictus *Parachinus angulosus*	Childs et al. (1982)	Childs et al. (1982)	Strickland et al. (1977a, 1977b) Strickland et al. (1978)	Strickland et al. (1980b)
Psammechinus miliaris	Wouters-Tyrou et al. (1976)	Schaffner et al. (1978)	Schaffner et al. (1978)	Wouters et al. (1978)
Strongylocentrotus purpuratus	Schaffner et al. (1978) Busslinger et al. (1980) Grunstein et al. (1981)	Busslinger et al. (1980) Sures et al. (1978)	Busslinger et al. (1980) Sures et al. (1978)	Schaffner et al. (1978) Busslinger et al. (1980) Sures et al. (1978)
8. Shark		Brandt et al. (1974)		
9. Fishes				
Carp		Hooper et al. (1973)		
Trout				
Salmo guirdnerii *Salmo frutta*	Winkfein et al. (1985)	Connor et al. (1984)	Winkfein et al. (1985) Kootstra and Bailey (1978)	Connor et al. (1984)

Table 4-3A. *Continued*

Organism	H4	H3	H2B	H2A
10. Amphibians				
Frogs				
Xenopus borealis	Turner and Woodland (1982)			
Xenopus laevis	Moorman et al. (1981) Turner and Woodland (1982)	Moorman et al. (1981)	Moorman et al. (1982)	Moorman et al. (1982)
11. Birds				
Chicken	Sugerman et al. (1983)	Brandt and von Holt (1974)	Van Helden et al. (1982) Grandy et al. (1982) Harvey et al. (1982)	Laine et al. (1978) Harvey et al. (1983)
12. Mammals				
Calf	DeLange et al. (1969a) Ogawa et al. (1969)	DeLange et al. (1972) Patthy and Smith (1975) Franklin and Zweidler (1977)	Iwai et al. (1972)	Yoeman et al. (1972)
Human	Hayashi et al. (1982) Sierra et al. (1983) Zhong et al. (1983)	Ohe and Iwai (1981) Zhong et al. (1983)	Ohe et al. (1979) Zhong et al. (1983)	T. Hayashi et al. (1980) Zhong et al. (1983)
Mouse	Seiler-Tuyns and Birnstiel (1981)	Sittman et al. (1983)	Sittman et al. (1983)	
Pig	Sautiere et al. (1971b)			
Rat	Sautiere et al. (1971a)		Martinage et al. (1979)	Laine et al. (1976)

Table 4-3B. Complete Sequences of Lysine-Rich Histones

Organism	H5	H1
1. Sea Urchins		
Parechinus angulosus		Strickland et al. (1980a)
Psammechinis miliaris		Schaffner et al. (1978)
Strongylocentrotus		Levy et al. (1982)
purpuratus		
2. Fishes		
Trout		MacLeod et al. (1977)
		Mezquita et al. (1985)
3. Amphibians		
Xenopus laevis		Turner et al. (1983)
4. Birds		
Chicken	Briand et al. (1980)	Sugerman et al. (1983)
	Ruiz-Carillo et al. (1983)	
Duck	Tonjes and Doenecke (1984)	
Goose	Yaguchi et al. (1979)	
5. Mammals		
Boar		Cole et al. (1984)
Rabbit		Cole (1977); see also
		Cole et al. (1984)

histone types and representing phyla ranging from the fungi to the vertebrates (see Table 4-3). I shall not discuss here the technicalities of protein sequencing; very complete descriptions of the methods employed and the difficulties peculiar to the histones are given by DeLange (1978) and Hsiang and Cole (1978). Until recently, most primary structures were determined from the proteins themselves, but in the past few years the availability of cloned histone genes has led to an increasing use of DNA sequencing. While this method is rapid and convenient, two limitations should be noted: First, the DNA sequence will not reveal certain modifications (i.e., methylation, some acetylation) that are stable to the protein sequencing technique. Second, histone genes almost always exist in multiple, nonidentical copies (see *Histone Variants,* below), not all of which may be transcribed. A gene picked at random from a clonal library may represent a minor variant (or even a silent pseudogene), rather than the major cell product. Identification of the cloned gene may require cell-free synthesis of the gene product. Of course, in those instances where the clone was prepared by reverse-transcriptase from mRNA (c-DNA clones), we know that the

Fig. 4-5. Representative H4 sequences from a number of phyla. Only differences from the calf sequence are shown for other organisms. Code for this and following figures: A = ala, C = cys, D = asp, E = glu, F = phe, G = gly, H = his, I = ileu, K = lys, L = leu, M = met, N = asn, P = pro, Q = gln, R = arg, S = ser, V = val, Y = tyr. For sources of data, see Table 4-3A. The sea urchin sequences are from "early" *S. purpuratus* genes, in all cases.

H4 Sequences

	1	5	10	15	20	25	30
Calf	S	G R G K G	G K G L G	K G G A K	R H R K V	L R D N I	Q G I T

Chicken —— Identical to calf ——→
Urchin —— Identical to calf ——→
Yeast | (position ~20)

	31	35	40	45	50	55	60
Calf	K	P A I R R	L A R R G	G V K R I	S G L I Y	E E T R G	V L K V

Chicken —— Identical to calf ——→
Urchin —— Identical to calf ——→
Yeast | V A S

	61	65	70	75	80	85	90
Calf	F	L E N V I	R D A V T	Y T E H A	K R K T V	T A M D V	V Y A L

Chicken —— Identical to calf ——→
Urchin | S L
Yeast | S (C) L

	91	95	100	102
Calf	K	R Q G R T	L Y G F G G	102

Chicken —— Identical to calf ——→ 102
Urchin —— Identical to calf ——→ 102
Yeast —— Identical to calf ——→ 102

S (under position ~93)
S (under position ~97)

gene in question represents at least *one* of the variants expressed in the cell.

The histone sequences that I have located in a systematic search up to January 1985, plus a few more recent additions, are all listed in Table 4-3. In Figures 4-5 through 4-9, a number of representative sequences from diverse phyla are given. While we should not be concerned with details at this point, a cursory examination of these data reveals some important generalizations. Figures 4-5 through 4-9 and Table 4-4 illustrate these points:

1. All histones are basic proteins, containing relatively large amounts of lysine and arginine. The lysine/arginine ratio varies considerably, from a high of about 20 in the "lysine-rich" histones such as H1 to less than 1 in the "arginine-rich" H3 and H4. But in every case there is a sufficient excess of basic over acidic residues to give histones a substantial positive net charge at physiological pH. (See Table 4-4.)

2. The positive charge is distributed quite unevenly in the histone sequences. In all cases, the N-terminal region contains a high concentration of basic residues; in H2A and H3, there are similar domains near the C terminus. The lysine-rich histones (H1 and its many variants) have a very long, positively charged C-terminal domain in addition to a lysine-rich N-terminal region (Fig. 4-9). Apart from these end regions, the remainder of each histone molecule is roughly balanced in acidic and basic residues and contains most of the hydrophobic amino acids. The composition of this portion approximates that of a "normal" globular protein. As will be shown in later sections, it is probable that each of these domains has both structural and functional significance.

3. The histones have been *generally* conserved in evolution. This was first brought out by the striking similarity between calf and pea H4 (DeLange et al., 1969b), which differ at only two sites in 102 residues. While further studies have confirmed this general trend to conservatism (see, for example, Wilson et al., 1977; Isenberg, 1979), it is now clear that the situation is far more complex than first believed. In the first place, not all histones are so highly conserved. As Figure 4-10 illustrates, H2A and H2B are not nearly so conservative in sequence as H3 and H4, and H1 is a quite variable molecule as compared to many other proteins. Second, the evolutionary changes appear to be unevenly distributed within the sequence of each histone. In H2A, H2B, and H1, the lysine–arginine-rich terminal domains show the greatest evolutionary divergence, whereas the "core" of each molecule seems much more highly conserved. The terms *variable region* and *constant region* are sometimes used to denote these portions of the histone sequences. However, the data available on H3 and H4 divergence (which are limited because of the slow evolution of these histones) show a much more uniform distribution of amino acid substitutions. This striking

H3 Sequences

	1				5					10					15					20					25					30
Calf	A	R	T	K	Q	T	A	R	K	S	T	G	G	K	A	P	R	K	Q	L	A	T	K	A	A	R	K	S	A	P
Chicken	← Identical to calf →																													
Urchin	← Identical to calf →																													
Yeast																						S								

	31				35					40					45					50					55					60
Calf	A	T	G	G	V	K	K	P	H	R	Y	R	P	G	T	V	A	L	R	E	I	R	R	Y	Q	K	S	T	E	L
Chicken	← Identical to calf →																													
Urchin	← Identical to calf →																													
Yeast	S											K												F						

	61				65					70					75					80					85					90
Calf	L	I	R	K	L	P	F	Q	R	L	V	R	E	I	A	Q	D	F	K	T	D	L	R	F	Q	S	S	A	V	M
Chicken	← Identical to calf →																													
Urchin																					E									
Yeast																													I	G

	91				95					100					105					110					115					120
Calf	A	L	Q	E	A	C	E	A	Y	L	V	G	L	F	E	D	T	N	L	C	A	I	H	A	K	R	V	T	I	M
Chicken						S																								
Urchin						S							?	?	?															
Yeast					S	V						S								A										Q

	121				125					130					135	
Calf	P	K	D	I	Q	L	A	R	R	I	R	G	E	R	A	← 135
Chicken	← Identical to calf →															← 135
Urchin	← Identical to calf →															← 135
Yeast	K				K					L					S	← 135

Fig. 4-6. Representative H3 sequences from a number of phyla.

H2B Sequences

```
            1       5            10           15           20           25        30
Calf      P  E  P        A  K  S  A  P  A  P  K  K  G  S  K  K  A  V  T  K  A  Q  K  K  D  G  K  K  R  K
Chicken                  -  -                                               T                          G
Urchin       K           -  -                                               T                          G
Yeast  S A K A           Q  V  A  K  K  A  E  K  K  P  A  K  K  T  S  T  S  A* G  D                    N  S

            31      35           40           45           50           55        60
Calf      R  S  R  K  E  S  Y  S  V  Y  V  Y  K  V  L  K  Q  V  H  P  D  T  G  I  S  S  K  A  M  G
Chicken   K                                                                                      G
Urchin    K  A           -  -  -                          T                             R  S        V
Yeast     K  A     T                                      T                       Q     R  S        S

            61      65           70           75           80           85        90
Calf      I  M  N  S  F  V  N  D  I  F  E  R  I  A  G  E  A  S  R  L  A  H  Y  N  K  R  S  T  I  T
Chicken   S
Urchin    S        T              -                          S                    K  Q     S  K  S  T
Yeast     L        T              -  -        S              A                    K  A     S        T

            91      95           100          105          110          115       120
Calf      S  R  E  I  Q  T  A  V  R  L  L  L  P  G  E  L  A  K  H  A  V  S  E  G  T  K  A  V  T  K
Chicken   A                          -                                              R
Urchin              |———— Identical to calf ————                                   
Yeast     A                          -                                              R

            121     125
Calf      Y  T  S  S  K ———|  125
Chicken   |——— Identical to calf ———|  125
Urchin    T ———|  123
Yeast     S        T  Q  A |———  130
```

Fig. 4-7. Representative H2B sequence from a number of phyla. *Indicates insertion of an unidentified residue in *S. purpuratus* H2B. Yeast sequence is for variant 1.

H2A Sequences

	1				5					10					15					20					25					30
Calf	S	G	R	G	K	Q	G	G	K	A	R	A	K	A	K	T	R	S	S	R	A	G	L	Q	F	P	V	G	R	V
Chicken																S														
Urchin						–	S					T																		
Yeast			G*			A		S	A		K		S	Q	–	S			A	K				T						

	31				35					40					45					50					55					60
Calf	H	R	L	L	R	K	G	N	Y	A	E	R	V	G	A	G	A	P	V	Y	L	A	A	V	L	E	Y	L	T	A
Chicken	← Identical to calf →																													
Urchin			F								K				G						M									
Yeast						R					Q		I		S							T							A	

	61				65					70					75					80					85					90
Calf	E	I	L	E	L	A	G	N	A	A	R	D	N	K	K	T	R	I	I	P	R	H	L	Q	L	A	I	R	N	D
Chicken	← Identical to calf →																													
Urchin																S											V			
Yeast	← Identical to calf →																													

	91				95					100					105					110					115					120
Calf	E	E	L	N	K	L	L	G	K	V	T	I	A	Q	G	G	V	L	P	N	I	Q	A	V	L	L	P	K	K	T
Chicken	← Identical to calf →																													
Urchin									G																					
Yeast	D								N													H	Q	N						S

	121				125					130		
Calf	E	S	H	H	K	A	K	G	K	⊢←		129
Chicken	D		–					A		⊢←		128
Urchin	A	K	S	S	⊢←	—	—	—	—			123
Yeast	A	K	A	T			S	Q	E	L	⊢←	131

Fig. 4-8. Representative H2A sequence from a number of phyla. Yeast sequence is variant 1. *The dipeptide KG is inserted between positions 3 and 4 (calf numbering).

(a)

```
                    5        10A      15        20        25
Trout   c A E V  A P A P A P A A A  P A K A  P K K A A A A K P K K
Rabbit  c S E A P A E T A  A P A P A P A E K S  P A K  K K A A K K P G
                         10        15        20        25

                30            40      45        50
Trout   S G  P A V G E L  A G K A V A A S K E R  S G V  S L A A
Rabbit  A G A A K R K A A G  P P V S E  L I T  K A V A A S K E R N  G L S  L A A
                      35        40      45        50        55        60

        55            60        65        70        75        80        85
Trout   L K K S L A A G G Y D V E  K N N S R V K  I A V K S L V T K G T L V E
Rabbit  L K K A L A A G G Y D V G  K N N S R I K  L G L K S L V  S K G T L V E
              65        70        75        80        85        90        95

        90            95        100        105        110        115
Trout   T K G T G A S G S F K L N K K A V  E A K  K P A K K A A A P K A
Rabbit  T K G T G A S G S F K L D K K A A S G E A K P K P  K K A G A A K P
                  100        105        110        115        120        125

        120        125        130        135        140        145
Trout   K K V A A K K P A A A K K P K K V A  A K K A V A A K K S P K K A
Rabbit  K K P A G  A T P K K P K K A A G A K K A V  K K T P K K A P
               130        135        140        145        150        155

        150        155        160        165        170
Trout   K  K P  A  T P K K A A K S P K K A T K A A K P K A
Rabbit  K P K A A A K P K V A K P K S  P A K V A K S P K K A  K A V K P K A
              160        165        170        175        180        185        190

        175        180        185        190
Trout   A K P K K A A K S P K K V K  K P A A A K K
Rabbit  A K P K  A P K  P K A A K A K T A A K K K
              195        200        205        210
```

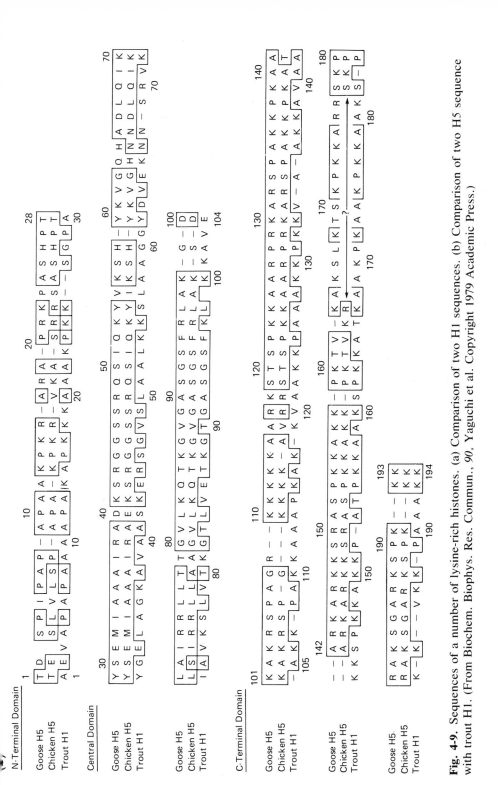

Fig. 4-9. Sequences of a number of lysine-rich histones. (a) Comparison of two H1 sequences. (b) Comparison of two H5 sequence with trout H1. (From Biochem. Biophys. Res. Commun., **90**, Yaguchi et al. Copyright 1979 Academic Press.)

Table 4-4. Some Properties of Histones

Histone[a]	Molecular weight	No. of residues	Mol% Lys	Mol% Arg	Estimated[b] net charge	Extinction[c] coefficient $\times 10^{-3}$ (cm^{-1} M^{-1})
H1	22,500[d]	224[d]	29.5[d]	1.3[d]	+58[d]	1.35
H2A	13,960	129	10.9	9.3	+15	4.05
H2B	13,774	125	16.0	6.4	+19	6.70
H3	15,273	135	9.6	13.3	+20	4.04
H4	11,236	102	10.8	13.7	+16	5.40

[a]Data are for calf thymus histones, except for H1 data indicated by [d].
[b]At pH 7.5 histidine residues are assumed to be nonprotonated.
[c]At 275 nm for H1, 275.5 for the inner histones. H1 value is from Smerdon and Isenberg (1976a), inner histone data from D'Anna and Isenberg (1974d).
[d]The complete sequence of calf thymus H1 is not available. These values are for rabbit thymus H1.3, as given by Cole et al. (1984).

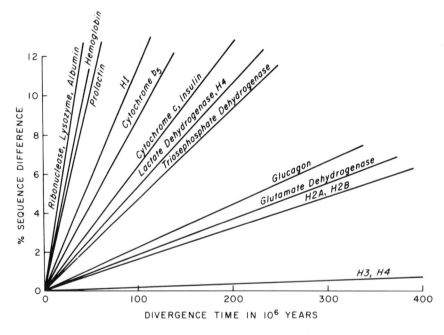

Fig. 4-10. Evolutionary rates of the histones as compared to other proteins. From Isenberg, 1978. The figure is schematic and does not include data from more recent sequences, but provides an approximate comparison for evolutionary rates in higher eukaryotes. (From The Cell Nucleus, H. Busch, Ed. vol. 4, pp 135–154, Isenberg. Copyright 1978 Academic Press.)

difference may reflect different roles for the domains in H3 and H4 as compared to the other histones. In particular, the data suggest that whereas wide variation in N and C domains can be tolerated in H2A and H2B, these regions have a much more defined role in H3 and H4.

The determination of the yeast and *Tetrahymena* sequences has done much to negate the earlier idea that the H3 and H4 primary structures were inviolate. However, there remains one respect in which the conservatism of H3 and H4 is remarkable. As Table 4-3 shows, nearly thirty H4 histones and over twenty H3 histones have so far been sequenced, ranging from yeast to man. Every one of these H4s contains *exactly* 102 residues, and all but one of the H3s contains 135 residues. The sole exception is an unusual, highly divergent mouse variant H3, which has 134 amino acid residues (Sittman et al., 1981). This constancy is striking and must reflect the critical role played by H3 and H4 in determining nucleosome structure (see Chapter 6). Such restrictions do not seem to apply to H2A, H2B, and H1, each of which exhibits considerable variations in chain length.

4. No known histone contains tryptophan, and the tyrosine and phenylalanine contents are relatively low. As a result, histones have much lower absorbance in the near ultraviolet than most other proteins. Values of the appropriate extinction coefficients are listed in Table 4-4. As a consequence of the low UV absorptivity, coupled with the very high absorbance of DNA and RNA in this spectral range, histone concentrations are often determined by chemical methods, the Lowry technique being commonly used. However, it should be noted that histones give somewhat anomalous results in Lowry analysis when calibrated against proteins of more usual composition; the use of histones themselves as primary standards is therefore recommended.

5. It seems likely that the unusual amino acid composition of histones is also responsible for their anomalous behavior in SDS gel electrophoresis. It is found that these proteins migrate more slowly than the comparably sized "normal" proteins often used as standards. Thus, molecular weights of histones will be overestimated by this technique, unless histones of known size are run as standards.

Histone Variants

General Considerations

In most organisms, the histone genes are present in multiple copies. For example, the complete set of histone genes is reiterated about ten- to twentyfold in mouse, twenty- to fiftyfold in *Xenopus*, about a hundredfold in *Drosophila*, and several hundredfold in some sea urchins. In some cases, these genes are arranged in tandem arrays, each repeat carrying all five histone genes. However, this is not universally so; sometimes the histone genes are irregularly clustered or even dispersed. The ordering of the

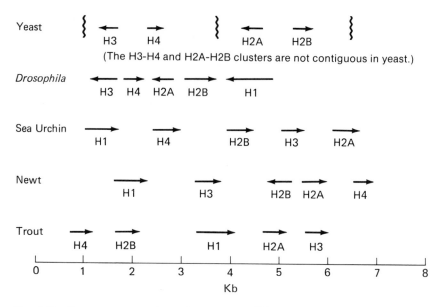

Fig. 4-11. Some representative arrangements of histone gene clusters in a number of different organisms. Data for yeast, *Drosophila*, newt, and trout are adapted from figures in the book *Histone Genes* (Stein, Stein, and Marzleff, Eds., 1984). The sea urchin (*S. purpuratus*) data are from Kedes (1979).

genes, their directions of transcription, and the arrangement of spacer sequences all seem to vary greatly. More than one arrangement can sometimes be found in a single organism. Some examples are shown in Figure 4-11; for more details, the reader is referred to the book *Histone Genes: Structure, Organization, and Regulation* (Stein, Stein, and Marzleff, Eds., 1984) or to several excellent reviews (Kedes, 1979; Hentschel and Birnstiel, 1981; Old and Woodland, 1984).

The existence of multiple copies of histone genes in a given organism suggests the possibility of *nonallelic* variation. It has been demonstrated that even inbred, homozygous individuals may carry genes coding for several different primary sequence variants of a given histone. A great deal of research has centered on the identification of such variants and attempts to understand their differential function.

The first clear indication of the existence of histone variants appears to have been provided by Kincade and Cole (1966). They showed that calf thymus H1 could be fractionated by ion exchange chromatography into a number of electrophoretically distinct fractions. Moreover, they demonstrated that these fractions differed slightly but significantly in amino acid composition and in their peptide maps. This was soon followed by Panyim and Chalkley's (1969a, 1969b) introduction of new gel techniques that allowed the facile detection of such heterogeneity. Panyim and

Chalkley found that one variant (H1°) contained histidine, unlike other H1 components, and thus could not possibly be a modification. Such evidence is especially important in this field, for it does *not* suffice simply to show electrophoretic resolution of multiple forms of a histone. Differences in mobility can be produced by any of a number of kinds of posttranslational modification (see *Histone Modifications*). Therefore, it is the obligation of the experimenter to provide proof of *sequence* difference, if the existence of true variants is to be claimed.

There are a number of ways in which this evidence can be obtained, all of which have been used by one group or another:

1. *Comparison of amino acid and/or peptide composition:* These were the original methods of Kincade and Cole and the only techniques available at the time. The first suffers from the disadvantage that amino acid compositions may differ only slightly, or not all, and hence variants may go undetected. Simple "finger-printing" of peptides is not adequate, for the mobilities of peptides can be changed by posttranslational modification. Nor is the observation of a difference in amino acid composition necessarily sufficient proof. In some cases, histones are specifically proteolyzed in vivo so as to yield somewhat shorter molecules. These would show up as different in composition, but are most certainly not variants in the sense used above. A case in point is the conversion of $H3^S$ to $H3^F$ in *Tetrahymena,* which will be discussed below. On the other hand, if peptides are isolated following digestion, a compositional difference in any specific peptide is sufficient evidence for the existence of variants.

2. *Isolation and in vitro translation of multiple mRNAs for a given histone:* This method was employed, for example, by Newrock et al. (1978) to establish that the electrophoretically separable forms of sea urchin histones were in fact variants. If two or more unique mRNA molecules can each be shown to produce a single histone that migrates in electrophoresis together with one, and only one, of the forms observed in vivo, the presumptive evidence that these are sequence variants is strong. It should be noted, however, that the mere presence of multiple and separable mRNAs is *not* in itself evidence for the existence of histone variants. As an example, it has been shown that there are several distinguishable human H4-mRNAs, all of which produce the same histone molecule (Lichtler et al., 1980, 1982). Presumably, differences in the noncoding region of the mRNA and/or silent codon substitutions account for the different mobilities of these RNA molecules.

3. *Isolation, cloning, and sequencing of the histone genes:* This technique was employed by Wallis et al. (1980) and Rykowski et al. (1981) in their elegant studies of the histones of yeast. Since it is always possible that certain copies of a multiple gene will be silent, it is essential in this method to demonstrate that the genes in question are actually tran-

scribed and translated in vivo. The two yeast H2B genes, which differ significantly in coding sequence, could be shown to correspond to two electrophoretically resolvable components (Wallis et al., 1980). But the yeast H2A genes differ only in the transposition of two amino acids at residues 89 and 90. Such protein variants cannot be resolved by any known electrophoretic technique. The fact that both are expressed was demonstrated genetically by Kolodrubetz et al. (1982), who produced mutations that inactivated the products of one or the other of the genes and showed that the organism could survive with the one remaining functional gene. This example also points out the possible existence in other organisms of such "hidden" variants, which would not be detected by the usual electrophoretic techniques.

4. *Isolation and sequencing of the variant proteins:* This method has been used extensively by von Holt and co-workers (see Brandt et al., 1979; Rodrigues et al., 1979, for examples) and is certainly the most direct, albeit laborious way to identify variants. Sequencing need not be complete to demonstrate the existence of variants; *any* difference in sequence in any peptide is sufficient. Furthermore, if the sequencing is careful, complete separation of the histone variant mixture is not essential in some cases. For example, sequencing of whole H3 from pea embryo revealed an ambiguity at residue 96, which could only be resolved by the assumption that 60% of the molecules had alanine, 40% serine at this location (Patthy et al., 1973).

A very large number of histone variants have been reported, encompassing many cell types from both animal and plant phyla. However, in

Table 4-5. Examples of Histone Variants in a Fungus, a Protist, and an Invertebrate

	H1	H2A	H2B	H3	H4	References
Yeast (*Saccharomyces cerevisiae*)	no H1 identified	H2A.1 H2A.2	H2B.1 H2B.2	H3	H4	Wallis et al. (1980) Grunstein et al. (1984)
Tetrahymena thermophila macronucleus	H1	H2A^S H2A^F hv1	H2B	H3 hv2		Vavra et al. (1982) Allis et al. (1980a)
Tetrahymena thermophila micronucleus	α β γ^a δ^a	H2A^S H2A^F	H2B	H3	H4	Vavra et al. (1982) Allis et al. (1980a) Allis et al. (1984)
Sea urchin embryo (*Strongylocentrotus purpuratus*)	H1$_{CS}$ H1 α H1 β H1 γ	H2A$_{CS}$ H2A α H2A β H2A γ H2A δ	H2B$_{CS}$ H2B α H2B γ H2B δ	H3	H4	Newrock et al. (1978) Childs et al. (1979)

[a]Recent evidence indicates that γ is probably not a true variant, but rather the product of specific proteolysis of α, itself derived from a larger precursor. Further processing of γ results in the fourth, minor component, δ. Allis et al. (1984).

Table 4-6. Core Histone Variants in the Mouse[a]

Variant	Source[b]	Type[c]	Range[d]
H2A.1	S	SRD	V
H2A.2a	S	PRD	V
H2A.2b	S	PRD	V
H2A.3	S	RI	M
H2A.4	S,T[e]	NM	M
MI[f]	S	NM	M + others
H2B.1	S	PRD	V
H2B.2	S	SRD	m
H2B.S	T		m
H3.1	S	SRD	M
H3.2	S	SRD	V
H3.3	S	RI	V
H3.S	T		m (+ rat)
H4	S,T		V

[a]Data from Zweidler (1980); and private communication.
[b]S = somatic cells, T = testes (spermatocytes).
[c]Code: SRD = strictly replication-dependent, PRD = partially replication-dependent, RI = replication-independent, NM = noncompeting minor protein.
[d]M = mammals, m = mouse spec., V = vertebrate.
[e]Minor in somatic cells, increases greatly in spermatocytes. Originally called H2A.S.
[f]Equivalent to H2A.Z reported in other organisms.

many cases the demonstration that these are true variants is equivocal. Rather than list all of the data, I shall describe in some detail a few of the better studied systems. Data from these are summarized in Tables 4-5, 4-6, and 4-7.

Before considering the specific systems, it is worthwhile to note some generalizations that appear immediately from inspection of these tables. Variance is by no means uniform among histone types. H4 seems almost

Table 4-7. Lysine-Rich Histones in Mouse Somatic Cells[a]

	H1a	H1b	H1c	H1d	H1e	H1⁰

	H1a	H1b	H1c	H1d	H1e	H1[0]
Synthesized in dividing cells	+	+	+	+	+	
Synthesized in nondividing cells	?	?	+	+	+	+
Present in nondividing, nonlymphoid cells			+	+	+	+
Present in nondividing lymphoid cells	+	+	+	+	+	
Unnecessary for cell division		+		+		+

[a]Adapted from Lennox and Cohen (1983).

immune to variation. Only two reports of nonallelic variants of H4 are known to the author. Two H4s have been claimed to exist in the urchin *Parechinus* (Schwager et al., 1983), on the basis of amino acid compositional difference. In *Tetrahymena,* two H4s have been sequenced (Hayashi et al., 1984b). Variants of all other core histones are numerous, with H2A and H2B showing the greatest variability. There are *many* H1 variants; and the differences in sequence appear to be most extreme with this histone. Indeed, H1 variants found in some species and cell types are so different from "normal" H1 in composition, solubility, or electrophoretic behavior that it becomes difficult to decide whether to classify them as variants or different histones. A classic example is histone H5 of erythrocytes. Although it could be considered an extreme H1 variant, it has been traditionally classified as a separate histone. A list of these types of lysine-rich histones is given in Table 4-8, and some sequences are given in Figure 4-9. A minireview has been presented by Cole (1984). These proteins clearly form a class; each has some variation of the domain structure associated with H1, and all seem to play a similar role in binding to linker DNA. H1° and H5 are both concentrated in cells that are terminally differentiated, but H5 is specific for the transcriptionally inactive erythrocytes of fish, amphibians, reptiles, and birds. It has been found that the H1° and H5 types exhibit sequence similarities that distinguish them from other members of the H1 family (Pehrson and Cole, 1981). Indeed, at the level of amphibians, it becomes very difficult to distinguish between the

Table 4-8. Lysine-Rich Histone Types

Histone type	Found in	Multiple variants	Comments
H1	Almost all eucaryotic phyla	Yes	The most widespread class of lysine-rich histones
H1°	Mammals, possibly birds,[a] amphibians[b]	Yes	Most concentrated in nonreplicating terminally differentiated cells; some similarities to H5
H5	Birds, reptiles, amphibians,[b] fish	Yes	Specific to reticulocytes and erythrocytes; increases as cells mature and become transcriptionally inactive
AK	Mammals	?	Major differences in amino acid composition, sequence, and solubility as compared to other types;[c] status as member of class uncertain
H1t	Mammals	Not reported	Specific to maturing spermatocytes in testes
H1-S	Urchins, some other invertebrates	Yes	Specific to sperm that retain nucleosomal structure

[a]Srebreva et al. (1983) report an H1° in terminally differentiated chicken liver cells.
[b]Yasuda et al. (1984) claim that the putative H5 in bullfrog is actually an H1°.
[c]See Jenson et al. (1980).

putative "H5" and an H1° histone (Moorman et al., 1984). There is evidence that the replacement of H1 by one or another of these "H1-like" proteins is accompanied by changes in the higher order structure of chromatin (see Chapter 7). Classification of the lysine-rich histones is further complicated by the fact that some *types* (such as H1 and H1°) have been demonstrated to themselves exist as multiple *variants* in some cells. For example, Dupressoir and Sautiere (1984) report five variants of H1 in chicken erythrocytes. Whether such individual variants of the different lysine-rich histone types play distinct functional roles in chromatin organization is as yet uncertain, although there exist two systems that give us indication that they may indeed do so. First is the progressive replacement of H1 variants during the development of invertebrate embryos, which will be described in detail below. Second, there has been discovered a remarkable series of exchanges of lysine-rich histones during spermatogenesis in mammals. During this process, spermatagonia are first transformed into spermatocytes, which undergo meiosis, finally maturing into spermatids. In the mature spermatid, histones have been replaced by small, arginine-rich protamine-like proteins. However, before protamine replacement occurs, there has already been a major replacement of some somatic histones with "testes-specific" forms. Seyedin and Kistler (1980) isolated and characterized a testes-specific H1 in rats, which has been termed H1t. This has been shown to be present in testicular tissues of a wide variety of mammals, including humans (Seyedin et al., 1981; Seyedin and Kistler, 1983). The precise role of H1t has been clarified by Bucci et al. (1982). As shown in Figure 4-12, they find that H1t is present specifically in the late leptotene and pachytene periods of meiotic prophase. Even the spermatagonia exhibit an H1 distribution distinct from that of most somatic cells; the variants H1a and H1c dominate, whereas H1b, H1d, and H1e are most common in most cells (Lennox and Cohen, 1984b; Lennox, 1984). Lennox and Cohen suggest that H1a and H1c may not be capable of stabilizing the higher order structure of chromatin, thus leaving it "open" for the genetic recombination that occurs at pachytene.

The histone variant switching that occurs in embryogenesis and spermatogenesis is by no means confined to the linker histones. As will be seen, there exist developmental variants of the inner histones as well.

Let us now turn to a more detailed discussion of variants in a few specific systems, covering a wide range of eukaryotic phyla.

Yeast *(Saccharomyces cerevisiae)*

This organism is unique in its low multiplicity of histone genes. Only two gene copies of each of the four inner histones are present (Hereford et al., 1979; Wallis et al., 1980; Smith and Murray, 1983; Smith, 1984; Grunstein et al., 1984). Furthermore, the H2A and H2B genes occur in quite different regions of the genome than do the H3 and H4 genes (see Fig. 4-11). The products of the two H4 copies, and likewise of the two H3 copies,

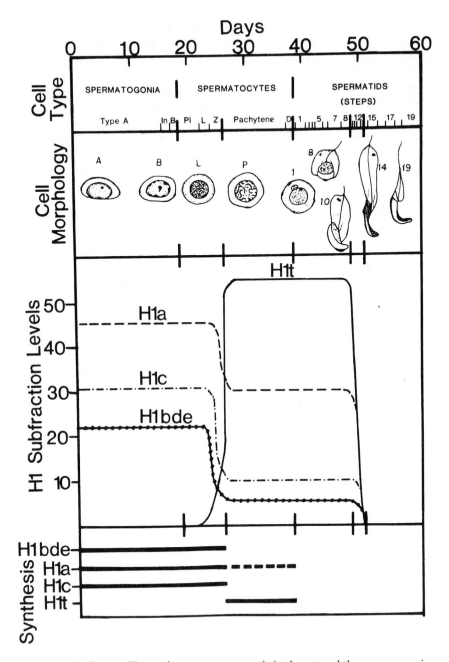

Fig. 4-12. A diagram illustrating spermatogenesis in the rat and the accompanying changes in H1 variant levels and synthesis. Shown are changes in cell type, morphology (not to scale), variant levels, and times of synthesis. (From Experimental Cell Research, vol. 140, pp. 111–118, Bucci et al. Copyright 1982 Academic Press.)

are identical; thus, there is only one H4 and one H3 in yeast. On the other hand, as mentioned above, there are two variants each of H2A and H2B. The low copy number of yeast histone genes, together with the ease of carrying out site-directed mutagenesis with this organism, have made possible some truly elegant and penetrating studies of histone function. Grunstein and co-workers have specifically inactivated one or another of the gene copies by frameshift mutation. This first allowed them to show that either copy of either H2A or H2B was dispensable, but that at least one of each was necessary for viability (Rykowski et al., 1981; Kolodrubetz et al., 1982; Grunstein et al., 1984). Furthermore, by using these mutants in which only *one* functional gene copy remains, they have been able to explore the effects of modification of the only histone variant of a given type available to the cells. Remarkable changes can be tolerated. For example, the entire region between amino acid residues 3 and 22 in H2B can be deleted without affecting viability (Wallis et al., 1983). Since this comprises essentially the whole amino-terminal tail of the molecule, we are led to conclude that this region plays either *no* role in yeast histone function or that its role is exceedingly subtle.

Such studies severely shake our preconceptions as to the significance of histone structure in the life of the cell!

Tetrahymena

While the putative histone variants in this protist have not been characterized as thoroughly as those in some other organisms, the system has some especially interesting features. *Tetrahymena* contains two nuclei, a vegetative macronucleus and a micronucleus in which the chromatin is highly condensed and transcriptionally inactive. Gorovsky and co-workers have examined the histone composition of these two kinds of nuclei (see Allis and Gorovsky, 1979; Allis et al., 1980a, 1980b; Bannon and Gorovsky, 1984; Allis and Wiggins, 1984b). The results are summarized in Table 4-5. Considering first the macronucleus, we find two major forms of H2A plus a minor variant (hν1) that bears considerable resemblance to H2A. This latter variant first appears in new macronuclei at about the time that these commence RNA synthesis, suggesting that it may be a "transcription" variant (Wenkert and Allis, 1984; Allis et al., 1986). H3 is also accompanied by a minor variant, termed hν2. The micronucleus contains what appear to be the same two major H2A variants, but not the minor form hν1. The micronuclear H3 seems to be identical to that in the macronucleus, but it is accompanied by an electrophoretically faster band that was at first believed to be a sequence variant. It has been found, however, that this H3^F is actually a proteolytic product of H3, in which six residues have been cleaved from the N terminus (Allis et al., 1980a). This does not represent a proteolysis artifact, as might be assumed. Rather, the trimmed H3 is made by a physiologically controlled process, and its appearance is associated with maturation of new micronuclei following conjugation of *Tetrahymena* (Allis and Wiggins, 1984a). A curious parallel to this modification is observed in mammalian cells infected with foot-and-mouth disease virus (Grigera and Tisminetsky, 1984). In such cells, the H3 is selectively degraded to produce a product that runs on SDS gels a bit more slowly than H4. A specific protease seems to be involved, for treatment with cycloheximide immediately after infection blocks the degradation.

The two nuclei of *Tetrahymena* also display pronounced differences in lysine-rich histones. Macronuclei contain a single H1, but micronuclei have four "H1-like" proteins, denoted α, β, γ, and δ. Evidence suggests that α, γ, and δ are all produced by successive, specific proteolytic cleavage of a precursor X. Whether β is produced in the same process or is coded by a separate gene is presently unclear (Allis and Wiggins, 1984b).

Micronuclei and macronuclei of *Tetrahymena* are each formed anew after conjugation. Apparently, old micronuclei de-differentiate by losing the micronuclear-specific histones; redifferentiation then yields both new micronuclei and macronuclei. The details of the process are complex, but the fact that both transcriptionally inactive and transcriptionally active nuclei are generated make this an exceptionally interesting system for further study (see Allis and Wiggins, 1984b; Allis et al., 1986).

Sea Urchins

While a number of laboratories had long been concerned with aspects of the role of histones in sea urchin development, the full importance of urchin histone variants in this process only became apparent with a 1975 paper from Leonard Cohen's laboratory (Cohen et al., 1975). These workers, utilizing a new triton-urea gel technique together with pulse labeling of the histones, were able to demonstrate specific "switches" in the variant pattern during embryonic development in *Strongylocentrotus purpuratus*. A long series of subsequent papers from this and other laboratories have amplified and refined these results, with the consequence that this is now among the best-known examples of changes in histone variant pattern accompanying development (see, for example, Newrock et al., 1977; Kunkel and Weinberg, 1978; Newrock et al., 1978; Childs et al., 1979; Hieter et al., 1979; Shaw et al., 1981; Poccia et al., 1981; Newrock et al., 1982; Herlands et al., 1982; Mauron et al., 1982; Maxson and Wilt, 1982; Harrison and Wilt, 1982; Weinberg et al., 1983; Richards and Shaw, 1984). The developmental program is described in Figure 4-13 and the recognized variants are listed in Table 4-5. For histones H1, H2A, and H2B, several variants have been identified, and their distribution is found to change during development of the embryo. Each of these histones has a specific "cleavage stage" variant that apparently exists in a maternal pool in the egg (Poccia et al., 1981) and predominates during the first few cell divisions. These are replaced at about the 8- to 16-cell stage by the "α" variants that persist to approximately the mesenchyne blastula stage, whereupon

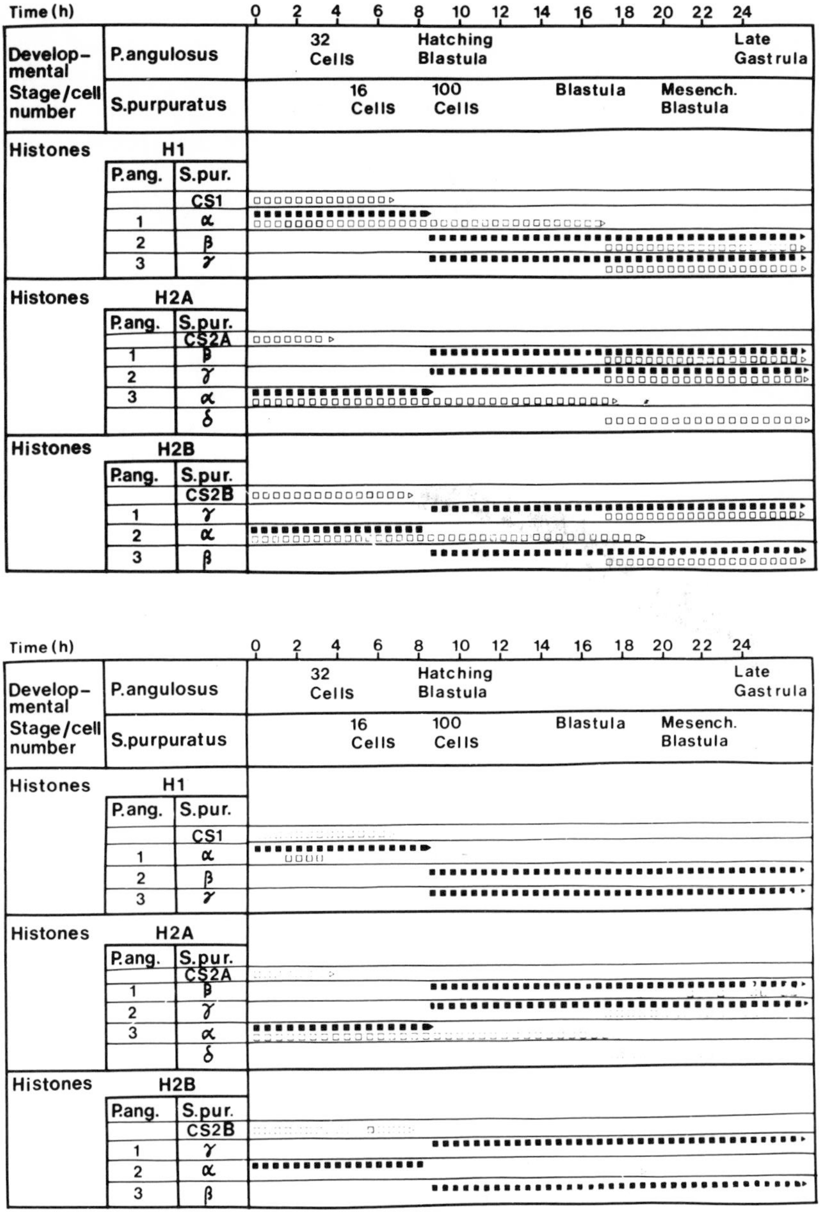

Fig. 4-13. Developmental pattern of sea urchin embryo histones in *P. angulosus* (■ ■) and *S. purpuratus* (□ □). (From Histone Genes, Stein, Stein and Marzluff, Ed., pp. 65–105, von Holt et al. Copyright © 1984 John Wiley & Sons, Inc. Reprinted by permission of John Wiley & Sons, Inc.)

they are in turn replaced by a new set of "late" histones. The histones H3 and H4, on the other hand, are represented by only one molecular type each, throughout development. It must not be assumed, however, that these embryonic histone variants complete the list. *Strongylocentrotus purpuratus* also has H1 and H2B variants unique to the sperm (Carroll and Ozaki, 1979) and a fourth type (H1λ) present only in adult somatic tissues (Pehrson and Cohen, 1984). It is quite possible that further variants remain to be discovered.

The existence of a "developmental program" of histone variant switching is by no means unique to *S. purpuratus*. The urchin *Parechinus angulosus* has been extensively studied in the laboratory of C. von Holt (see, for example, Brandt and von Holt, 1978; Brandt et al., 1979; De Groot et al., 1983; von Holt et al., 1984). This urchin exhibits a developmental "switching" program that is similar to, but distinct from, that of *S. purpuratus* (see Fig. 4-13). In addition to the embryonic forms listed in the figure, specific somatic cell variants have been identified in this organism. The sperm shows further unique histone types; this same laboratory has isolated—and sequenced—no less than three sperm H2Bs from this organism (Strickland et al., 1977a, 1977b, 1978). Further studies, using individual animals, have shown that different individuals in the same population can express different combinations of these sperm H2B variants (Strickland et al., 1981). There are also sperm-specific H1 and H2A. All together, this group counts at least 24 histone variants in *P. angulosus* (Schwager et al., 1983).

It is evident that the repertoire of histone variants available to sea urchins is extensive. Although I have concentrated, for brevity, on two organisms, parallel studies on other species have yielded very similar results. The reader should compare, for example, the studies by Gross and co-workers on *Arbacia* and *Lytechinus* (Ruderman and Gross, 1974; Senger et al., 1978; Arceci and Gross, 1980). Histone variants in a number of other urchin species have been reported by DePetrocellis et al. (1980). The histone switching phenomenon in invertebrates is by no means restricted to urchin development; Mackay and Newrock (1982) have demonstrated quite similar processes in the snail, *Ilyanassa* (see also Flenniken, 1984). Evidence for histone switching has also been obtained for the clam, *Spisula* (Gabrielli and Baglioni, 1975, 1977). Franks and Davis (1983) report switching in embryos of the marine worm *Urechis caupo*. Despite the very extensive studies of histone changes during invertebrate embryogenesis, the physiological function of the phenomenon remains obscure. Newrock's laboratory has studied the effect of removal of the polar lobe in *Ilyanassa* embryos (see Flenniken, 1984; Flenniken and Newrock, 1987). In this case, both the normal pattern of histone replacement and normal embryonic development are perturbed. The data suggest that cell-cell interactions may participate in the histone-switching mechanism. A rather different insight into histone switching has been provided by Knowles and Childs

(1984). They have examined the utilization of "early" and "late" H3 and H4 genes in *Lytechinus*. Although these code for identical proteins, they are arranged very differently in the genome. The early H3 and H4 genes exist with other histone genes in tandem arrays reiterated many hundreds of times. The late genes are not in tandem arrays, are not closely linked to other histone genes, and are reiterated only about 8 to 10 times. This is reminiscent of the histone gene arrangements in vertebrates. Clearly, there can be no change in *protein* function associated with this transcriptional switch. Rather, the existence of the highly reiterated early genes seems to be an adaptation to facilitate the great burst in histone synthesis required in early urchin development. This raises the question: Might *all* switching in early embryonic development simply be a mechanism to provide different rates of mRNA formation? Perhaps the only reason why multiple H1, H2A, and H2B products are observed lies in the weaker evolutionary pressures on these proteins—it may not *matter* to the organism that the products of these late and early genes are slightly different. However, it should be noted that Simpson (1981) has observed differences in stability in nucleosomes containing early and late urchin histones.

Mouse
The histone variation in mouse is probably the most thoroughly studied of that in any higher organism. Much of this work comes from the laboratories of Dr. A. Zweidler and L. Cohen at the Fox Chase Cancer Center. Zweidler's studies of the core histone variants are succinctly summarized in a review (Zweidler, 1984); for recent work on the lysine-rich histones, see Lennox and Cohen (1983, 1984a, 1984b) and Lennox (1984). The results are of especial interest, for not only are they very complete, but they suggest a quite different role for histone variation in the mouse, as compared to that implicated in invertebrate development.

The mouse histone variants presently recognized fall into two main classes: those present in somatic cells, and a special set of "testes" histones found only in developing spermatocytes (Seyedin and Kistler, 1979a; Zweidler, 1980; Meistrich et al., 1985). In a pattern common to many vertebrate species, these testes-specific histone variants replace the somatic cell variants during the meiotic prophase of spermatogenesis. As the spermatocyte further develops into the mature sperm, all (or nearly all) of these histones are then replaced by protamines. The mammalian spermatocyte-specific H1 histones have been discussed above. In addition, there exist similar variants of the core histones, as listed in Table 4-6. It should be noted that this pattern is by no means universal in the animal kingdom; many invertebrate sperm contain histones rather than protamines and retain a nucleosomal structure. Even in such cases, some of the sperm histones are distinct from their somatic counterparts (see, for examples, the earlier discussion of sea urchin histones).

The somatic histones of the mouse (Fig. 4-14) present a complex picture,

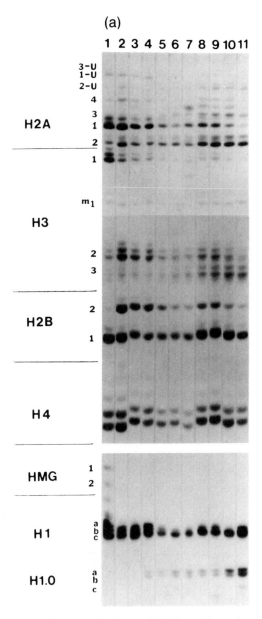

Fig. 4-14. (a) Histone variants and their modified forms in various mouse tissues, as resolved on acid–urea–triton gels. Tissues: (1) calf thymus (reference), (2) mouse thymus, (3) mouse spleen, (4) intestinal mucosa, (5) mammary gland, (6) lung, (7) heart, (8) Kupffer cell of liver, (9) salivary glands, (10) kidney, (11) polypoid hepatocytes. (b) A similar gel, showing the changes occurring in mouse liver histones

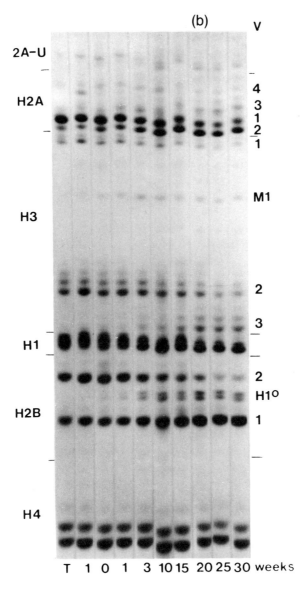

during maturation. Lane (T) is a mouse thymus standard; the remaining lanes range from one week before birth (− 1) to 30 weeks of age. (From Histone Genes, Stein, Stein and Marzluff, Ed., pp. 339–371, Zweidler. Copyright © 1984 John Wiley & Sons, Inc. Reprinted by Permission of John Wiley & Sons, Inc.)

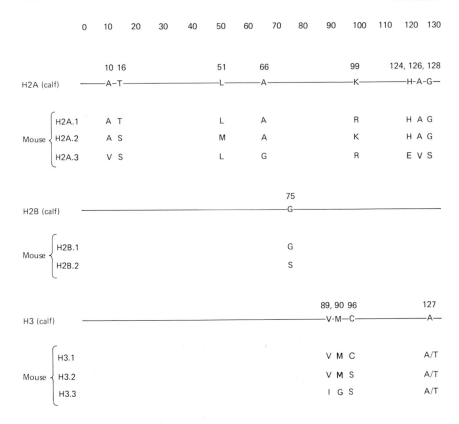

Fig. 4-15. Sequence differences in the somatic core histone variants of the mouse. Most of these variants are found throughout the vertebrates. Data from F. Zweidler.

which may be taken as a paradigm for mammalian histone diversity. Figure 4-14a depicts high-resolution gel electrophoresis of the histones in a number of adult mouse tissues. Even if we disregard the modified forms, the number of variants is impressive, and there are clear tissue differences. The presently identified core histone variants are listed in Table 4-6. The corresponding sequence differences are shown in Figure 4-15. The mouse tissues in Figure 4-14a are arranged (left to right) in order of decreasing rate of DNA replication in the adult; the accompanying changes in histone variant distribution are obvious.

Zweidler and his colleagues have also studied the changes in variant distribution during development, from embryo implantation to the adult (see Franklin and Zweidler, 1977; Zweidler et al., 1978; Zweidler, 1980). Kaye and Wales (1981) have conducted comparable studies on preimplantation embryos. Careful examination of such data indicates that histone variants play a quite different role in mammalian cells than they do in invertebrate development. There is little evidence for the dramatic "histone switching'' observed in sea urchin embryos. Almost all of the somatic variants seem to be present at all developmental stages from neurula to the adult. However, there are significant changes in the relative amounts of some of the variants as tissues mature (see Fig. 4-14b). This developmental change, together with the tissue specificities in the adult, led Zweidler (1980) to postulate the existence of at least two classes of variants: those associated with cell replication, and those utilized in replacement in nondividing cells. Tests of this hypothesis have involved either partial hepatectomy (in which resting cells are induced to replicate) or stimulated differentiation of erythroleukemia cells, which leads to a cessation of cell division. Such experiments have supported the general concept, but led to a somewhat more sophisticated classification of variants, based on the relationship of their synthesis to DNA replication (see Zweidler, 1984; Grove and Zweidler, 1984).

1. *Strictly replication-dependent:* Induced at the beginning and repressed at the end of DNA synthesis. These include H2A.1, H2B.2, H3.1, and H3.2.
2. *Partially replication-dependent:* Induced at the beginning of DNA synthesis, but not wholly repressed at the end. H2A.2 and H2B.1 are examples.
3. *Replication-independent:* Can be continually expressed at a low rate even in nondividing cells. Such variants (H2A.3 and H3.3) tend to accumulate in differentiated, nondividing cells.
4. *Noncompeting:* Minor variants that normally occur in somatic cells at low levels, but can be induced under special circumstances. These include H2A.4 and M1.

Figure 4-16 depicts the behavior of the major variants following partial hepatectomy: The first three classes are clearly delineated. In Table 4-6, all of the core histone variants have been classified according to this scheme. As Figure 4-16 shows, it is precisely those variants labeled ''replication-independent'' that accumulate as the tissue matures.

Turning to the lysine-rich histones of the mouse, one finds equal or greater complexity. Lennox and Cohen (1983) have identified five H1 variants (a–e), plus H1°. According to Zweidler, there are in fact two H1° variants in the mouse, a pattern common to many mammals (Harris and Smith, 1983). Some of the distinguishing features of the H1 variants, as noted by Lennox and Cohen, are presented in Table 4-7. These authors feel that the behavior of these variants is too complex to be wholly encompassed by Zweidler's classifications. Nevertheless, it would appear that H1a and H1b exhibit some of the characteristics of class 1 whereas H1e resembles class 3. The behavior of H1° is more complex. In some respects it resembles class 3, for it exhibits a tendency to accumulate in nondividing cells (see, for examples, Panyim and Chalkley, 1969a; Marks et al., 1975; Seyedin and Kistler, 1979b; Pehrson and Cole, 1980; Pieler

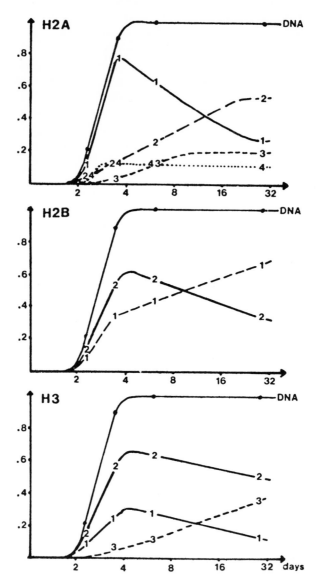

Fig. 4-16. Changes in the newly synthesized DNA and histone variants after partial hepatectomy in 4-month-old mice. Numerical symbols indicate the changes in specific variants. (From Histone Genes, Stein, Stein and Marzluff, Ed., pp. 339–371, Zweidler. Copyright © 1984 John Wiley & Sons, Inc. Reprinted by permission of John Wiley & Sons, Inc.)

et al., 1981). On the other hand, it rises rapidly in response to certain stimuli. Both Keppel et al. (1977) and Osborne and Chabanas (1984) have noted a very rapid accumulation of H1° in mouse erythroleukemia cells subsequent to chemical stimulation to differentiation. Larue et al. (1983) have examined H1° expression during development and differentiation of tissues in the rat liver. They observe distinct steps of increase in this variant that are claimed to correlate with developmental stages.

This rapid and specific response to differentiation has led Zweidler to classify H1° as a type 4 variant of the H1 class of histones. Many of the specific histone variants detected in the mouse are found to be common to other mammals and, in some cases, to nonmammalian vertebrates (see Table 4-6 and Zweidler, 1984). This might be thought to argue for common roles, but it has been pointed out that H2A.2, for example, behaves like a replacement variant in mammals, but not in the chicken. Zweidler has suggested that some of these variants may have arisen via neutral mutations, and later acquired specific roles in certain organisms.

Yet there are curious examples of variant preservation, even between distantly related phyla. For example, Wu et al. (1982) report remarkable similarity in peptide maps between one minor mouse variant (H2A.Z = M1) and a sea urchin H2A (see also West and Bonner, 1984). Clearly, nonallelic variants not only exist, but are evolutionarily conserved. This would suggest that some, at least, must play more than a passive role in chromatin function.

Possible Molecular Roles of Histone Variants

A major problem in assessing the physiological function of histone variation is that we have virtually no knowledge as to what effect such variation has at the molecular level. The sequence differences in most inner histone variants are by no means dramatic, usually involving only a few amino acid substitutions (see Fig. 4-15). However, as Figure 4-15 shows, many of these changes occur in the globular regions of the histones, where they might be expected to influence histone–histone interactions, and thus the stability of the nucleosome (cf. Simpson, 1981). Zweidler has suggested that the "replication" variants might be associated with a more open and dynamic chromatin structure, appropriate to replicating chromosomes. But as we shall see in *Histone–Histone Interactions*, below, Isenberg and co-workers have found that the histone–histone interactions are remarkably uniform, even when histones from widely different phyla are mixed. This does not rule out the possibility of some "fine tuning" of interactions by histone variation, but there is virtually no data with which to test such a hypothesis. In yeast it has been shown that the H2A variants and H2B variants can interchangeably associate to yield viable cells (Kolodrubetz et al., 1982). However, the relative interaction strengths have not been determined. Careful comparative studies of the stability of histone–histone interactions involving histone variants would be welcome.

At the nucleosome level, there is an almost equal dearth of information. Simpson and Bergman (1980) have investigated a rather extreme example, the stability of the sperm nucleosomes of the urchin *S. purpuratus*. In this case, the H2B variant is quite radically different; it contains about 20 additional residues on the N terminus, as compared to somatic H2B. The H2A of the sperm also differs considerably in sequence from the somatic variant. These nucleosomes exhibited some increased stability toward thermal denaturation, as well as an increased resistance to DNase-1 digestion. Otherwise, the nucleosomal properties were much like those from other cells (see also Simpson, 1981).

It has been suggested that the length of the H2B chain, which varies quite widely among variants and organisms, could play a role in determining the linker length and hence the nucleosome repeat (Spadafora et al., 1976; Zalenskaya et al., 1981). This speculation was founded on the unusual length of both the H2B molecule (145 residues) and the nucleosomal repeat in sea urchin sperm (237 bp). However, it does not seem very convincing when a wide variety of organisms are considered. For example, *Tetrahymena* macronuclei have an H2B 119 residues long and a repeat of 205 bp; yeast has a 130-residue H2B and a 165-bp repeat. In general, there does not seem to be *any* correlation between repeat length and H2B chain length. This is not to assert that H2B (and possibly H2A as well) may not play some role in determining linker size, but if they do, features more subtle than mere chain length of the histone must be involved.

It would seem more likely that H1 variants might help determine linker size, since these proteins interact directly with that part of the nucleosome. Indeed, a progressive change in nucleosome spacing has been observed in developing avian erythroid cells, as H5 replaces H1 (Schlegel et al., 1980a). Similarly, the very long nucleosome repeats in invertebrate sperm seem to be associated with specific sperm H1 types. However, that these may be special cases is suggested by studies of H1 variant distribution in neuronal and glial cells of calf cerebellum (Harris et al., 1982). These cell types exhibit very different nucleosome repeats (165 bp and 193 bp, respectively). Yet Harris et al. find only a small difference in the $H1°/H1$ ratios in these nuclei (0.135 and 0.099, respectively). Since there is only about one $H1°$ per 10 nucleosomes, it is difficult to see how this variant could account for the repeat difference. In many other cases, $H1°/H1$ ratios have been observed to change with no reported difference in spacer length. Complicating the picture is the fact that H1 and $H1°$ are apparently nonrandomly distributed (Delabar, 1985).

It remains possible that differences in H1 variant composition have a major effect on nucleosome–nucleosome interaction, and hence on the higher order structures of chromatin. In support of this hypothesis, Liao and Cole (1981a, 1981b) have shown that different fractions of bovine H1 differ dramatically in their abilities to aggregate dinucleosomes or superhelical DNA. How this may correlate with in vivo function remains unknown.

The truth of the matter is that we do not know, at this time, exactly what it is that histone variation accomplishes in *any* organism. We do not know why there are as many variants as there are—or so few, depending on one's point of view. We do not understand why a certain set of variants is highly conserved, at least among the mammalia. Much is yet to be revealed, but it is becoming evident from these studies that histones may play roles much more sophisticated than the simple packaging of eukaryotic DNA. In a certain sense, the Stedman hypothesis (see Chapter 1) seems to be vindicated. There *are* many kinds of histones, and they *do* exhibit some cell-type specificity. It is probable that these differences play a functional role, but we do not, as yet, understand that role at a molecular level.

Histone Modifications

The preceding section has described in some detail nonallelic variations in the *sequences* of the principal histone classes. However, as Figure 4-14 indicates, histone heterogeneity is much more complicated than can be accounted for by this mechanism alone. There exists a set of post-translational covalent modifications of these proteins which, in its full complexity, can only be described as baroque. These modifications include acetylation, methylation, phosphorylation, ADP-ribosylation, glycosylation, and covalent addition of peptides; their basic characteristics are summarized in Table 4-9. Evidence has accumulated that such alteration of the histones may have important effects on chromatin structure and function. For that reason, an enormous number of papers on this topic have appeared in recent years. It would be impracticable to describe all of this work, much of which is fragmentary and repetitive in any event. In this section, I shall briefly summarize salient features of the known forms of modification, leaving for later chapters much of the discussion and speculation concerning their supposed roles in altering nucleosome or chromatin structure and behavior. The reader looking for more detail should consult the reviews listed in Table 4-9. It must be kept in mind that despite the abundance of research effort in this field, some subtle or evanescent forms of histone modification may have yet escaped our notice. The modification reactions, in the main, occur in the nucleus and are in most cases accompanied by reactions that can remove the covalently bound groups. Thus, turnover of some modifications is rapid, and the spectrum of modified forms observed after nuclear isolation may not accurately reflect the distribution in vivo. Attempts to circumvent the problem by studying these reactions by in vitro catalysis with the appropriate enzymes and purified histones have been found to be of limited usefulness, for the accessibility of histone sites is often markedly different in situ (see

Table 4-9. Types of Histone Modifications

Modification	Group added	Acceptor group	Reviews
Acetylation	O $\parallel$ CH_3C-	N-terminal serine, lysine	Allfrey (1977, 1980) Doenecke and Gallwitz (1982)
Methylation	CH_3-	Lysine; *possibly* histidine, arginine, glutamic and aspartic acids	Allfrey (1977) Paik and Kim (1980)
Phosphorylation	$-PO_3=$	Serine, threonine, histidine, *possibly* lysine and arginine	Smith (1982) Langan (1978) Johnson and Allfrey (1978) Hohmann (1983)
ADP-ribosylation	ADP-ribose (or polymer of)	Glutamic acid, carboxyl terminal; *possibly* arginine	Hayashi and Ueda (1977) Smulson and Sugimura (1984)
Ubiquitin addition	The protein ubiquitin, through C terminus	ε-amino groups of lysine, via an amide bond to carboxy terminus of ubiquitin moiety	Goldknopf and Busch (1978) Busch and Goldknopf (1981)
Glycosylation	Fucose, mannose (?)	Unknown	Levy-Wilson (1983)[a]

[a] While not a review, this is, to my knowledge, the only paper reporting such modification.

Garcea and Alberts, 1980, for examples). Accordingly, some rather sophisticated in vivo methods have been developed, including the use of radiotracer pulses, selective inhibition of certain enzymes, and cell cycle synchronization. In the following pages, I shall outline what these and other methods have revealed about each of the above-mentioned kinds of modification. Even though modification of histones probably is to some extent interdependent, the only way to comprehensibly describe such a complex matter is to consider each of the categories listed in Table 4-9 separately.

Acetylation

Histone acetylation was first noted by Phillips (1963). Two distinct types of modification occur in vivo. First, a number of histones are acetylated on the α-amino moieties of N-terminal serine residues (Phillips, 1963, 1968; Allfrey et al., 1964). In most cells H1, H2A, and H4 are found to contain such blocked N termini (see Fig. 4-17). This modification appears to be an early, cytoplasmic event in histone synthesis (Liew et al., 1970; Ruiz-Carrillo et al., 1975). Jackson et al. (1976) present evidence that H4 is

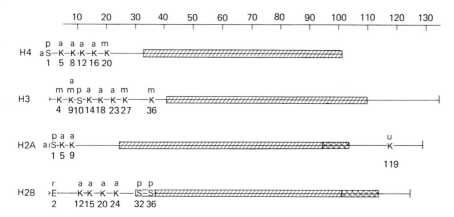

Fig. 4-17. Sites of covalent modification in the core histones. Code: a = acetylation, m = methylation, p = phosphorylation, r = ADP-ribosylation, u = ubiquitination. Only sites known to be modified in vivo are illustrated. The "globular" domain of each histone is indicated by shading, with both more and less conservative estimates shown.

diacetylated within a few minutes after synthesis. One of the acetyl groups is then removed, with a half-life of about 5 minutes, leaving the single N-terminal acetyl. That acetylation may be directly associated with histone deposition is supported by more recent studies by Allis et al. (1985), who have studied acetylation of H3 and H4 in the transcriptionally inactive micronuclei of *Tetrahymena*.

The second kind of acetylation occurs within the nucleus and involves addition of acetyl groups to the ε-amino groups of selected lysines in all four of the "core" or "inner" histones, but not in H1 (Candido and Dixon, 1972a). Figure 4-17 identifies acetylation sites on calf histones. It is noteworthy that of all the lysine sites potentially available for modification, only certain ones are used, and these are invariably in the N-terminal domains of the inner histones. The pattern shown by calf thymus histones seems to be maintained in other cell types and organisms as well. For example, studies of trout histones (Candido and Dixon, 1971, 1972a, 1972b) show these same sites in H4 and H3 to be acetylated; there are only minor differences in H2A and H2B. Even in such distantly related organisms as *Physarum* and *Tetrahymena,* the acetylation sites in H3 and H4 are conserved (Waterborg et al., 1983; Waterborg and Matthews, 1983a, 1983b; Chicoine et al., 1986). While knowledge of the precise loci of acetylation is not available for most other organisms, the phenomenon has been observed in all eukaryotic phyla investigated, including fungi, plants, and animals.

It seems probable that the N-terminal domains of histones play some special role in chromatin organization (see Chapters 6, 7, and 8). If so, nuclear reactions that modify histones in these regions could be of great significance to chromatin function. As many authors have noted, maximal acetylation of these sites would markedly reduce the net positive charge and therefore DNA-binding strength of the N-terminal domains. That the precise location of these modifications may be important to their role in chromatin function is suggested by the fact that *these* lysine residues, in otherwise evolutionarily unstable domains, are highly conserved (Isenberg, 1979; see also Figs. 4-5 through 4-8).

The in vivo source of acetyl groups for histone acetylation seems always to be acetyl coenzyme A. The transfer of acetyl groups to histone lysine residues is catalyzed by a number of *acetyltransferases*. Such enzymes have been purified from numerous sources; a partial listing is given in

Table 4-10. Histone Acetyltransferases[a]

Acetylase type	Reported in	Localization	Histones acetylated in vitro
A	Rat thymus (1)[b] Rat liver (2) Rat hepatoma (3) Calf liver (4) Calf thymus (5) Bovine lymphocytes (6) Artemia (7) Yeast (8)	Nuclear; bound to rapidly sedimenting fraction	All free histones, including H1; all core histones on particles
B	Rat thymus (1) Rat liver (2) Calf liver (4) Calf thymus (5) *Drosophila* (9) Rat hepatoma cells (10)	Mainly cytoplasmic; does not bind to chromatin	H4, and H2A to a lesser extent; inactive on isolated nucleosomes
C	Calf thymus (5)	Nuclear	All free histones, including H1
DB	Bovine lymphocytes (6) African green monkey cells (SV-40 minichromosomes (11)) Artemia (7)	Nuclear, chromatin bound	All free histones except H1; all core histones on core particles

[a]Most data derived from Doenecke and Gallwitz (1982).
[b]References:
1. Gallwitz and Sures (1972).
2. Gallwitz (1973); Libby (1980).
3. Belikoff et al. (1980); Garcea and Alberts (1980).
4. Libby (1978).
5. Sures and Gallwitz (1980).
6. Böhm et al. (1980b).
7. Estepa and Pestana (1983).
8. Travis et al. (1984).
9. Weigand and Brutlag (1981).
10. Garcea and Alberts (1980).
11. Otto et al. (1980).

Table 4-10. It is clear that there exist a number of acetyltransferases exhibiting different substrate specificities, which can be grouped into several major classes. However, the complexity of the situation is probably greater than Table 4-10 would imply. It should not be assumed that all of the enzymes grouped under a given category in Table 4-10 are in fact identical, or even necessarily homologous. There are numerous apparent discrepancies in the literature concerning the substrate specificities of presumably homologous enzymes, and indirect evidence that there may exist enzymes which have not as yet been isolated. Consider the results of Garcea and Alberts (1980), who studied the specificity of a purified preparation of acetyltransferase A from rat hepatoma cells. Garcea and Alberts observed marked differences in histone reactivities when the enzyme was incubated with nucleosomes or free histones. Further, neither of these patterns corresponded to the distribution of acetylation observed in vivo; in particular, H3 was poorly acetylated by exogenous enzyme. Thus, these workers suggested that either a special cofactor or a still unidentified acetyltransferase is required for the efficient acetylation of H3 observed in living cells. More recent studies (Estepa and Pestana, 1981, 1983) indicate that the acetylase DB (Table 4-10) may be responsible and that spermine may activate the enzyme.

The acetylation of histones in nuclei is a dynamic process, and there exist enzymes that catalyze the *deacetylation* reaction (see, for example, Vidali et al., 1972; Fujimoto and Segawa, 1973; Kaneta and Fujimoto, 1974; Hay and Candido, 1983a, 1983b). As a consequence of the presence of both acetyltransferases and deacetylases in the nucleus, the acetyl groups on at least some histones in some portions of the genome are in rapid turnover (see Moore et al., 1979; Chestier and Yaniv, 1979; Covault and Chalkley, 1980; Perry and Chalkley, 1982; Duncan et al., 1983; and further discussion in Chapters 7 and 8). Furthermore, whereas acetylation reactions are slowed during nuclear isolation (presumably through loss of acetyl Co A), the deacetylation reaction continues. As a result, isolated nuclei will probably contain histones at lower levels of acetylation than would be typical of the in vivo steady-state.

The discovery that butyrate has an inhibiting effect on the deacetylases was therefore of great importance. Riggs et al. (1977) first noted that addition of low levels of butyrate to cultured cells led to accumulation of what appeared to be highly acetylated forms of the histones. The mechanism for this effect was established in the following year, when three laboratories demonstrated almost simultaneously that the effect was due to inhibition of deacetylation (Candido et al., 1978; Sealy and Chalkley, 1978a; Vidali et al., 1978). The use of butyrate to enhance levels of acetylation has since been extended to many kinds of cells and has proved to be a very useful technique. A quantitative study of its effect on deacetylases is provided by Cousens et al. (1979). Figure 4-18 illustrates the effect of butyrate treatment on the HeLa histones. Enhancement of ace-

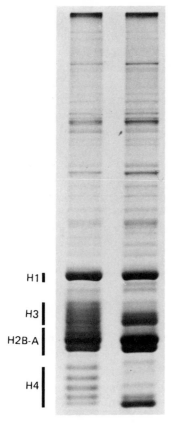

H1

H3

H2B-A

H4

Fig. 4-18. Comparison of histone acetylation in butyrate-treated (left) and nonbutyrate-treated (right) HeLa cells. Note the very clear difference in modified forms of H4. From J. Ausio.

tylation of histones H4 and H3 is clearly demonstrated. Table 4-11 gives some quantitative data on the effects of butyrate treatment on observed levels of H4 acetylation in HeLa cells. The use of this method has permitted the isolation of highly modified histones, nucleosomes, and chromatin. The effects of such modification will be discussed in Chapters 6,

Table 4-11. Acetylation of H4 in Hela Cells As a Function of Butyrate Treatment[a]

	% of total histone as acetylated form				
Sites acetylated: Butyrate (mM)	0	1	2	3	4
0	56	32	8	3	1
10	7	23	23	22	25
15	8	20	22	20	30

[a]Data of D'Anna et al. (1980). Growing cells were treated with indicated concentrations of butyrate for 24 hours.

7, and 8. It should not be presumed, however, that the high levels of acetylation induced by butyrate necessarily represent the in vivo distribution of modifications. One-half of a dynamic process has been blocked, and ongoing acetylation during butyrate treatment may in fact be producing a "hyperacetylated" state (see Sealy and Chalkley, 1978a). This view is disputed, at least with respect to Ehlich ascites tumor cells, by Multhaup et al. (1983). They find no difference in histone acetylation between experiments in which butyrate was added to growing cells and those in which the inhibitor was utilized only during isolation. Therefore, they argue, the so-called hyperacetylated state is the normal state for these cells. It is important to recognize that butyrate has other effects, in addition to its inhibition of the deacetylase. Some cell types are blocked in G-1 phase by this reagent (D'Anna et al., 1980; Darzynkiewicz et al., 1981; Littlefield et al., 1982). Perhaps in consequence of this, modifications in the degree of phosphorylation and methylation are also observed (Whitlock et al., 1980; Boffa et al., 1981). Thus, although the use of butyrate has made possible many studies on the effects of histone acetylation, it must be kept in mind that cells so treated are significantly perturbed and may exhibit unexpected anomalous behavior.

It should already be clear that the superposition of varying degrees of posttranslational modifications upon the range of histone subtypes available through nonallelic genetic variation can produce an enormous variety of histone forms. That the situation may hold even subtler complexities is indicated by several recent observations. First, as Table 4-12 shows, different variants of a given histone exhibit marked differences in acetylation. Second, a careful study by Pantazis and Bonner (1981) shows

Table 4-12. Acetylation of Histone Variants in Non-butyrate-Treated CHO Cells[a]

Acetylation: Variant	% of total histone type as given acetylated variant					
	0	1	2	3	4	Σ
H2A.1	33.5	14	0.24	—	—	47.74
H2A.2	13.3	5.9	0.23	—	—	19.43
H2A.2ox	12.6	3.8	0.12	—	—	16.52
H2A.X	13.4	2.5	0.05	—	—	15.95
H2B.1	54.1	12.3	0.4	—	—	66.8
H2B.2	11.8	3.3	0.9	—	—	16.0
H2B.3	13.3	3.2	0.3	—	—	16.8
H3.1	3.3	3.2	0.57	0	0	7.07
H3.2	31.8	13.5	2.6	0.24	0	48.14
H3.3	22.6	20.5	1.1	0.6	0	44.8
H4	47.2	36.1	13.3	3.3	0	100

[a]Data of Joseph et al. (1983). Cells were in exponential growth.

that some of the sequence variants of H2A differ in the number and positions of sites *available* for acetylation. Finally, there are studies which clearly indicate that such modifications as acetylation and phosphorylation are not independent; the presence of one type may modify the propensity for another (see Whitlock et al., 1983). If these turn out to be general rules, full elucidation of the mechanisms and consequences of posttranslational modification may require many years of careful effort. However, more powerful techniques continue to be developed. To take one example: Sterner and Allfrey (1982) have shown that the acetate analog 2-mercaptoacetate can be used to introduce mercaptoacetyl groups into histones in duck erythrocytes. The availability of such a modification allows the *selective* isolation of acetylated histones, through mercury-affinity chromatography. Such techniques should prove to be of great aid in unscrambling the complex interrelationships between different kinds of modification and variation.

Although the physiological role of histone acetylation has been the subject of enormous interest, it is still imperfectly understood. There appear to be at least three important processes in which this histone modification is involved:

1. *Transcription:* Soon after the discovery of histone acetylation, Allfrey et al. (1964) suggested that this modification might promote the transcription of chromatin. A great deal of evidence has accrued in the following years to support this hypothesis. It will be, however, most appropriate to defer discussion of this topic to Chapter 8.
2. *Histone replacement in spermiogenesis:* In those organisms in which histones are replaced by protamines during spermatid maturation, a complex series of events occurs. In the early 1970s, G. Dixon and coworkers noted that in the trout this process was accompanied by marked increases in acetylation (see Sung and Dixon, 1970; Candido and Dixon, 1972a; Christensen et al., 1984). Other workers have reported similar results for a number of other organisms. Examples include mouse (Bouvier, 1977), rooster (Oliva and Mezquita, 1982), and rat (Grimes and Henderson, 1983, 1984a, 1984b). The latter studies are especially complete and indicate that high levels of acetylation occurs on all of the core histones in a brief period during spermatid maturation. This is just before the final removal of histones for protamine replacement. Many workers have suggested that the function of this acetylation is to somehow "loosen" chromatin structure to allow replacement.
3. *Replication:* As noted above, there is evidence that some newly synthesized histones are partially acetylated in the cytoplasm prior to import into the nucleus. Since histone synthesis and DNA replication are strongly linked in most cells, this implies that there should be specific acetylation associated with S-phase. First noted by Candido and Dixon (1972b), this acetylation has been the object of a detailed study by Waterborg and Matthews (1982, 1983a, 1983b, 1984) using the slime

mold *Physarum*. Cell cycle synchronization in this organism is easy and exact, allowing fine discrimination. Labeling with [^{3}H]-acetate and using high-resolution gels allowed the discrimination of two partially superimposed patterns of modification. A "G-2 phase" acetylation, involving addition of up to four acetyl groups on H3 and H4, appears to correlate with transcription (see above). The acetylation peculiar to S-phase occurs on all histones, but involves mainly mono- and diacetylation of H3 and H4. Figure 4-19 shows how these patterns have been resolved in the case of H4.

If the analysis of Waterborg and Matthews can be extended to other organisms, it may clarify much confusion regarding both the pattern of histone acetylation and acetyl turnover. The idea that diacetylation of H4 is specifically associated with the incorporation of new histone into chromatin is supported by the observation of Woodland (1979) that H4 stored in *Xenopus* oocytes is so modified. Chambers and Shaw (1984) have noted very high levels of diacetylation during periods of rapid cell division in sea urchin embryos, again consistent with the Waterborg–Matthews hypothesis. Finally, recent studies of acetylation of *Tetrahymena* histones point to a similar pattern (Vavra et al., 1982; Allis et al., 1985; Chicoine et al., 1986). But the pattern of acetyl turnover in somatic cells of higher organisms seems complex (see, for example, Cousens et al., 1979; Brotherton et al., 1981; Perry and Chalkley, 1982; Leiter et al., 1984). And it is not immediately evident how all of the data can be fitted into so simple a picture as Waterborg and Matthews present.

Nonetheless, a general pattern is beginning to emerge. Histone acetylation, particularly the higher levels of modification of H3 and H4, appear to be associated with a general relaxation of the structure of chromatin. As we shall see in later chapters, there is only incomplete evidence concerning the molecular mechanisms involved. Acetylation seems, however, to play other roles as well, as exemplified by the modifications of H4 prior to incorporation into chromatin. Much of the confusion regarding acetylation may be a consequence of the existence of multiple roles.

Methylation

Protons of the ε-amino groups of lysine side chains in proteins can be replaced by methyl groups to yield mono-, di-, and trimethylated derivatives (see Paik and Kim, 1980, for a general overview of these reactions). The presence of such modifications in histones was first demonstrated by Murray (1964). The fact that at least some of the methyl groups are stable under commonly used methods of protein sequencing soon led to the recognition of specific locations for methylation in a number of histones. In calf thymus H4, for example, lys 20 was found to be the sole methylation site (DeLange et al., 1969a; Ogawa et al., 1969). In most H4 molecules, the dimethyl derivative predominates, with a smaller fraction of the mono-methylated species. Calf H3 is methylated at two sites (lys 9, lys 27), and mono-, di-, and trimethylated derivatives are observed (DeLange et al., 1973). These sites are shown in relation to other modification locations in Figure 4-17. The specificity of methylation has been quite strongly preserved through evolution as is illustrated in Table 4-13. In almost all organisms examined to date, H4 is methylated at residue 20. The sole exceptions are pea (DeLange et al., 1969b) and *Tetrahymena* (H. Hayashi et al., 1980), in which species this histone is wholly unmethylated. In all H3 molecules in which methylation has been examined up to now, lysines 9 and 27 are modified; in some cases there is additional modification at sites 4 and 36.

A special feature of methylation is that a large fraction of the molecules in any sample are actually modified at the sites listed in Table 4-13. Thus methylation, at least of histones H3 and H4, is a general (rather than localized) feature of mature chromatin. This stands in sharp contrast to the situation with most other forms of histone modification, wherein only a small fraction of the genome is in the modified form at any moment. This high level of modification does not necessarily mean that the turnover of H3/H4 methylation in vivo is slow. Although a number of experiments

Table 4-13. Known Histone Methylation Sites in H3, H4[a]

Organism (tissue)	H4	H3	Reference[b]
Human (spleen)	20[c]	9, 27, 36	Hayashi et al. (1982) Ohe and Iwai (1981)
Calf (thymus)	20	9, 27	DeLange et al. (1969b, 1973)
Chicken (erythrocytes)	20	9, 27, 36	Urban et al. (1979) Brandt and von Holt (1974)
Trout (testes)	20	4, 9, 27 (36)[d]	Honda et al. (1975b)
Sea urchin (sperm)	20	9, 27, 36	Wouters-Tyrou et al. (1976) von Holt et al. (1979)
Pea (embryos)	none	9, 27	Patthy et al. (1973)

[a]Data are included for organisms (tissues) in which H3 and H4 have both been studied.
[b]When two references are given, first is to H4 data, second to H3.
[c]Numbers refer to lysine residues at which methylation has been reported.
[d]Questionable.

Fig. 4-19. Resolution of S-phase and G_2 phase acetylation of *Physarum* H4. (a) The S-phase pattern. (b) The G-2 phase (—) scan of the stained gel; (---) scan of a [³H]-acetate fluorogram. Bands are numbered according to number of acetyl lysines/molecule. (c) Scan of the S-phase fluorogram, and its resolution into S-type and G_2-type bands. From Waterborg and Matthews, 1984.

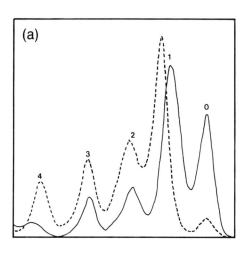

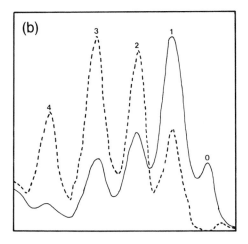

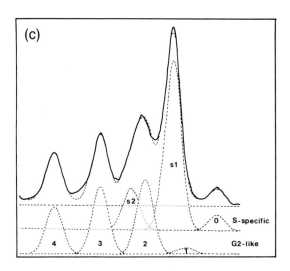

have indicated little turnover (see Borun et al., 1972; Honda et al., 1975b; Thomas et al., 1975), other workers have obtained contrary results. Hempel et al. (1979) report half lives of only a few days for methyl groups on H3 and H4 in cat kidney. Since these cells divide only rarely and have very low levels of histone turnover, the data clearly indicate displacement. Hempel et al. emphasize that kidney tissues are known to be high in demethylase activity (see below) and that cell types may in fact vary greatly in their rates of methyl turnover.

The extent to which histones other than H3 and H4 are methylated is a controversial question. Reports are to be found claiming methylation in *every* histone. (For example: H1 and H5, Gershey et al., 1969; Byvoet et al., 1972; all core histones, but not H1, Borun et al., 1972; Marsh and Fitzgerald, 1973). However, there are reasons to be cautious in interpretation of many of the earlier studies. Some of these experiments were based upon the detection of methylated amino acids in hydrolysates of supposedly purified histones. Contamination from other histones or non-histone proteins may have caused confusion. The same problem can arise even when radiolabeling of methyl groups is utilized. For example, Honda et al. (1975a) note that contamination of H2B fractions with a small amount of H3 gave an initial false impression that trout H2B was methylated. Earlier studies used radiolabeled methionine as a methyl donor; there is in some cases the possibility that methionine itself was incorporated into newly synthesized histones. More recent in vivo experiments, in which protein synthesis has been blocked (i.e., Sung et al., 1977) as well as in vitro experiments using isolated nuclei and the more discriminating S-adenosyl-methionine as a methyl donor (i.e., Duerre et al., 1977; Branno et al., 1983) tend to report mainly methylation of H3 and H4. However, there are some solid experiments that point to some level of methylation in other histones. For example, Levy-Wilson (1983a) finds *only* H2A methylated in in vitro studies with *Tetrahymena* nuclei. Since other studies (see above) indicate no H4 methylation in *Tetrahymena*, this may suggest that methylation patterns in protists are quite different from those in higher organisms. Even in multicellular organisms, methylation of histones other than H3 and H4 may occur under *some* conditions. Thus, Shepherd et al. (1971) reported incorporation of methionine label into methylated lysines of H2B during the S phase of synchronized CHO cells. Since most H3/H4 methylation occurred much later in the cell cycle, it is hard to rationalize this result in terms of contamination from the arginine-rich histones. Again, Camato and Tanguay (1982) and Arrigo (1983) have observed that heat shock or arsenite treatment of cultured *Drosophila* cells induces methylation of H2B, even under circumstances where the synthesis of new H2B has been blocked.

We have no information concerning the sites of methylation on the more lysine-rich histones. Sequencing of the polypeptide chains, which routinely uncovers methylated groups in H3 and H4, regularly fails to find these in other histones. In all of the studies listed in Table 4-13, H2A and H2B have also been sequenced, but no evidence for methylation was found. This suggests that either (1) only H3 and H4 are methylated in most organisms, and results to the contrary are either artifactual or represent very special situations, or (2) the methylation of other histones is at much lower levels, so as to go unnoticed in sequencing studies. Alternatively, it is conceivable that methylation of H1, H2A, and H2B occurs extensively in vivo, but turns over rapidly, and is mostly removed by demethylases active during nuclear isolation, to leave only traces detectable by radiometric methods. Such a situation would be somewhat analogous to that found for acetylation; but in this case, we do not have available a specific inhibitor for the hypothesized demethylase.

The existence of demethylases has been documented, and one such enzyme has been studied in detail. Paik and Kim (1973) discovered an enzyme in nuclei and mitochondria capable of removing ε-methyl groups from histone lysine residues. The level of this activity varies greatly among rat tissues, being especially high in kidney. A year later, these same workers showed that the enzyme was identical to the previously known protein, ε-alkyllysinase (EC1.5.3.4), which was known to remove methyl groups from the free amino acid. The process is oxidative, and the product is formaldehyde (Paik and Kim, 1974).

There has been quite extensive study of the companion enzymes, the protein *methylases*. Table 4-14 lists the characteristics of three recognized classes. The fact that all three types will catalyze the methylation of histones in vitro need not suggest that these proteins are, in fact, the in vivo substrates. In the first place, the type II methylases have been reported only in cytoplasm, and the type I enzymes are found mainly therein. There are no reports, to my knowledge, of cytoplasmic methylation of histones.

Table 4-14. Protein Methylases[a]

	Methylase type		
	I	II	III
Tissue source	Brain	Brain	Thymus
Localization	Cytoplasm and nucleus	Cytoplasm	Nucleus (chromatin)
In vitro substrates	Histones, others	Many, including histones	Histones
Residue modified	Arg	Glu/asp	Lys
pH optimum	7.0	6.0	9.0
K_m(s-adenosyl methionine)	$2.1 \times 10^{-6}M$	$0.9 \times 10^{-6}M$	$3 \times 10^{-6}M$

[a]Data for calf enzymes; abbreviated from Paik and Kim (1980). For information on purification and assay procedures for the various enzymes, see: type I: Durban et al. (1978); type II: Kim and Paik (1978); type III: Nochumson et al. (1978).

However, neither is there proof that even type II enzymes are *excluded* from nuclei in vivo. The situation is complicated by the fact that the products of type II methylation are acid-unstable and could well be lost during histone isolation (see Byvoet et al., 1972, for indirect evidence suggesting that such histone methylation might occur in vivo).

The type I methylases present an even thornier problem. These enzymes *are* found in nuclei, and there are a number of reports of methylarginine in histone hydrolysates (see Byvoet, 1971; Byvoet et al., 1972; Borun et al., 1972). Methyl histidine has been reported as well (Gershey et al., 1969; Byvoet, 1971). However, it should be kept in mind that some of the earlier histone preparations could have been contaminated with nonhistone proteins. To further complicate the situation, Gupta et al. (1982) have recently reported a "histone-specific" type I methylase in wheat germ.

Despite these various hints at other types of methylation, it must be emphasized that *only* lysine methylation has ever been identified as an in vivo product by location of site-specific modifications during sequencing. Presumably the other kinds of methylation, if they exist on histones, are rare, transient, or labile. Thus, it is clear that most of the *well-defined* methylation of histones must be the work of type III methylases. Although such an enzymatic activity was first detected many years ago (Kaye and Sheratsky, 1969), it is only recently that the nuclear enzyme(s) have been purified and studied. The available data suggest that there must be more than one such activity, for Sarnow et al. (1981) find that the enzyme they have purified from calf thymus acts *only* on H4 in vitro. This would suggest that there may exist at least one other enzyme to account for the methylation of H3, a conclusion in accord with the observation by Thomas et al. (1975) that the kinetics of in vivo methylation of H3 and H4 point to different mechanisms. Sarnow et al. could detect no activity toward H4 incorporated into nucleosomes, and thus proposed that methylation of this histone must occur before nucleosome assembly. An alternative explanation might be found in the work of Duerre et al. (1982), who observed pronounced ionic effects on histone methylation in isolated nuclei. These experiments suggest that under the ionic conditions used by Sarnow et al., the nucleosomal conformation might not be appropriate for enzymatic methylation of H4.

The role of histone methylation in chromatin structure and function is unknown. Cell cycle studies, although carefully conducted, have not been easy to interpret. A number of experiments have indicated that methylation begins in late S phase and extends through G_2 and even into mitosis (see Borun et al., 1972, for example). In a very careful kinetic study, Thomas et al. (1975) find that the data can be explained by the fact that histone methylation is a slower process than histone synthesis. Thus, although histone synthesis peaks in S phase, some histones require much of the cell cycle to become fully methylated. Such a slow and progressive methylation would seem to argue against a temporal role for the modification.

Furthermore, there is one unique aspect of histone methylation that indicates, at least in the case of H3 and H4, that this modification is not involved in dynamic cell processes. The methylation of these histones is virtually complete in most nondividing cells. For example, Duerre et al. found only a very small fraction of H3 and H4 unmethylated in the brain cells of 12-day-old rats. With adult rats, no nonmethylated sites could be detected. Similarly, Sung et al. (1977) found no ongoing methylation of H3 and H4 in mature avian erythrocytes, even though acetylation of these same histones was actively proceeding. The completeness of methylation of H3 and H4 would seem to argue against any selective role in chromatin function. Rather, it seems likely that methylation may simply be a part of the maturation of nucleosomes, once they are incorporated into the chromatin structure. On the other hand, it must be remembered that there are suggestions of a more evanescent methylation of other histones, and the roles for such modification might be rather different.

Phosphorylation
Although the presence of phosphorylated proteins in the nucleus had been recognized for many years (see, for example, Johnson and Albert, 1953), it was only in 1966 that two groups succeeded in demonstrating that these phosphoproteins included histones. Ord and Stocken (1966) showed that H1, as well as other unspecified histones, could be labeled covalently with ^{32}P in rat liver and thymus. The existence of phosphoserine in H1 was demonstrated. Independently, Kleinsmith et al. (1966) found incorporation of phosphate into phosphoserine residues of H1 and H3 in lymphocytes. These experiments initiated long programs of studies in many laboratories concerning the mechanisms of phosphorylation and dephosphorylation, the sites of reactions, and the possible roles of such modification in nuclear processes.

It should be noted at the outset that there exist two distinct modes of histone phosphorylation. Predominant and most thoroughly studied is that phosphorylation which takes place on the hydroxyl moieties of serine and threonine. This kind of phosphorylation has been extensively reviewed (see, for example, Gurley et al., 1978b; Johnson and Allfrey, 1978; Allfrey, 1980; Hohmann, 1983) and will be discussed below in some detail. Such phosphorylation is catalyzed by both cyclic nucleotide-dependent and -independent kinases using ATP as a phosphate donor, and these phosphate residues can be removed in vitro by alkaline phosphatase. Most important, the phosphate-ester linkages to threonine or serine are stable to even strongly acidic solutions, and thus survive the usual techniques of histone isolation and purification.

The second class of histone phosphorylations involves P–N linkages to lysine, histidine, and possibly arginine residues. This mode of phosphorylation was first observed in histones by R.A. Smith and collaborators and has been mainly studied by this group (Smith et al., 1973, 1974; Chen et al., 1974, 1977; Smith et al., 1978; Bruegger et al., 1979; Fujitaki et al., 1981). These P–N linkages are exceedingly acid-labile; Smith et al. (1978) assert that they possess half-lives of the order of only minutes in $10^{-2}\,M$ mineral acid at room temperature. Thus, it is to be expected that such modification would have gone wholly undetected in histones prepared by the older acid-extraction techniques. So far, two distinct enzymatic activities that catalyze such phosphorylation in vitro have been found; one forms ε-N phospholysines in H1, the other phosphorylates both of the histidine residues (18 and 75) in H4 (Bruegger et al., 1979). These researchers also report the presence of phosphoarginine in histone H1. Evidence has been presented that the modifications catalyzed by these enzymes actually occur in vivo (Chen et al., 1974, 1977).

Since most studies of histones and their interactions have been carried out with acid-extracted material, the existence of these acid-labile modifications may be cause for concern. Does it imply that most experimenters have utilized irreversibly damaged histones? Isenberg (1979) argues that it does not. This contention is strongly supported by two studies (Bidney and Reeck, 1977; Beaudette et al., 1981) that show identical physical properties, including CD spectra, for core histone complexes prepared by salt extraction or by reconstitution from acid-extracted monomeric histones. The latter workers also showed that the complexes could be dissociated at low pH and then reassociated, with no change in properties (see also Greyling et al., 1983). However, there is some evidence suggesting that the situation might be different in the case of H1. Brand et al. (1981) have reported irreversible changes in the CD spectrum and sedimentation coefficient of salt-extracted H1 after exposure to low pH. It is not clear, however, whether the conditions used would have been sufficient to cleave the P–N bonds, and no analysis for phosphate was reported. In any event, the question of possible irreversible changes in histone behavior as a consequence of acidic dephosphorylation should be regarded as still unsettled.

Whatever role phosphorylation of basic residues may have, it is clear that the majority of sites phosphorylated on histones are serines and threonines, and mainly the former. This kind of phosphorylation has received a great deal of attention, as have the *histone kinases* that catalyze such modification. The known kinases can be divided into two main categories, as is shown in Table 4-15: those dependent on cyclic nucleotide monophosphates for activation, and those kinases whose activity is independent of the presence of cyclic mononucleotides. This table lists only those kinases for which specific sites of in vitro phosphorylation have been identified. A number of other kinase activities have been found, the sites of action of which are unknown. Three notes of caution should be directed toward Table 4-15. First, it is by no means clear that all enzymes in a given class are identical or even homologous. For example, Martinage et al. (1979) show evidence that the cAMP-dependent kinases from rat pancreas and from pig brain exhibit somewhat different specificities toward

Table 4-15. In Vitro Sites of Serine and Threonine Phosphorylation of Histones

Kinase type	Source(s)	Phosphorylation sites reported (S = serine, T = threonine)	
I. Cyclic nucleotide mono- phosphate-dependent			
(a) c-AMP-dependent (HK-I)[a]	Many mammalian tissues; silkworm (9)	H1: H5:	S37[b] (1–4, 7, 9, 16)[c] S45, 49, 91, or 92 (8) S22, 29, 145, 166 (2)
		H4:	S47 (6)
		H3:	S10 (4, 6, 14)
		H2B:	S14, 36 (3) S32, 36 (2, 4, 5, 9) S32, 36, 87, 91 (2)
		H2A:	S1, S19 (2, 3)
(b) c-GMP-dependent	Bovine lung (5, 7) Silkworm (9)	H1: H2A:	S37 (7, 9, 16) S32, 36 (5, 9)
II. Cyclic nucleotide mono- phosphate-independent:			
(a) HK-II	Many animal tissues (1, 2)	H1:	S103 (1, 16)
(b) GR (growth associated)	Rapidly growing cells (1)	H1:	S16, T136, T153, S173, S180 (1, 2, 16)
(c) H3-specific	Calf thymus (10)	H3:	T3 (10)
(d) H4-kinase I	Mouse lymphosarcoma (11)	H4:	S47 (11, 15)
H4-kinase II	Mouse lymphosarcoma (11)	H4:	S1 (11, 15)
(e) miscellaneous	Calf thymus (12) Mouse plastocytoma (13)	H2B: H5:	S32 (12) S3, 7, 104, 117, 148 (13)

HK-I is the designation given by Langan (1) to the cAMP-dependent kinase that phosphorylates site 38 in calf H1. It has not been established that *all* of the activities listed here are identical, or even homologous to this enzyme.

Numbering of Langan (1) based on rabbit RTL-3 (H1.3) sequence. In calf thymus H1, this corresponds to serine 38.

References:
1. Langan (1978).
2. Martinage et al. (1980).
3. Shlyapnikov et al. (1975).
4. Taylor (1982).
5. Glass and Krebs (1979).
6. Martinage et al. (1981b).
7. Zeilig et al. (1981).
8. Kurochkin et al. (1977).
9. Hashimoto et al. (1976).
10. Shoemaker and Chalkley (1980).
11. Masaracchia et al. (1977).
12. Romhanyi et al. (1982).
13. Martinage et al. (1981a).
14. Paulson and Taylor (1982).
15. Eckols et al. (1983).
16. Hohmann (1983).

H2B. Second, it has not been established in every case that the kinase used was a *nuclear* enzyme. Just because an enzyme extracted from a cell homogenate will catalyze the phosphorylation of a histone in vitro does not prove that it is located so as to function in this way in vivo. Finally, note that the histone phosphorylations listed in Table 4-15 are, in the main, those found upon incubation of *isolated* histones with the enzymes. We should not assume that these sites will be available when the proteins are compacted into nucleosomes or chromatin fibers. Indeed, in a number of cases where such comparisons have been made, site reactivities in chromatin-bound and free histones were found to differ considerably (see, for example, Knippers and Böhme, 1978; Walton and Gill, 1981). It is likely, therefore, that the sites of phosphorylation given in Table 4-15 include some that are never found in the living cell. Careful studies, including site-determination, on chromatin phosphorylated in vivo are much less frequent in the literature than experiments with isolated histones. It should also be noted that attempts to determine in vivo sites of phosphorylation on extracted histones may be compromised by phosphatase or other hydrolytic activity during isolation. The question of *what* is phosphorylated in vivo is not easy to answer. In order to relate the data in Table 4-15 to the in vivo studies, it is necessary to say a few words about each histone in turn.

H4: The only specifically demonstrated site of in vivo phosphorylation of H4 is the N-terminal serine residue. This reaction apparently occurs in the cytoplasm shortly after histone synthesis, but dephosphorylation and some rephosphorylation may take place on entry into the nucleus (Ruiz-Carillo et al., 1975; Jackson et al., 1976). Enzymes have also been identified that phosphorylate serine 47 (Masaracchia et al., 1977; Martinage et al., 1981b), and others specific for histidines 18 and 75 (Bruegger et al., 1979). However, proof for in vivo phosphorylation at serine 47 is lacking, and whether one or the other (or both) of the histidines is actually phosphorylated in the cell is still unclear.

H3: There is very good evidence that H3 is phosphorylated at ser 10 both in vivo (Paulson and Taylor, 1982; Ajiro and Nishimoto, 1985) and in vitro by a cAMP-dependent kinase (see references in Table 4-15). The reaction will also proceed when H3 is incorporated into the (H3·H4)$_2$ tetramer (Martinage et al., 1981b). Although serine 10 appears to be the only firmly established cellular site, in vitro phosphorylation has also been reported on serine 28 (see Johnson and Allfrey, 1978) and at threonine 3. Curiously, the latter seems to be the *specific* site for a chromatin-bound enzyme isolated by Shoemaker and Chalkley (1980). Thus, as in the case of H4, there exists an identified activity that is specific for a reaction site which has never been shown to be phosphorylated in vivo. Another kinase specific for H3 has been identified by Whitlock et al. (1980). This is clearly different from the enzyme of Shoemaker and Chalkley, for it is calcium-dependent and phosphorylates an (unidentified) serine residue in the N-terminal region (Whitlock et al., 1983).

H2B: While phosphorylation has been *claimed* at serine sites 6, 14, 32, 36, 87, and 91 (Table 4-15), there seems to be no definitive evidence as to which sites are modified in vivo. Both the cAMP-dependent and the cGMP-dependent kinases will phosphorylate serines 32 and 36, with the cGMP enzyme preferring the latter, the cAMP enzyme the former (Hashimoto et al., 1976; Glass and Krebs, 1979). Since these enzymes are nuclear and widely distributed, and since there is much evidence in the literature for H2B phosphorylation, it is probably safe to presume that at least these two residues are modified in vivo. The other sites listed are either noted as "minor" sites or reported only in a limited number of cases.

H2A: The phosphorylation of the N-terminal serine in trout testes has been demonstrated (Sung and Dixon, 1970). In addition, the cAMP-dependent kinase has been found to catalyze modification at serine 19 in vitro. In the protist *Tetrahymena,* Fusauchi and Iwai (1984) have discovered a remarkable series of C-terminal phosphorylations. Serines 122, 124, and 129 were modified in H2A.1, and serines 122 and 128 in H2A.2. These appear to be unique.

H1: Phosphorylation of this histone has been the subject of extensive investigation, with major contributions by T. Langan and his colleagues. It is the subject of an excellent review by Hohmann (1983). Description and comparison of the results is complicated by our incomplete knowledge of H1 sequences and the extreme variability of these, even between H1 variants in a single cell type. For simplicity, I shall adopt, as have other writers, a site numbering based on the known sequence of the rabbit H1 variant, HLT-3 (Cole, 1977). However, the reader is cautioned that most experiments have been done using other (i.e., calf) H1s, and exact positions (and even existence) of particular serine and threonine sites will differ from one H1 variant to another. For example: A frequently studied serine site lies at residue 38 in calf H1. The corresponding position in the rabbit variant HLT-3 would be residue 37, but this histone has an alanine at this position and no serine anywhere nearby. Authors may refer to this position as either serine 38 or serine 37. Following Langan (1978) and Hohmann (1983), a schematic of the H1 sequence is shown in Figure 4-20. Serine 37 is phosphorylated by the cAMP-dependent histone kinase (HK-I) (Lake, 1973; Langan, 1978) whereas serine 16, threonine 136, threonine 158, serine 173, and serine 180 are substrates for the "growth-dependent" kinase (see below). Modification at sites corresponding to each of these positions has been demonstrated in histones extracted from cells (see Langan, 1978; Hohmann, 1983, for details). There is another enzyme, HK-II, which specifically phosphorylates serine 103 in vitro, but the corresponding in vivo product is yet to be demonstrated. Hohmann (1983) gives specific sites for phosphorylation of other H1 variants. It is by no means certain that

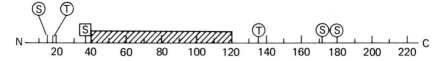

Fig. 4-20. Approximate sites of in vivo phosphorylation of mammalian H1. The positions are from a combination of data on rabbit thymus H1 variants, as summarized by Hohmann, 1983. The exact locations may vary from one mammalian variant to another. The symbol S denotes a serine, T a threonine site. The cyclic–AMP-dependent site is shown by a square symbol, "growth-associated" sites by circles. The shaded area indicates the approximate extent of the folded domain in H1.

the above list of sites phosphorylated in H1 is complete. Other kinase activities have been identified that have unknown reaction sites. These include a calcium–phospholipid-dependent kinase (for which there also exists a specific phosphatase—Iwasa et al., 1980; Sahyoun et al., 1983) and a unique kinase in *Physarum* (Chambers et al., 1983).

The specific phosphorylations of different sites in H1 probably play important roles in the regulation of cellular processes. The cyclic–AMP-dependent phosphorylation of the serine 37/38 site is under hormonal control and may play a role in the regulation of transcription (see Chapter 8). On the other hand, the sites accessible to the "growth-dependent" kinase appear to be relevant to the condensation of chromatin that takes place during mitosis. Further evidence for the roles of H1 phosphorylation in cell cycle regulation will be discussed below.

H5: Table 4-15 lists numerous sites of phosphorylation that have been reported from in vitro studies. Of these, Sung and Freedlender (1978) have confirmed that serines 3 and 7, as well as two or more C-terminal sites, are modified in chicken erythrocytes. The phosphorylation pattern appears to be quite different from that in H1 and, probably, plays a rather different role. Sung (1977) has shown that H5 becomes progressively more phosphorylated as avian erythroid cells mature and is then dephosphorylated just before the final inactivating condensation of the chromatin.

From these data on the localization of modified sites, certain generalizations can be drawn.

1. Phosphorylation in each histone is quite specific, not only with respect to serine and threonine residues phosphorylated, but also in terms of those that are *not* so modified. There have been a number of attempts to seek common features of the phosphorylated sites. Williams (1976) has argued that there is a strong propensity for phosphorylation at residues separated by one amino acid from either a lysine or arginine residue. Indeed, the statistics in favor of such a rule are strong, not only in the examples pointed out by Williams, but also in histone phosphorylation sites that have subsequently been discovered. However, sev-

eral points argue against so simple a rule for kinase specificity. First, it is not absolute; there are phosphorylated sites that do not obey the Williams rule, and there are nonmodified serines and threonines that should, according to this model, be reactive. Second, the rule does not explain the specificities exhibited by the various kinases. Finally, the data lose much of their power to convince when it is recognized that most histone phosphorylation occurs in the terminal domains of the histones (see below), which are in all cases extremely rich in lysine and arginine.

An alternative suggestion has been put forward by Small et al. (1977), who point out that in a number of instances the phosphorylation sites fall within, or at the borders of, predicted β-turn regions in the histone structure. Whether such secondary structure can in fact be determining seems questionable, however, since many of the in vitro studies have been made under circumstances where the histone molecules or at least the regions that are phosphorylated are probably in a random coil conformation.

2. There is a very strong tendency for phosphorylation sites to be concentrated in the nonglobular "tails" of the histone molecules (see Figs. 4-17, 4-20). Since it is the folded, globular portions of the histones that are presumed to be primarily involved in histone–histone interactions in the nucleosome (see *Histone–Histone Interactions*, below, and Chapter 6), it might therefore be assumed that histone phosphorylation would play a minor role in modulating such interactions. The experimental results of Szopa et al. (1980) are consequently very surprising. They find that phosphorylation of serine 10 in H3 decreases the association constant for formation of the $(H3 \cdot H4)_2$ tetramer about 30,000-fold! Curiously, addition of two phosphate residues to H2B (at S32, S36 within the globular region) only halves its affinity for H4, and reduces the $(H2A \cdot H2B)$ dimerization constant only tenfold. There seems, then, to be no obvious correlation between the propinquity of phosphorylation sites to the globular region and the influence of phosphorylation at these sites on histone–histone interaction.

Such considerations lead us to the primary question: What is the role of histone phosphorylation in vivo? While we do not as yet have a complete picture, strong suggestions have arisen from the numerous studies that follow phosphorylation and dephosphorylation through the cell cycle. Indications of a cell-cycle correlation of histone phosphorylation are to be found in a number of early papers (see, for example, Cross and Ord, 1970; Lake et al., 1972; Oliver et al., 1972; Marks et al., 1973). However, it is from two series of investigations that the most detailed and complete information has been gleaned. The first of these has utilized the slime mold *Physarum polycephalum,* which can readily be grown in synchronized culture (Bradbury et al., 1973b, 1974a, 1974b; Mitchelson et al., 1978; Glotov et al., 1980; Chambers et al., 1983; Jerzmanowski and Maleszewski,

1985). This work has been particularly suggestive concerning the role of
H1 phosphorylation in mitosis. The second major investigation, instigated
by L. Gurley and collaborators, has used CHO cells, synchronized by a
variety of methods. While the synchrony obtained with these cells is not
so exact as in the *Physarum* studies, I shall choose the CHO work as an

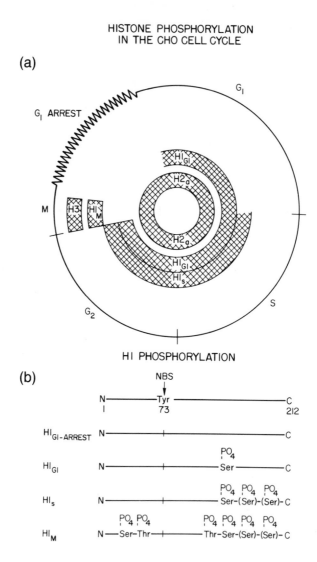

Fig. 4-21. (a) The pattern of histone phosphorylation in the CHO cell cycle. (b)
H1 phosphorylation at various stages of the cell cycle. The regional locations of
the phosphorylation sites were located by cleaving H1 at tyrosine 73. (From Cell
Cycle Regulation, Jetar et al. Eds., Gurley et al. Copyright 1978 Academic Press.)

example, for the phosphorylation of *all* histones has been followed throughout the cycle. For a review of the earlier work in this area, see Gurley et al. (1978b).

The basic experimental protocol is straightforward. Proliferating CHO cells were synchronized and pulse-labeled with $^{32}PO_4^{-3}$ at various stages in the cell cycle. It was found that of all five histones, only H1, H3, and H2A showed major incorporation of ^{32}P. The temporal patterns of modification of these three histones were quite different as is shown schematically in Figure 4-21a. Each will be considered in turn.

H1: As has also been shown in many other cell types, H1 goes through a complex series of phosphorylations and dephosphorylations during the cell cycle. When CHO cells arrested in early G_1 are released, phosphorylation on one serine site in the C-terminal region of the molecule occurs during passage through G_1 (Gurley et al., 1973, 1975). During S phase, two more sites in this same region are phosphorylated. As the cells prepare to enter mitosis, a burst of additional phosphorylation occurs. Two sites in the N-terminal portion of the molecule are now reacted; one of these is a serine, the other a threonine. An additional threonine site in the C-terminal portion of the molecule is also phosphorylated at this time. This burst of modification preceding mitosis has been termed *superphosphorylation,* not only in reference to the number of sites involved, but also because a large fraction of the H1 molecules are so modified. However, it should be noted that in CHO cells, the two major H1 variants are phosphorylated to different degrees, one accepting a total of six phosphates, the other only four (Hohmann et al., 1976; Gurley et al., 1978b). The overall program of H1 phosphorylation of CHO cells is outlined in Figure 4-21b. This pattern of H1 phosphorylation is qualitatively similar to that observed in a number of other systems. For example, the mitotic superphosphorylation in CHO bears a marked resemblance to the "growth-associated" phosphorylation of calf H1 first noted by Lake et al. (1972) and explored so thoroughly by Langan and his collaborators. It would be of great interest to compare the H1 loci phosphorylated in the CHO program with the sites identified by Langan (1978; see above). However, such attempts are not yet possible, for we do not know the sequence of any Chinese hamster H1.

A number of studies have indicated that histone subtypes can differ greatly in their sites of and capacity for phosphorylation (see Ajiro et al., 1981a, 1981b; Langan et al., 1981; Wilkinson et al., 1982; Langan, 1982). In any event, an indication of the difficulty of the problem of comparing H1 phosphorylation in different organisms is exemplified by consideration of the S37 site. This has been shown to be a significant site of *interphase* phosphorylation in a number of cell types (Langan, 1971, 1978). According to Gurley et al., no phosphorylation occurs anywhere in the N-terminal region of CHO H1 until the beginning of mitosis. As first pointed out by Langan, some histone variants do not have a serine residue corresponding to S37 (Langan, et al., 1971). This may be the case in CHO cells.

A different approach to the cell cycle problem has been taken by Kurochkin et al. (1977) who have studied the in vitro phosphorylation of calf thymus H1 by kinases isolated from different stages in the first cell cycle of *Strongylocentratus intermedius* embryos. They find the following results: With G_1-phase kinases, serine 38 (S37 in the rabbit numbering) is phosphorylated, followed by a site analogous to rabbit S114 during S and G_2 phases. Using mitotic-stage enzymes, four more sites, three serine and one threonine, are phosphorylated in the C-terminal portion of the molecule. The special interest of these experiments derives from the fact that they seem to indicate that it is the enzymatic activities present, rather than the state of H1, that determine phosphorylation sites. It is clear that there are some differences between these results and the studies described above, but whether this stems from substrate and enzyme differences or the use of in vitro reactions remains uncertain. The various studies of cell-cycle dependent H1 phosphorylation seem consistent only in the broadest sense; there is specific, but limited phosphorylation through much of the cell cycle, followed by a burst of phosphorylation immediately before mitosis.

As cells divide and enter a new G_1 phase, there is rapid loss of histone phosphorylation. Both the premitotic rise and postmitotic decline are nicely illustrated by the data of D'Anna et al. (1981) shown in Figure 4-22. It should be noted that, at least with the CHO cells used in this work, the phosphorylation levels of H1 and H1° change in a completely parallel manner.

The loss of phosphorylation is probably catalyzed by histone-specific phosphatases. Such activities have been identified (Tamura et al., 1978; Sahyoun et al., 1983) and at least one has been purified (Tamura and Tsuiki, 1980). Furthermore, D'Anna et al. (1978) have shown that isolated metaphase chromosomes contain phosphatases which remain active during chromosome isolation. Since activity of the endogenous kinase will decrease during isolation because of ATP depletion, this may result in artifactually low values for phosphorylation of chromosomes. Paulson and Taylor (1982) find that such phosphatase activity can be countered by supplying ATP to fuel the endogenous kinases. Such observations suggest a much more dynamic state in the metaphase chromosome than had been hitherto imagined.

H3: A very specific phosphorylation event occurs on the H3 of CHO cells during, and only during, mitosis. In HeLa cells, this has been shown to occur at serine 10 (Paulson and Taylor, 1982), and because of the conservatism of H3 sequences in mammals, we may guess that this is also the site in CHO cells. Curiously, this is the site where phosphorylation was shown by Szopa et al. (1980) to grossly modify H3/H4 interactions. Does this hint at something unusual about nucleosome structure in metaphase chromosomes? As with H1, phosphorylation of H3 is largely lost

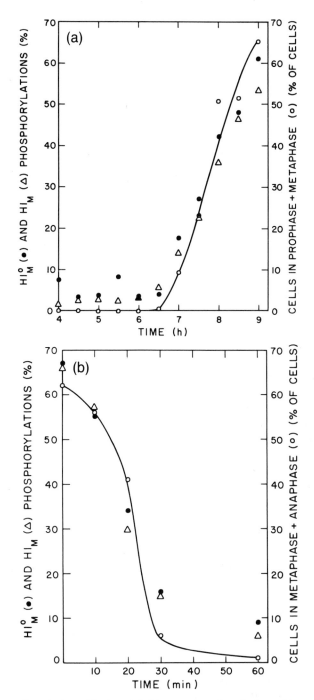

Fig. 4-22. (a) Correlation of lysine-rich histone phosphorylation with entry of CHO cells into metaphase. The solid circles (●) show phosphorylation of HI°, the triangles (△) the phosphorylation of HI. (b) Correlation of the dephosphorylation of HI° (●) and HI (△) with exit of cells from mitosis. (From Protein Phosphorylation, Rosen & Krebs, Eds., pp. 1053–1072, D'Anna et al. Copyright 1981 Cold Spring Harbor Laboratory.)

upon cell division, as the cells enter a new G_1 phase (Gurley et al., 1975, 1978a).

H2A: Phosphorylation of H2A appears to be largely constituitive in CHO cells. Although there are quantitative changes in the rate at certain stages, a generally high level is observed throughout the entire cell cycle, even after G_1-arrest in isoleucine-deprived cells (Gurley et al., 1978b). This observation, as well as the periodicity in H1 and H3 phosphorylation, have been confirmed in many cell types (see Gurley et al., 1978b; Prentice et al., 1982, for citations). In a quantitative study of H2A phosphorylation in both dividing and quiescent cells, the latter authors observed considerable phosphorylation (up to 20% of H2A) and evidence for rapid turnover. Such observations have led to suggestions that H2A phosphorylation is associated with transcriptional activity, rather than cell-cycle programming. Indeed, Prentice et al. (1982) find that mild DNase-1 digestion of chromatin in situ leads to preferential release of nucleosomes containing phosphorylated H2A. This can be taken as at least circumstantial evidence for such a role of H2A phosphorylation, since active chromatin has been shown to be preferentially sensitive to DNase-1 (see Chapter 8). Further support for this idea comes from the work of Allis and Gorovsky (1981) on *Tetrahymena*. They find that H2A phosphorylation occurs to a significant extent *only* in the transcriptionally active macronucleus; the inactive micronuclear chromatin shows little or no such modification. An alternative role for H2A phosphorylation has been proposed by Gurley et al. (1978b). From comparisons of the levels of H2A phosphorylation in different cells and organisms, they suggest that it is related to the maintenance of constituitive heterochromatin. Support for this view comes from the experiments of Marian and Wintersberger (1982) on sporulating yeast. They observe a major increase in H2A phosphorylation as the yeast enter sporulation and the chromatin becomes *inactive*.

It is truly difficult to see how all of these seemingly contradictory results can be resolved. One might attempt to explain the *Tetrahymena* data by arguing that H2A phosphorylation is required to maintain constituitive heterochromatin in otherwise active nuclei, but this idea seems in direct contradiction to the DNase-1 digestion results. Comparisons of *Tetrahymena* H2A phosphorylation with that in other organisms may be inappropriate, since the modification in the protists appears to be anomalous (see above).

As with other forms of histone modification, we remain in some uncertainty concerning the physiological roles of phosphorylation. There are, to be sure, strong hints in certain cases, but the overall pattern is elusive. It seems to me very possible that the histone modification program is a highly interactive one, with each kind of modification helping to determine not only the extent but also the effect of others. If this be so, a more holistic approach to the problem will be needed before real progress can be made.

ADP-Ribosylation

There are a number of cellular reactions in which nicotinamide adenine dinucleotide (NAD^+) is cleaved between the nicotinamide moiety and the adjacent ribose, with transfer of the adenosine diphosphate ribose (ADP-ribose) group to an acceptor molecule (see Fig. 4-23). Proteins can serve

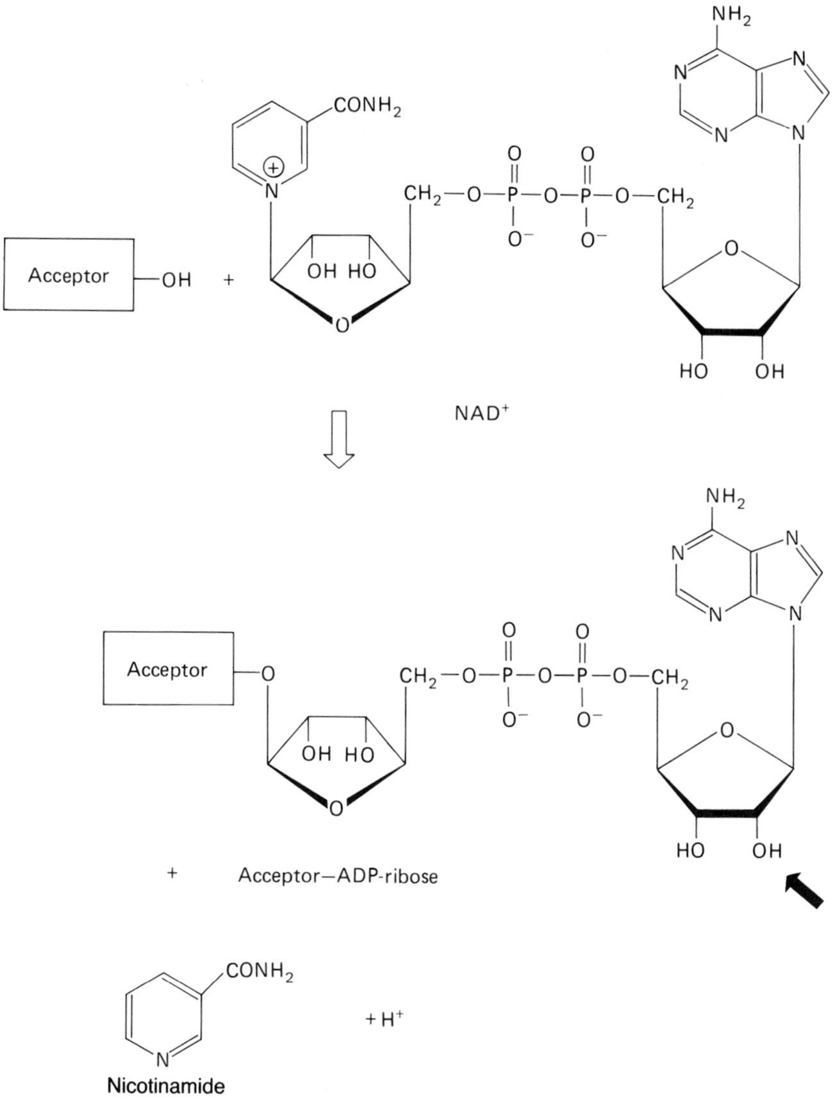

Fig. 4-23. ADP-ribosylation by the transfer of an ADP-ribose from NAD^+ to an acceptor molecule. The solid arrow indicates the point of addition of another ADP-ribose unit for the start of an poly(ADP-ribose) chain.

as acceptors, leading to the formation of *ADP-ribosylated proteins*. Furthermore, an ADP-ribose moiety (ADPR) thus conjugated to a protein can in turn act as an acceptor for another ADPR group; if this process is continued, a poly(ADP-ribose) chain will be built upon the acceptor protein (see Fig. 4-24). Protein modifications of these kinds are widespread. Poly(ADP-ribosylation) has been observed in many eukaryotic organisms, and the addition of single ADP-ribose units is not uncommon among the prokaryotes. The literature has been extensively reviewed; three excellent review papers (Hilz and Stone, 1976; Smulson and Shall, 1976; Hayaishi and Ueda, 1977) as well as a symposium volume are devoted to this subject (Smulson and Sugimura, Eds., 1980). In addition, Volume 106 of *Methods in Enzymology* (1984) contains a series of chapters concerning various aspects of ADP-ribosylation. The reviews deal not only with general features of ADP-ribosylation but also cover quite completely the earlier literature on such modification of histones. Accordingly, I shall concentrate on advances made since 1977, with only occasional references to the earlier work.

The first intimations that such reactions occur in nuclei came from the work of Chambon et al. (1963), who observed that incubation of chicken nuclei with NAD^+ stimulated the incorporation of ATP into a polymeric substance. They first believed this to be polyadenylic acid, but a few years later were able to show that the polymer was in fact poly(ADP-ribose) (Chambon et al., 1966). Using this technique, Nishizuka et al. (1968) demonstrated that if the NAD^+ were radiolabeled in the ribose moiety, the ADP-ribose units formed were linked to nuclear proteins, including H1, H2A, H2B, and H3. This appears to have been the first direct evidence for ADP-ribosylation of histones. The technique of incubating isolated nuclei with NAD^+, radiolabeled either in the ribose or adenine moiety, has been widely used in subsequent studies of ADP-ribosylation. The

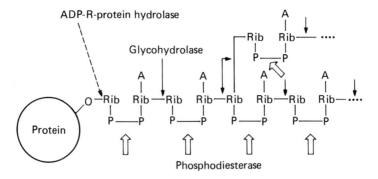

Fig. 4-24. A schematic representation of a poly (ADP-ribose) chain attached to a protein. A branch point is shown, as well as the places in which various degradative enzymes cleave the chain.

method is convenient and sensitive, but it should be noted that these are essentially in vitro studies; the environment in an isolated nucleus is not identical to that found in vivo. Therefore, as with all in vitro experiments on histone modification, we must be cautious in extrapolating results back to the living cell. The first unequivocal demonstration that histones are actually ADP-ribosylated in vivo was provided by Ueda et al. (1975b). Intraperitoneal injection of ^{14}C-ribose into rats was shown to lead to the incorporation of radioactivity into histones. The in vivo modification has since been confirmed by a number of other kinds of studies, including histone isolation and characterization (i.e., Adamietz et al., 1978) and immunological analysis (Wong et al., 1983a).

A number of experimenters have utilized the in vitro studies with isolated nuclei to assess the relative susceptibility to modification of different histones. (For examples, see Burzio et al., 1979; Lichtenwald and Suhadolnik, 1979; Ogata et al., 1980a; Poirier and Savard, 1980.) While there are some minor disagreements, all concur that H1 is most extensively modified, with 50 to 80% of the radioactivity being incorporated into this histone. Significant modification of H2B (Adamietz and Rudolph, 1984) and H3 (Lichtenwald and Suhadolnik, 1979) have also been reported, but only trace amounts of ADP-ribose appear to be associated with H2A and H4. However, great caution should be exercised in extrapolating these results in vivo, since Adamietz et al. (1978) have shown marked differences in comparing the in vivo data from HeLa cells with in vitro results obtained with HeLa nuclei. Their experiments also indicate that the amount of ADP-ribosylation in vivo is usually extremely small; only about 0.005 mol ADP-ribose were found per mole of H1 (see also Kreimeyer et al., 1984). However, when DNA is damaged, levels can run much higher; Adamietz and Rudolph (1984) report as many as 10% of H2B molecules to be ADP-ribosylated under such circumstances. The relationship of ADP-ribosylation to the repair of damaged DNA will be discussed below. Unless the in vivo studies are being compromised by lytic activity (see below) during nuclear isolation, the above data would indicate that ADP-ribosylation is a relatively rare event in unperturbed living cells. Most of the available data also indicate that the average ADP-ribose chain is relatively short, a few residues being typical. However, as we shall see below, longer chains are sometimes observed, and the average length observed may depend on relative polymerizing and degradative activities.

Poly(ADP-ribose) chains are frequently branched. Evidence for branching was first detected through in vitro studies by Miwa et al. (1979) and shown also to occur in vivo by Jaurez-Salinas et al. (1982). The structure of the branch points has been examined by Miwa et al. (1981) and is illustrated in Figure 4-24. Miwa et al. (1979) estimate about one branch point for 20 to 30 residues, suggesting that only very long ADPR chains are extensively branched. Isolated nuclei have also been used in studies of the locations of ADP-ribosylation sites on histone molecules. Ogata et al. (1980a) find that glutamate 2 is the modified residue in H2B. Riquelme et al. (1979) propose glutamate residues 2 and 116 in H1, but Ogata et al. (1980b) report glutamates 2 and 14, as well as the C-terminal carboxyl, to be the modified residues in this histone. With the exception of the unconfirmed residue 116 in H1, these results place all of the known ADP-ribosylation sites in histones within the basic "tail" regions of the molecules, and all on carboxyl groups.

However, there have been claims of ADP-ribosylation at other kinds of sites. In an early study, Smith and Stocken (1973) reported that a phosphoserine residue in H1 is an acceptor for ADP-ribosylation. This proposal has been disputed by Ogata et al. (1980a). More recently, Tanigawa et al. (1983) have found evidence for ADP-ribosylation on arginine in H1. The situation remains uncertain, particularly in regard to in vivo modification. The relative susceptibilities of in vivo and in vitro ADP-ribosylation to cleavage by dilute alkali and neutral hydroxylamine are quite different (Adamietz et al., 1978), suggesting that different kinds of linkages may be formed under different conditions. Hilz and Stone (1976) provide a detailed discussion of the significance of stability with respect to bonding mode.

An enzyme [poly(ADP-ribose) synthase, or poly(ADP-ribose) polymerase] that catalyzes the formation of poly(ADP-ribose) in nuclei was first isolated by Ueda et al. (1975a). Its subsequent history has been marked by rather massive confusion, which has only recently begun to be cleared away. Although Ueda et al. observed that DNA was required for and histones stimulatory to enzymatic function, these same workers later found that a purified enzyme preparation was incapable of ADP-ribosylating added histones (Okayama et al., 1977)! Instead, an "exogenous acceptor" that appeared to co-purify with the enzyme was the major recipient of ADP-ribosylation. In the same year, Yoshihara et al. (1977) presented evidence that the purified enzyme itself was capable of automodification. This result did not really clarify the situation, for the poly(ADP-ribose) polymerase has been found to have a molecular weight of about 120,000 (Ito et al., 1979; Ohghushi et al., 1980) whereas the "acceptor" found by Okamaya et al. migrated between histones H3 and H1 on SDS gels. Furthermore, Ueda et al. (1979) found that their enzyme preparation *was* capable of adding ADP-ribose units to a synthetically mono-ADP-ribosylated histone. This result raises a question that has been put forward by many authors: Are there in fact *two* enzymes, a "ligase" to attach the first ADP-ribose unit to the protein, and a polymerase to elongate the chain? Partial clarification has come from the work of Tanaka et al. (1979) and Yoshihara et al. (1981) who observed that the mode of activity of the enzyme preparation was markedly dependent upon the level of magnesium ion in the assay mixtures. In the absence of Mg^{2+}, extensive modification of H1 was observed; in the presence of this cation, histone modification was very limited, but automodification of the enzyme was now extensive.

The laboratory of Dr. M. Smulson has taken a rather different approach, which has added greatly to our understanding of this perplexing enzymatic activity. Rather than isolating the enzyme itself, they began studies on the poly(ADP-ribose) polymerase activity of chromatin fragments prepared by micrococcal nuclease digestion. It was found that whereas mononucleosomes were virtually inactive, oligonucleosomes exhibited increasing specific activity (see Fig. 4-25, and Giri et al., 1978a, 1978b). Evidently

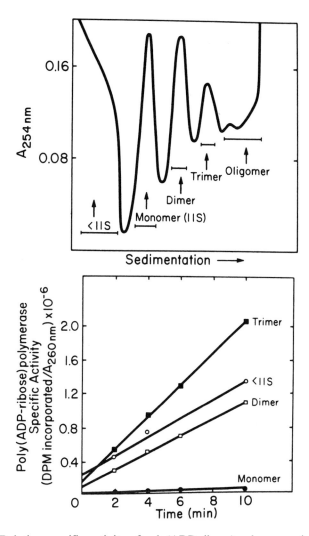

Fig. 4-25. Relative specific activity of poly(ADP-ribose) polymerase in HeLa nucleosome monomer, dimer, trimer, and in the fraction sedimenting at less than 11S. (Reprinted with permission from Biochemistry *17*, p. 3501, Giri et al. Copyright 1978 American Chemical Society.)

the enzyme is chromatin-bound and probably attached to linker DNA sites. In subsequent studies, this group has confirmed that the major acceptor for the purified enzyme is the enzyme itself; only when properly complexed into oligonucleosomes does it readily modify histones (Jump et al., 1979; Jump and Smulson, 1980; Jump et al., 1980; Butt and Smulson, 1980). By the use of immunoaffinity chromatography, employing sepharose-linked antibodies to poly(ADP-ribose), these researchers have been able to isolate virtually all of the polymerase activity on fragments corresponding to only about 10% of the chromatin (Malik et al., 1983; Smulson, 1984). This method allows both a purification of the activity in something resembling its native state for subsequent in vitro experiments, and the examination of the products of in vivo ADP-ribosylation.

One of the products has a most interesting structure. A number of years ago, Stone et al. (1977, 1978) discovered evidence for the synthesis of poly(ADP-ribose)-linked H1 dimers in isolated HeLa nuclei. Nolan et al. (1980) showed that the same reaction could occur on isolated oligonucleosomes. Using antibodies to APD-ribose and to H1, Wong et al. (1983a) have demonstrated that this product is formed in vivo. These adducts contain a poly(ADP-ribose) chain of about 15 units, covalently linked on at least one end to an H1 molecule. The mode of linkage to the second H1 is unknown, but it is probably covalent as well, since it is resistant to both detergents and urea. According to Wong et al. (1984), the poly(ADP-ribose) chain extends from the N-terminal region of one H1 molecule to the C-terminal domain of the other.

Such a reaction would have the potential to crosslink chromatin over considerable distances. Indeed, Butt and Smulson (1980) have observed the formation of aggregates of oligonucleosomes following ADP-ribosylation. Such results have suggested that a role of ADP-ribosylation may be to stabilize the higher order chromatin structure. However, an indication that the situation may be more complex is provided by claims (Poirier et al., 1982; Aubin et al., 1983) that ADP-ribosylation induces *relaxation* of chromatin structure. The amounts of ADP-ribosylation in these experiments were high (20 to 30% of H1 modified) and they may therefore not represent the overall picture in vivo. Yet if modification sites are clustered in chromatin (Butt and Smulson, 1980) these results might suggest that *local* unfolding would be a consequence of ADP-ribosylation. A local relaxation would also be consistent with several reports that ADP-ribosylated regions in chromatin are particularly prone to acetylation and phosphorylation (Wong et al., 1983b; Malik and Smulson, 1984; Wong and Smulson, 1984).

This leads us to an important and controversial question: What is the physiological function of ADP-ribosylation? Many roles have been suggested: inhibition of DNA synthesis, stimulation of DNA synthesis, both enhancement and repression of template activity, and the maintenance of chromatin structure during DNA repair have all been proposed, and evidence has been adduced for each. Unfortunately, in many cases contradictory experiments have been reported. The interested reader is referred to the review papers cited at the beginning of this section, wherein the evidences and counterevidences are cited in bewildering detail.

At the present time, it seems to me that a strong case can be made for the participation of ADP-ribosylation in DNA repair. This is supported by many observations of increased modification following DNA damage (see reviews, and Man and Shall, 1982; Thraves and Smulson, 1982; Kriemeyer et al., 1984; Adamietz and Rudolph, 1984). Ohghushi et al. (1980) and Benjamin and Gill (1980) have observed that the polymerase is stimulated by nicked or cleaved DNA. In support of this concept, Malik et al. (1983) have found that immunoaffinity-purified ADP-ribosylated oligonucleosomes contain many DNA nicks. Furthermore, Ohashi et al. (1983) note that DNA ligase is stimulated by ADP-ribosylation of chromatin. If ADP-ribosylation is clustered in damaged regions, a structural relaxation of the kind proposed by Poirier and his collaborators (see above) might facilitate repair.

It is not at all clear whether or not ADP-ribosylation is more extensive in transcriptionally active regions. One study, using a particular chromatin fractionation technique, states that it is (Mullins et al., 1977). Conversely, Yukioka et al. (1978), using a different method, could find no difference whatsoever between putatively "active" and "inactive" fractions. While the contradictions in these experiments may arise from deficiencies in one or another of the fractionation methods, it should also be noted that Malik and Smulson (1984), using the immunoaffinity method, could find no clear enrichment for active chromatin. The uncertainty is accented by studies of wheat embryos by Sasaki and Sugita (1982a, 1982b). They observed that ungerminated seeds, in which the chromatin is highly condensed and inactive, showed much higher levels of ADP-ribosylation than germinating seedlings. Furthermore, a *decrease* in ADP-ribosylation preceded the increase in transcriptional activity accompanying germination.

There *does* seem to be general agreement concerning the timing of changes in ADP-ribosylation of histones during the cell cycle. Both Kidwell and Mage (1976) and Kanai et al. (1981) find maximal levels of incorporation during the G_2 phase. The former also showed that the level of polymerase activity is maximal during this period.

The level of ADP-ribosylation in nuclei is probably under dynamic control, since the nucleus contains enzymes that degrade the polymer as well as the polymerase that forms it. There are three nuclear enzymes presently known that are involved in poly(ADP-ribose) degradation; their sites of cleavage are shown in Figure 4-24.

1. *Liver phosphodiesterase:* This enzyme has been purified from rat liver, among other sources (Futai and Mizuno, 1967; Futai et al., 1968; Hilz and Stone, 1976). The action is exolytic, from the AMP terminus. An

endolytic phosphodiesterase has also been reported (Ferro and Kun, 1976).

2. *Poly(ADP-ribose)glycohydrolase:* First purified by Miwa et al. (1974) from calf thymus nuclei, this enzyme has since been obtained from a variety of eukaryotic sources and extensively studied (see Hilz and Stone, 1976; Hayaishi and Ueda, 1977, for details). As Figure 4-24 shows, the glycohydrolase has the somewhat unusual property of cleaving the ribose–ribose linkage. It appears to be the major enzyme for degradation of poly(ADP-ribose) in nuclei.

3. *ADP-ribosyl-histone splitting enzyme:* Discovered by Okayama et al. (1978), this enzyme appears to be specific for the poly(ADP-ribose)-protein bonds but does not cleave the poly(ADP-ribose) chain itself. The products of this reaction have been characterized (Komura et al., 1983).

Very little is known concerning the control of these degradative enzymes. One might wonder at the utility of two or more different nuclear enzymes for degradation of the poly(ADP-ribose) chain, but as Hayaishi and Ueda point out, only the glycohydrolase cleaves the chain in a manner that will allow its re-elongation. Strong evidence for a role of the glycohydrolase in dynamic control of ADP-ribosylation is to be found in the work of Lorimer et al. (1977). Using a variety of mammalian nuclei, they found that the average chain length of poly(ADP-ribose) chains attached to H1 varied dramatically from one cell type to another. Furthermore, the average chain length was inversely proportional to the glycohydrolase activities of the various nuclei.

It seems possible that our knowledge of ADP-ribosylation in vivo is compromised by degradative activity during isolation procedures as in the cases of acetylation and phosphorylation. As NAD^+ is lost during isolation, synthase activity must decrease, but degradation can continue unabated. If chain elongation and cleavage are in dynamic balance in the living cell, the observed level of ADP-ribosylation will then be artifactually low. Studies in which either synthase activity is maintained or the degradative enzymes are inhibited seem essential at this point.

Ubiquitination

Some histones are modified by the covalent attachment of a small protein, ubiquitin. The discovery of this modification has a curious and interesting history. I shall recount it briefly here; for more details, the reader is referred to the excellent reviews by Goldknopf and Busch (1978) and Busch and Goldknopf (1981).

The first evidence for such modification appeared when Orrick et al. (1973) reported an unknown protein as one of many acid-soluble nuclear proteins appearing as spots on a two-dimensional gel electropherogram. They called it A-24. Although its nature was wholly unknown, A-24 attracted the further attention of H. Busch and co-workers by being one of a very few such proteins to change markedly in amount when rat liver nuclei were stimulated to RNA synthesis (Ballal et al., 1975). Purification and partial characterization of A-24 led to perplexing results. The protein was found to resemble histone H2A in composition. Surprisingly, the peptide map of A-24 included all the tryptic peptides of H2A as a partial subset (Goldknopf et al., 1975). Furthermore, A-24 was found to possess not only the acetyl-blocked N-terminal sequence of H2A, but in addition *another,* unblocked N terminus! Thus, the protein had to have a bifurcate polypeptide chain. Sequencing of 37 residues from this second terminus produced a sequence "not homologous to any known histone" (Olson et al., 1976).

But the sequence was *not* unknown. Hunt and Dayhoff (1977) pointed out the remarkable fact that these 37 residues were identical to the first 37 residues of the protein ubiquitin. Ubiquitin was quite unexpected as a component of chromatin. This small protein, first isolated by Goldstein et al. (1975) and sequenced by Schlessinger et al. (1975) was known initially as a polypeptide that stimulated thymocyte differentiation. It has been found, however, in every eukaryotic cell examined, hence the name. It now appears that ubiquitin may be involved in some very common cellular processes. More about these will follow.

Continuing the analysis, Goldknopf and Busch (1978) showed that A-24 consisted of a ubiquitin molecule, linked by its C terminus to the ε-amino group of lys 119 of H2A (see also Goldknopf et al., 1980a, 1980b; Goldknopf and Busch, 1980; and Busch and Goldknopf, 1981, for further details). The remarkable structure is depicted in Figure 4-26. More recently, it has been found that H2B can be modified in a similar way, although the levels of H2B ubiquitination are much lower (~1% *vis à vis* 10–20%) than the H2A modification in cells studied to date (West and Bonner, 1980). To provide a common nomenclature, ubiquinated histones are now designated by a prefix "u"; thus, A-24 now becomes uH2A and will be so designated in the remainder of this book.

Covalent attachment to ε-lysine residues appears to be relatively common for ubiquitin. The polypeptide has been found to participate in a widely distributed mechanism for protein degradation (Wilkinson et al.,

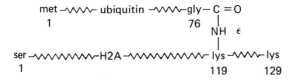

Fig. 4-26. A schematic of the structure of uH2A. For the whole sequence, see Busch and Goldknopf (1981).

1980; Ciechanover et al., 1980; Haas et al., 1982; see Ciechanover et al., 1984, for a review). In this process, ubiquitin is covalently attached, via an ATP-dependent reaction, to a protein to be targeted for proteolysis. A sequential series of three enzymatic reactions has been elucidated, and the enzymes involved purified (Ciechanover et al., 1982; Hershko et al., 1983). Just how much this mechanism has in common with the nuclear reaction with H2A and H2B is unclear, since the "marking" reaction appears to be much less specific than is histone modification and leads to multiple ubiquitination. That at least part of the same enzymatic system is involved is indicated by recent experiments described by Cienchanover and collaborators (Cienchanover et al., 1984; Finley et al., 1984). A cell line carrying a temperature-sensitive mutation in the first of the three enzymes was found to rapidly lose ubiquitin from uH2A at the nonpermissive temperature. Curiously, this mutant also shows a defect in H1 phosphorylation and is blocked from entering mitosis. Reversion of the mutant removes both defects. Therefore, there is a strong suggestion that ubiquitination and some H1 phosphorylation may be closely linked.

It has definitely been shown that uH2A (and probably uH2B) are present in nucleosomes. Several studies (Goldknopf et al., 1977; Martinson et al., 1979a; Levinger and Varshavsky, 1980) have demonstrated this, using a variety of tissues and techniques. Furthermore, uH2A binds to H2B to form dimers (Martinson et al., 1979a) and will crosslink to H1 in chromatin just as will H2A itself (Bonner and Stedman, 1979). Kleinschmidt and Martinson (1981) have prepared reconstituted core particles in which uH2A quantitatively replaced H2A. Surprisingly, these particles showed no significant differences from "normal" particles in digestion by micrococcal nuclease or DNase-1, or in their ability to bind the nonhistone proteins HMG-14 and HMG-17 (see Chapters 5 and 6). Thus, it seems that uH2A can handily replace H2A without significant effects *at the core particle level*.

What, then, is the biochemical role for the ubiquitination of histones? The evidence has been contradictory. The initial interest in ubiquitinated histones was stimulated by the observation that the content of uH2A decreased markedly in rat liver cells hyperinduced to RNA production by thioacetate administration or partial hepatectomy (Ballal et al., 1975). Therefore, it was suggested that such modification might inhibit transcription. Support for this idea was claimed by Goldknopf et al. (1978) who reported less uH2A in transcriptionally active fractions of chromatin. However, examination of these data shows that the overall level of histones is low in the "active" fraction, and the ratios of uH2A to histones are not compared.

A rather different role has been suggested by a number of other studies. Matsui et al. (1979) found that the ubiquitin moiety is lost from nucleosomes at the onset of metaphase and reappears early in G_1, indicating that the modification may hinder chromatin condensation (see also Goldknopf et al., 1980b). Such an effect does not seem unreasonable, since the ubiquitin moiety is attached to H2A and yet does not seemingly modify interactions of the core histones with DNA. Therefore, it probably lies on the outer surface of the histone core, in a position that might well hinder chromatin supercoiling. The idea that ubiquitination may "open" chromatin structure is an attractive one and is supported by the striking increases in uH2A content seen during spermatogenesis (Agell et al., 1983). However, as Finley et al. (1984) point out, the ts mutant in which ubiquitination is blocked is *also* incapable of entering mitosis, suggesting that ubiquitin removal *alone* is not sufficient to trigger chromatin condensation.

It now seems likely that ubiquitination must play other roles, in addition to whatever part it may play in mitosis or spermatogenesis. Goldknopf et al. (1980a) have shown that uH2A is formed continuously during interphase, and R.S. Wu et al. (1981) have demonstrated rapid turnover of bound ubiquitin in both dividing and nondividing cells. These results suggest a more dynamic role for this modification. In support of this concept is the finding of nuclear enzymes that cleave the isopeptide bond between ubiquitin and ε-amino groups of protein lysines (Anderson et al., 1981a, 1981b; Matsui et al., 1982). Another enzyme that cleaves H2A between residues 114 and 115 has been reported in nuclei (Eickbusch et al., 1976). The latter would produce much the same effect as the isopeptidase, since it removes the C-terminal region of H2A to which the ubiquitin is bonded. Together with the system of ubiquinating enzymes described above, these enzymes provide the mechanism for continual turnover.

If a function of ubiquitination is to open chromatin structure, one might expect it to *stimulate* transcription. Mention has been made of the early suggestions that ubiquitination inhibits transcription, but there are much more convincing data to indicate the converse. For example, Goldknopf et al. (1980b) have demonstrated that uH2A is converted to H2A during the general shutdown of transcription accompanying erythropoiesis in the chicken. Most impressive, perhaps, is the evidence by Varshavsky and co-workers. Levinger and Varshavsky (1982) find that the transcribed *copia* and heat shock (hsp 70) genes of *Drosophila* contain about one uH2A per two nucleosomes, whereas the ratio in an untranscribed satellite is about one per 25 nucleosomes. In further studies, transcribed dihydrofolate reductase genes were observed to contain very high levels of uH2A, especially near their 5' ends (Varshavsky et al., 1982). Results of this kind are entirely consistent with the idea that high levels of ubiquitination may disrupt higher order structure. A dynamic role in transcriptional regulation would be in accord with the observed turnover during interphase.

There is one perplexing problem that has received little attention. Outside of the nucleus, the major role for ubiquitination seems to be the "targeting" of proteins for proteolytic degradation. It seems that something else is happening in nuclei, for R.S. Wu et al. (1981) show that ubiquitination turns over more rapidly than does H2A; the H2A moiety is stable and cannot be entirely marked for degradation. Levinger and Varshavsky (1982) have suggested that rapidly transcribed genes may have lost nucleosomal structure via proteolysis after ubiquitin "marking." There is, at this writing, little further evidence to support this concept, and even if substantiated, it would seem that the evidence still requires at least two roles for ubiquitin in the nucleus. Mysteries remain.

Glycosylation

Many proteins, including some nuclear nonhistone proteins, are known to be glycosylated (see Chapter 5 for data on HMG proteins). Until recently, however, there was no evidence to indicate that histones might also be included in this category. Levy-Wilson (1983a) has reported that *Tetrahymena,* when fed radiolabeled fucose, will incorporate counts into histones. Since this organism does not metabolize fucose, the evidence for glycosylation is strong. This interpretation is supported by the observation that *Tetrahymena* histones will bind to lectins, from which they can be eluted with fucose. Such experiments also provided evidence for the presence of mannose residues. All of the histones, with the possible exception of H4, seemed to be glycosylated. The sites and extent of this modification are unknown. Nor is it clear whether glycosylation of histones is a general phenomenon or merely a peculiarity of *Tetrahymena.* It is possible that only a very small fraction of the histones are so modified, which might help explain why glycosylated histones have not been observed hitherto. It will be of considerable interest to see whether this novel observation can be confirmed in other organisms.

Secondary and Tertiary Structure of Histones

Introduction to the Problem

The histone core of a nucleosome is a compact object. The polypeptide chains therein must, therefore, be tightly folded, and it might seem logical to approach the problem of nucleosome structure through studies of individual histone conformations. Unfortunately, such attempts are beset with difficulties. First, we must consider, as detailed in the preceding sections, the almost infinite histone diversity allowed by sequence variation and posttranslational modification. In most studies of histone conformation carried out to date, such complications have been disregarded. In fact, the majority of these experiments have been carried out with calf thymus histones, without separation of variants or modified types. In only a few instances have there been attempts to study the effects of variation or modification on conformation; as will be shown below, the effects are not necessarily trivial.

A second and more fundamental problem arises from the considerable evidence that the conformation of a given histone can be modified in a major way by interaction with other histones and with DNA. We are here at face with a fundamental limitation of the conventional, reductionist approach to biochemistry: A *part* of a structure may be differently conformed when it is incorporated into the *whole*. But what is the "whole" in this case—the histone core complex, the nucleosome, the folded chromatin fiber, or the nucleus itself? We are dealing with a complex, interacting system. Very possibly, the "physiological" conformation of a histone molecule will ultimately be found to depend, in its details, not only on its incorporation into a chromatin structure, but also upon the functional state of the local chromatin region. From this point of view, studying the folding of an isolated histone molecule, even in an ionic environment approximating that of the nucleus, seems hopelessly naive. Yet the approach that has proved so successful in modern science is precisely this: to investigate the simplest system first, and then to use the insights so gained to move to a comprehension of the higher levels of complexity. For such reasons, considerable effort has been expended in histone conformation studies.

But even to elucidate the secondary and tertiary structures of individual histones is no easy task. These proteins have not been crystallized, and because of their peculiar properties, it seems unlikely that molecular crystals of individual histones can in fact be obtained. Therefore, conformational studies are limited to the more indirect techniques, such as circular dichroism, infrared spectroscopy, Raman spectroscopy, and nuclear magnetic resonance. Powerful as these methods may be, analysis of their results is forever beset with a certain level of ambiguity.

One more fundamental problem has caused a great deal of difficulty and confusion in attempts to delineate individual histone conformations. As early as 1969, it became clear that some individual histones (notably H3 and H4) tend to form large, aggregated structures at ionic strength levels comparable to those found in vivo (Edwards and Shooter, 1969). To be sure, each histone can be molecularly dispersed in dilute solution at very low ionic strength, but under these conditions all are found to exist largely in unfolded, random coil conformations (see Isenberg, 1979; and below). Conversely, there is no assurance that the individual histones, when aggregated at higher ionic strength, have conformations similar to those in nucleosomes. In fact, as we shall see, there is evidence to the contrary.

From the arguments given above, it should be clear that conformational studies on individual histones can provide only limited information relevant to the principal aim of this book—the description of *chromatin* structure. For this reason, I shall review these efforts only briefly here, principally as a background for more detailed discussion, to be taken up in following sections, concerning the folding of histones in histone complexes and nucleosomes.

To understand histone folding, it is necessary to consider first the major effects that ionic strength has upon the conformation of these histones. Therefore, I shall consider this topic first, before describing the details of individual structures.

The Effect of Ionic Strength on Histone Folding

As early as 1965, optical rotatory dispersion experiments indicated that
increasing salt concentrations had the effect of promoting secondary
structure formation in histones (Bradbury et al., 1965; Jirgensons and
Hnilica, 1965; see also Tuan and Bonner, 1969). However, the complexity
of the salt effects only became apparent through a series of elegant studies
in the laboratory of I. Isenberg (Li et al., 1972; Wickett et al., 1972; D'Anna
and Isenberg, 1972, 1973; this work has been reviewed by Li, 1977, and
by Isenberg, 1979). The principal result is illustrated by the example shown
in Figure 4-27. Following the addition of salt, changes in the fluorescence
anisotropy and circular dichroism of histone H4 occur in two steps: a
"fast" change (virtually instantaneous at room temperature), followed by
a "slow" step, which may be hours in duration. The nature of the slow
step is demonstrated by Figure 4-28; the large increase in light scattering,
which accompanies this step, indicates aggregation. From the anisotropy
and CD data, it is clear that conformational changes occur in both steps.
The CD spectrum of H4 in Figure 4-29 shows this clearly; in water, the
curve is essentially that of a random coil. A major change has occurred
just after the fast step is complete and change in the spectrum continues
through the slow step. More will be said about these changes below.

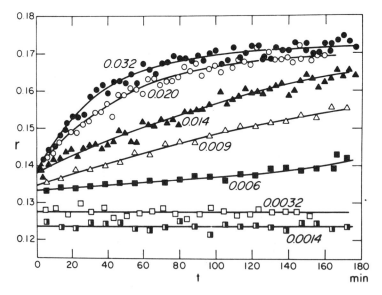

Fig. 4-27. The fluorescence anisotropy (r) of histone H3 as a function of time at
the salt (phosphate) concentrations indicated (pH = 7.0). The fast step, which
appears instantaneous on this time scale, is clearly distinguished from the slow
step. (Reprinted with permission from Biochemistry *13*, p. 4987, D'Anna and Isenberg. Copyright 1974 American Chemical Society.)

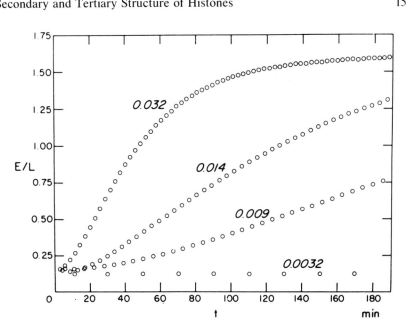

Fig. 4-28. Light scattering from H3 solutions at some of the salt concentrations indicated in Figure 4-27. This shows that the slow step, but not the fast step, involves an association process. (Reprinted with permission from *Biochemistry 13*, p. 4987, D'Anna and Isenberg. Copyright 1974 American Chemical Society.)

The nature of the long, fibrous polymers formed in the slow aggregation has been studied in some detail (Sperling and Amos, 1977; Wachtel and Sperling, 1979). Whereas H3 behaves in much the same manner as H4, H2B and H2A exhibit only the fast conformational change, showing no evidence for aggregation at low protein concentration. They can, however, aggregate if sufficiently concentrated (see *Histone–Histone Interactions*, below).

Both steps in the folding of H3 and H4 depend upon histone concentration; therefore, it would seem that even the fast process involves some histone–histone interaction; there is strong evidence that this is a dimerization (Smerdon and Isenberg, 1976a). The magnitude of the changes observed in the fast process depends upon the nature of the anion added, and from such data it is possible to deduce apparent binding constants. Tetrahedral ions, such as phosphate, sulfate, and perchlorate, bind more strongly and are much more effective in promoting the conformational change than are spherical ions such as chloride or fluoride (see Li, 1977).

The preceding discussion should make it clear that there is no simple way to define the "native" conformation of a histone molecule. The folding depends on salt concentrations and on the interactions with other histone molecules. It is not surprising that the experimental results have been

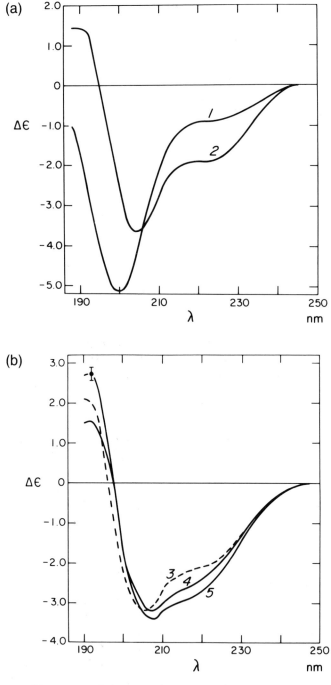

Fig. 4-29. (a) CD spectra of histone H3 at 1 mM HCl (curve 1) and 1.6 mM phosphate (curve 2). (b) Spectra after 24 hours at 3.2 mM phosphate (curve 3), 5.9 mM phosphate (curve 4), and 20 mM phosphate (curve 5). (Reprinted with permission from Biochemistry *13*, p. 4987, D'Anna and Isenberg. Copyright 1974 American Chemical Society.)

conflicting and confusing. Therefore, I shall take a somewhat unorthodox approach to the problem, first considering the *predictions* of folded structure that can be made from the amino acid sequences, and then asking how well the results of experiments can be reconciled with these predictions.

Predictions of Secondary Structure in Histones

While a number of investigators have considered this problem for specific histones, the most thorough and careful analysis appears to be that of Fasman et al. (1976). Using the methods of Chou and Fasman (1974), they have predicted regions of α-helix, β-sheet, and β-turn in each of the five major histones. However, the analysis has been extended well beyond a blind application of the Chou–Fasman rules. Fasman et al. have considered both the possibilities of α→β transitions in some regions, and the important role that electrostatic repulsion will play in destabilizing secondary structure, especially at low ionic strength. The latter is particularly important in histones, for many of the predicted α-helical regions are rich in lysine residues. Lysine is normally a strong helix former, but a locally dense constellation of lysines and arginines should be destabilized by repulsion. As a consequence, Table 4-16 lists both maximal and minimal amounts of α-helix and β-sheet. The former are considered to apply under conditions where electrostatic repulsion is minimized (i.e., in nonaqueous solvents, or perhaps when the positive charges are compensated, as in chromatin). The minimal values are then to be expected at very low ionic strength in aqueous solution. A third estimate, giving the absolute minimum amount of α-helix, also includes the possibility of α→β transitions in some helical regions with high concentrations of β-formers. The differences are very large and point up the complexity of the problem. Fasman et al. also list the results of a number of predictions by other researchers.

Of course, the predictive method used does not merely estimate percentages; it specifies regions for each conformational type. The diagrams (for maximal secondary structure) are reproduced in Figure 4-30; for de-

Table 4-16. Prediction of α-Helix and β-Sheet Structure in Histones[a]

Histone	Predicted by Chou–Fasman rules		Predicted if charge repulsion destabilizes helices		Predicted with both charge repulsion and α → β transition	
	%α	%β	%α	%β	%α	%β
H4	28	31	11	31	0	37
H3	39	15	16	15	12	25
H2B	35	20	21	20	13	27
H2A	40	19	22	19	5	34
H1	55	5	16	5	16	5

[a]Adapted from Fasman et al. (1976).

H4

(a)

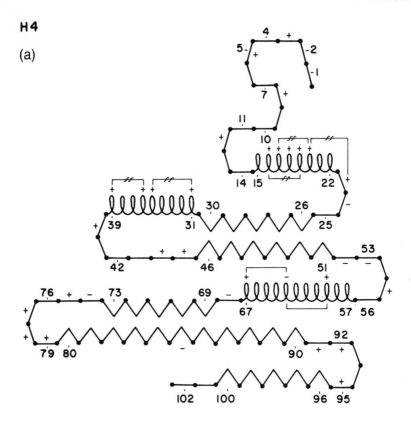

Fig. 4-30. Predicted secondary structure of histones. Regions of α-helix are denoted by spirals, β-sheet by zig-zag lines, and β-turns at 3-point corners. Helices denoted by (–//–) contain sufficient charge repulsion to be disrupted at low ionic strength. Shown are calf H4 (a), H3 (b), H2B (c), and H2A (d), and rabbit H1 (e). (From *Molecular Biology of the Mammalian Genetic Apparatus*, Ts'o, Ed., pp. 1–52, Fasman et al. Copyright 1976 Elsevier Science Publishers.)

H3

(b)

H2B

(c)

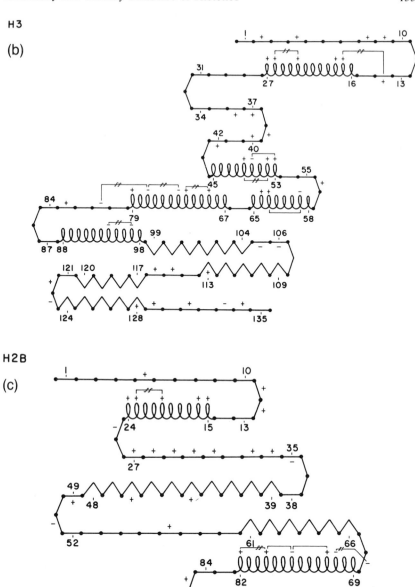

H2A

(d)

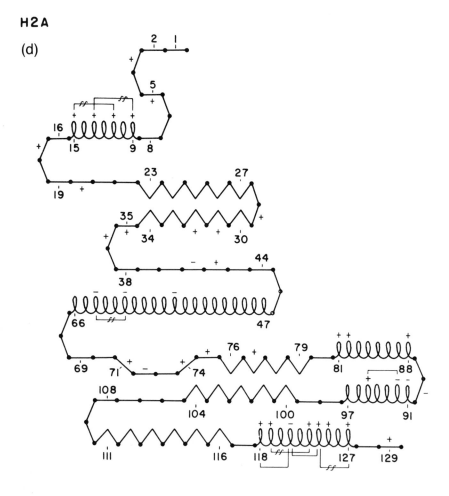

H I

(e)

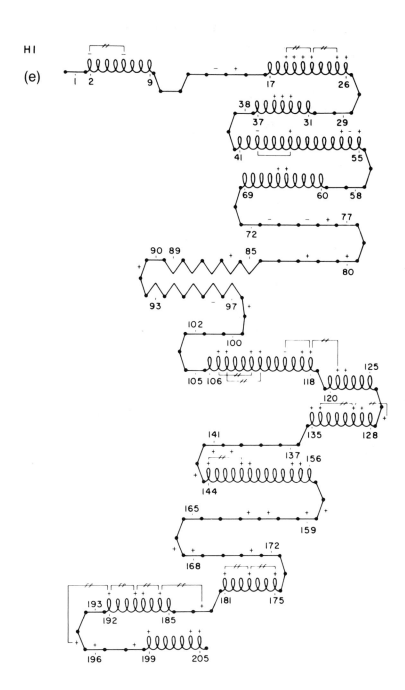

tails, the reader is referred to the original paper. We shall make reference to some of these specific regions in the following sections.

Conformations of the Individual Histones: Experimental Results

Before considering each of the histones in turn, a few general remarks are in order. First, there has been an extensive use of NMR methods, particularly by E.M. Bradbury and his associates, to delineate those regions of the histones that are not folded into defined structure. Each histone can, by this method, be divided into portions that have a definite tertiary structure and freely moving random coil sections. There may well be segments within the tertiary domains that are not describable in terms of the simple secondary structure types, but by NMR evidence these regions are nevertheless relatively immobile. The truly "random coil" domains are invariably N- or C-terminal regions. As comparison of Table 4-17 and Figure 4-30 shows, parts of these terminal domains are frequently described as helical in the "maximum" predictions of Fasman et al. This is, as expected, an indication that the latter are overestimates.

A comment should be made concerning the conformations of histones in pure water or *very* dilute salt solutions. There exist in the literature some very marked disagreements concerning this point, some workers claiming wholly random coil conformation, others seeing evidence for some secondary structure.

I shall not enter into this controversy, for it does not concern us. Such an environment is *wholly* unlike that which the proteins experience in any physiological situation. The existence of such contradictory analyses, however, should serve as a warning about *all* of the experimental results. The conformation of histones has been studied by a variety of techniques, and the results do not always agree. Even circular dichroism data can be

Table 4-17. Structured Regions of Histones, Based on NMR Data[a]

Histone[b]	Chain length (res.)	Structured region	Reference
H4	102	33–102	Bradbury and Rattle (1972)
H3	135	42–110	Bradbury et al. (1973a)
H2B	125	31–102	Bradbury et al. (1972)
		37–114	Moss et al. (1976b)
H2A	129	25–113	Bradbury et al. (1975b)
		25–95	Moss et al. (1976b)
H1	213	36–121	Hartmann et al. (1977)
H5	189	22–100	Aviles et al. (1978)

[a]Adapted from Fasman et al. (1976) with addition of more recent data.
[b]Inner histones and H1 from calf thymus (H1 numbering based on rabbit thymus H1); H5 from chicken erythrocytes.

variously interpreted, depending on which reference spectra are used, particularly for the "random coil" form. Finally, widely different histone and salt concentrations have been used by different investigators, and, as *The Effect of Ionic Strength on Histone Folding*, above, has emphasized, these parameters can play critical roles in determining conformation. Consequently, all of the following results on individual histones should be regarded skeptically, at least in their quantitative aspects.

H4: According to the NMR studies of Bradbury et al., all but the first 32 residues of H4 are in a structured conformation in salt solution. Pekary et al. (1975) have claimed that the structured domain extends from only about residues 70–80 to the C terminus, but this is contraindicated by the observation that the peptide (1–53) contains some α-helix (Crane-Robinson et al., 1977b). The amounts of helix and β-sheet predicted by Fasman et al. in the region between residues 31–102 corresponds, respectively, to about 20% and 32% of the whole molecule.

In their studies of the fast step, Isenberg's group detected the formation of about 15 to 16% α-helix and no β-sheet (Wickett et al., 1972; Baker and Isenberg, 1976). The slow step was seen to be accompanied by the formation of 20 to 30% of β-sheet structure. The final conformation has been roughly confirmed by IR (Shestopalov and Chirgadze, 1976 [10% α; 32% β]), by Raman spectroscopy (Pezolet et al., 1980 [20% α; 40% β]), and by Crane-Robinson et al. (1977b) [25% α; 27% β]), using a variety of techniques. The latter workers have also studied proteolytic fragments of H4 and conclude that one 13-residue α-helix lies between residues 50–67 in the sequence, with more somewhere toward the C-terminal end. All of these results are in quite good agreement with the theoretical predictions of Fasman et al. (1976). Furthermore, they suggest that in moderately concentrated salt solutions, H4 adopts nearly its maximum predicted structure, except in the N-terminal domain.

H3: According to the NMR studies, H3 in salt solutions is structured in a central region, with a 41-residue N-terminal and a 25-residue C-terminal random coil tail. Partial confirmation of this range is provided by the spectroscopic studies of Palau and Padros (1975), who observe that all three tyrosines (at positions 41, 54, and 99) become "buried" when the salt concentration is increased. In this structured central domain, the calculations of Fasman et al. predict about 41 residues of α-helix and 8 to 11 residues of β-sheet. These correspond to about 30% and 6 to 8% of the entire molecule, respectively.

D'Anna and Isenberg (1974c) found that H3 contained about 15% α-helix (and no β-sheet) after the fast folding. This was later revised (based on better methods of CD analysis) to 12.5% α (Baker and Isenberg, 1976). Upon slow aggregation, about 30% β-sheet is formed. Somewhat different results have been reported by Morris and Lewis (1977), who find 15% α-helix, 9% β-sheet in the fast folding, and an additional 4% helix and 10% sheet formed during aggregation. Analysis of the behavior of CNBr peptides indicated that most of the secondary structure lay between residues 1 and 90. It must be remembered, however, that the interpretation of such peptide studies is potentially misleading; the tertiary folding of the whole protein may modify the secondary structure of particular regions. Pezolet et al. (1980) find that results from Raman spectroscopy depend strongly on the method of analysis, but estimate roughly 20% α, 30 to 44% β.

H2B and H2A: Since these two proteins do not exhibit a slow aggregation, a more direct comparison of results from different laboratories is possible. These are summarized in Table 4-18. As can be seen, there is general agreement for about 10 to 15% α-helix at moderate salt concentrations, but considerable discrepancy in the amount of β-sheet reported. As Fasman et al. (1976) point out, this may be in part a consequence of the CD reference spectra employed. When, as for H3 and H4, we compare the observations with predictions of α and β structure within the structured domains, we find fair agreement, especially at higher salt concentrations.

H1 and H5: The folding of the lysine-rich histones has been the object of very extensive investigation, especially by E.M. Bradbury and his as-

Table 4-18. Secondary Structure in the More Lysine-Rich Histones

Histone	Method	Solution conditions	%α	%β	Reference
H2B	CD	10 mM NaCl	11	0	Moss et al. (1976b)
	CD	26 mM phosphate	14	0	D'Anna and Isenberg (1972)
					Baker and Isenberg (1976)
	CD	0.14 M NaF	14	29	Adler et al. (1974)
	Predicted for structured domain		19–26	15	Fasman et al. (1976)
H2A	CD	3.3 mM phosphate	13	0	D'Anna and Isenberg (1974)
			11	0	Baker and Isenberg (1976)
	IR	5 mM NaCl	10	15	Shestopalov and Chirgadze (1976)
	CD	10 mM NaCl	17	0	Moss et al. (1976b)
	CD	0.14 M NaF	15	31	Adler et al. (1975)
	CD	1 M KF	23	18	Garel et al. (1975)
	CD	2 M NaCl	28	0	Bradbury et al. (1975b)
	Predicted for structured domain		27	~11	Fasman et al. (1976)
H1	CD	0.14 M NaF	4	15	Fasman et al. (1976)
	IR	0.15 M NaCl	5	N.D.	Shestopalov and Chirgadze (1976)
	RD	1.0 M kNaCl	15	0	Bradbury et al. (1975a)
	CD	Extrapolated to high salt	5–7	6–10	Smerdon and Isenberg (1976b)
	Predicted for structured domain		18	5	Fasman et al. (1976)
			22	0	van Helden (1982)
H5	CD		16	ND	Crane-Robinson et al. (1977a)
	Predicted for structural domain		22	0	van Helden (1982)

sociates. The data in Table 4-18 reveal only a small part of the story. Both H1 and H5 can be cleaved by proteases so as to divide the molecules into three domains: an unfolded N-terminal region, a central globular region, and an extended C-terminal domain. The globular portion extends roughly from residues 40 to 120 in H1, and from 20 to 100 in H5 (Chapman et al., 1978; Aviles et al., 1978). This region is much more compact by hydro-dynamic measurements than is the whole protein, and it contains virtually all of the organized secondary structure of the molecule (Hartman et al., 1977; Aviles et al., 1978; Barbero et al., 1980; Tiktopulo et al., 1982; Crane-Robinson and Privalov, 1983). A ^{13}C NMR study by Saito et al. (1982) indicates that the major helical segments are probably located between residues 41–55 and 60–69 in H1, in good agreement with both the estimates of overall helicity (Table 4-16) and with predictions for this region from the sequence (Fasman et al., 1976; van Helden, 1982). It should be noted that the *overall* predictions of helicity for H1 are much higher, since they include regions in the N- and C-terminal domains that have sequences rich in helix formers. It seems likely that the high charge density in these regions vitiates the predictions, at least in aqueous solution. In a non-aqueous solvent such as chloroethanol, the overall helicity of H1 does in fact approach 50% (Saito et al., 1982).

It may be an oversimplification to assume that the N- and C-terminal arms of H1 adopt a freely moving random coil conformation. Hydrody-namic studies by Lewis and Reams (1983) indicate an *exceedingly* ani-sotropic structure, even in moderate salt concentration. Most significantly, addition of 6 *M* guanidine hydrochloride (which should yield a wholly random coil molecule) resulted in a *decrease* of the frictional coefficient. The arms may be highly extended by electrostatic repulsion.

Impressive as these studies of H1 may be, a word of caution is in order. Virtually all of the experiments have been carried out with histones isolated by acid extraction. In an important study, Brand et al. (1981) have com-pared the properties of chicken erythrocyte H1 prepared by acid ex-traction and salt extraction at neutral pH. They find that the salt-extracted H1 has a higher sedimentation coefficient ($S_{20,w}$ = 2.2 S vs. 1.5 S) and a much more negative ellipticity at 222 nm than the acid-extracted material. Furthermore, brief exposure of the salt-extracted histone to low pH *ir-reversibly* modifies the properties to those exhibited by the acid-extracted material. There is some evidence (see *Phosphorylation*, above) that H1 contains acid-labile phosphate residues. Accordingly, one is left with the discomforting suspicion that a great deal of effort may have been expended on a molecule irreversibly altered from its native state!

It should be emphasized that similar questions cannot be raised con-cerning the inner histones. In an exactly parallel study, Beaudette et al. (1981) have shown that the salt-extracted core histones can be taken to low pH, with no irreversible change in properties when the pH is brought back to neutrality.

The erythrocyte-specific histone H5 appears to exhibit a structure resembling that of H1. Crane-Robinson et al. (1977a) showed that virtually all of the folded structure is contained in the first 99 residues of this 189-residue protein. A more detailed study by the same group (Aviles et al., 1978) further localized the folded domain to the region between residues 22 and 100 and showed that this portion of the molecule behaves, in hydrodynamic and NMR experiments, as a globular protein. In agreement with this, 25 of the 29 helical residues were found to lie in this domain (see also Chan et al., 1984).

Despite their superficial similarity, there is evidence that the globular domains in H1 and H5 are quite differently conformed. Spectroscopic and NMR studies by Cary et al. (1981b) note substantial differences between the globular domains of these two proteins. H5 was found to be conformationally more similar to H1°. These relationships are emphasized by the observation that antibodies against the globular region of H5 would not cross-react with H1, but would with H1° (Allan et al., 1982b).

Despite the many complications that have been mentioned in the foregoing discussion, analysis of the conformational states of the histones leads to an internally consistent overall picture. Each of the histones contains a central or C-terminal domain that is readily folded in salt solutions. In each case, the maximal folding under these conditions agrees quite well with that predicted from sequence analysis. There remain, in each histone, regions that are potentially capable of further folding under conditions of charge neutralization. In subsequent pages, we shall see that the histones do, in fact, seem to adopt very highly organized structures when present in heterotypic association in the nucleosome. It may well be that this potentiality for further organization is a driving force in chromatin organization.

At this point, it is intriguing to compare the limited structural information presently available with the patterns of histone modification. This is illustrated in Figure 4-17, where the putative structured domains have been shaded. Virtually all of the known sites for in vivo modification fall outside of these regions. On the other hand, as we shall see in the following section, the portions of the chains believed to be responsible for histone–histone interaction in the nucleosome fall *within* the structured domains. This suggests that the major role of histone modification may not be modulation of the stability of the histone core but, rather, adjustments of the interaction of the core with DNA, with nonhistone proteins, or with other nucleosomes.

Histone–Histone Interactions

There are two ways in which histones can interact so as to form associated products. *Homotypic* association occurs between histone molecules of the same kind. These interactions have been mentioned in preceding sections as impediments to the study of histone structure in solution. They are probably not very relevant to the structure of chromatin. The *heterotypic* interactions, which involve two or more different types of histones, form the basis for the formation of the histone core of the nucleosome. Accordingly, they have been studied extensively.

To begin with, we must recognize that either type of histone complex is formed only in the presence of appreciable salt concentration. As was emphasized in *Secondary and Tertiary Structure of Histones,* above, histones in distilled water or in salt solutions of exceedingly low ionic strength are stable as denatured, loosely coiled monomers. It appears that the addition of salts allows the condensation of parts of these highly charged molecules into compact, globular structures that can then interact with one another. The reduction of intermolecular electrostatic repulsion probably also plays a role in this process. The globular structures are quite possibly stabilized by their interactions with other histone molecules. Thus, association is expected to be a complex, cooperative process. But it is clear, as the early kinetic experiments of Isenberg and co-workers showed, that *partial* folding of histones *precedes* their interaction to form complexes. Likely, it is a necessary prelude.

Homotypic association appears to occur with every histone (save possibly H5) given sufficiently high histone and salt concentrations. A summary of reported cases is given in Table 4-19. With the exception of the H3 and H4 fiber formation, these are mainly weak interactions. We shall not consider them further here; the interested reader is referred to the citations in the table.

Heterotypic associations can occur between histones so as to form simple complexes of defined stoichiometry. For the sake of clarity, I shall begin by describing the interactions between *pairs* of different histones and then turn to the more complex associations that build the histone core of the nucleosome.

Table 4-19. Homotypic Histone Interactions[a]

Histone	Aggregates observed	Reference
H1	Dimer (weak)	Sperling and Bustin (1975)
		Diggle et al. (1975)
		Roark et al. (1976)
H5	None	Diggle et al. (1975)
H2A	Dimer, tetramer, and some larger aggregate	Roark et al. (1976)
H2B	Dimer, tetramer (weak)	Diggle et al. (1975)
		Roark (1978)
H3	High molecular weight aggregates	Sperling and Bustin (1975)
		Diggle et al. (1975)
H4	High molecular weight aggregates	Sperling and Bustin (1974, 1975)
		Diggle et al. (1975)

[a]Adapted from Roark et al. (1976), with later data added.

Pairwise Interactions

All of the possible pair interactions between the inner histones are listed in Table 4-20. The parameters given derive mainly from data of D'Anna and Isenberg (1974d). Although all possible interactions seem to occur to some extent, some are much stronger and more important than others. In the table, free energies of formation are also given per monomer unit; these values allow the assignment of a hierarchy of reaction strengths. It is clear that the strongest interactions are found in the formation of the $(H3 \cdot H4)_2$ tetramer and in the $(H2A \cdot H2B)$ dimer. Association of H3 and H4 to form the tetramer appears to proceed through a heterotypic dimer intermediate (see Roark et al., 1974). Comparison of the first two entries in Table 4-20 shows that most of the free energy change is developed in this first step. The $H2A \cdot H2B$ and $H3 \cdot H4$ bondings are sufficiently strong to hold these complexes together even in very dilute solution, at physiological pH and ionic strength. Furthermore, the stability is such that the formation of some of the heterotypic complexes competes effectively against the tendency of histones to form homotypic polymers, as demonstrated by D'Anna and Isenberg (1973).

These interaction patterns are by no means unique to animal cell histones; very similar results have been obtained with histones from species as diverse as yeast (Mardian and Isenberg, 1978), *Tetrahymena* (Glover & Gorovsky, 1978), and pea (Spiker and Isenberg, 1977a). Even *mixtures* of histones from sources as different as yeast and calf have been shown to yield very similar interaction energies (see, for example, Spiker and Isenberg, 1977b; Mardian and Isenberg, 1978; Isenberg, 1979). These studies show that the standard state free energy change for the formation of $H2A \cdot H2B$ pairs is almost insensitive to the source of the histones; in most cases, the difference in $\Delta G°$ is only a few tenths of a kilocalorie; the largest difference is 1.6 kcal. While it should be remembered that a difference in $\Delta G°$ of 1.6 kcal corresponds to nearly a 15-fold factor in the

Table 4-20. Heterotypic Pairwise Interactions Between Core Histones

Reaction	$\Delta \alpha^a$	$\Delta G°$ (kcal/mol product)	$\Delta G°$ kcal/mol histone	Reference[b]
$2H3 + 2H4 \leftrightarrows (H3 \cdot H4)_2$	9	-28.1	-7.0	1
$2 (H3 \cdot H4) \leftrightarrows (H3 \cdot H4)_2$	ND	-7.2	-1.8	2
$H2B + H2A \leftrightarrows (H2B \cdot H2A)$	15	-8.1	-4.0	1
$H2B + H4 \leftrightarrows (H2B \cdot H4)$	8	-8.1	-4.0	1
$H2A + H3 \leftrightarrows (H2A \cdot H3)$	0	$-6.7 \rightarrow (-8.1)$	$-3.4 \rightarrow (-4.0)$	1
$H2A + H4 \leftrightarrows (H2A \cdot H4)$	1	-6.2	-3.1	1
$H2B + H3 \leftrightarrows (H2B \cdot H3)$	0	Weak	—	1

[a] Increase in number of α-helical residues per molecule of product. Data of D'Anna and Isenberg (1974c).
[b] 1. D'Anna and Isenberg (1974d).
  2. Roark et al. (1974).

equilibrium constant, the free energy differences are certainly small in comparison to the magnitude of most noncovalent interactions between macromolecules. Consequently, Isenberg (1979) has argued that the binding surfaces must have been conserved to an extraordinary degree throughout evolution.

We now know something about those binding surfaces. NMR studies of defined peptides (Böhm et al., 1977) have indicated that the regions between residues 38 and 102 in H4 and 42 and 120 in H3 are involved in the H3·H4 interaction. Similarly, residues 31 to 95 in H2A and 37 to 114 in H2B are asserted to participate in the H2A·H2B association (Moss et al., 1976b). As Figure 4-17 shows, these segments all fall within the "globular," "invariant" regions of these histones. Since these experiments have utilized only those peptides available from defined cleavages, they may overestimate the extent of the binding domains. Thus, it is probably not surprising that the regions given above for the interaction of H2B with H2A and H4 with H3 overlap the regions defined by contact-site cross-linking studies as being involved in H2B·H4 interaction in the nucleosome (see Chapter 6). Despite their present limitations, the NMR experiments seem to rule out an important contribution of the positively charged histone "tails" in such interactions. This conclusion is at least indirectly supported by neutron scattering experiments by Carlson (1984).

The formation of these complexes apparently involves a considerable reorganization of the secondary structure of the participant molecules, as judged by CD (Baker and Isenberg, 1976; Spiker and Isenberg, 1977a) and NMR (Moss et al., 1976a, 1976b). There is an increase in α-helix content, which must be distinguished from that which occurs during the "fast" folding of histones, which occurs immediately upon addition of salt, and in the absence of heterotypic interactions. Estimates of the helix increase upon association are included in Table 4-20.

Further changes in the conformation of these complexes may occur at higher salt concentrations. For example, Khrapunov et al. (1984) have detected a "compaction" of the (H2A·H2B) dimer above 0.5 M NaCl that involves the screening of tyrosine residues from solvent contact. Such changes may be a necessary prelude to the higher order interactions discussed below.

Further Association: Formation of the Octamer

The pairwise interactions predominate in dilute histone solution at ionic strengths of 0.2 and below. At higher salt concentrations or in concentrated histone solutions, further association is possible. Two modes of such association can be detected. First, at high histone concentrations, condensation of each kind of pair unit can occur, ultimately leading to the formation of long fibrous polymers. This phenomenon has been extensively studied by R. Sperling and her associates (see Sperling and Bustin, 1974, 1975; Wachtel and Sperling, 1979; also Diggle et al., 1975). While the fibers are well-defined structures, it is not clear what their relationship may be to the nucleosomal organization of chromatin.

The second mode of association leads to the octamer and is observed even in dilute histone mixtures at sufficiently high ionic strength. Sedimentation equilibrium and gel filtration experiments in 2 M NaCl (Eickbusch and Moudrianakis, 1978b) as well as osmotic pressure studies (Stein and Page, 1980) show that octamer formation is a two-step process:

$$(H3 \cdot H4)_2 + (H2A \cdot H2B) \rightleftharpoons (H3 \cdot H4)_2(H2A \cdot H2B) \tag{1}$$

$$(H3 \cdot H4)_2(H2A \cdot H2B) + (H2A \cdot H2B) \rightleftharpoons (H3 + H4)_2(H2A \cdot H2B)_2 \tag{2}$$

The very careful sedimentation equilibrium analysis by Godfrey et al. (1980), coupled with the calorimetric experiments of Benedict et al. (1984), have provided a nearly complete thermodynamic analysis of these reactions. The results are summarized in Table 4-21. It should be noted that the free energy and entropy changes listed in the table are "intrinsic" values; they have been corrected for the fact that there are two possible ways to add the first dimer to make a hexamer, but only one way to complete the octamer. Thus, the difference of almost 1 kcal/mol in $\Delta G°$ indicates a positive cooperativity in the addition of the second dimer to form the octamer. This, in turn, suggests a conformational change between the hexameric and octameric states. The negative values of both the enthalpy and entropy changes argue against a major role for hydrophobic bonding. Indeed, Eickbush and Moudrianakis (1978b) suggest, from the effect of temperature, urea, and pH on the reactions, that histidine–lysine or histidine–tyrosine hydrogen bonds may be involved in stabilization of the octamer. The participation of histidine is supported by the observation (Benedict et al., 1984) that some group with pKa $\simeq$ 6.7 must participate in the association. In accord with this idea, Butler and Olins (1982) have found that the octamer is unstable below pH $\simeq$ 6. An involvement of tyrosine residues is strongly suggested by the spectroscopic studies of Michalski-Scrive et al. (1982). They find that of 8 tyrosines in an H2A·H2B dimer, 5 are inaccessible to solvent. In the (H3·H4)$_2$ tetramer, all 14 ty-

Table 4-21. The Thermodynamics of Octamer Assembly in 2 M NaCl[a]

	Reaction (1)	Reaction (2)
$\Delta G°_{int}$ (kcal/mol)	-7.3	-8.2
$\Delta H°$ (kcal/mol)	-30.8	-22.5
$\Delta S°_{int}$ (cal/mol·K)	-76	-48

[a]Reaction (1) and reaction (2) are given in the text. See the text for references to the original data.

rosines are accessible. Yet in the octamer, *none* can be detected as solvent-accessible (see also Kleinschmidt and Martinson, 1984).

Hydrodynamic characterization of the octamer has proved difficult. Thomas and Butler (1977) reported a sedimentation coefficient of 4.77 S, which would correspond to a quite asymmetric structure (f/fo = 1.43). However, Philip et al. (1979) observed that the sedimentation coefficient of the histone octamer was strongly dependent on rotor speed; the data suggested a pressure-dependent dissociation in high centrifugal fields (see also Jamaluddin and Philip, 1982). When data were extrapolated to zero angular velocity, a value of $s^{\circ}_{20,w}$ = 6.6 S was found. This value, combined with their experimentally determined diffusion coefficient, gave a molecular weight of about 105,000, indicating the octamer to be intact under these conditions. The value of f/fo obtained from these data is 1.23, a more reasonable value for such a presumably compact structure.

It is possible to crystallize the histone octamer from concentrated ammonium sulfate. Using such crystals, Burlingame et al. (1984, 1985) have obtained resolution of 3.3 Å. The resulting model differs strikingly from the octamer structure that might be expected on the basis of crystallographic studies of nucleosomal core particles (Richmond et al., 1984; Uberbacher and Bunick, 1985a; see Chapter 6). In particular, the octamer structure obtained by Burlingame et al. can be described as prolate ellipsoid approximately 110 Å long, whereas the core particle studies would suggest a more symmetrical structure with a maximum dimension of about 90 Å. The apparent discrepancies have generated rather heated controversy (see Klug et al., 1985; Moudrianakis et al., 1985a, 1985b; Uberbacher and Bunick, 1985b). Uberbacher and Bunick (1985b) suggested that the discrepancy might be more apparent than real; the elongated structure observed by Burlingame et al. could be a consequence of both the absence of DNA and the high salt conditions and in crystallization.

In a subsequent study, Uberbacher et al. (1986) employed low-angle neutron scattering to examine the shape of the histone octamer in concentrated salt solutions. They examined the particles in both 2 *M* NaCl and the ammonium sulfate conditions used by Burlingame et al. There is a complication that should be brought to the attention of readers of this paper. In their 1986 paper, Uberbacher et al. state the ammonium sulfate concentration in the solutions they used (made by the protocol given by Burlingame et al., 1984) to be 3.5 *M*. Apparently this is in error; the actual concentration seems to have been about 2.3 *M* (Uberbacher et al., via private communication from Moudrianakis et al.). So we must consider that Uberbacher et al. have compared the particle structure in 2.0 *M* NaCl and 2.3 *M* $(NH_4)_2SO_4$. It should be noted that the latter is still a *much* higher ionic strength than the former, because of the presence of the divalent sulfate ions. In any event, length distribution analysis from the low-angle neutron scattering leads to the conclusion that the maximum particle dimension is about 90 Å in 2 M NaCl and about 120 Å in the ammonium sulfate solution.

The controversy is not likely to be quickly settled to the satisfaction of all, but I would interpret the available data to indicate that the dimensions of the octamer core are quite strongly dependent on solvent conditions. For this reason, comparisons of a free octamer in high salt to an octamer condensed into a nucleosome at low salt seem specious. This is basically the position that Uberbacher et al. have taken.

One conclusion to be drawn from all of these studies is that the histone octamer, in the absence of DNA, is a fragile object. It cannot, for example, exist free in solution under physiological conditions. Therefore, the assembly of nucleosomes in vivo must either require some kind of "carrier" to deliver intact octamers, or must proceed in a stepwise fashion, involving first the attachment of the very stable $(H3 \cdot H4)_2$ tetramer to the DNA. We shall encounter both concepts again in Chapter 9.

Noneukaryotic Histone-Like Proteins and the Evolution of the Histones

Previous chapters have emphasized that a nucleosomal structure has been observed in chromatin of eukaryotes ranging from yeast to man. This structure is highly conserved and invariably involves the four inner histones, and in most cases H1 or H1-like proteins as well. Consideration of the entire range of life, however, shows that this is a specialized feature. No convincing evidence for such structure has been obtained for any eubacterium, archaebacterium, or organelle. Figure 4-31 depicts a recent proposal for an overall phylogeny; nucleosomal chromatin is restricted to one particular branch.

Nonetheless, there do exist DNA-binding proteins in the eubacteria, archaebacteria, chloroplasts, and mitochondria. If, as is supposed, all living organisms have evolved from one original line, might not "histone-like" proteins be expected throughout all kingdoms? If such exist, can their structures reveal anything about the evolution of the histones and eukaryotic chromatin structure?

Prokaryotic DNA-Binding Proteins: Histone Precursors?

Among the prokaryotes, *E. coli* has been studied in by far the greatest detail. Its genome is folded into a compact, supercoiled structure termed the *nucleoid* that may be isolated as a more or less intact entity by a number of procedures (Stonington and Pettijohn, 1971; Worcel and Burgi, 1972; Kornberg et al., 1974; Varshavsky et al., 1977a, 1977b). Without question, this is a nucleoprotein complex; depending on its method of preparation, many different kinds of proteins are found therein (see, for example, Figure 6 in Varshavsky et al., 1977b; and Lossius et al., 1984). Furthermore, electron microscopy of the partially unfolded bacterial nu-

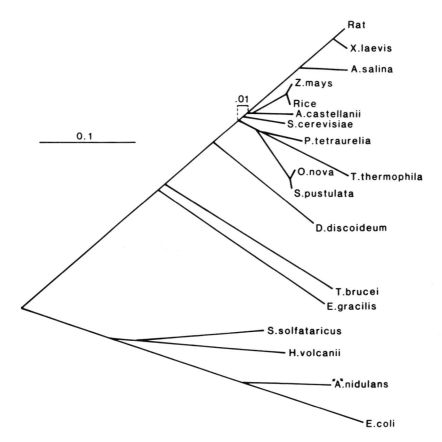

Fig. 4-31. An overall phylogeny deduced from the sequences of small-subunit ribosomal RNAs. The whole small subunit RNAs were used. Histones and nucleosome structure are observed *only* in the upper branch. (From Proc. Natl. Acad. Sci. USA *83*, pp. 1383–1387, Sogin et al.)

cleoid shows, under certain conditions, a beaded structure remarkably similar to the eukaryotic chromatin fiber (Griffith, 1976). "Particles" about 130 Å in diameter are found to be spaced closely along the fiber.

Despite these apparent similarities to eukaryotic chromosomes, digestion of the *E. coli* nucleoid has never been reported to yield the kind of "ladder" of DNA fragments so typical of eukaryotic chromatin. Correspondingly, neither *E. coli* nor any other prokaryote has been shown to contain histones or proteins even closely resembling them in amino acid sequences.

The acid-soluble, "histone-like" proteins that *have* been reported in *E. coli* and other prokaryotes are listed in Table 4-22, along with appropriate references. Dominant in the *E. coli* nucleoid are two very similar low molecular weight proteins: HU-1 and HU-2 (also designated as HLP-IIa,

Table 4-22. "Histone-Like" Proteins of Prokaryotes

Organism/organelle	Common symbol for protein	Molecular weight	Comments	Reference
1. Archaebacteria				
Halobacterium halobium	—	~10,000	Isolation, characterization:	Obha and Oshima (1980)
Sulfolobus acidocalderius	—	~14,500	Isolation, characterization:	Green et al. (1983)
Thermoplasma acidophilum	HTa	10,065 (90 residues: also form with 89 res.)	First isolation: Purification, characterization: Sequence: Forms DNA complex: Structure of complex:	Searcy (1975) DeLange et al. (1981a) DeLange et al. (1981b) Searcy and Stein (1980) Notbohm (1982)
2. Eubacteria				
Bacillus globigii	—	~11,000	Isolation, comp., immunology:	Imber et al. (1982)
Bacillus staerothermophilus	HBst	9,716	First isolation, characterization: Sequence: 3-dimensional structure:	Dijk et al. (1983) Kimura and Wilson (1983) Tanaka et al. (1984)
Bacillus subtilis	—	~9,000	First isolation; composition:	Nakayama (1980)
Escherichia coli	HU1, HU2 (also called NS1, NS2)	9,232 9,643	First observation: Sequences of both: Forms DNA complex in vivo:	Rouviere-Yaniv and Gros (1975) Mende et al. (1978) Laine et al. (1980) Rouviere-Yaniv (1978) Rouviere-Yaniv et al. (1979)

Table 4-22. *Continued*

Organism/organelle	Common symbol for protein	Molecular weight	Comments	Reference
	H1	~15,300	Nature of complex: First observation:	Jacquet et al. (1971)
			Identification of 3 forms (a,b,c); purification of (a):	Spassky et al. (1984)
			Partial sequence:	Laine et al. (1984)
	HLP-1	~17,000	First observation:	Schafer and Zillig (1973)
			Identified as product of fir gene:	Lathe et al. (1980)
	H	~28,000	Isolation, composition, immunology:	Hübscher et al. (1980)
Pseudomonas aeruginosa	HPa	~10,000	Isolation, characterization, partial sequence:	Hawkins and Wootton (1981)
Rhizobium meliloti	HRm	9,303	Isolation, characterization:	Laine et al. (1982)
			Sequence:	Laine et al. (1983)
3. Cyanobacteria				
Anabaena sp.	—	~10,000	Isolation, characterization, composition, immunology:	Haselkorn and Rouviere-Yaniv (1976)
Apanocapea sp.	—	~10,000	Isolation, characterization, composition, immunology:	Haselkorn and Rouviere-Yaniv (1976)
Synechocystis sp.	—	~10,500	Isolation, partial sequence:	Aitken and Rouviere-Yaniv (1979)
4. Chloroplast (spinach)	—	~17,000	Isolation, immunology:	Briat et al. (1984)
5. Mitochondrion (yeast)	—	~20,000	Isolation, characterization:	Caron et al. (1979)

HLP-IIb and NS-1, NS-2 by some authors). These proteins have been sequenced and show considerable homology to one another but no significant resemblance to any histone (Laine et al., 1980). The proteins HU-1 and HU-2 form a heterotypic dimer. They are claimed to be capable of superhelical coiling of DNA, with the formation of nucleosome-like structures (Rouviere-Yaniv et al., 1979; Broyles and Pettijohn, 1986). In addition to the HU-proteins, three other acid-soluble proteins have been reported to be associated with the *E. coli* DNA (see Table 4-22). The protein H has an amino acid composition similar to that of the histone H2A and reacts with anti-H2A antibody (Hübscher et al., 1980). Unfortunately, the sequence of this interesting protein does not seem to have been determined. If it is closely related to H2A, it must represent a gene doubling, since its molecular weight is about twice that of the histone. No proteins resembling any of the other eukaryotic histones have been reported.

If the quantities claimed for each of these proteins are added, the total ratio of mass of acid-soluble proteins to DNA in the bacterial nucleoid can be estimated to be between about 0.5 and 1.0. Unfortunately, this value is not very precise, since data from different laboratories are not consistent. Nevertheless, a picture emerges of the *E. coli* DNA being associated with a significant amount of protein, at least some of which has the capacity to produce supercoiling and something resembling nucleosome structure. However, this seems to be accomplished without histones, and while the chromatin structure may under certain circumstances resemble that of eukaryotes, it is clearly very different. It is of exceptional interest that the HU–DNA interaction studied by Broyles and Pettijohn (1986) appears to be very weak at physiological ionic strength. This suggests that the structure may be a dynamic one and could explain the lack of a nucleosome "ladder."

Proteins similar to HU have been reported in other eubacteria, cyanobacteria, and archaebacteria (see Table 4-22). Figure 4-32 depicts an alignment of some of the sequences that have been determined. Clearly, these form a moderately conserved group; they have been collectively termed "Type II DNA binding proteins." Two more of these proteins are of particular interest.

The HTa Protein of *Thermoplasma acidophilum*

An archaebacterial DNA binding protein, HTa is present in appreciable amounts—up to 0.6 g/g DNA (Searcy and Stein, 1980). It appears to protect DNA fragments of about 40 to 80 bp from micrococcal nuclease digestion. Low-angle X-ray diffraction studies (Notbohm, 1982) support the electron microscope evidence of Searcy and Stein for the formation of nucleoprotein particles about 55 Å in diameter. The amino acid sequence of this protein shows considerable similarity to that of the HU proteins (DeLange et al., 1981b; see Fig. 4-32). Of greater interest is the fact that some correspondence seems to exist between the HTa sequence and some histone se-

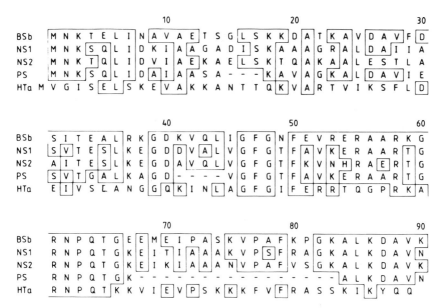

Fig. 4-32. Comparison of the amino acid sequences of a number of prokaryotic DNA binding proteins. Included are the completed sequences of the *E. coli* HU-1 and HU-2 (here designated NS-1 and NS-2), the *Bacillus staerothermophilus* protein (BSb), the *Thermoplasma acidophilum* protein (HTa), and a partial sequence of the DNA protein from *Pseudomonas aeruginosa* (PS). Residues identical with BSb are boxed. From Kimura and Wilson, 1983.

quences. The correspondence is closest with H2A and occurs in two regions of H2A that have been shown to exhibit internal homology (see below). A possible alignment is included in Figure 4-35. Interestingly, most of the modest number of identities occur at positions in the H2A sequence conserved over the entire eukaryotic range. It should also be noted that whereas HTa can be aligned with both H2A and HU, similarities between the latter two proteins are hard to find. Thus, HTa seems to occupy an intermediate position.

HB St, The DNA-Binding Protein from *Bacillus staerothermophilus*

The three-dimensional structure of this protein has been determined (Tanaka et al., 1984). Considering the sequence conservation among the type II proteins, the structure can probably be considered an archetype, and it is a remarkable one. As shown in Figure 4-33, the protein exists as a dimer, with two extended "arms" that together enclose one turn of B-form DNA. The dimers can pack together in such a fashion as to supercoil DNA about a helical protein core (Fig. 4-34). Such a complex cannot yield a *complete* description of the bacterial nucleoid, for two

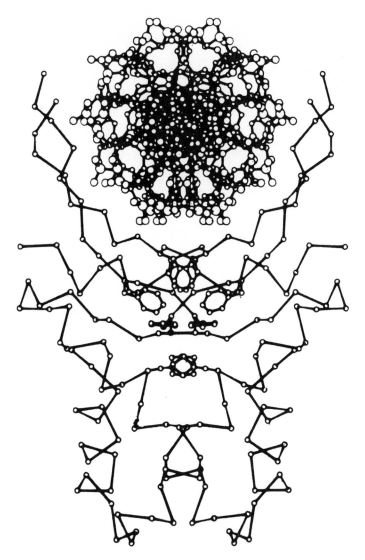

Fig. 4-33. The three-dimensional structure of the DNA-binding protein of *Bacillus staerothermophilus*. A view of the dimer of the protein, showing its putative interaction with DNA from an end-on view. (Reprinted by permission from *Nature,* Vol. 310, p. 376. Copyright (c) 1984 Macmillan Journals Limited.)

reasons. First, the type II protein, in itself, would seemingly lead to the formation of a continuous, nonbeaded supercoil, contrary to observation. Indeed, it is amusing to note that the kind of structure proposed is very reminiscent of the "uniform supercoil" model (Chapter 1). Second, we know (see above) that there are other proteins associated with bacterial

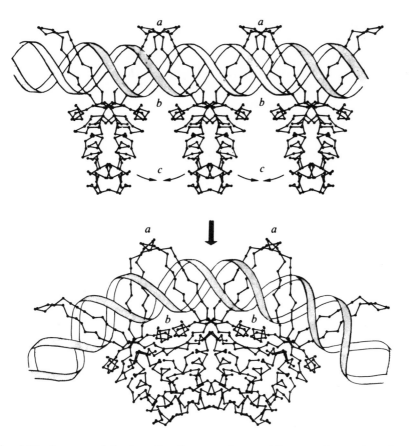

Fig. 4-34. A proposal for how the *B. staerothermophilus* protein can introduce supercoiling of DNA. This representation is diagramatic and does not reflect a detailed structural analysis. The important interactions are felt to be (a) contact between "arms" of adjacent molecules, (b) contact between C-terminal helices, and (c) electrostatic interactions between the wedge-shaped molecule. (Reprinted by permission from *Nature,* Vol. 310, p. 376. Copyright (c) 1984 Macmillan Journals Limited.)

DNA. While the overall structure will doubtless prove complex, these results give the first insight into the organization of a prokaryotic chromatin.

We observe, then, in the contemporary prokaryotes, both eubacterial and archaebacterial, certain "histone-like" proteins that not only bind DNA but compact it in ways faintly resembling the nucleosomal structure of eukaryotic chromatin. But between these proteins and histones, or between nucleoid and chromatin structure, there are enormous differences. For the nucleosome to develop, a whole set of histones had to evolve. Once evolved, they and the structure they produce seem to have become fixed. What can we hypothesize as to how and when this major evolutionary step occurred?

Evolution of the Histones: A Hypothetical Scenario
McCarroll et al. (1983) and Sogin et al. (1986) have utilized the sequences of small-subunit ribosomal RNAs to examine phylogeny over vast periods of time. According to such studies, protists such as *Euglena gracilis* represent the most primitive eukaryotes (see Fig. 4-31). We know that *Euglena* has a nucleosomal chromatin structure (Magnaval et al., 1980). Therefore, evolution of the histones must have taken place after the divergence of the eukaryotic line from the eubacterial–archaebacterial stem and before the divergence of the protists. As we have seen, there are hints of common, archaic elements of structure in the sequences of DNA-binding proteins from all three contemporary kingdoms. Therefore, one can postulate the existence of a primitive DNA-binding protein (or proteins) in the ancient organisms that were the predecessors of all three lines of descent. Such proteins could evolve independently to produce the HU protein of eubacteria, the HTa-type proteins of the archaebacteria, and the histones of the eukaryotes.

A weak case can be made that a single DNA-binding protein was ancestral to both the type II proteins and the histones, based on the apparent homologies cited above. If this were the case, the ancestral molecule must have been shorter than the contemporary type, for a portion of HTa shows homology with *both* halves of H2A, each about 60 residues in length. A gene doubling must be postulated in the evolution of H2A from this hypothetical ancestor.

How, then, did the other histones evolve? Three groups, examining the sequences of the core histones, have claimed to find significant sequence homologies (Wuilmart and Wyns, 1977; Reeck et al., 1978; Brown, 1983). There is, not surprisingly, considerable disagreement concerning the relationships. Wuilmart and Wyns postulate the evolution of H2A and H2B from a proto-H4; they find no evident relationship between H3 and the others. Reeck et al. claim homologies between all four core histones (see Fig. 4-35). Furthermore, the regions claimed to be homologous by these two groups are not the same! Brown contends that all four histones show evidence for a primitive tetrapeptide repeat, with ancestral features best seen in H3. Clearly, more information from more histone sequences is required to resolve these questions, if resolution is indeed possible. On a *structural* basis, an argument could be made for H3 and H4-like proteins as elements of the primitive nucleosome. As will be described in Chapter 6, this pair of histones alone can form nucleosome-like structures with DNA; H2A and H2B, either singly or in combination, are ineffectual. We may hypothesize, then, that at some point between the eukaryotic–prokaryotic divergence and the divergence of the protists, duplication of a proto-type II gene gave rise to two proteins that could form the octameric

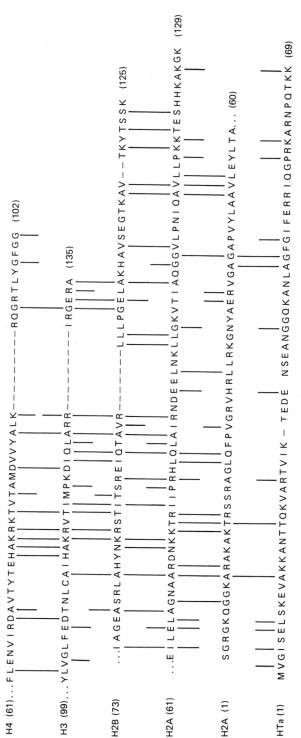

Fig. 4-35. Alignments of portions of the core histones, according to Reeck et al., 1978. The C-terminal portions of H4, H3, and H2B are employed. Both an N-terminal and C-terminal region of H2A are aligned. I have added the N-terminal region of HTa, the DNA-binding protein from *Thermoplasma acidophilum*. Vertical lines show correspondences. Gaps necessary for alignment are shown by dashes. Numbers in parentheses give the first and last residues shown. Sequences that continue are indicated by dots.

core of a primitive nucleosome. Further duplication and independent evolution then produced the kind of structure seen today. Partial support for this idea may come from observations by Rizzo et al. (1985) on the histones of the unicellular alga, *Olisthodiscus luteus*. They find that this organism contains H3 and H4, but a *single* protein fulfilling the roles of H2A and H2B. (See also Rizzo, 1985.)

The relationship of H1 to the other histones is obscure. So far as we can tell, the H1 sequences seem unrelated to either other histone sequences or those of prokaryotic proteins. This may, of course, simply be a consequence of the rapid evolution of this protein, which has obscured its origins; alternatively, H1 may have evolved from an entirely different protein.

When one considers further evolution of the histones, some difficult problems arise. According to the phylogeny depicted in Figure 4-31, plants, animals, and fungi diverged from the ancestral eukaryotic branch at almost the same time. Yet it has been noted (see *The Primary Structures of Histones,* above) that plant and animal histones show great sequence homology and are quite different from those of yeast and *Tetrahymena.* Indeed, on the basis of the apparent evolutionary rates of plant and animal histones Brandt and von Holt (1982) have calculated that the fungi must have diverged from the other phyla about 4 billion years ago! While this value, which is comparable to the presumed age of the earth, is clearly too large, it points up the paradox. One cannot escape the difficulty by postulating that histone structure became "fixed" only in the animal phyla. If so, why are cow and pea histones so similar? Only two possible explanations are obvious: Either the phylogeny based on ribosomal RNA is grossly wrong, or the evolutionary "clock" has run at very different rates in yeast and *Tetrahymena,* as compared to plants and animals.

Even if the problems described above can be reconciled, there still remains one group of organisms that is exceedingly difficult to fit into the overall picture. These are the *dinoflagellates.*

The Peculiar Problem of Dinoflagellate Chromatin
The free-living dinoflagellates share traits in common with both prokaryotes and eukaryotes (see Loeblich, 1976; Hinnebusch et al., 1981). There is a nucleus, complete with nuclear membrane. The chromosomes are highly condensed, and in most species remain so throughout the life cycle. These organisms share such eukaryotic features as a large genome size and the presence of large amounts of repetitive DNA (Allen et al., 1975; Hinnebusch et al., 1980). Dinoflagellates have been thought of as very primitive eukaryotes (i.e., Kubai and Ris, 1969; Loeblich, 1976), as "mesokaryotes" intermediate between prokaryotes and eukaryotes (Dodge, 1966), and as a degenerate eukaryotic line (Hinnebusch et al., 1981). A very thorough discussion of dinoflagellate phylogeny and evolution is provided by Taylor (1980), who reviews the paleontological record as well.

The gist of the matter is that we do not know when dinoflagellates first evolved, nor their phylogenetic affinities with any certainty.

Despite the above-mentioned similarities to eukaryotes, dinoflagellates have a very different chromatin. Attempts to observe a periodic pattern of protection of dinoflagellate chromatin to nuclease digestion have been uniformly unsuccessful (van Holde et al., 1979; Herzog and Soyer, 1981; Shupe and Rizzo, 1983). In accord with this, a number of electron microscopic studies of spread chromosomes have failed to show nucleosomal structure, revealing instead smooth chromatin fibers about 55 to 65 Å in thickness (Hamkalo and Rattner, 1977; Rizzo and Burghardt, 1980; Herzog and Soyer, 1981). Thus, it is not surprising to find that dinoflagellate nuclei contain little basic protein and *no* protein that can be identified as a histone. The acid-soluble proteins that have been reported have molecular weights of about 12,000 to 13,000 (Herzog and Soyer, 1981; Rizzo, 1985). We do not know much about these proteins. They resemble neither histones nor the prokaryotic DNA-binding proteins in amino acid composition (Rizzo and Nooden, 1974; Herzog and Soyer, 1981; Rizzo and Morris, 1983). The ratio of basic proteins to DNA is small, of the order of 0.1 g/g DNA in all of the free-living dinoflagellates so far examined (Rizzo and Morris, 1983; Rizzo et al., 1984). However, Herzog and Soyer also report the presence of considerable amounts of acid-insoluble protein associated with dinoflagellate DNA. It seems unlikely that either the unusual chromatin structure or the absence of histones can be explained as artifacts of the unusual difficulties in isolating nuclei from these armored organisms. An internal control is provided by the fact that some dinoflagellates (i.e., *Peridinium balticum*) have two nuclei: a typical "dinokaryotic" nucleus and a eukaryotic nucleus, which is probably endosymbiotic. It has been shown that the latter has both histones and a beaded chromatin structure, whereas co-isolated dinokaryotic nuclei do not (Rizzo and Cox, 1977; Rizzo and Burghard, 1980; Rizzo, 1982).

How are we to relate the dinoflagellates to other organisms? Hinnebusch et al. (1980) have sequenced portions of the 5 S and 5.8 S RNAs from *Cryptothecodinium cohnii*. Their interpretation of the 5 S RNA data places the dinoflagellates as considerably closer to the eukaryotes than to the prokaryotes. Furthermore, a phylogenetic tree based on these data places the divergence of dinoflagellates at about the same point as the divergence of plants and animals, and *considerably later than the divergence of fungi!* Thus, we are confronted with the remarkable possibility that yeast, whose ancestors (according to this model) diverged from the eukaryotic stem before dinoflagellates, possess a nucleosomal structure, whereas the latter do not. The problem is only partially resolved by a recent determination of the small-subunit rRNA sequence from a dinoflagellate, *Prorocentrum micans* (Herzog and Maroteaux, 1986). According to their analysis, divergence of the dinoflagellates from the main eukaryotic line occurred only slightly earlier than that of *Dictyostelaum discoidum*, the slime mold.

In an attempt to reconcile such problems, Hinnebusch et al. (1980) suggest that the free-living dinoflagellates represent a degenerate eukaryotic line, which has *lost* histones and the accompanying nucleosomal structure. It is truly difficult to imagine how such a profound rearrangement could occur. Presumably, both transcription and replication mechanisms must differ in nucleosomal and nonnucleosomal chromatin. It is hard to envision how such a loss could occur without a massive and coordinated reordering of the machinery for these fundamental cellular processes.

Escape from this conundrum is not easy. One can, of course, assume that *all* of the proposed phylogenies are incorrect, at least insofar as the dinoflagellates are concerned; certainly, they are not wholly consistent. Perhaps the dinoflagellates diverged even earlier than yeast or *Euglena* from the eukaryotic line and represent an evolutionary "dead end" in nuclear organization. Alternatively, one is tempted to wonder if the large amount of DNA in the dinoflagellate nucleus may be obscuring the true picture. Some dinoflagellates have as much as 200 picograms of DNA/cell, as compared to a few picograms for a typical eukaryote and about 0.03 picograms for yeast. Only a fraction of that total should be needed for transcription of necessary messages; and that small fraction might be a nucleosomal euchromatin, undetected in the nucleus because of a large excess of nonnucleosomal, tightly folded chromatin. It is perhaps noteworthy that Sigee (1984) has observed that transcription in dinoflagellates seems to occur on "peripheral fibers" attached to the condensed chromosomes.

Until we know more about their nuclear structure, or can fix their evolutionary position more exactly, the dinoflagellates remain anomalous.

5

The Proteins of Chromatin. II. Nonhistone Chromosomal Proteins

It has been argued in previous chapters that the histones determine the fundamental structural features of chromatin. However, it must be emphasized that chromatin is a dynamic structure, constantly changing during transcription, replication, mitosis, meiosis, and DNA repair. The histones themselves undergo both partial replacement and major modification during the cell cycle. All of these processes must require the participation of innumerable other proteins in the nuclear milieu.

It has long been recognized that the histones represent only a part of the protein component of chromatin. The remaining "nonhistone" proteins might be defined as those proteins other than histones that are associated with chromatin in vivo. While admirably simple, such a definition is useless. Even if we could better define the vague phrase "associated with," no techniques are currently capable of clearly ascertaining what is associated with what within the intact nucleus in vivo.

Therefore, researchers tend to fall back on more operational definitions, such as: "Nonhistone chromosomal proteins (NHCPs) are those, other than histones, attached to isolated chromatin." But again there are difficulties. The many ways of isolating chromatin yield different (though overlapping) sets of "nonhistone" proteins. To understand why this should be so, we must consider briefly the structure of the eukaryotic nucleus. For further details, the reader should consult reviews by Agutter and Richardson (1980), Hancock (1982), and Bouteille et al. (1983).

The nucleus is bounded by an *envelope*, which consists of three layers: an *outer membrane*, an *inner membrane*, and a proteinaceous, fibrous *lamina*. The envelope is pierced at many points by *nuclear pores*. These are large enough to allow small protein molecules to pass through, a fact taken advantage of in studies where chromatin is digested by nucleases diffused into intact nuclei. Therefore, the nucleus is to some extent open to the cytoplasm, and some proteins can presumably exchange between these compartments in vivo. It may be expected, then, that vigorous washing of nuclei will extract some proteins that are normally bound, but only weakly so, to the nucleoprotein strands.

Many believe that the nucleus contains a fibrillar protein structure called the *nuclear matrix*. This is said to contain 10 to 20% of the total nuclear protein (see Berezney and Coffey, 1975; Long et al., 1979; and Chapter 7). Although the exact composition is difficult to fix, most matrix preparations seem to include some components of the lamina (the *lamins*) as well as specialized proteins such as topoisomerases. It should be noted, however, that a number of scientists remain skeptical as to the significance or even the reality of the "nuclear matrix." For one such view, see Bouteille et al. (1983).

This, then, is the environment in which the histone–DNA complex, with its associated proteins, is found. There is evidence (see Chapter 7) that the chromatin fibers are themselves attached to both the matrix and the lamina. Bathing the chromatin and matrix is presumably a *nuclear sap,* a concentrated solution of proteins that associate only weakly (or not at all, in some cases) with other components.

Almost all methods for isolating chromatin begin with the preparation of nuclei. The first problem is evident: Does one wash the nuclei in dilute buffer or not? If nuclei are not sufficiently washed, they are likely to be grossly contaminated with cytoplasmic proteins, and even some noninteracting nuclear sap proteins will remain. Such proteins may artifactually condense upon chromatin in later steps. Alternatively, if nuclei are extensively washed, not only nuclear sap proteins but also some of the more weakly bound nonhistone chromosomal proteins will be removed. For example, both Comings and Harris (1976) and Hyde et al. (1979) found that repeated washing with quite low salt buffers removed a large fraction of the nuclear proteins. (See Table 5-1.) In order to study chromatin and the proteins associated with it, the nuclei must then be lysed and the chromatin released and separated from the residual nuclear structure. Again, there are many ways in which this can be done, and differing results may be obtained, depending upon whether detergents are or are not used, what extent of nuclease digestion is employed, etc.

Finally, we are faced with the likelihood that some proteins will be attached to and isolated with the nucleohistone fiber only at certain stages in the cell cycle. An example is provided by the studies of Adolph (1984) comparing the proteins of interphase chromatin and whole metaphase chromosomes of HeLa cells. While 90% of the antigenic determinants of interphase chromatin could be identified in the metaphase chromosomes, the latter also contained a substantial number of additional proteins. On the other hand, *chromatin* prepared from the metaphase chromosomes

Table 5-1. Nonhistone Protein Content of Rat Liver Nuclei and Chromatin[a]

Isolation stage	% total DNA	gm nonhistone protein / gm DNA
I. Nuclei, unwashed	100	1.38 ± 0.12
II. Nuclei, after 3 cold washes in buffer A,[b] one supplemented with 0.2% Triton	~100	1.00 ± 0.06
III. Nuclei, incubated at 37° in buffer A supplemented with divalent ions		
A. without nuclease	96	0.72 ± 0.08
B. with micrococcal nuclease		0.75 ± 0.04
C. with Hae III		0.73 ± 0.04
IV. Lysed nuclei, sedimented to yield "soluble" chromatin and "insoluble" fraction		
A. soluble	79	0.3–$0.5 \rightarrow {\sim}0.2$[c]
B. insoluble	17	${\sim}0.2 \rightarrow {>}3.0$[c]

[a]Adapted from Hyde et al. (1979).
[b]Buffer A: 65 mM KCl, 65 mM NaCl, 0.5 mM spermidine, 0.15 mM spermine, 0.2 mM EDTA, 0.2 mM EGTA, 1 mM phenylmethylsulfonyl fluoride, 5 mM β-mercaptoethanol, 10 mM Tris, pH = 7.4
[c]Amount in soluble fraction decreased, and amount in insoluble fraction increased with increasing digestion.

seemed much more similar to that from interphase (see also Wray et al., 1980).

Given all of these complications, and the fact that a standard method for chromatin preparation does not exist, it is not surprising that the values given for "nonhistone proteins" in Table 4-1 vary so greatly, even when the same tissue is examined in different laboratories. For a specific, contemporary view of the problem, consider the data of Hyde et al. given in Table 5-1. They found 1.1 g of histone per gram of DNA, in accord with most current estimates. The unwashed nuclei contained 1.38 g of other proteins in toto, but this value was reduced to 1.00 g upon washing in physiological saline. This step presumably removed cytoplasmic contaminants, nuclear sap proteins, and probably some of the more weakly bound NHCPs as well. A further incubation in saline at 37° reduced the nonhistone nuclear protein level to about 0.75 g/g DNA. Upon lysis, after light nuclease digestion, about 0.3 to 0.5 g nonhistone protein remained attached to the soluble chromatin, but this amount decreased dramatically upon further digestion. The implication is that some proteins were displaced from the linker regions in the chromatin and rebound to the matrix/envelope fraction, which constitutes the "insoluble" chromatin. Thus, even the *extent* of digestion may have a marked effect on the amount of NHCP reported! Support for the idea that aqueous washing may remove weakly associated proteins comes from the studies of Kuehl et al. (1984). Using a nonaqueous technique for nuclear isolation, they find much higher levels of HMG proteins (see below) than have usually been reported.

I conclude that we do not have, at the present time, the basis for even a good *operational* definition for nonhistone chromosomal proteins. About all one can say is that chromatin, isolated by any of the usual procedures from washed nuclei, without excessive nuclease digestion, is usually found to contain between 0.3 and 0.8 g of nonhistone protein per gram of DNA. Considering the interconnections between entities in the eukaryotic nucleus, and its communication with the cytoplasm, an exact definition of nonhistone chromosomal proteins is probably not possible. Nonetheless, it is clear that there is a significant amount of nonhistone protein that remains bound even when chromatin is isolated by the most drastic methods.

There certainly are very *many* such proteins. This was demonstrated in the pioneering studies of Elgin and co-workers (Elgin and Bonner, 1972; Elgin et al., 1973), but the true complexity became evident only with the development of two-dimensional electrophoretic methods.

An example is shown in Figure 5-1, taken from the careful study by Peterson and McConkey (1976). They report more than 450 distinguishable spots. In a similar examination of rat brain NHCPs, Fleischer-Lambropoulos and Pollow (1978) claim to have found over 1200 components! At least some of the multitude of spots recorded on such gels must correspond to modified derivatives, variants, and possibly degradation products. Nonetheless, the nucleus undoubtedly contains a very large number of kinds of nonhistone proteins. A corollary of this is that some must be present at relatively small copy numbers.

The functions of most of these proteins are wholly unknown. This has resulted in a certain aimlessness in much research on NHCPs. The literature is replete with studies that indicate changes in amounts, or chemical modification of one or more proteins in response to cell cycle, differentiation, hormone administration, etc. There are many, many papers comparing NHCP distributions in different cell types. Unfortunately, most such studies are uninterpretable at the present time. In the rare cases where dramatic changes in a particular component can be identified, its isolation in significant quantities remains a formidable task. Even if this can be accomplished, one is usually left with a protein whose function must still be sought.

How can progress be made? The most reasonable course, which has met with some success, is to begin with those chromatin-associated propH

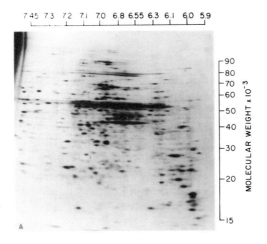

Hela Cell Chromatin

pH

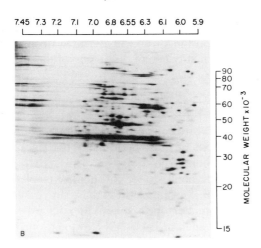

Hela Cell Nucleoplasm

Fig. 5-1. Autoradiograms of the nonhistone chromosomal proteins from HeLa cells. Cells were labeled with [^{35}S]-methionine. The two-dimensional gel technique involved isoelectric focusing in the first dimension and SDS-gel electrophoresis in the second. (From Peterson and McConkey, 1976. Reprinted with permission.)

teins whose functions are clearly identifiable or which are present in substantial amounts. It is these that I shall discuss in this chapter. The recognized categories include:

1. *Chromatin-bound enzymes:* These are easily identifiable by their activities and thus constitute a well-studied group.
2. *Transcription-regulating proteins:* There is a small but growing list of proteins that bind at specific DNA sites so as to regulate transcription in eukaryotes. They are usually identified by "footprinting" methods.
3. *Hormone-receptor proteins:* A number of identified proteins are involved in the regulation of transcription in response to hormones. Radiolabeled hormones can be used to identify them.
4. *The high mobility group (HMG) proteins:* These are small, relatively abundant NHCPs that can be isolated and purified by simple procedures. Consequently, they have received extensive study.
5. *Chromosomal scaffold proteins:* There exists a class of proteins that bind very tightly to DNA, and thus can be isolated after removal of all other proteins. Some of these seem to be involved in maintaining chromosome integrity and higher-order structure.

Finally, there is conflicting evidence as to whether certain contractile proteins, which have cytoplasmic counterparts, are bona fide components of the nucleus, or are merely contaminants.

Even a list of all of the identified members of the above classes leaves the vast majority of NHCPs unaccounted for. Although each of the remainder may be of importance, and many have been investigated, the data are so fragmentary that coherent discussion of them seems impossible at the present time.

Enzymes Associated with Chromatin

The chromatin of eukaryotic cells is an exceedingly dynamic material. It is replicated, and portions of it are continually undergoing transcription. Concomitantly, it is modified in a host of ways. Therefore, it is to be expected that among the nonhistone, chromatin-associated proteins will be found many of the enzymes involved in these processes. This is indeed the case, and a large number of such activities have been identified. A selection of these is presented in Table 5-2. However, one must be cautious in evaluation of much of the enzymological literature in this field. The difficulty lies in ascertaining which enzymes are truly chromatin-bound, which are present as nonchromosomal nuclear proteins (i.e., matrix proteins or nuclear sap proteins), and which are primarily cytoplasmic proteins that may only adventitiously be isolated with the nucleus. Many studies have been conducted on whole cell (or whole tissue) homogenates, and it has sometimes been assumed that if chromatin components are substrates, the corresponding enzymes must be chromatin-associated. In many other instances, whole nuclei have been used as a starting material for

Table 5-2. Enzymes Associated with Chromatin

Enzyme	Organism; tissue	Comment	Reference
I. Enzymes of DNA replication			
A. DNA polymerase	Rat liver	Examined "active" and "inactive" fractions	Chan et al. (1977)
	Rat ascites hepatoma	Both cytoplasmic and chromatin-bound	Tsuruo and Ukita (1974)
	Mouse ascites	Associated with nucleosomes	Schlaeger et al. (1978)
	Bovine lymphocytes	Associated with nucleosomes	Schlaeger et al. (1978)
	Sea urchin	About 80% activity in chromatin	Loeb (1970)
B. DNA ligase	Rabbit bone marrow	Some chromatin-bound	Gaviez and Kuzin (1973)
C. Terminal deoxynucleotidyl transferase	Calf thymus	Probably chromatin-bound	Wang (1968)
	Tobacco	From chromatin	Srivastava (1972)
II. Enzymes of DNA repair			
A. AP endodeoxyribonuclease	Rat liver	On both cores and linkers	Bricteau-Gregoire and Verly (1983)
B. Deoxyribonuclease IV	Rat liver	Together with above enzyme, ligase, and polymerase β, forms system to repair apurinic or apyridimic sites	Goffin and Verly (1984)
C. O⁶-ethylguanine transferase	Rat liver	Chromatin-bound	Renard and Verly (1983)
III. Deoxyribonucleases	Rat liver	Alkaline endonuclease; chromatin-bound	O'Connor (1969)
	Rat liver	Ca^{2+}-Mg^{2+} dependent endonuclease, chromatin-associated	Machray and Bonner (1981) Ishida et al. (1976)
	HeLa cells	Four chromatin-associated endonucleases detected	Urbanczyk and Studzinski (1974)
	HeLa S₃ cells	Isolated 22 K endonuclease; probably on chromatin	Fischman et al. (1979)
	Papovavirus (SV-40, polyoma)	Associated with viral minichromosome	McGuire et al. (1976; Waldeck and Saver (1981)

Table 5-2. *Continued*

Enzyme	Organism; tissue	Comment	Reference
IV. Ribonucleases			
A. Nuclear exoribonuclease and endoribonuclease	Rat liver and tumor cells	Exonuclease is 3'-5'	Lazarus and Sporn (1967)
B. Nucleolar RNase I	Ehrlich ascites cells	Cyclizing endonuclease	Eichler and Tatar (1980)
C. Nucleolar RNase II	Ehrlich ascites cells	Endonuclease	Eichler and Eales (1982)
D. Nucleolar RNase III	Ehrlich ascites cells	5' 3' exonuclease	Lasater and Eichler (1984)
		Note: *none* of the above have been proven to be chromatin-bound	
V. RNA polymerases			
A. RNA polymerase I	Coconut milk	Obtained from nuclei or crude chromatin	Ganguly et al. (1973)
B. RNA polymerase II	Wheat seedlings	Only activity detected; not certain II	Sasaki and Tazawa (1973)
	Drosophila salivary glands	Detected by immunofluorescence in polytene chromosomes	Jamrich et al. (1977)
	Rat liver	Both free and chromatin-bound	Tata and Baker (1974)
			Yukioka et al. (1979)
			Hatayama et al. (1982)
			Okai (1984)
	Hen oviduct	At least 10% is chromatin-bound	Cox (1973)
	Maturing chicken erythrocytes	Nuclear, but not proved chromatin-bound	van der Westhuyzen et al. (1973)
VI. Enzymes of ADP-ribose metabolism			
A. Poly (ADP-ribose) synthetase	Rat liver	Chromatin-bound	Ueda et al. (1975a)
			Okayama et al. (1977)
			Mullins et al. (1977)

Table 5-2. *Continued*

Enzyme	Organism; tissue	Comment	Reference
	Bovine thymus HeLa cells	Bound to oligonucleosomes	Yoshihara et al. (1978) Mullins et al. (1977) Giri et al. (1978a, 1978b)
B. Poly (ADP-ribose) glycohydrolase	Wheat embryo Rat liver	Chromatin-bound Chromatin-bound	Sasaki and Sugita (1982) Miyakawa et al. (1972)
	Calf thymus	Not proved to be chromatin-bound in this tissue	Miwa et al. (1974)
C. ADP-ribosyl-histone splitting enzyme	Rat liver	20–30% in nuclei; not proved to be chromatin-bound	Okayama et al. (1978)
D. NMN adenyltransferase	HeLa cells	Nuclear, but localization uncertain	Uhr and Smulson (1982)
VII. Other enzymes of histone modification			
A. Histone acetyltransferase			
Acetyltransferase A	Calf thymus Bovine lymphocytes Artemia	Acetylates all 5 histones Assoc. with large chromatin Resembles "A" of mammals	Belikoff et al. (1980) Böhm et al. (1980b) Estepa and Pestana (1983)
Acetyltransferase DB	Bovine lymphocytes	Found on mono- and oligonucleosomes; binds to DNA	Böhm et al. (1980b)
B. Histone deacetylase	SV-40 Calf thymus HeLa cells	Bound to minichromosomes Probably chromatin-bound Not definitely chromatin-bound; might be matrix protein	Otto et al. (1980) Kaneta and Fujimoto (1974) Hay and Candido (1983a)
C. Histone methylase	Calf thymus and bovine lymphocytes	At least portion chromatin-bound; specific for H4	Sarnow et al. (1981)
D. Kinases and phosphatases			
cAMP independent histone kinase	Mouse, several tissues SV-40 transformed rat cells	One of three kinases chromatin-bound One major, one minor kinase in NHCPs	Schlepper and Knippers (1975) Segawa and Oda (1978)

Table 5-2. *Continued*

Enzyme	Organism; tissue	Comment	Reference
	Rat liver	At least 88% in nuclei, not shown to be chromatin-bound	Thornburg et al. (1978)
	Soybean hypocotyl	At least part is chromatin-bound	Murray et al. (1978)
			Lin and Key (1980)
	Wheat seeds	Obtained from chromatin	Sasaki and Sugita (1982a)
	Hepatoma cells	Kinase associated with chromatin	Kitzis et al. (1980)
	HeLa cells	Kinase on metaphase chromatin	Paulson and Taylor (1982)
	Physarum	3 nuclear kinases	Chambers et al. (1983)
	Bovine thymus	Chromatin-bound	Shoemaker and Chalkley (1978, 1980)
H3 specific kinase	Walker 256 cells	At least two nuclear activities	Smith et al. (1974)
Histidine-lysine kinase	Several mammalian cells	2 activities from chromatin	Stahl and Knippers (1980)
Acidic protein kinase	CHO cells	Associated with metaphase chromosomes	D'Anna et al. (1978)
Histone phosphatase			
VIII. Proteases			
	Rat liver	Chromatin-bound; prefers histone substrate	Garrels et al. (1972)
	Rat liver	3 major proteins in nuclei and chromatin react with diisopropylfluorophosphate	Carter et al. (1976)
	Rat liver	Histone protease; possibly same as that of Garrels et al.	Ramponi et al. (1978)
	Rat liver	Neutral protease, chromatin-bound	Chong et al. (1974)
	Rat peritoneal macrophages	Neutral protease, 89% in chromatin	Suzuki and Murachi (1978)

Table 5-2. *Continued*

Enzyme	Organism; tissue	Comment	Reference
	Rat, several tissues	Alkaline protease; only found in nuclei	Hagiwara et al. (1980)
	Yeast	Histones protease associated with nuclear homogenate and chromatin	Ruggieri and Magni (1982)
	Calf thymus	Specifically cleaves H2A at res 119; present on mononucleosomes	Watson and Moudrianakis (1982)
IX. Miscellaneous enzymes			
A. Cyclic-AMP phosphodiesterase	Rat liver	Chromatin-bound	Tanigawa et al. (1981)
B. Nucleotide-phosphohydrolyzing enzyme	Pea cotyledon	Catalyzes UTP hydrolysis	Hirasawa et al. (1978)
C. Topoisomerase I	*Drosophila*	Detected in polytene chromosomes by immuno techniques	Howard et al. (1980)
		Associated with active chromatin	Fleischmann et al. (1984) Gilmour et al. (1986)
D. Topoisomerase II	*Sacchromyces pombe* HeLa cells	In isolated nuclei Associated with scaffold	Uemura and Yanagida (1984) Earnshaw and Heck (1985); Earnshaw et al. (1985) Gasser et al. (1986)
	Drosophila	Associated with matrix	Berrios et al. (1985)

enzyme preparations, but tests for cytoplasmic contamination have not been presented. Clearly, in neither situation are we warranted in assuming that the enzyme in question is actually a nonhistone chromosomal protein. An example will be useful here. At least three different acetyltransferase activities have been identified in mammalian cells. These are designated A, B, and DB (see Böhm et al., 1980b). All three are active in histone acetylation, with somewhat different specificities and preferences. However, only acetyltransferase DB can be, at this point, *clearly* identified as chromatin-bound. Not only is it found in nuclei, but it can also be obtained on purified oligonucleosomes, and it binds to DNA (Böhm et al., 1980b; Otto et al., 1980). Neither of the other enzymes meet these criteria. Acetyltransferase B appears to be primarily a cytoplasmic enzyme (Belikoff et al., 1980). Acetyltransferase A is a nuclear enzyme, but it is not found to be associated with small oligonucleosomes. Rather, it can be sedimented with high molecular weight fractions from lysed nuclei (Böhm et al., 1980b). It is possible that this enzyme associates only with higher-order chromatin structures. It does not appear, by the usual definitions, to be a matrix protein, for extensive nuclease digestion promotes its release.

The marked differences in localization of these three enzymes, all of which catalyze the acetylation of histones, underlines the need for caution. Accordingly, I have included in Table 5-2 *only* those enzymes isolated from nuclei or chromatin. Where direct evidence for chromatin-association has been presented, it is noted. I have excluded the many examples of putative nuclear enzymes that have been obtained only from whole cell homogenates.

Even with such cautious selection, the list is impressive. It is surely far from complete, and many more activities will undoubtedly be added. But even as it stands, it includes the essential enzymes for all of the major dynamic processes in chromatin. The polymerases for both DNA replication and transcription are recognized as nonhistone chromosomal proteins. As we shall see in later chapters, these processes can be carried out, at least in part, in isolated chromatin. The entire set of enzymes necessary for at least one kind of DNA repair are present (see Goffin and Verly, 1984). Enzymes for all of the major histone modifications—acetylation, methylation, phosphorylation, and ADP-ribosylation—are included.

In fact, a major problem arises because so *many* enzymes appear to be associated with chromatin. How are their diverse (and sometimes contradictory) functions regulated? To take one example: Both acetyltransferases and deacetylases are known to be chromatin-bound. Yet acetylation of histones appears to be a carefully controlled process in living cells. Histone molecules of a given kind appear to be differently acetylated in different chromatin regions, and this modification varies during the cell cycle (see Chapter 4). Alternatively, consider the enzymes of ADP-ribosylation. Not only the synthetase, but also a number of enzymes involved in various modes of degradation of poly (ADP-ribose) have been shown to be present in nuclei. How are these activities controlled so as to produce the correct amount of ADP-ribosylation at the right place and the right time?

Most perplexing of all, perhaps, are the deoxyribonucleases and histone proteases listed in Table 5-2. These enzymes have the potential capacity to destroy the basic chromatin structure. Indeed, it is a common experience of all who have worked with chromatin that DNA digestion and proteolysis of histone and nonhistone proteins can be annoying problems during chromatin isolation. What roles do these enzymes play in vivo, and how are their activities kept in check?

One gains the impression that there exists a vast, shadowy network of interrelated activations and inhibitions of chromatin enzymes. We have, as yet, only the faintest glimpse as to how this may function. The indications of interrelations between different kinds of histone modification mentioned in Chapter 4 may be a part of this control. But to unravel the whole pattern may prove exceedingly difficult. In the first place, we may have to understand much better than we do now the interactions between chromatin, its associated proteins, and the nuclear matrix. Indeed, as suggested above, that distinction may itself be meaningless, at least on a functional basis. Moreover, it is highly unlikely that the whole pattern of control of chromatin-associated enzymes can ever be understood through studies of isolated nuclei, for the nucleus and cytoplasm are in constant communication.

Transcription-Regulating Proteins

For some years, it has been recognized that significant changes in chromatin organization occur within the 5'-flanking (promoter) regions of genes that are activated for transcription. The appearance of nuclease hypersensitive sites (or, more precisely, regions) is a frequent aspect of these changes. (See Chapter 8 for a detailed description.) More recently, it has been found that a number of such regions bind specific nonhistone proteins. These proteins are necessary co-factors for RNA polymerase II, which seems unable to recognize promoter sites in their absence. The function of these proteins has been reviewed by Dynan and Tijan (1985).

The first to be discovered is the factor Sp1, which interacts with multiple GGGCGG sites in the nucleosome-free promoter region of SV-40, to facilitate the expression of both early and late genes (Dynan and Tijan, 1983a, 1983b; see also Sassone-Corzi et al., 1985; Gidoni et al., 1985; and Wildeman et al., 1986). Furthermore, it has been discovered that Sp1 binds to similar sites adjacent to a number of other genes, and GGGCGG tracts have been detected in still other promoters. These are probably also Sp1 sites (see Dynan and Tijan, 1985, for a summary of these results).

Nor is Sp1 the only known protein of this type. A heat-shock transcription factor has been isolated from *Drosophila* that appears to bind to upstream sequences of heat-shock genes (Parker and Topol, 1984a, 1984b; reviewed by Pelham, 1985). This is a nonhistone protein of approximately 70 Kd. Recently, Jones et al. (1987) have reported the purification of a family of polypeptides (NF-1) of about 52 to 66 Kd that bind to a sequence (CCAAT) common to many promoters, including the globin and heat-shock promoters.

A particularly interesting system in which regulatory proteins have begun to be characterized is the β-globin gene cluster in chicken erythroid cells. Emerson and Felsenfeld (1984) found a factor in nuclear extracts of these cells that would bind to the 5'-flanking region of the β^A gene and generate a particular hypersensitive site in vitro. Subsequent studies (Emerson et al., 1985; Plumb et al., 1985, 1986) have revealed multiple binding sites and multiple protein factors for both the β^A gene and the β^H ("hatching") gene.

The field of polymerase II transcription factors is in very rapid development at the present time, and to attempt to thoroughly review it would be premature. It appears that there may be many such factors, and the fact that molecular weights ranging from 20 Kd to 200 Kd have been reported (see Plumb et al., 1986) suggests that they may be a diverse group of nonhistone chromosomal proteins.

Hormone Receptors

A number of hormones appear to elicit control of transcription by binding to certain nonhistone chromosomal proteins. These receptor proteins have been objects of considerable study, for several reasons. First, like the enzymes, they possess a substrate (the hormone) that can be used to identify them and, in some cases, to aid in their purification. Second, the fact that hormone administration often elicits a specific transcriptional response holds forth the possibility that these may provide model systems for the more general problem of transcriptional control. Finally, elucidation of their role is an important part of the whole field of hormone-related research.

Progress in this field has been rapid in recent years and appears to be at the edge of major breakthroughs (see Yamamoto, 1985; Ringold, 1985; and Gorski et al., 1986, for contemporary reviews). It now appears that the receptor proteins for thyroid hormones, estrogens, progestins, and glucocorticoids may constitute a related family of proteins with similar functional mechanisms.

In the past, a major problem in working with these proteins has been their low abundance. Consider the thyroid hormone receptor, which will be discussed in more detail below. Jump and Oppenheimer (1983) point out that there are only about 4.5×10^3 receptors in each nucleus of rat liver, a responsive tissue. This corresponds to only about 4×10^{-16} g per nucleus. Put another way, the receptor represents only about 0.002% of the total nuclear proteins. It is difficult to work with such a material. The consequence is that few of the receptor proteins have been purified to anything approaching homogeneity. On the other hand, recent success in cloning the cDNAs for a number of these proteins has provided detailed information concerning their primary structures. In some cases, site-directed mutagenesis has been used to discern the function of different domains within receptors (see below).

Major advances have centered on a relatively small number of the numerous receptor proteins that must exist. I shall concentrate the discussion on these.

The Thyroid Hormone Receptor

Thyroxine, produced in the thyroid gland, is mono de-iodinated in peripheral tissue to yield triiodothyronine (T_3). Receptors for this hormone have been detected in substantial quantities in a number of tissues and are most abundant in the anterior pituitary and liver. Current knowledge of T_3 receptors is very nicely summarized in reviews by Samuels et al. (1982) and Jump and Oppenheimer (1983).

The free receptor is a nonhistone protein with a mass of about 50,000 daltons, $s_{20,w} = 3.5$ S. It can be released from chromatin by salt extraction or by digestion with either DNase I or micrococcal nuclease. The latter method releases not the protein itself, but particles with $s_{20,w}$ of 5.5 S or 12.5 S (Jump et al., 1981; see also Gruol, 1980). The former contains DNA. The latter is probably a nucleosome to which the receptor is attached. The present view is that the receptor attaches not to the core particle, but to the linker region adjacent to it. Of course, neither the receptor protein nor these particles have been isolated in anything like pure form; rather, they are detected by the presence of radiolabeled T_3. The receptor itself has been purified only up to 500-fold (Latham et al., 1981; Apriletti et al., 1984); a 40,000-fold purification would be required to approach homogeneity (see above).

There now exists good evidence that T_3 administration to cultured cells elicits a marked increase in specific mRNA and growth hormone synthesis (see Jump and Oppenheimer, 1983, for details). As might be expected from its putative function, the T_3 receptor appears to be concentrated in chromatin fractions associated with transcriptional activity (see, for example, Levy-Wilson et al., 1979; Jump and Oppenheimer, 1980). However, chromatin structural differences associated with receptor binding must be either subtle or localized, since experiments comparing euthroid and hypothyroid liver nuclei (Jump et al., 1981) or liver and cerebral cortex nuclei (Silva, 1983) failed to reveal *any* differences in the nuclease sensitivity of receptor binding. Clearly, further work on this interesting system is warranted.

Steroid-Receptor Proteins

Many tissues respond in diverse ways to steroid hormones, and there exists a vast and diffuse literature on the subject. It would be quite out of place to even begin to summarize it here. Rather, I shall touch upon only one question: Are there specific nonhistone chromosomal proteins involved in this response?

It has long been known that there exist, in cells which are targets for steroid hormones, proteins that act as steroid receptors. Early evidence suggested that these proteins were abundant in the cytoplasm, and a model was proposed in which the receptor molecule, when occupied by hormone, migrated to the nucleus (Jensen et al., 1968, 1982; Gorski et al., 1968). In recent years, however, evidence has accumulated to indicate that the estrogen receptor molecule, even in its unoccupied state, is a weakly bound nonhistone chromosomal protein, mainly concentrated in the nucleus (see Gorski et al., 1986, for a comprehensive review of the evidence for and against this new model). It should be noted, however, that none of the data rule out the existence of a *low* concentration of unoccupied receptor in the cytoplasm. What appears to happen, as judged by phase-partitioning experiments, is that binding of the hormone to the receptor causes a conformational change in the latter, which markedly decreases its hydrophobicity (Hansen and Gorski, 1985) and might well increase its preference for a nuclear environment. In addition, it should be noted that most receptors, even when unoccupied, exhibit weak *nonspecific* DNA binding.

To what does the occupied receptor bind in the nucleus? Virtually every component of the nucleus has been suggested by one group or another as the site of binding (see Spelsberg et al., 1983, for an excellent review). As will be detailed below, there is strong evidence for site-specific DNA binding. However, in the past few years evidence has developed to indicate the participation of additional proteins in the binding of steroid-receptor complexes. Hiremath et al. (1980) reported isolation of an "acceptor" for the dihydroxytestosterone receptor in rat prostate. The protein, which was obtained by 0.35 M salt extraction of chromatin, has a mass of 14,000 daltons. Spelsberg et al. (1983, 1984) have shown that a particular fraction of nonhistone chromosomal proteins is capable of enhancing the binding of the progesterone-receptor complex in chicken oviduct. In fact, the reconstitution of sites on specific DNA sequences is claimed. The procedure used by Spelsberg and co-workers to isolate their "acceptor" activity is quite different from that employed by Hiremath et al. In fact, the "acceptor" was obtained from proteins residual on DNA after extraction of chromatin with 3.0 M NaCl and 4.0 M guanidine hydrochloride—thus, it shares certain properties in common with "scaffold" or "matrix" proteins. There have, in addition, been numerous reports that the nuclear matrix of target cells is rich in steroid *receptors* (reviewed by Barrack and Coffey, 1982). While it is clear that labeled steroids do bind to matrix preparations, and perhaps to receptor proteins contained therein, there is as yet little evidence to show that these represent the in vivo sites. It is quite possible that the receptor or acceptor molecules have been simply displaced onto the matrix during the salt extraction and/or nuclease digestion steps in matrix preparation. (The evidence of Hyde et al., 1979, for protein translocation during nuclease digestion of nuclei must be kept in mind.) On the other hand, the possibility of a matrix-associated site has interesting implications. As will be shown below, there is now evidence that gene-flanking regions may be associated with nuclear "scaffold" proteins, which in turn are probably at least a part of the nuclear matrix. Such a site would be an ideal location for proteins involved in gene control.

Nor should the identification of "acceptor" proteins be taken as *prima facie* evidence that hormone receptors bind to other nonhistone proteins *rather* than to DNA. Both could occur, or the role of the additional protein factors could be fundamentally an accessory one. Recent studies of the glucocorticoid response of the promoter in the long terminal repeat of mouse mammary tumor virus (MMTV) support this idea. After MMTV DNA is synthesized following infection, it becomes integrated into the host cell genome. The expression of MMTV genes then comes under glucocorticoid control. Payvar et al. (1983) found that the purified receptor reacts selectively with MMTV DNA at a number of sites. In subsequent studies, Zaret and Yamamoto (1984) have observed changes in chromatin structure accompanying this binding. More recently, Cordingly et al. (1987) have found that glucocorticoid stimulation leads to the binding of two other transcriptional factors to promoter sites. One of these has been identified as NF-1 (see above). In other words, receptor binding appears, in this case, to *recruit* the binding of other transcription-regulatory proteins to the promoter.

In the past few years, cDNA cloning of receptor proteins has allowed their sequence determination, with spectacular results. At this writing, sequences have been determined for the chicken and human estrogen receptors, the human glucocorticoid receptor, and the chicken progesterone receptor (see Krust et al., 1986, for appropriate references). Site-directed mutagenesis of the human estrogen receptor has allowed the identification of a steroid-binding domain and a putative DNA-binding domain (Kumar et al., 1986).

Studies utilizing "footprinting" methods have allowed the identification of some receptor-specific binding sequences in DNA. While these experiments strongly support the presumption that receptors do indeed bind to the DNA itself in vivo, the "concensus sequences" derived to date for different hormones show little in common (Yamamoto, 1985). In vitro determinations of the selectivity of such sites (specific site affinity/nonspecific affinity) range from 5 to about 10^3. Such values seem very low to account for the biological selectivity in vivo. Perhaps the accessory transcription factors mentioned above may enhance site selectivity in a complete system.

Despite the remarkable advances that have been made in recent years, a number of perplexing problems remain with respect to receptor–chromatin interactions. These include:

1. A given hormone may elicit either positive or negative responses in transcription. For example, estrogen stimulates lactotroph prolactin gene transcription in the pituitary, but *decreases* gonadotroph transcription of follicle stimulating hormone and luteinizing hormone genes.
2. A given hormone may regulate different genes in different tissues. For example, estrogen stimulates yolk protein synthesis in avian *liver,* but egg white protein production in the oviduct.

Such observations are not easily explained by any *simple* model of receptor–DNA interaction. Rather, a multifactor interaction with promoter elements seems now to be essential.

The High Mobility Group (HMG) Proteins
Discovery, Nature, and Distribution
In the early 1960s, E.W. Johns and his collaborators were working out their pioneering methods for purification of the histones (see Chapters 2 and 4). In isolating H1 by perchloric acid extraction of chromatin, it was observed that this fraction contained significantly larger amounts of acidic amino acids than the H1 fraction obtained from whole histones by Crampton et al. (1957). This led Johns to suspect that PCA extracts were contaminated by some unknown chromosomal proteins (see Johns, 1964). Johns showed that the contaminating proteins could be removed and that they exhibited unusual amino acid composition, with high levels of both basic and acidic amino acid residues. However, the major interest at the time centered on histones, and these nonhistone chromosomal proteins were put aside for many years. Serious investigation began only in the early 1970s, when G.H. Goodwin, working in Johns' laboratory, began a long and comprehensive study. By this time, Johns and Forrester (1969) had observed that extraction of chromatin with 0.35 *M* NaCl yielded some of the same nonhistone proteins that had contaminated the early H1 preparations. Thus, by this date were established the two methods that have since been used for the preparation of this class of nonhistone protein: extraction with 5% perchloric acid followed by fractionation from the H1-class, or extraction with 0.35 *M* NaCl. The latter method yields a large number of proteins, with a wide range of mobilities on SDS gels; in addition, there is contamination with RNA. Goodwin et al. (1973) found that addition of 2% trichloroacetic acid to the 0.35 *M* NaCl extract precipitated both the RNA and most of the higher molecular weight ("low mobility group") proteins. What remained in the TCA supernatant were the "high mobility group" (HMG) nonhistone proteins.

This history, which is recounted in more detail in two excellent reviews (Goodwin et al., 1978a; Johns, 1982b), illustrates most clearly the vagaries of nonhistone protein research. A small group of proteins has been selected almost by chance, by the circumstance of isolation procedures, from the host of chromatin-associated proteins. To be sure, the HMGs have turned out to be an exceptionally interesting class, which fact has stimulated much of the subsequent research. Nonetheless, there must exist other constellations of equally interesting nonhistone proteins, which, if they could be isolated in so simple a manner, would command equal attention.

The early isolation of HMG proteins yielded, by gel electrophoresis, the initial identification of 16 and later 20 bands (Goodwin et al., 1973, 1975, 1980). From the first, it was realized that not every one of these bands necessarily represented a separate protein of the HMG class; some of them could well be degradation products of others, and some might represent quite unrelated proteins. Both suspicions have turned out to be correct. Proteins 3 and 8, for example, which figure predominantly in early studies, turn out to be degradation products of HMG 1 and histone H1, respectively (see Goodwin et al., 1978b). Band 20 has been identified as ubiquitin (Walker et al., 1978).

When HMG proteins are isolated under conditions where proteolysis is minimized, only four major components remain: 1, 2, 14, and 17 (see Fig. 5-2). These proteins fall into two distinct groups: HMGs 14 and 17 are similar in sequence (see below) and have molecular weights of about

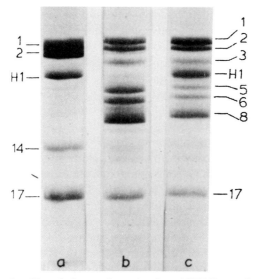

Fig. 5-2. Polyacrylamide gel electrophoresis of total HMG proteins. (a) Perchloric acid-extracted HMG proteins from calf thymus tissue frozen in liquid nitrogen. (b) HMG proteins extracted from calf thymus chromatin by 0.35 M NaCl. (c) Perchloric acid-extracted HMG proteins from calf thymus *chromatin*. (From Biochim. Biophys. Acta *519*, p. 233, Goodwin et al. Copyright 1978 Elsevier Science Publishers.)

10,000 g/mol. HMGs 1 and 2 form another homologous pair, but they are considerably larger (about 29,000 daltons). All four are soluble in 5% perchloric acid or 0.35 M NaCl. Each contains a large fraction of acidic and basic residues and is low in hydrophobic residues. These proteins have been very thoroughly studied, and an entire book is devoted to their distribution, isolation, and properties (Johns, 1982a).

Since these early studies, two more components of the 0.35 M salt extract have been investigated, corresponding to bands 18 and 19B (Goodwin et al., 1980). Band 18 shows much similarity to histone H5 and is probably not closely related to the HMG proteins. Band 19B resembles HMG 17 and may be a genuine, although minor, HMG protein.

Quite recently, what appears to be a third class of HMG proteins has been discovered. These proteins bind strongly to α-satellite DNAs, exhibiting a specificity in binding not found in other HMGs (Levinger and Varshavsky, 1982; Strauss and Varshavsky, 1984; Solomon et al., 1986).

Most of the early studies of HMG proteins were carried out with calf thymus as the protein source. There have been serious difficulties in defining what is and what is not an HMG protein in other organisms. The question has been considered in some detail by Johns (1982b), who has presented a list of criteria based on solubility and amino acid composition. However, even these criteria fail for some proteins (such as the HMG-T of trout; see below) that nevertheless clearly exhibit strong sequence homology with the other HMGs. A full understanding will probably only come when we can understand the *evolutionary* relatedness of these proteins. It is becoming increasingly clear that the older kinds of protein classification, based on solubility, composition, or function, are faulty. Closely related proteins may behave quite differently, and proteins with similar solubility, composition, or function may in fact be wholly unrelated. Eventually, we may expect that a genealogy of nonhistone proteins will be established which will render such classifications as ''HMG-proteins'' obsolete.

Such are the problems encountered when one attempts to assess the distribution of HMG proteins among the eukaryotic phyla. Table 5-3 summarizes reports of such proteins, ranging from fungi and protists to mammals. As is made clear by the table, the major HMG proteins (1, 2, 14, 17) seem to be present in mammals, birds, and probably amphibians, albeit with definite species and phylal differences. As one progresses to the lower vertebrates and invertebrates, however, the trail becomes muddled. Certainly there *are* HMG-like protein in fishes; the HMG-T of trout, for example, exhibits considerable homology to HMGs 1 and 2. With insects, plants, fungi, and protists, the relationships become wholly unclear. Proteins can be extracted from the chromatin or nuclei of these organisms that·satisfy some of the criteria for HMG proteins, in terms of solubility or amino acid composition, but it remains very difficult, in most cases, to make a 1:1 correspondence between such proteins and the major mam-

Table 5-3. HMGs and HMG-Like Proteins

Organism, tissue	Proteins	Comments	Reference
I. Mammals			
Calf thymus	HMG 1,2,14,17	These were the first four shown to be individual proteins, and not degradation products	Goodwin et al. (1973, 1975)
Fetal calf thymus	HMG 1,2,14,17,18,19A,19B	Latter three are minor proteins	Goodwin et al. (1980)
Pig thymus	HMG 1,2,14,17		Levy-Wilson and Dixon (1978)
Rabbit thymus	HMG 1,2,14,17		Goodwin et al. (1978a)
Mouse brain	HMG 1,2,14,17		Goodwin et al. (1977a)
	HMG 1,2,14,17		Levy-Wilson and Dixon (1978)
Mouse myeloma	HMG 1,2	Associated with nucleosomes	Jackson and Rill (1981)
Rat liver	HMG 1,2,17		Craddock and Henderson (1980)
	E	Early survey of nonhistone proteins; component E is probably HMG1 + HMG2	Elgin and Bonner (1972)
Rat, various tissues	HMG 1,2	Study directed toward tissue and cellular distribution of HMG 1/2	Seyedin and Kistler (1979b)
II. Cultured mammalian cells			
HeLa	HMG 1,2	Study mainly concerned with synthesis of proteins in cell cycle	Goldknopf et al. (1980a)
African green monkey	HMG 14,17,α	α is associated with α-satellite DNA	Strauss and Varshavsky (1984)
Rat hepatoma	HMG 1,2	Termed NH-1 and NH-2 originally	Bidney and Reeck (1978)
Mouse ascites, L1210	HMG 1,2,14,17	Both cell types contain all, but in differing relative amounts	Saffer and Glazer (1980)
Mouse L	HMG A→H	Of these, HMGA = HMG 1, HMGE = HMG 17, HMG-G = HMG 14	Bakayev et al. (1979)
Mouse Friend	HMG 14,17	Found in "active chromatin" fractions	Reeves and Candido (1980)
CHO	HMG 1,2,14,17		Schröter el al. (1980)

Table 5-3. *Continued*

Organism, tissue	Proteins	Comments	Reference
III. Birds			
Chicken erythrocyte	HMG 1,2,14,17	Proteins almost same, but not identical to mammalian	Rabbani et al. (1978)
	HMG 1,2,14a,14b,17 and E	14a and 14b are variants of 14; HMG-E resembles HMG 1/2 and appears to be erythrocyte specific	Isackson et al. (1980a)
Chicken thymus	HMG 2a and 2b	HMG 2b migrates with calf HMG 2; HMG 2a is faster	Mathew et al. (1979)
Chicken, various tissues	HMG 2A and 2B	HMG 2A probably = HMG 2b, and HMG 2B probably = HMG 2a	Gordon et al. (1980)
Chicken, oviduct and erythrocyte	HMG-Y	Is similar in composition to 14 and 17, but lower molecular weight	Goodwin et al. (1981)
Chicken oviduct	HMG 1,2,14,17, + 95 K protein	Tissue specific; extracts with 0.35 M NaCl, composition like HMG; not released by nuclease digestion	Teng et al. (1979)
Duck erythrocyte	HMG 1,2,E		Vidali et al. (1977) Sterner et al. (1978)
Rooster testis	HMG 1,2	Similar in composition to erythrocyte HMG 1/2	Chiva and Mezquita (1983)
IV. Amphibians			
Xenopus laevis ovary	HMG 1,2,17,18	Proteins are similar in electrophoretic mobility to mammalian HMGs, but have not been positively identified	Mayes (1982)
V. Fishes			
Trout testis	HMG-T	Resembles HMG 1 and HMG 2	Watson et al. (1977) Peters et al. (1979)

Table 5-3. *Continued*

Organism, tissue	Proteins	Comments	Reference
	HMG-T1 and HMG-T2	Appear to be variants of HMG-T	Brown et al. (1980) Bhuller et al. (1981)
Trout liver	H6	Significant homology with HMG 17	Watson et al. (1979)
	HMG-T, H6, Protein C and D	Proteins C & D exhibit homology with HMGs 14 and 17	Rabbani et al. (1980) Walker (1980)
Winter flounder testis	HMG-T, H6	H6 of flounder similar to, but distinct from that of trout	Kennedy and Davies (1980)
VI. Sea urchin *Strongylocentrotus purpuratus* embryos	P2, P4?	Two potential HMGs; P2 resembles mammalian HMGs in composition; status of P4 as HMG uncertain	Katula (1983)
VII. Insects *Drosophila melanogaster*	HMG 1,2,14,17	Putative HMG 14/17 have similar mobilities to mammalian, those of putative 1/2 somewhat different	Howard et al. (1980)
	D1	Similar in composition to HMG 1/2; appears to be concentrated in satellite chromatin	Rodriguez-Alfagame et al. (1980) Levinger and Varshavsky (1982)
Ceratitis capitata	Cla1, Cla2, Clb	All resemble D1 and mammalian HMG 1/2 in composition	Marquez et al. (1982)
VIII. Plants Wheat germ	HMG a→d	These four proteins have both resemblances to and differences from both HMGs and 1 and 2 and HMGs 14 and 17 of mammals	Spiker et al. (1978)
	HMG's A→D	A may not be HMG; other three have composition like mammalian HMGs; relation to proteins of Spiker et al. ambiguous	Mayes (1982) Mayes and Walker (1984)

Table 5-3. *Continued*

Organism, tissue	Proteins	Comments	Reference
IX. Fungi			
Saccharomyces cerevisiae	HMG's a→i	Mobilities unlike calf HMG's; composition of total mixture resembles calf	Petersen and Sheridan (1978)
	HMGa	Not same as HMGa of Petersen and Sheridan. Does not extract in 5% PCA or 0.35 M NaCl, but in 0.25 M HCl	Spiker et al. (1978)
	S_1, S_3, S_4	Compositions resemble but are distinctly different from mammalian HMGs; mobilities quite different	Weber and Isenberg (1980)
X. Protista			
Tetrahymena pyriformis (macronucleus)	LG-1, LG-2	Composition bears some resemblance (but is significantly different from) mammalian HMGs	Hamana and Iwai (1979)
Tetrahymena thermophila (macronucleus)	A,B,C,D	A,B resemble HMG 1/2, C,D are like HMG 14/17; all are nucleosome-bound.	Levy-Wilson et al. (1983)

malian HMGs. This problem is discussed carefully and at length by Mayes (1982). As he points out, the HMG proteins do not appear to be as highly conserved as the inner histones, which makes seeking their counterparts in distantly related organisms much more difficult.

The HMG proteins are generally considered to be minor but not insignificant components of chromatin. According to Goodwin et al. (1977a) and Isackson et al. (1980a), they are present in an amount of a few percent of the DNA weight. This is to be contrasted with histones, which are present in approximately equal weight to the DNA. As Goodwin et al. have shown, HMGs seem to be associated with nucleosomes in the nucleus, but their data would indicate that only a small fraction of nucleosomes can carry these proteins. A very different view is given by the data of Kuehl et al. (1984), who measured concentrations of HMG proteins in nuclei isolated by a nonaqueous technique. They find in several rat tissues the following numbers of moles of HMGs per nucleosome: HMG 1, 0.28; HMG 2, 0.18; HMG 17, 0.46. Kuehl et al. argue that much of the HMG content of nuclei is lost by conventional isolation techniques.

It does appear that at least some HMG proteins can move between the nucleus and cytoplasm. In an elegant study, Rechsteiner and Kuehl (1979) microinjected labeled HMG 1 into the cytosol of bovine fibroblasts and HeLa cells. They observed that in the former case at least 90% and in the latter case at least 70% of the protein was rapidly transported into the nucleus. That the process is not irreversible, however, was shown by cell fusion experiments; in 12 to 24 hours, radiolabeled HMG 1 in one nucleus had equilibrated between both nuclei. Thus, at least some HMG 1 must be able to escape from nuclei. Isackson et al. (1980b), using an immunochemical technique, could demonstrate the presence of HMG 1 in both the cytosol and nucleus of cultured rat hepatoma cells. However, there may be differences in such distribution for different HMGs, for Bhuller et al. (1981) were able to detect very little trout HMG-T in trout cell cytosol.

There is evidence that levels of HMG proteins in nuclei may vary from tissue to tissue and during the cell cycle. Gordon et al. (1980) have carefully analyzed the distributions of HMGs 1 and 2 in different chicken tissues. Representative data are presented in Table 5-4. Clearly the differences are well beyond experimental error. In cell cycle studies with synchronized HeLa cells, Goldknopf et al. (1980a) found that HMG 1 and 2 were most rapidly synthesized in early G_1 phase. Marked changes in the levels of some HMG proteins have also been observed during spermatogenesis (see Chiva and Mezquita, 1983, for data on roosters; Kennedy and Davies, 1980, for studies in the winter flounder). The significance of any of these observations is presently unknown and must be considered to be questionable if serious losses of HMGs are in fact occurring during isolation (cf. Kuehl et al., 1984).

Table 5-4. Distribution of HMG 1/HMG 2 in
Some Chicken Tissues[a]

	% of:		
Tissue	HMG 1	HMG 2A	HMG 2B
Muscle	53	27	20
Liver	76	4	20
Brain	67	9	24
Thymus	54	39	7
Erythrocytes	42	15	43

[a]Selected data from Gordon et al. (1980). Uncertainties
in values average about ±4%.

Primary Structure and Modifications

Complete amino acid sequences of calf and chicken HMG 17, the complete
sequence of HMG 14 of calf, and a partial sequence of chicken HMG 14
have been determined. The homologous trout protein H6 has also been
completely sequenced. Among the HMG 1/2 class, complete sequences
are available for the calf HMG 1 protein and for the analogous HMG-T
of trout. Sequences of the calf and trout proteins are presented in Figure
5-3. The chicken HMG 14 and 17 show only small differences from their
calf counterparts. Two other trout proteins (C and D; Table 5-3) have
been partially sequenced and show considerable homology to HMGs 14
and 17. In addition, fragments of the sequences of HMG 18 and 19B have
been reported. Details of these incomplete sequences can be found in the
review by Walker (1982).

The primary structures of HMGs 14 and 17 and of H6 are characterized
by a high concentration of basic residues in the N-terminal region and an
excess of acidic amino acids near the C terminus. In the former respect,
they resemble the histones, but no histone contains a comparable acidic
region.

The sequences of the HMG 1/2 class are considerably more interesting.
In the first place, both HMG 1 and 2 from calf contain, in the C-terminal
region, a long stretch comprised entirely of aspartic and glutamic acid (see
Fig. 5-3). At first, it was believed that such a sequence might not be present

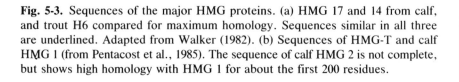

Fig. 5-3. Sequences of the major HMG proteins. (a) HMG 17 and 14 from calf,
and trout H6 compared for maximum homology. Sequences similar in all three
are underlined. Adapted from Walker (1982). (b) Sequences of HMG-T and calf
HMG 1 (from Pentacost et al., 1985). The sequence of calf HMG 2 is not complete,
but shows high homology with HMG 1 for about the first 200 residues.

(a)

```
HMG 17  P K R K – – – A E G D A K G D K A K V K D E P Q R R S A R L S A K P A P P

HMG 14  P K R K V S S A E G A A K – – – – – – – E E P K R R S A R L S A K P A P A

H 6     P K R K – – S A T – – – K G D – – – – – E P A R R S A R L S A R P V P –
```

```
HMG 17  K P E P K P K K A P A – K K G E K V P K G K K G K A D A G K D G N N P A E

HMG 14  K V E T K P K K A A G K D K S S D K K V Q T K G K R G A K G K Q A E V A N

H 6     K P A A K P K K A A A P K K A V K G K K A A E N G D – A K A E – A K V Q A
```

```
HMG 17  N G D A K T N Q – A E K A E G A G D A K

HMG 14  Q – E T K E D L P A E N G E T K N E E S P A S D E A E E K E A K S D

H 6     A G D – – – – – – – – – – – G A G N A K
```

(b)

```
HMG-T   P G K D P N K P K G K T S S Y A F F V A T S R E E H K K K H S G A K V N G S E S

HMG 1   G K G D P K K P R G K M S S Y A F F V Q T S R E E H K K K H P D A S V N F S E F
```

```
HMT-T   S K A C G K S P R D S K A P K R W R T M G A K E K V K F E D M A K G D K V R Y D

HMG 1   S K K C S E – – – – – – – – R W K T M S A K E K G K F E D M A K A D K A R Y E
```

```
HMG-T   K D M K T Y I P P K G E K A A G K R K K D P N A P K R P – S A F F G Y E S A E R

HMG 1   R E M K T Y I P P K G E T – – K K K F K D P N A P K R P P S A F F L F A S E Y R
```

```
HMG-T   A A R I K A D H P G M G I G D I S K Q L G L L W G K Q S S K D K L P H E A K A A

HMG 1   – P K I K G E H P G L S I G D V A K K L G E M W N N T A A D D K Q P Y E K K A A
```

```
HMG-T   K L K E K Y E K C V A A Y K P K G G A A A P A R E R V D K A K G T A G A T A K H

HMG 1   K L K E K Y E K D I A A Y R A K G K P D A A K K G V V K A E K S K K K K – – –
```

```
HMG-T   G P G V P A V G K P K A A P M D D D D D D D D D E E E E E D D D E E E D D D D D

HMG 1   – – – – – – – – – – – – – – E E E E D E E – D E E D E E E E E D E E D E E
```

```
HMG-T   D D D

HMG 1   E D D D D E
```

in HMG-T (Cary et al., 1981), but more recent studies indicate a structure very much like HMG 1/2 (See Fig. 5-3 and Pentacost et al., 1985). An additional insight into HMG 1/2 structures is provided by Reeck et al. (1982), who have detected a "repeat" within the N-terminal portion of each sequence: Residues 1 through 92 show a very strong homology to residues 98 through 176. Indeed, there is other evidence for a well defined domain structure in HMG 1 and 2 (see below). These results suggest that this class of proteins may have evolved from a much smaller protein, comparable in size to the smaller histones or the prokaryotic HU proteins (see Chapter 4). However, there is no convincing evidence for any sequence homology between any HMG and any of these proteins.

Like the histones, HMG proteins are subject to postsynthetic modification. The subject has not been studied nearly so completely, however, as has histone modification. The rather scanty data known to the writer are summarized in Table 5-5. Most of the kinds of modification are similar to those of histones, i.e., acetylation at sites near the N terminal and phosphorylation at serine or threonine residues. However, the only methylation detected to date in HMG proteins appears to occur on arginine rather than lysine residues (Boffa et al., 1979). Details of all but the most recent data are given in the review by Allfrey (1982). It is clear that modification is not confined to the mammalian HMGs. In a detailed study of the *Tetrahymena* proteins, Levy-Wilson et al. (1983) find evidence for both phosphorylation and ADP-ribosylation. We understand even less concerning the role of HMG modification than we do concerning modification of the histones. One peculiar observation has, however, emerged.

Table 5-5. Modification of Vertebrate HMG Proteins

Protein	Acetylation	Methylation	Phosphorylation	ADP-ribosylation	Glycosylation
HMG 1	+,2,11,a[a]	+,c	−,d,e	+,f	+,f,h
HMG 2	+,a	+,c	−,d,e	+,f	+,f,h
HMG-T				+,g	
HMG 6				+,g	
HMG 14	+,2,4,b;−e		+,d,e	+,f	+,f,h
HMG 17	+,2,4,10,b;−,e		+,d,e	+,f	+,f,h

[a]A (+) indicates evidence for the modification, a (−) that the modification should have been observed if present, but was not. Numbers following give sites, where known. The letter gives the reference.

References:
a. Sterner et al. (1979).
b. Sterner et al. (1981).
c. Boffa et al. (1979).
d. Saffer and Glazer (1982).
e. Levy-Wilson (1981b).
f. Reeves et al. (1981).
g. Levy-Wilson (1981a).
h. Reeves and Chang (1983).

Reeves and Chang (1983) find that glycosylated HMGs bind to the nuclear matrix much more strongly than do the unmodified protein. The significance of this intriguing result awaits clarification.

Secondary and Tertiary Structure

HMG 14, HMG 17, and H6 are unusual proteins in that they exhibit little or no secondary or tertiary structure. This might be expected on the basis of their amino acid sequences, which are rich in hydrophilic residues and proline and unusually low in hydrophobic residues. This prediction has been confirmed by experimental studies of these proteins employing a variety of physical techniques. This work has been compactly reviewed by Bradbury (1982) and need not be repeated here. It can be summarized by stating that the largest amount of secondary structure detected in any of these proteins under any conditions studied is about 3% α-helix. As Bradbury points out, this value is so small as to be meaningless. Thus, HMG 14 and 17 and H6 may be regarded as unstructured proteins, at least when free in aqueous solution.

The conformations of the other class major of HMG proteins, 1, 2, and T, present a more complicated picture. As pointed out in Bradbury's review, both Chou–Fasman predictions and earlier physical studies indicated a substantial amount of secondary structure in these proteins. A detailed investigation by Cary et al. (1983, 1984) demonstrates that HMG 1 and HMG 2 can each be described in terms of three conformational domains (see Fig. 5-4). Under physiological conditions, domains A and B appear to be compactly folded, whereas domain C (which contains the long sequence of acidic residues) remains randomly coiled. Study of the individual domains is made possible by the fact that A and B may be isolated by controlled proteolysis (Reeck et al., 1982; Cary et al., 1983). Domain B is found to contain about 30% α-helix, domain A about 50%. Neither shows evidence for significant β-sheet, although frequent β-turns are predicted from the sequence.

The similarity of domains A and B is far from coincidental. Reeck et al. (1982) pointed out that there is extensive sequence homology, and even

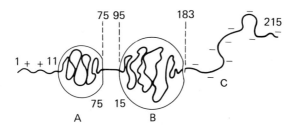

Fig. 5-4. A schematic view of the domain structure of HMG 1/2. After Cary et al., 1983, 1984. Numbers indicate the approximate beginnings and ends of domains.

more striking homology in predicted secondary structure, in these two regions. Domain C appears to be unrelated.

The tertiary structures of these proteins bear a certain weak resemblance to that of H1. Each contains a central globular region, with extended N-terminal and C-terminal tails. But the sequence has no obvious similarity to that of H1, and in other respects the two classes differ greatly. All of the H1 proteins have a lysine-rich, positively charged C-terminal tail; in HMG 1 and HMG 2, this region is highly acidic. The N-terminal tail, although positively charged in the HMGs, is very much shorter than in any H1.

A recent study by Kohlstaedt et al. (1986) suggests that care should be taken in evaluating in vitro studies of HMG 1 and 2. They observe that two of the four -SH groups in these proteins are readily oxidized to an -SS-bridge under conditions often employed for isolation.

The third HMG protein of this class, HMG-T, appears to represent a further variant on the same structural theme. Its overall helical content is much lower than that of HMG 1 and HMG 2 (Cary et al., 1981a). As Cary et al. point out, this difference is paralleled by the fact that HMG-T, like H1, can cause aggregation of DNA upon binding; this does not happen with HMGs 1 and 2. Such differences are surprising, considering the considerable homology between HMG-T and HMG 1 shown in Figure 5-3.

Binding of HMG Proteins to DNA and Histones
There has been, as one might expect, considerable interest in the interactions of the HMG proteins with the constituents of chromatin. The binding of these proteins to chromatin itself and their probable locations thereon will be discussed later, in the chapters dealing with nucleosomes and higher-order structure. For the moment, I shall restrict the discussion to in vitro studies of HMG binding to purified DNA and histones. The work up to 1982 has been thoroughly reviewed in *The HMG Chromosomal Proteins* (Johns, 1982a; see Chapters 5 and 8 in particular). Therefore, I shall only briefly summarize the earlier studies herein.

HMG 14 and 17 and H6
The first studies of interactions of this class of proteins with DNA appear to have been carried out by Javaherian and Amini (1977, 1978), who observed that HMG 17 and 14 would precipitate double-strand DNA. Circular dichroism measurements showed that the DNA conformation was unchanged, although it was stabilized against melting. There followed an extensive series of studies by P.D. Cary and collaborators, from the laboratory of E.M. Bradbury (see Chapters 5 and 8 of Johns, 1982a, for references). These experiments, which utilized proton NMR as the major tool, showed that the basic N-terminal half of each of these proteins con-

tained the primary interaction sites for DNA. The more acidic C-terminal region seems to remain as an unbound random coil. It has been observed that HMG 14 and 17 are preferentially released from chromatin by intercalating agents such as ethidium bromide (Schröter et al., 1985).

In vitro binding of this class of HMG proteins with free histones appears to have been little studied. They most certainly do interact with nucleosomes, as will be discussed in Chapter 6. HMG 14, but not HMG 17, is reported to form a 1:1 complex with H1 (Espel et al., 1983).

HMG 1, HMG 2, and HMG-T
These proteins interact with DNA in a more complex and interesting fashion than do the HMG 14/17 class. In the first place, they exhibit a preference for single-strand DNA and act as helix-destabilizing proteins (Javaherian, 1977; Javaherian et al., 1978, 1979; Isackson et al., 1979; Yoshida and Shimura, 1984; Makiguchi et al., 1984). The binding of HMG 1 and HMG 2 to supercoiled DNA leads to a definite unwinding: 22° for HMG 1, 26° for HMG 2 (see Javaherian et al., 1979). Curiously, in the absence of salt, HMG 1 and 2 *stabilize* double-helical DNA (Yu et al., 1977; Javaherian et al., 1979; see also Butler et al., 1985).

There is definitive evidence for in vitro interaction of HMG 1 with the inner histones. Bernúes et al. (1983) report crosslinking between HMG 1 and H2A·H2B dimers, and with the (H3·H4)$_2$ tetramer. HMGs 1 and 2 also exhibit a most interesting interaction with the lysine-rich histones. First noted by Shooter et al. (1974), the interactions were studied in detail by Smerdon and Isenberg (1976c). Figure 5-5 depicts the results of sedi-

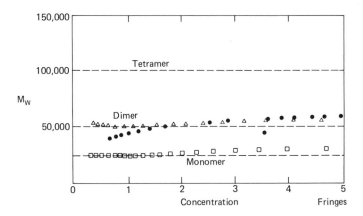

Fig. 5-5. Sedimentation equilibrium studies of the interaction of HMG 1 with subfractions of calf H1. Symbols: (●) fraction lb; (△) fraction 2, (□) fraction 3b. All in 0.02 M phosphate, pH 7.6. A concentration of 4 fringes correspond to approximately 1 mg/ml. (Reprinted with permission from Biochemistry *15*, p. 4242, Smerdon and Isenberg. Copyright 1976 American Chemical Society.)

mentation equilibrium studies of mixtures of HMG 1 with three different variants of calf thymus H1. Clearly, there are strong differences in the association constants. HMG 2 was found to associate only with variants 3a and 3b, and then only weakly. It seems probable that these interactions involve the acidic C-terminal domain of the HMGs (see Carballo et al., 1983), but the reasons for the strong selectivity are not at all clear, considering the high degree of homology between HMG 1 and 2.

The Satellite-Binding Proteins
The α-protein from African green monkey cells and the D1 protein from *Drosophila* exhibit strong binding preference for double-strand, A–T-rich satellite DNA (Levinger and Varshavsky, 1982; Strauss and Varshavsky, 1984). The α-protein binds at specific sites in the African green monkey α-satellite 172 bp repeat. In the major phasing frame, nucleosomal DNA folding will bring these sites into mutual proximity. Recently, Solomon et al. (1986) have shown that the α-protein will bind to any stretch of six A–T base pairs. This site selectivity is unusual among HMG proteins, but solubility, amino acid composition, and primary structure seem typical of HMGs (Solomon et al., 1986).

Summary
The HMGs are a well-studied, interesting group of nonhistone chromosomal proteins. As will be shown in later chapters, the localization in chromatin structure of the two major groups (HMG 14, 17, H6 vs. HMG 1, 2, T) is quite different. Although the functional role of neither group is clearly elucidated as yet, many suggestive experiments indicate a fundamental difference in this respect as well. There is no obvious sequence similarity between these groups. Why, then, should they be even considered together? The answer seems to be a rather trivial one: All are extracted from chromatin together by the same techniques, and all are relatively small nonhistone proteins that run rapidly on SDS gels. Some day, when a rational classification of nonhistone chromosomal proteins finally emerges, the presently recognized subclasses of HMG will probably be placed in quite different categories.

Chromosomal Structural Proteins

There appear to be at least two classes of "structural" nonhistone proteins associated with chromatin. The first of these is represented by those proteins that remain firmly attached to eukaryotic DNA even after extraction with very high salt, polyelectrolytes, or strong detergents. These are sometimes referred to as chromosomal "scaffold" proteins, or "residual" proteins. A second and more problematic class includes contractile and structural proteins similar to those found elsewhere in cells and tissues; i.e., actin, myosin, and tubulin. The role of these latter as nonhistone proteins or even their *presence* in nuclei of living cells has been a matter of considerable controversy.

Scaffold Proteins

A number of workers have used high salt concentration (2 *M* or greater), sometimes augmented by 5 *M* urea, to remove the majority of proteins from either interphase or metaphase chromatin (see Pedersen and Bhorjee, 1975; Benyajati and Worcel, 1976; Adolph et al., 1977; Laemmli et al., 1977; Gates and Bekhor, 1979a, 1979b; Lebkowski and Laemmli, 1982a, for examples). Typically, such treatments leave a few percent of the total protein, or about 10% of the nonhistone protein, remaining firmly attached to the DNA. Furthermore, if such protein removal is carried out under conditions where damage to the DNA is minimal, the residual proteins are found to constitute a "scaffold" of the metaphase chromosome from which enormous loops of DNA are extended (Paulson and Laemmli, 1977; see Fig. 5-6). Similar structures can be obtained from metaphase chromosomes by acid extraction of the histones (Jeppesen et al., 1978).

When *interphase nuclei* are subjected to this kind of treatment, "eukaryotic nucleoids" with DNA highly extended from a nuclear "skeleton" are observed (see Chapter 7, particularly Fig. 7-20). If the DNA is then digested, a "nuclear matrix" structure is left.

The structures obtained upon high salt extraction of either chromosomes or nuclei can be further relaxed by proteolysis (Adolph et al., 1977; Nakane et al., 1978) or by reduction with β-mercaptoethanol (Nakane et al., 1978; Lebkowski and Laemmli, 1982a, 1982b). The latter reaction appears to actually *release* proteins. Lebkowski and Laemmli (1982b) have followed the process in some detail and find that after extraction of HeLa nuclei with 2 *M* NaCl or dextran sulfate/heparin, about 10 to 15% of the nonhistone proteins remain attached to the DNA. Reduction with mercaptoethanol reduces this to about 3 to 5%, accompanied by a further relaxation of the chromatin structure. The principal proteins remaining at this point depend upon the starting material but have been reported to include lamins and topoisomerase II (Gasser et al., 1986). Further details are given in Chapter 7; see particularly Table 7-6.

If the unfolded chromosomes are digested with nucleases, DNA fragments are found to remain associated with the scaffold proteins. The first attempts to determine the sequence-nature of these fragments were inconclusive (see Mirkovitch et al., 1984, for references and discussion). Mirkovitch et al. have found that this confusion probably results from sliding of the DNA relative to the scaffold at high salt and that this artifact can be eliminated by the use of lithium diiodosalicylate as an extractant. Under these conditions, they find attachment of the scaffold to regions flanking sets of coordinately expressed genes. The implications of this remarkable observation for chromatin higher-order structure and transcription will be discussed in later chapters. For our present purposes, it

Fig. 5-6. Electron micrograph of a histone-depleted metaphase chromosome from a HeLa cell. Proteins have been extracted with dextran sulfate and heparin. The dark, remaining protein scaffold is surrounded by loops of DNA that emerge from it. (From Cell *12,* p. 817, Paulson and Laemmli. Copyright 1977 M.I.T. Press.)

signifies that the scaffold proteins may be a very important subset of the nonhistone chromosomal proteins.

Identification of these proteins, while still incomplete, points up the difficulties in classifying nuclear proteins. As mentioned above, the lamins are constituents of the "nuclear scaffold," and yet they are clearly proteins of the nuclear envelope. Other scaffold proteins appear to be identical to previously identified matrix proteins (see Lewis et al., 1984). Yet by their persistent attachment to the DNA fiber, we should certainly class all of these as NHCPs. What is becoming clear is that the interphase chromosome cannot be thought of as an independent entity, floating freely within a nuclear envelope. Rather, it seems to be connected to both the envelope and to an internal fibrillar structure, the matrix. Any isolation of "chromatin" from interphase cells must in some way break these connections, and certain proteins will be classed as "envelope," "matrix," or "chromatin" proteins depending only upon how the connections are broken and where these proteins happen to end up. It may well be that some of the scaffold proteins play a major role in the control of processes such as transcription and replication. If so, researchers may have to abandon attempts to further *analyze* function in a reductionist manner. The entire interphase nucleus may have to be considered as a functional entity.

Contractile Proteins

The question of whether proteins such as actin, myosin, or the tubulins are normal nuclear constituents has been the subject of considerable debate. Early studies have been reviewed by Douvas and Bonner (1977) and LeStourgeon (1978). Interest in the subject increased dramatically following the publication of a careful study by LeStourgeon et al. (1975) of *Physarum* nonhistone proteins. These authors claimed the presence of actin and myosin as major constituents and included control studies to argue against the possibility of cytoplasmic contamination. Such contamination is always a concern when a protein abundant in the cytoplasm is also claimed to be a nuclear protein. Many, however, have been skeptical; for example, Goldstein et al. (1977) found that in *Amoeba,* actin could readily exchange between nucleus and cytoplasm. In careful studies of the constituents of mouse liver nuclei, Comings and Harris (1976) found that actin, though a major component in unwashed nuclei, was largely removed in the first wash. They conclude that actin, myosin, tubulin, and tropomyosin are very minor components, at best, of chromatin itself.

Studies of this kind would suggest that actin in nuclei is either adventitious or not a chromatin-bound protein. However, more recent experiments tend to support the idea of a specific role or roles for actin in the nucleus and its probable association with chromatin. For example, Clark and Merriam (1977) hand-isolated nuclei from *Xenopus* oocytes, a technique claimed to minimize or eliminate cytoplasmic contamination. They found actin to represent 6% of the total nuclear protein. About 75% of this amount was diffusible, but even "nuclear gels," stripped of their envelopes, retained 25%. The work of Bremer et al. (1981) is also of considerable interest. They found that whereas actin was not extracted from Novikoff hepatoma chromatin by 0.6 M NaCl, it was released by 5 M

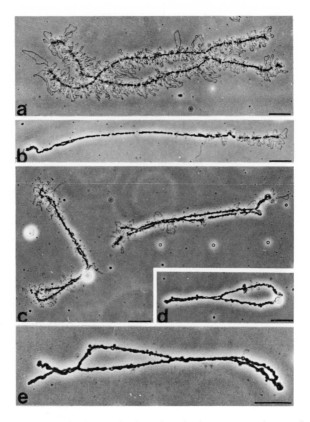

Fig. 5-7A. Retraction of salamander lampbrush chromosome loops after injection of antibodies to actin into living oocytes. (a) Chromosomes isolated 1 hour after injection of 5 nl of 0.2 mg/ml purified anti-actin IgG. (b–d) Representative chromosomes after 4.5 hours. (e) A chromosome 4.5 hours after injection of 10 nl.

urea. Most important, this actin, while similar to cytoplasmic actin, was clearly distinguishable from it in amino acid composition and peptide maps. Thus, it appears that even if cytoplasmic actin can exchange with the nucleus, there also may exist a chromatin-specific form.

A remarkable role for nuclear actin has been suggested by the work of Scheer et al. (1984). Using salamander nuclei, which contain lampbrush chromosomes, they have demonstrated that microinjection of either actin antibodies or actin modifying proteins causes the retraction of the lampbrush loops (see Fig. 5-7A). On the other hand, if transcription itself is inhibited by actinomycin, but no anti-actin agents are added, the retraction

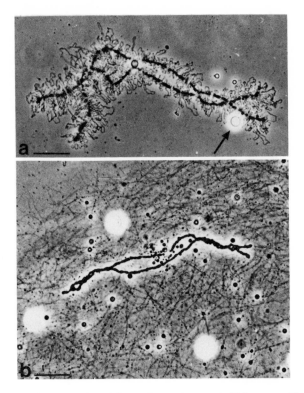

Fig. 5-7B. Retraction of lampbrush loops by actinomycin D leaves a fibrillar network. (a) Control chromosome. (b) Chromosome 6 hours after incubation of oocytes in medium containing 50 μg/ml actinomycin. Loops are retracted, but a fibrillar network is left. (From Cell *39*, p. 111, Scheer et al. Copyright 1984 M.I.T. Press.)

of the loops reveals a filamentous structure surrounding the chromosome (Fig. 5-7B). The implications of these studies, which remain to be worked out in detail, are that actin filaments are somehow involved in the extension of the lampbrush chromosome loops.

   While the functional presence of actin in interphase nuclei now seems probable, the situation regarding tubulin is less clear. Anachkova et al. (1977) reported α- and β-tubulin among the nonhistone proteins of rat liver, but only β-tubulin in brain. Similar results were reported by Farr et al. (1979), who found α- and β-tubulin (as well as actin) as prominent constituents of lymphocyte chromatin, and by Menko and Tan (1980) using 3T3 cells. A related result comes from Villasante et al. (1981), who have detected a microtubule-associated protein that interacts specifically with centromeric repetitive DNA sequences. Considering the well-established role of microtubules in mitosis, such specific interaction is not unexpected. However, whether such components remain associated with the DNA throughout the cell cycle is presently unclear.

6

The Nucleosome

The preceding three chapters have described the chemical constituents of chromatin: DNA, histones, and nonhistone chromosomal proteins. We are now in a position to ask how these interact to form the complex, functional structure of chromatin. It is logical to begin with the elemental unit, the nucleosome, and to consider first the ubiquitous component of all nucleosomes, the core particle.

The Core Particle

Isolation

If either nuclei or chromatin are digested with micrococcal nuclease, the first products are a series of oligomers of nucleosomes. As digestion proceeds, these are cleaved to yield mononucleosomes and simultaneously "trimmed" to produce core particles (see Chapter 2). However, such a procedure is inefficient for the purification of the core particle. If the digestion is stopped too soon, both nucleosome oligomers and larger mononucleosomes carrying lysine-rich histones remain (see Fig. 6-1, lanes 12, 13). On the other hand, if the digestion is carried far enough to wholly remove these intermediates, many of the core particles are degraded by internal cuts by micrococcal nuclease (Fig. 6-1, lanes 3–5, 8–10). Furthermore, long digestion will produce a further artifact. Micrococcal nuclease has a preference to cut in A–T-rich regions, and prolonged digestion of a random population tends to concentrate as survivors those nucleosomes rich in G–C sequences, and particularly those that have G–C pairs in "exposed" positions on the core particle surface (McGhee and Felsenfeld, 1983).

For these reasons, a single direct digestion is not to be preferred for

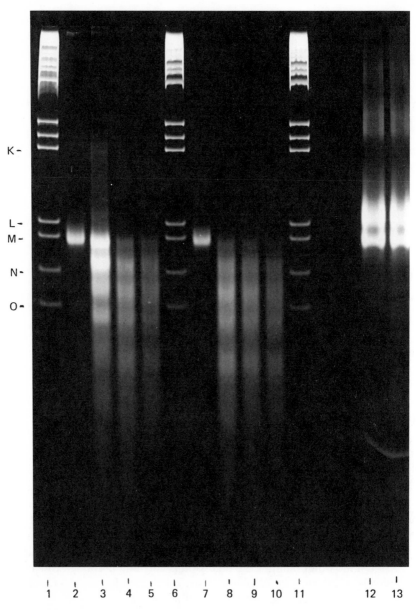

Fig. 6-1. Preparation of monocleosomes by various methods. This is a polyacryl-amide gel showing double-strand DNA fragments. Lanes (left to right): Lanes 1, 6, 11—Hae III restriction fragments of PM-2 DNA. Lanes 2, 7—core particle DNA prepared by digestion of H1-depleted chromatin. Lanes 3–5, 8–10—further digestion of such particles, showing formation of subnucleosomal DNA. Lanes 12, 13—DNA from nucleosomes prepared by digestion of whole nuclei. Both 145 bp and ~165 bp fragments are present, plus a small amount of dimer. (Reprinted with permission from Biochemistry *16*, p. 5295, Tatchell and Van Holde. Copyright 1977 American Chemical Society.)

core particle preparation, and more complex schemes have been devised. One approach is that taken by Olins et al. (1976). This utilizes a relatively mild digestion and separates core particles from H1-containing mononucleosomes and oligonucleosomes on the basis of solubility properties. Figure 6-2 shows that mononucleosomes are partially precipitated in the ionic strength range between 0.03 M and 0.3 M. If DNA and histones are extracted from the precipitate and the supernatant fractions, the result shown in Figure 6-3 is obtained. Whereas the precipitated particles contain longer DNA (often about 170 bp) and some of the lysine-rich histones, the supernatant particles are enriched in DNA of the core size (140 to 150 bp) and are devoid of H1 or H5. While the method is simple, it does not appear to yield core particles as homogeneous as some described below, nor does it give high yield. However, variations of this technique have proved very useful for isolation of the larger particles with their accompanying lysine-rich histones (see Simpson, 1978b; and *Organization of the Nucleosome Above the Core Particle Level,* below).

Since it appears that the binding of lysine-rich histones (and probably nonhistone proteins as well) to the linker regions inhibits micrococcal nuclease digestion, it would seem reasonable that removal of these proteins *prior* to digestion should facilitate "trimming" of the DNA to the core before substantial numbers of intracore cuts are made. A number of techniques have been used for this purpose. Extraction of chromatin at salt concentrations in the range of 0.6 M (Ohlenbusch et al., 1967; Tatchell and van Holde, 1977) will efficiently remove these proteins. After diges-

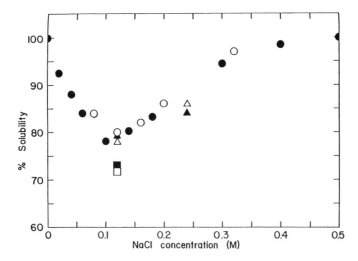

Fig. 6-2. Solubility of nucleosome preparations like those shown in lanes 12 and 13 of Figure 6-1, as a function of salt concentration. The soluble portion at 0.12 M salt is mainly core particles. Different symbols represent different preparations. From Tatchell (1978).

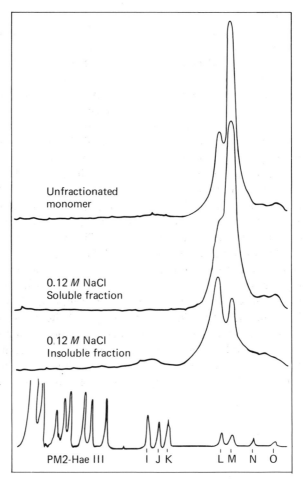

Fig. 6-3A. Scans of electrophoretograms of the DNA from soluble and insoluble fractions (Fig. 6-2).

tion, core particles are obtained that have the very narrow DNA distribution shown in Figure 6-1, lanes 2 and 7. A danger in this method, however, lies in the fact that the histone cores appear to be quite mobile on the DNA strand at ionic strengths above 0.5 M (Steinmetz et al., 1978; Weischet, 1979; Spadafora et al., 1979). In fact, the act of digestion itself tends to produce core movement in H1-depleted chromatin at ionic strengths of about 0.35 M or greater (Weischet and van Holde, 1980). For many purposes, this may be of little consequence, but it is to be avoided in studies in which one wishes to keep the histone cores in contact with those particular DNA sequences with which they were associated in vivo. Therefore, methods in which the lysine-rich histones can be removed at

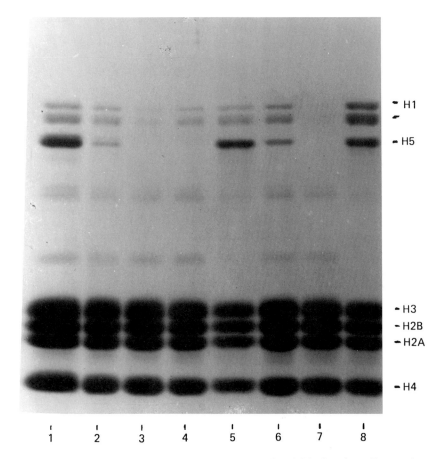

Fig. 6-3B. Histone composition of the soluble and insoluble fractions. Lanes: 1—chicken erythrocyte histone standards; 2—whole nucleosomes; 3—0.12 M NaCl soluble; 4—0.24 M NaCl soluble; 5—0.24 M NaCl insoluble; 6—whole nucleosomes; 7—0.12 M NaCl soluble; 8—0.12 M NaCl insoluble.

lower salt concentrations are often to be preferred. A number of such techniques have been developed; some are listed in Table 6-1. The method of Libertini and Small (1980) is attractive, for quantitative removal is accomplished at quite low ionic strength (50 mM) with a good yield of pure core particles. It has been demonstrated (Beard, 1978; Spadafora et al., 1979) that at ionic strengths this low *no* nucleosome movement is detectable even after long incubation at temperatures as high as 37°C.

As Figure 6-1 shows, careful digestion by micrococcal nuclease after removal of H1 and other linker proteins produces core particles with quite homogeneous DNA size. Many laboratories have reported preparations in which the breadth of the size distribution did not exceed a few base

Table 6-1. Some Methods for Removing Lysine-Rich Histones from Chromatin

Solvent medium	Recipient substance	Histones removed	Comments	Reference
0.4–0.5 M NaCl; 0.01 M Tris, pH 7.5	None	H1	Also give conditions for removal of other histones	Ohlenbusch et al. (1967)
1 mM MgCl$_2$ or 40 mM NaCl; pH 8.0	tRNA	H1	If salt omitted, get transfer of some other histones as well	Ilyin et al. (1971)
0.5–0.6 M NaCl; 0.05 M phosphate buffer, pH 7.0	Ion exchange resin AG50W-X2	H1 + H5	Slightly higher salt needed to remove H5 than H1; addition of urea and/or ethanol yields extraction of some other histones	Bolund and Johns (1973)
0.05 M Na phosphate + 0.1 M NaCl, pH 7.0	Ion exchange resin AG50W-X2	H1	Lower NaCl concentration led to incomplete removal of H1	Thoma and Koller (1977)

Table 6-1. *Continued*

Solvent medium	Recipient substance	Histones removed	Comments	Reference
0.5–0.65 M NaCl, 0.01 M Tris, 0.2 mM EDTA, pH 7.4	None	H1 + H5	Two-step swelling of chromatin gel in 0.5, then 0.65 M NaCl	Tatchell (1978)
0.05 M NaCl, 0.05 M Na phosphate, pH 7.0	Ion exchange resin AG50W-X2	H1	Two steps: first in 0.05 M phosphate buffer, then in buffer + 0.05 M NaCl; claim Thoma and Koller procedure yields some free DNA, this does not	Modak et al. (1980)
0.05 M NaCl, 0.01 M Tris, pH 7.5	Carboxy-methyl sephadex, C25-120	H1 + H5	Procedure allows production of core particles in high yield	Libertini and Small (1980)
0.08 M NaCl 1 mM phosphate pH 6.8, 0.2 mM EDTA	AG50W-X2	H1	Quantitative removal of H1 from chicken erythrocyte chromatin; complete retention of H5	Muyldermans et al. (1980a)

pairs. It seems likely that the limits to the homogeneity are imposed by the nuclease itself. Micrococcal nuclease prefers to cut between A–T pairs; Cockell et al. (1983) find that almost all of the 5' termini of DNA from random-sequence nucleosomes are T or A residues. Thus, the precise cutting point must depend upon the propinquity of A–T sequences to some "resisting point" in the nucleosome.

Protein Composition and Protein–Protein Interactions
in the Core Particle

Although the existence of a histone octamer was one of the basic premises of the first models of nucleosome structure, experimental confirmation that this is the *exact* stoichiometry has been remarkably slow in coming. It is easy enough to make a convincing argument, based on gel electrophoresis studies like those shown in Figure 6-3, that there are *approximately* equal weights of all four histones. However, exact quantitation of gel staining is difficult. More recently, a few papers have appeared in which radioactive labeling of histones in vivo has been employed to yield unambiguous results. A most detailed and careful study is that of Albright et al. (1979). These workers double labeled mouse tissue culture cells with [³H]-lysine and [¹⁴C]-arginine. Nuclease digestion products were then separated by gel electrophoresis, and the relative amounts of proteins in three resolvable dimer bands and three resolvable monomer bands were measured. The data are summarized in Table 6-2. Note that not only the four core histones but the protein uH2A, as well, were determined. As described in Chapter 4, uH2A is a covalent adjunct of H2A and the protein ubiquitin. The data in Table 6-2 show that the amount of H2B is virtually identical to that of H4 for all particles. Histone H3 appears a little high, but the authors argue that this is likely due to contamination by co-migrating nonhistone proteins. Note that the H3/H4 ratio for the core particle MI is 1.00. The content of H2A seems low, but this is partially made up for by uH2A, which had already been shown to sometimes substitute for H2A in nucleosomes. It is of interest that the data show that those particles preferentially digested to cores under the conditions used by Albright et al. do not contain uH2A. This should not be taken to imply that core particles cannot contain this modified histone. Indeed, several groups have detected uH2A as a component in core particles (see *Histone Variants* in Chapter 4), and Kleinschmidt and Martinson (1981) have reconstituted core particles with uH2A as the sole form of this histone.

Even with the ubiquitinated form included, the amount of H2A is about 10% below the expected value. In a "Note added in Proof," Albright et al. suggest that this may be accounted for by minor histone variants found in mouse tissue by Franklin and Zweidler (1977; also see Table 4-6). It appears that H2A may, to a limited extent, represent a "variable" position in the histone core.

These and other less detailed studies are convincing evidence that core

Table 6-2. Relative Histone Composition of Nuclei, Chromatin, and Nucleosomes[a]

Fraction[b]	DNA length (bp)	H1A/H1B mole ratio	Moles per mole of histone H4						Core histone mass relative to MI
			H1 (A + B)	uH2A	H3	H2B	H2A	H2A + uH2A	
Nuclei		2.3	0.32	0.10	1.11	1.01	0.82	0.92	
Chromatin		1.7	0.30	0.08	1.08	0.99	0.85	0.93	
D3	367	1.7	0.44	0.12	1.21	1.06	0.87	0.99	
D2	342	1.7	0.41	0.08	1.05	0.99	0.83	0.91	2.06
D1	305	2.1	0.20	0.11	1.07	0.99	0.81	0.92	1.94
MIV-V	212	1.3	0.36	0.07	1.28	0.92	0.70	0.77	
MIII	172	1.4	0.41	0.06	1.12	0.97	0.73	0.79	1.02
MI	150	—	<.001	<.007	1.00	1.01	0.89	0.89	(1.00)

[a]Data of Albright et al. (1979), on mouse mastocytoma cells.
[b]Code to electrophoretic fractions:
D3, D2, D1 are dimers with differing DNA lengths.
MIV-V are monomeric nucleosomes with approximately repeat-length DNA.
MIII corresponds approximately to the chromatosome.
MI is the core particle.

particles do contain equimolar amounts of the four core histones, with the exception of some substitution of H2A by minor variants. Since there must then be 4 n histone molecules in the particle (n integer), the presence of an *octamer* is demonstrated. The molecular weights of the core particle and the DNA, together with the histone/DNA ratio, allow no other possibility.

Are any other proteins ever found as integral parts of the histone core? There have been numerous reports that nucleosomes contain nonhistone proteins. However, on examination it is found that most of these studies have utilized mononucleosomes obtained by sucrose gradient fractionation of single-step chromatin digests. Particles so prepared will generally contain either additional DNA, lysine-rich histones, or both, which may act as binding sites for nonhistone proteins. The sole case in which significant amounts are claimed for carefully fractionated materials is that of Chan and Liew (1979). These workers claim to find as much as 10% of the total protein in an electrophoretically isolated "MN1" fragment (devoid of H1) to be nonhistone proteins. However, I do not believe that such results compromise the basic simplicity of core particle composition. There is now good evidence that core particles can *bind* certain nonhistone proteins, specifically, HMG 14 and 17 (see Mardian et al., 1980; Sandeen et al., 1980; Shik et al., 1985). This is quite a different matter, and it will be discussed in a later section.

In summary: It appears that the core particle contains exactly two copies each of H4, H3, H2B, and H2A, with occasional substitution of uH2A or other proteins for H2A. Despite this basic simplicity in composition, there is an enormous potential for variability in the core particle. This arises from the existence of histone variants and modifications. Consider, for example, the mouse histone variants listed in Table 4-6. Taking all possible combinations of just the major somatic variants, one calculates 2025 compositionally different core particles! If the multitudinous possible combinations of modifications are included, the number becomes astronomical. What fraction of these variations may be functionally significant is quite unclear. Differences in variant composition do not appear to strongly affect chromatin digestibility, for Bafus et al. (1978) could find no evidence for differing distributions of histone variants in mono, di, and trinucleosomes from mouse chromatin.

We turn now to a consideration of the contacts and interactions of proteins within the core particle. That the eight histone molecules are in reasonable propinquity has been demonstrated by the fact that reagents such as dimethyl suberimidate can crosslink the entire octamer (Kornberg and Thomas, 1974; Stein et al., 1977; Carter et al., 1980). While such results reinforce the octamer concept, they do not tell us much about internal arrangement. This is because "long" crosslinkers such as dimethyl suberimidate span distances of the order of 10 to 20 Å. They can reach sufficiently far across the particle to tie together virtually any pair of histones.

Of greater usefulness in determining histone propinquity are the "contact site" crosslinkers, which make either direct or very short crosslinks between reacting groups. A representation of such reported crosslinks is given in Figure 6-4. Several points should be noted:

1. The principal contacts are between H2A and H2B, H2B and H4, and H3 and H3. These correspond closely to the "strong" interactions found in solution by Isenberg and co-workers (see Chapter 4; also Carter, 1981).
2. The existence of contact-site crosslinks between H3 and H3, but not between H4 and H4, suggests that the (H3·H4)$_2$ tetramer is held together

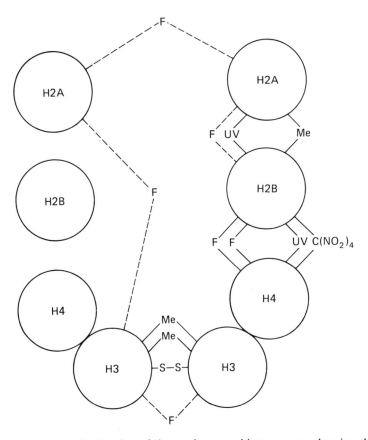

Fig. 6-4. A schematic drawing of the nucleosome histone core, showing the interactions detected by contact-site crosslinkers. Contacts in one heterotypic tetramer are shown and are presumed to exist symmetrically in the other as well. Strong contacts are indicated by solid lines, weaker ones by broken lines. Code: F, formaldehyde; Me, methylacetimidate; UV, ultraviolet light-induced; C(NO$_2$)$_4$, tetranitromethane. See the text for further discussion and references.

mainly through H3·H3 interactions. Even some "long" crosslinkers do not succeed in producing H4·H4 dimers (Hardison et al., 1977; Bonner, 1978). However, Chung and Lewis (1986) report eximer fluorescence when met 84 in H4 molecules in nucleosomes are labeled with a fluorescent dye, indicating the existence of at least some points of close contact.

3. Although cysteine 110 of H3 seems to be unoxidized in the majority of native core particles, these residues must lie close together, for they can be oxidized to form an -S-S-bridge in situ (Camerini-Otero and Felsenfeld, 1977b) and such dimers can be employed in core particle reconstitution (Lewis and Chiu, 1980).

4. Homotypic contacts (other than H3–H3) are not generally observed, leading to the picture of the histone core as a bifurcate structure. This idea is supported by the dimethyl suberimidate studies of Suda and Iwai (1979), who found the major dimers to be heterotypic.

5. A number of these points of histone–histone contact have been mapped quite precisely on the histone sequences (see, for example, Camerini-Otero and Felsenfeld, 1977b; DeLange et al., 1979; Martinson et al., 1979b; McGhee and Felsenfeld, 1980a; Callaway et al., 1985).

In almost every instance, these contacts lie within the "globular" portions of the histone molecules and in strongly conserved regions. Thus, suggestions that these portions of the histones are essential for core particle integrity seem to be borne out. The fact that in some cases two or more different crosslinks have been established between the same pair suggests that there are in fact surfaces of contact between histone molecules. Spiker and Isenberg (1977b) have emphasized that such surfaces must be highly conserved, arguing the disruption of even one interaction might be sufficient to severely weaken the noncovalent bonding. Most remarkably, it has been found that the crosslinks established between H2A and H2B by UV radiation or tetranitromethane are the same in leek and calf nucleosomes. It is even possible to crosslink interkingdom histone pairs in this way (Martinson and True, 1979b). It is noteworthy that the H2B region involved (near tyrosine residues 37, 40, 42) is highly conserved. Kleinschmidt and Martinson (1984) find that iodination of these residues is blocked in nucleosomes. Iodination on the free histone does *not* block reconstitution.

It appears that the histones in the core particle are very high in α-helix content. Thomas et al. (1977) have used both Raman spectroscopy and circular dichroism to analyze the secondary structure in the histone core. For DNA free cores, in 2 M NaCl, both techniques yield about 50% α-helix and very little β-sheet. The result gains in significance from the observation of Thomas et al. that the Raman spectrum and CD spectrum of the DNA-free core in 2 M NaCl are indistinguishable from the corresponding spectra of the core particle in dilute salt, once DNA contributions are subtracted from the latter. This means that the values given above can be applied to proteins in the core particle and that there is no significant change in protein secondary structure when the DNA wraps around the core. On the other hand, there may well be changes in tertiary structure and compaction (see Chapter 4).

As far as the *globular* regions of the core particle proteins are concerned, the value given above for α-helix is probably an underestimate. This is because the N-terminal regions appear, from a number of criteria (see Chapter 4), to lack defined secondary structures. Thomas et al. estimate that as much as 80% of the *globular* portions of the protein in the histone core may be α-helical. This would make these histones among the most highly ordered of globular proteins.

In summary, the picture of the histone core that emerges is of a tight, highly ordered complex of the globular regions of the histones with highly conserved surfaces of interaction. This leaves out of consideration, however, the disposition and roles of the N-terminal (and to some extent C-terminal) portions of the histones. I shall defer consideration of this point until the arrangement of the DNA and its interactions with the histones have been described.

The DNA in the Core Particle

What kinds of nucleic acids can interact with histone octamers to form core particles? This question has been approached by numerous reconstitution studies, with the results shown in Table 6-3. They can be summarized as follows: Histone cores will *not* accept double-strand RNA, RNA–DNA hybrid, or even a moderate inclusion of RNA bases in the double helix. Also unacceptable are the complementary deoxyribo-homopolymers poly dG·poly dC or poly dA·poly dT. Short A/T "runs" can be accommodated, but apparently only in certain regions of the nucleosome. Stem-loop structures cannot be incorporated. There exists, at this writing, a controversy as to whether or not Z-form polydeoxyribonucleotides can form nucleosomes (see Table 6-3). In my opinion, the evidence that they can is weak, a conclusion supported by very recent studies (Ausio et al., 1987).

Nevertheless, the histone octamer will apparently form nucleosome-like structures with some very peculiar polynucleotides, including double-strand poly U, single-strand DNA, and the glycosylated DNA of T4 phage. It is clearly difficult to draw *any* general conclusions from these results, and we shall probably have to learn much more about the details of histone–DNA interaction before a pattern becomes clear.

On the other hand, the core particles formed with various kinds of double-strand DNA are remarkably uniform objects. Early studies indicated that these all contained about 140 bp of DNA (see Chapter 2). More recently, improvements in particle isolation and gel electrophoresis techniques have led to a more precise, slightly larger figure. Table 6-4 collects some data of high precision. Only those experiments are included in which

Table 6-3. Nucleosome Reconstitution on Specific Types of Nucleic Acids

Nucleic acid	Results	Reference
Double-strand RNA	No reconstitution	Kunkel and Martinson (1981)
Double-strand (?) poly U	Reconstitution; beaded strand in EM	Klump and Falk (1984)
RNA–DNA heteroduplex	No reconstitution	Dunn and Griffith (1980)
Ribosubstituted DNA	No reconstitution if >5%	Hovatter and Martinson (1987)
Single-strand DNA	Reconstitution; obtain 10S particles	Palter et al. (1979) Caffarelli et al. (1983, 1985)
Poly(dA·dT)·poly(dA·dT)	Reconstitution; particles digest to 146 bp length DNA	Simpson and Kunzler (1979)
	Identical results	Bryan et al. (1979)
	Reconstitution	Rhodes (1979)
	Reconstitution to 200 bp repeat	Prevelige and Fasman (1983)
Poly(dA)·poly(dT)	No reconstitution	Simpson and Kunzler (1979)
	No reconstitution	Rhodes (1979)
Poly(dA)·poly(dT) inserts in plasmid DNA molecules:		
long	No reconstitution	Kunkel and Martinson (1981)
97 bp	Reconstitution only extending about 60 bp into insert	Prunell (1982)
20 bp	Reconstitutes, but 20 bp insert lies near "ends" of nucleosomal DNA	Prunell (1982)
Poly(dG·dC)·poly(dG·dC)	Reconstitution; very similar to (dA·dT) (dA·dT) particles	Simpson and Kunzler (1979)
	No reconstitution	Bryan et al. (1979)
	No reconstitution	Rhodes (1979)
	Reconstitution, 200 bp spacing	Prevelige and Fasman (1983)
Poly(dG)·poly(dC)	No reconstitution	Simpson and Kunzler (1979)
	No reconstitution	Rhodes (1979)
Poly(dG-m⁵dC)· poly(dG-m⁵dC) (B-form)	Reconstitution; normal DNase I pattern	Nickol et al. (1982)
	Reconstitution, 11S particles	Miller et al. (1985)
Poly(dG-m⁵dC)· poly(dG-m⁵dC) (Z-form)	No reconstitution, using Co $(NH_3)_6^{3+}$ to convert to Z-form	Nickol et al. (1982)
	Reconstitution, using trout testes assembly factor	Miller et al. (1985)
Glucosylated (T4) DNA	Reconstitution; normal 11S particles obtained	McGhee and Felsenfeld (1982)
DNA containing stem-loop structure	No reconstitution with stem-loop in nucleosome	Nickol and Martin (1983)

Table 6-4. Precise Measurements of the Core DNA Length

Tissue, cell	Length of DNA	Method of measurement	Reference
Yeast	148 ± 3 bp	Comparison with Hae III PM2 fragments calibrated against several series of sequenced restriction fragments	Lohr and van Holde (1979)
Pea	145 bp	Calibrated vs. Hae III PM2 fragments	Grellet et al. (1980)
Chicken erythrocyte	145 ± 3 bp	Comparison with sequenced restriction fragment of almost identical size	Shindo et al. (1980)
HeLa cells; "MNI" fraction of nucleosome	146 ± 3 bp	Comparison with Hae III fragments of SV-40 DNA	Levinger and Varshavsky (1980)
Drosophila embryos	145 ± 3 b	Comparison with φX174 sequenced fragments	Shick et al. (1980)
Rat liver	146 b	Comparison with sequenced plasmid fragments	Prunell et al. (1979)
mouse Ehrlich ascites tumor	145 ± 3 b	Comparison with sequenced φX174 fragments	Mirzabekov et al. (1978)
MOPC 21 tissue culture cells	145.9 ± .3 b	Band counting on high resolution gels; band center chosen for core length	Lutter (1979)
Core particles reconstituted from chicken erythrocyte histone and poly (dA·dT)·poly (dA·dT)	146 b	Band counting to highest band	Bryan et al. (1979)
Core particles; chromatin reconstituted from core histones and poly (dA·dT)·poly (dA·dT)	146 ± 1 bp	By comparison with sequenced Hae III φX174 RF fragments	Simpson and Kunzler (1979)
Core particles; chromatin reconstituted from core histones and poly (dG·dC)·poly (dG·dC)	146 bp	By comparison with sequenced Hae III φX174 RF fragments	Simpson and Kunzler (1979)

sequenced DNA fragments have been used for size calibration in electrophoresis, or individual bands were counted on high resolution gels.
The latter technique deserves comment. As we shall discuss later, DNase
I can make single-strand nicks on DNA in the core particle. The product
of a light digestion, analyzed on a denaturing gel of high resolution, will
contain a set of bands with 1 b spacing, extending up to the core particle
size (see Fig. 6-5). The bands need only be counted. In this way, Lutter

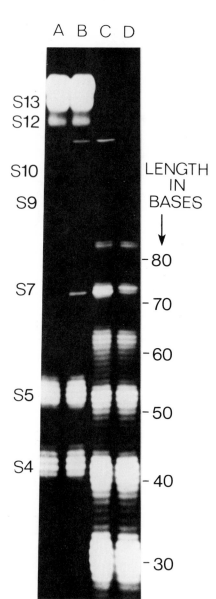

Fig. 6-5. A high-resolution electro-
phorogram of single-strand DNA fragments produced by DNase I digestion
of core particles. (From Nucleic Acids
Research, vol. 6, issue 1, p. 47, Lutter.
Copyright 1979 IRL Press.)

(1979) has obtained an average value of 145.9 b for core particles from mouse tissue culture cells, and Bryan et al. (1979) find 146 b for reconstituted particles containing the synthetic DNA, poly(dA·dT)·poly(dA·dT). For the sake of precision, it should be noted that these latter techniques, in which the DNAs are denatured, are measuring the lengths of each strand. By the nature of micrococcal nuclease cutting, the core particle excised by this enzyme should have a two base extension at each 5' end. That is, the core particle produced by micrococcal nuclease digestion is envisioned as containing two 146 base strands, offset by two bases. Extensive digestion may result in trimming of some of these projecting ends. These complications make measurement of core DNA size under non-denaturing conditions a trifle fuzzy. Summarizing the data in Table 6-4, we conclude that the size of the core DNA appears to be the same for particles containing a wide variety of DNA or histone types. The best estimate of the size now looks to be very close to 146 bp. This corresponds to almost exactly 14 turns of DNA if the nucleic acid is in the "solution-B-form" (Chapter 3), with 10.5 bp/turn. However, as will be discussed below, there is some controversy concerning the twist of DNA in the core particle. If it is, as some hold, more nearly 10.0 bp/turn, then there are about 14.5 turns of the DNA.

The source of the histones in the core seems to have very little effect on the length of the core DNA. While this fact is generally implied by the results shown in Table 6-4, it is further emphasized by Simpson and Bergman's (1980) observation that sea urchin sperm also yield a 145 bp core particle. This is of special significance, for the urchin sperm contains a different H2B and a considerably larger H2A than do other tissues. The universality of this DNA length seems truly remarkable, especially when we consider that it corresponds to nothing more than a "halting point" during the nuclease digestion. It must represent some feature of the histone core that severely and precisely limits further digestion, for Riley and Weintraub (1978) have found that exonuclease III will also trim nucleosomal DNA to the same core particle length.

What fundamental biological requirement has kept this DNA length so strongly conserved throughout evolution? While there is no real evidence, the following suggestion could be put forward: If there have evolved methods for precisely positioning nucleosomes on DNA (and there is much evidence for this—see Chapter 7), an organism that accepted histone mutations which led to a different core length would find these to "misfit." Even a few base pairs' difference in linker length could have significant effects, for the orientation of one nucleosome with respect to the next may be very sensitive to linker size (see Chapter 7). Thus, the organization of the whole genome would be disrupted. Therefore, such mutations cannot be tolerated, for the same reason that mutations in the genetic code are strongly suppressed. Either would have a "global" effect. It may be that the impressive conservation of the globular regions of the core histones and the chain lengths of H3 and H4 (see Chapter 4) are explained in turn by the necessity to maintain a constant core DNA length.

We now turn to the question of how this DNA is arranged on, or in, the core particle. That it must be on the surface was argued in the very first models (Chapter 2). After all, it is not easy to see any other way in which a 500 Å long piece of DNA could be accommodated in a particle with diameter of about 100 Å. Evidence came somewhat later. First, there was the observation by Noll (1974b) that digestion of chromatin by the enzyme pancreatic DNase I led to extensive nicking of the core DNA. This enzyme produces single-strand nicks under most conditions. When the fragments produced are electrophoresed under denaturing conditions, a remarkable "ladder" of bands is produced (see Figs. 2-3, 6-5). The distances between points of maximal cleavage are multiples of *approximately* 10 bases. Noll's explanation, which is now generally accepted, is that the DNase I preferentially nicks a strand whenever the phosphodiester backbone of that strand is maximally exposed to approaching nucleases. The fact that the entire length of the core particle DNA is accessible to such nicking (see Fig. 6-5) means that the whole DNA must lie on the surface. Support for this interpretation has come from studies in which DNA deposited on crystal surfaces is found to be cleaved by DNase I in a similar periodic fashion (Liu and Wang, 1978; Rhodes and Klug, 1980).

Noll's argument, while persuasive, was somewhat indirect. More compelling evidence that the DNA lies on the histone surface came from neutron scattering in solution. The rationale for these experiments is provided by the fact that the intensity of neutron scattering from an object bathed in a solvent depends upon the "contrast" in neutron scattering between the object and the surrounding solvent. This is true of scattering phenomena in general, but with neutron scattering this feature has particular power because the scattering intensity from different atoms (nuclei) varies greatly. Thus, DNA scatters more strongly than does protein, and (most important) D_2O and H_2O differ by a large amount in scattering power. It is, in fact, possible to make D_2O/H_2O mixtures that "match" the scattering of either protein or DNA. The situation is then as depicted in Figure 6-6. By choosing the proper solvent mixture in which to suspend the particles, one can make either DNA or protein "disappear." The angular dependence of the low-angle scattering can then be used to measure a radius of gyration for either the protein or DNA portion of a core particle.

Several groups have used this approach (see, for examples, Pardon et al., 1975, 1977a, 1977b; Suau et al., 1977; Braddock et al., 1981). The results from different laboratories are all in quite good agreement; the radius of gyration of the histone core is about 30 to 35 Å, whereas for the DNA a value of approximately 45 to 50 Å is found. The fact that the radius of gyration of the DNA is greater than that of the protein portion establishes that the DNA is on the outside. By considering additional features of the solution scattering curve, further information about the shape of the core

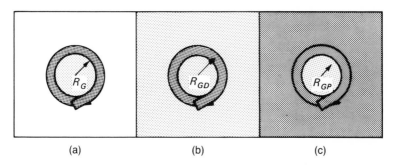

(a)   (b)   (c)

Fig. 6-6. The principle of contrast-matching in neutron scattering. A nucleosome, which has DNA (dark) wrapped about a protein core (lighter), is depicted schematically. (a) Solvent of low scattering power; the entire particle contrasts, and the radius of gyration R_G is for the whole structure. (b) Solvent matched to protein, the R_{GD} measured is that of the DNA. (c) Solvent matched for DNA; the DNA "disappears" and the R_{GP} of the protein core is measured. (From K.E. van Holde, PHYSICAL BIOCHEMISTRY, 2/E, © 1985, p. 229. Reprinted by permission of Prentice-Hall, Inc., Englewood Cliffs, New Jersey.)

particle can be obtained (see references above, especially Braddock et al., 1981). However, such information cannot compare in detail to that obtainable from diffraction studies of *crystals* of the core particles. These will be described in a following section.

Attempts to deduce the secondary structure of the DNA in the core particle have been beset with ambiguities and seeming contradictions. One of the most striking features of core DNA, noted very early, is its unusual CD spectrum (Sahasrabuddhe and van Holde, 1974; for more extensive studies see Cowman and Fasman, 1980). As Figure 6-7 shows, the CD spectrum of core particles at wavelengths above 250 nm is very different from that of free B-form DNA. The difference cannot be attributed to spectral contributions from the histones, for these proteins contain no tryptophan and little tyrosine, and consequently absorb very weakly in this spectral region. The unusual CD spectrum could admit of at least two possible explanations: (1) the DNA is not in the solution B form, or (2) the DNA is in the B form but its CD spectrum is perturbed by a vicinal effect due to the wrapping of more than one turn of DNA tightly about the histone core.

There are a number of indirect pieces of evidence to suggest that the DNA remains in something at least *close* to the solution B form when it wraps about the histone core. For example, Thomas et al. (1977) found that the Raman spectrum of core particles is very close to the sum of contributions from free B-DNA and histone cores. More direct evidence comes from the studies of Goodwin and Brahms (1978), who have identified Raman bands at 807 cm^{-1}, 835 cm^{-1}, and 865 to 870 cm^{-1} as corresponding to A, B, and C forms of DNA, respectively. They show that nucleosomes,

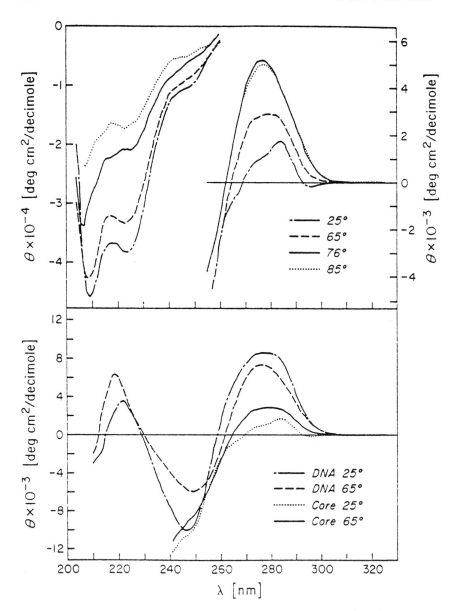

Fig. 6-7. The circular dichroic spectra of chicken erythrocyte core particles and DNA under various circumstances. In the top figure, data for core particles in the DNA region of the spectrum (right) and the protein region (left) are shown at various temperatures. In the bottom figure, core particles and purified DNA are contrasted at two temperatures. (From Nucleic Acids Research, vol. 5, issue 1, p. 146, Tatchell. Copyright 1978 IRL Press.)

oligonucleosomes, and chromatin all have the 835 cm^{-1} band and lack the 807 cm^{-1} and 865 to 870 cm^{-1} band. NMR experiments yield similar conclusions. For example, studies of ^{31}P NMR, which should be quite sensitive to DNA conformation, show little or no difference between solution DNA and core particle DNA (Cotter and Lilley, 1977; Shindo et al., 1980). A similar result was obtained by Feigon and Kearns (1979) from proton NMR experiments.

All of these studies suggest that the DNA conformation is similar to, if not identical with, the solution state, which has ~10.5 bp/turn. It has been suggested by Baase and Johnson (1979) that the difference observed in the CD spectrum could be explained by a relatively small change in the DNA winding (from ~10.5 bp/turn to ~10.3 bp/turn) when the polynucleotide is bound to the histone core. Alternatively, Shih and Lake (1972) pointed out that the CD-difference spectrum between chromatin and free DNA resembled the spectrum of the highly condensed (ψ) form of DNA. Fasman (1979) has noted the same type of difference in spectrum between core particles and free DNA. It is suggested, then, that the propinquity of DNA coils on the nucleosome corresponds to a "local" ψ-like condensation of the polynucleotide. However, there have been, to my knowledge, no theoretical analyses of this concept.

It would seem reasonable that a very precise mapping of points of maximum accessibility of the nucleosomal DNA to nuclease cleavage might provide information concerning the DNA periodicity. A number of very careful studies have been conducted; the major ones are summarized in Table 6-5. At first glance, most of the results shown in the table would seem to indicate that the periodicity is very nearly the same as in solution—about 10.4 bp/turn. However, closer examination of the problem reveals complexities. Each of the *endonucleases* must approach the DNA coil from the side, and the spacing of the most accessible points will vary along the DNA coil, as a consequence of steric hindrance to the enzymes. In fact, it is argued (Prunell et al., 1979; Klug and Lutter, 1981) that the observed *cutting* periodicity is most consistant with a DNA periodicity of 10.0 bp/turn in the nucleosome. According to this view, the apparent shift in periodicity observed through the nucleosomal DNA (see Table 6-4) is a consequence of the different angle of attack of the nucleases near the ends and near the center. The observation that the average spacing is >10 bp would then be a result of this bias. Interestingly, more recent results from exonuclease III digestion (Prunell, 1983) are generally consistent with this interpretation. This enzyme does not attack DNA from the sides but rather from the ends. Its points of "pausing" during digestion appear to be spaced about 11 bp apart near the ends, but after two periods shift to a 10 bp spacing.

It is probably wise not to overinterpret these data. All of the experiments described in Table 6-4 [with the exception of those on poly(dA·dT)· poly(dA·dT) particles] were carried out on random-sequence core particles

Table 6-5. Some Precise Studies of the Cleavage of Nucleosomal DNA by Various Nucleases

Enzyme used	Source of core particles used	Average periodicity (bases)	Comment	Reference
DNase I	MOPC21 tissue culture cells	10.4	Periodicity seems to vary along DNA; about 10.0 near ends, 10.7 near center	Lutter (1979); also see Lutter (1981); Klug and Lutter (1981) for further interpretation
	From chromatin reconstituted with poly(dA·dT)·poly(dA·dT) and chicken erythrocyte histones	10.5	About 10.25 near ends, 10.67 near center	Bryan et al. (1979)
DNase II	Beef kidney	10.4	About 10.0 bases near ends, 10.7 bases near center	Lutter (1981)
Micrococcal nuclease	Beef kidney	10.4	Periodicities very close to those seen for DNase I, DNase II (see above)	Cockell et al. (1983)
Exonuclease III	Rat liver	Not given	Spacing of pause points about 11 bp near ends, 10 bp in interior	Prunell (1983)

derived by cleavage of whole chromatin. In a more recent study of DNase I digestion of a nucleosome obtained by reconstitution on a defined sequence DNA, Simpson and Stafford (1983) find that cleavage periodicities are nonuniform and different on the two strands.

The question of whether the DNA periodicity is 10.0 or 10.5 bp/turn on the nucleosome has direct bearing on the so-called linking number paradox. Well-established models of the core particle (see below) indicate that the DNA makes about 1.6 to 1.8 left-hand superhelical turns about the histones. Yet direct measurement of the change in linking number made in circular SV-40 DNA by reconstitution with histone cores gave a value of $\Delta Lk \cong -1.25$ per nucleosome formed (Germond et al., 1975). More recent studies, using reconstituted phased nucleosomes, give $\Delta Lk = -1.01 \pm .08$ (Simpson et al., 1985). The results are inconsistent with the model *unless* the DNA changes in twist. The changes in twist corresponding to the difference between the solution periodicity of 10.5 bp/turn and a nucleosomal value of 10.0 bp/turn would correspond to a total, over 146 bp, of about 0.5 turns of DNA, very close to the amount needed to reconcile the discrepancy (Crick, 1976). Finally, it should be recalled from Chapter 3 that Levitt's (1978) theoretical calculations of DNA secondary structure predicted 10.5 bp/turn for free DNA. This same analysis, when applied to a DNA molecule bent to the radius of curvature expected in the nucleosome, predicts 10.0 bp per turn. Thus, although conclusive experimental evidence is not yet at hand, the weight of theoretical prediction is in favor of a more tightly wound DNA in the nucleosome.

As with so many scientific controversies, however, the debate as to the number of base pairs per turn may be an oversimplification, masking a more complex reality. As was emphasized in Chapter 3, the DNA molecule is flexible in its conformation, easily adapting to local influences. We may expect that when interacting with the histone core it may adopt many different local conformations. Indeed, the X-ray diffraction results hint strongly at this (see below), indicating the possibility of local irregularities. (See also McMurray et al., 1985, for NMR evidence.)

DNA–Histone Interactions in the Core Particle

We turn now to some basic questions: How is the DNA bound to the histone core? Is binding uniform, or localized in certain regions? How are the histones placed with respect to the DNA sequence? Although it seems likely that these questions will ultimately be answered by diffraction studies, present X-ray resolution is not sufficient. Therefore, a number of less direct techniques have been employed. These will be described and results summarized in an attempt to develop a coherent picture.

Reconstitution of Core Particles

A number of early experiments had given indications that DNA and histones, if dissociated from one another by high salt ($\geq 2 M$), could be made to reassociate by gradual lowering of the ionic strength. In some cases, 4 M urea was also present in the starting mixtures (Gilmour and Paul, 1970; Germond et al., 1975; Camerini-Otero et al., 1976). However, the first detailed critique of such reconstitution of core particles was provided by Tatchell and van Holde (1977), who obtained particles by dialyzing histone cores together with core-size DNA. The dialysis proceeded stepwise from 2 M NaCl, 10 mM Tris, to 10 mM Tris. Yields of up to 80% of 11S particles were obtained. These were found to be identical to "native" particles with respect to hydrodynamic behavior, digestion by nucleases and proteases, and thermal denaturation (see below). Of particular importance is the fact that DNase I digestion, using particles with ^{32}P end-labeled DNA, indicated that the DNA had attached in "correct" register (see *Nuclease Digestion of End-Labeled DNA*, below, for a description of this technique). Others, using a variety of characterization methods, have also found that core particles spontaneously formed in this manner have properties identical to those obtained by nuclease digestion of chromatin (see for example Brahms et al., 1981; Sibbet et al., 1983; Greyling et al., 1983). The technique has been used for many other purposes: to study the reassociation of DNA with partial complements of histones, to study reassociation with specific sequence DNA fragments, etc. These applications will be considered later. Of concern at the moment is that such experiments show that the interaction of 146 bp DNA with the histone core is a spontaneous process at low ionic strength. No special "assembly factors" are required, although such may be involved in vivo (see Chapter 9). The dynamics of this process have been studied in some detail by Stein (1979).

Nuclease Digestion of End-Labeled DNA

It has been noted above that DNase I will nick the core particle DNA at intervals of roughly 10.4 bases, and that this may be used to assay DNA periodicity. In interpreting DNA–histone interactions, it is important to know whether this nicking takes place with equal facility at every cutting site or if there are local differences in accessibility. An ethidium bromide-stained gel such as that shown in Figure 6-8A is quite uninformative, for as Figure 6-8B shows, fragments of a given length can arise from many different pairs of cleavages. The fluorescent label used to display the bands (ethidium bromide) cannot distinguish between these different contributors. A way to surmount this problem was devised by Simpson and Whitlock (1976). They labeled the 5′ ends of core particle DNA with ^{32}P, using the T4 polynucleotide kinase. When the DNA is then cleaved with DNase I, and the fragments electrophoresed, an *autoradiogram* of the gel will exhibit only those fragments that have an intact 5′ end (see Figs. 6-8C and 6-8D). Thus, the relative intensities of bands on the autoradiogram will measure the relative frequency of cutting at the different sites. The technique can be extended by the use of 3′ end labels (Lutter, 1978) and

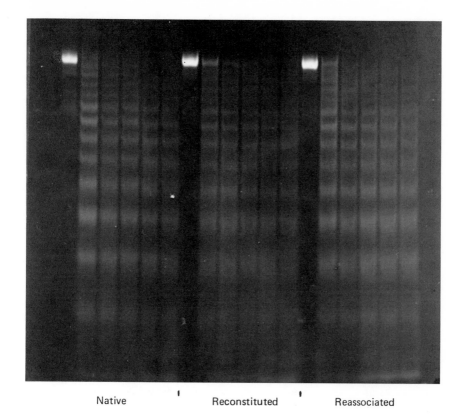

Fig. 6-8A. Gel electrophoresis of single-strand fragments obtained by DNase I digestion of chicken erythroctye core particles. Data for native, reconstituted, and reassociated particles are compared. The gels have been stained with ethidium bromide. Times of digestion in each set (left to right) are: 0, 30 sec, 2 min, 5 min, 10 min, 20 min. (Reprinted with permission from Biochemistry *16*, p. 5295, Tatchell and van Holde. Copyright 1977 American Chemical Society.)

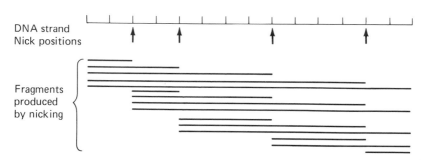

Fig. 6-8B. A schematic diagram to show how the fragments that appear in a gel such as that shown in Fig. 6-8A are generated. Note that the same fragment size can arise from many different sets of nicks.

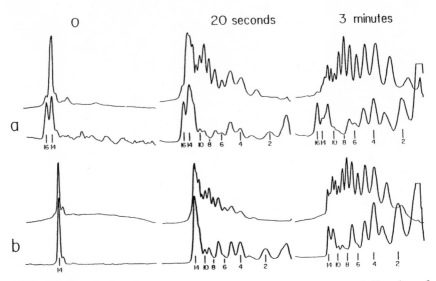

Fig. 6-8C. Scans of electrophororadiograms obtained upon DNase I digestion of core particles with 5′ DNA ends labeled with ^{32}P. Data for both native (a) and reconstituted (b) core particles are shown (cf. Fig. 6.8A). In each case scans of the corresponding ethidium bromide-stained gel are shown above. Note the presence of obvious "nulls" in the end-labeled pattern, near bands 3, 6, 8, and 11. (Reprinted with permission from Biochemistry *18*, p. 2871, Tatchell and van Holde. Copyright 1979 American Chemical Society.)

used with nucleases other than DNase I (see, for example, Whitlock et al., 1977; Lutter, 1981; Cockell et al., 1983).

The most detailed studies of DNAase-1 digestion are those of Lutter (1978), who has measured psuedo first-order rate constants for the appearance of bands of different lengths. The result is shown diagrammatically in Figure 6-9a; the general pattern can be observed in Figure 6-8C. Quantitatively similar results have been obtained in a number of laboratories, using core particles derived from a number of cell types, as well

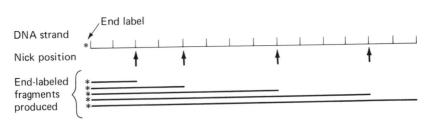

Fig. 6-8D. The source of the difference in pattern produced by end labeling. Only those fragments carrying the 5′ end label will appear on the autoradiograph of the gel.

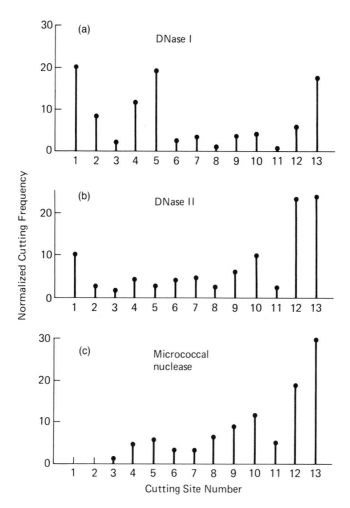

Fig. 6-9. Diagrams showing the relative frequency of cutting at different points in the nucleosome by various nucleases. (a) DNase I (Lutter, 1978). (b) DNase II (Lutter, 1981). (c) Micrococcal nuclease (Cockell et al., 1983).

as reconstituted particles (see, for example, Simpson and Whitlock, 1976; Noll, 1977; Tatchell and van Holde, 1977; Simpson and Kunzler, 1979). It is quite clear from Figure 6-9a that not all sites are cleaved with equal probability. In particular, sites approximately 30, 60, 80, and 110 bases from the 5' end (sites 3, 6, 8, and 11) are cleaved very slowly. This has been confirmed by studies in which the 3' end is labeled (Lutter, 1978). Clearly, *something* inhibits DNase I cleavage at these particular points. Is this a property of the nuclease or some inherent feature of core particle structure? Figure 6-9 compares cutting frequencies for DNase I with those for DNase II and micrococcal nuclease. Certain features appear to be common: All three nucleases cleave slowly at sites 3, 6, and 11, but the situation at site 8 is less clear. DNase II and micrococcal nuclease cut more frequently near the 3' end of the strand; DNase I shows, if anything, an opposite preference. Numerous attempts have been made to derive significance from these patterns. For example, the existence of resistance to nuclease cutting at sites 3 and 11, 0 and 8, 6 and 14, each separated by about 83 bp, has been used to support the idea that one turn of DNA on the nucleosomes corresponds to 83 bp of DNA (Lutter, 1978). Yet the micrococcal nuclease data of Cockell et al. (1983) do not show site 8 to be strongly protected.

The very existence of nulls in the cutting pattern obtained with 5'-labeled core particles has been used as an argument for psuedo-dyad symmetry. It is argued that the DNA at *either* (or both) 5' ends will be end-labeled in a typical preparation. If the particle were not symmetrical, one would expect the protected sites on the two strands to lie at different distances from the respective 5' ends. This would have the effect of obliterating any nulls, for what was not cleaved from one direction could be from the other. But the studies of defined sequence nucleosomes by Simpson and Stafford (1983) yield a very different result: The relative cutting frequencies at corresponding points on the two strands are very different.

Furthermore, in both these experiments and in similar studies by Ramsay et al. (1984) it is found that *one* strand is preferentially digested. The reason for the preference is obscure, but its existence casts doubt on many earlier interpretations of such experiments.

The most extreme explanation for the DNase I cutting pattern is that of Trifonov and Bettecken (1979). They point out that if the DNA has a particular, nonintegral periodicity (10.3–10.4 bp/turn), bonds at 30–31, 62, 82, and 113–114 will be so oriented as to be minimally accessible. The explanation is ingenious, but it suffers from two difficulties: (1) The periodicity must be uniform and lie within a very narrow range of values; and (2) since the explanation depends entirely on properties of the DNA, it cannot explain different cutting patterns by different enzymes. It seems likely, as Cockell et al. point out, that the cleavage pattern depends on properties of *both* the core particle and the individual nucleases. The observations of Riley (1980) further complicate the interpretation of DNase I cleavage data. He provides evidence that DNase I digestion of nucleosomes produces considerable single-strand DNA and contends that the interpretation usually given to such data is incorrect. In any event, what initially seemed to be a clarifying picture has become more obscure through further experimentation. It still seems likely that the pattern reflects *some* nonuniformity in the DNA conformation. As will be shown in a later section, the most recent X-ray diffraction studies suggest that the DNA is maximally distorted at just those points where DNase I digestion is least. In the view of Richmond et al. (1984), it is this distortion, rather than steric protection by histones, that may account for the pattern. This hypothesis has the advantage that it can account, in principle, for differences between nucleases; different nucleases may respond to distortion in different ways.

Thermal Denaturation Experiments
A quite different kind of evidence for the existence of distinguishable domains of histone–DNA interaction can be derived from thermal denaturation studies. The DNA in core particles is hypochromic and exhibits about 35% increase in absorbance when a solution of these particles is heated to temperatures approaching 100°C. This value is comparable to but slightly less than that of free DNA (cf. McMurray et al., 1985). The experiment is especially revealing when combined with parallel measurements of circular dichroism and calorimetry. Figure 6-10 displays such data, as derivative plots, for core particles from chicken erythrocytes (see also Weischet et al., 1978). Two wavelengths have been used for the CD measurements; as has been mentioned in *The DNA in the Core Particle,* above, the ellipticity at 273 nm is sensitive only to the DNA conformation, whereas that at 223 nm is dominated by the high α-helix content of the proteins.

The overall progress of the denaturation can be analyzed as follows: The hyperchromicity curve (d A 260/dt) is clearly biphasic; about 30% (or 42 bp) of the DNA melts at about 58°C, the remainder melting sharply at 74°C. Both of these temperatures are above the melting point of core-length DNA in this solvent (45°C). The nature of these transitions is elucidated by the CD studies; a part of the DNA shows a conformational transition before the first melting phase; the remainder increases in ellipticity to reach the value corresponding to free, double-strand DNA at about 72°. It then exhibits a small decrease (corresponding to melting) at about 75°C. The protein conformation is wholly unperturbed in the first transition, but denatures sharply at about 72°, just before the last of the DNA melts. The parallel calorimetric experiments also show that the first transition is a pure DNA melting process, since the enthalpy change per base pair (~8 kcal/mol base pairs) is equal to that for free DNA.

There are a number of reasons to believe that the 40 to 45 bp which melt first are at the ends of the core DNA. First, it has been shown in subsequent experiments using native and reconstituted particles containing DNA of different lengths that the magnitude of this first transition is proportional to the amount of DNA in excess of 100 bp (Tatchell, 1978; van Holde et al., 1980; Cowman and Fasman, 1980). More convincing are the elegant experiments of Simpson (1979), who utilized reconstituted core particles containing the synthetic DNA, poly(dA·dT)·poly(dA·dT). These particles undergo the first transition at a lower temperature, about 40 to 50°. This is preceded by a change in the DNA circular dichroism centered at 34°C, which is accompanied by a decrease in the sedimentation coefficient of about 10%. If particles carrying 5′ ^{32}P end labels are digested

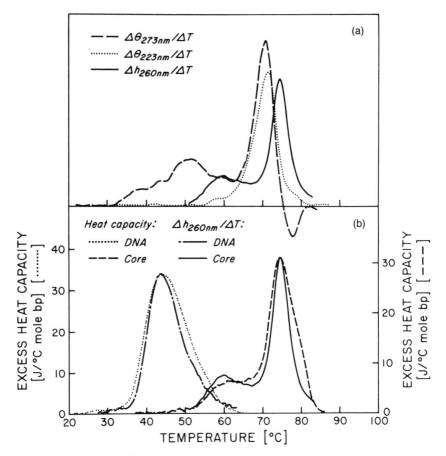

Fig. 6-10. Thermal denaturation of chicken erythrocyte core particles and core particle DNA. (a) As measured by circular dichroism at 273 nm ($\triangle\theta$ 273 nm; DNA region ——) and at 223 nm ($\triangle\theta$ 223 nm; protein region ------). Both are shown as derivative plots. From Tatchell (1978). (b) Measurements by hyperchromicity ($\triangle$h 260 nm) (shown as a derivative plot) and heat capacity.

with exonuclease III at 45°, the DNA is reduced in length to about 105 bp, and *all end label is lost*. Since exonuclease III will only attack single-strand DNA, this proves that melting in the first transition is from the ends.

Poon and Seligy (1980) have carried out careful electron microscopic studies of the thermal denaturation process of native core particles. As expected from the experiments described above, they observe no major change in particle structure until about 70°C. Above this temperature, the particles begin to open up into ring-like structures, some of which exhibit an apparent eightfold symmetry.

The model derived from all of these studies is depicted in Figure 6-11. Perhaps the most significant aspect is the division of the core particle DNA into three regions of differing interaction with the histones: two labile end regions of about 20 to 25 bp each, and a central region of about 105 bp that is much more firmly bound.

This concept is reinforced by other kinds of experiments. For example, Riley and Weintraub (1978) have observed that high levels of exonuclease III will digest into core particle DNA at 37°C with a strong "halting point" at about 100 bp, again suggesting that the end regions are more labile.

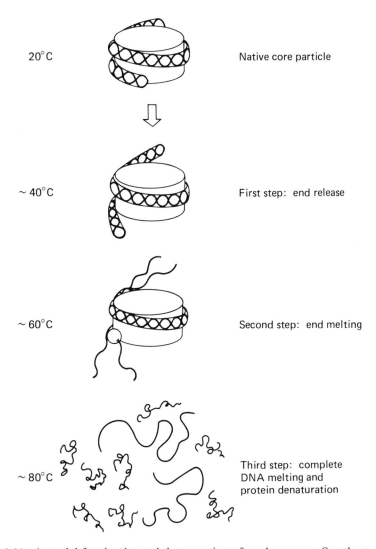

Fig. 6-11. A model for the thermal denaturation of nucleosomes. See the text.

Similarly, Smerdon and Lieberman (1981) find that in the digestion of H1-depleted chromatin by micrococcal nuclease there is a strong pause at 102 ± 4 bp. Shindo et al. (1980) report that the ^{31}P NMR of core particles shows a temperature dependence of the line width that parallels the hyperchromicity curve. Furthermore, at 65° (i.e., at the midpoint of the first transition), the NMR spectrum shows evidence for two states of the DNA.

The ionic strength dependence of the first transition has been utilized by McGhee and Felsenfeld (1980b) to estimate the number of phosphate groups that interact with the histones in this region. The interpretation of the data depends on the idea that approximately 0.88 Na^+ ions will be bound to each phosphate when the DNA is released from the histones (Record et al., 1978). While the calculation involves a number of assumptions and is acknowledged by the authors to be quantitatively imprecise, it is the order of magnitude of the answer that is surprising. Only about 15% of the phosphate groups in this region appear to be involved, corresponding to roughly one phosphate per turn of the DNA chain. It is as if each chain interacted with the histone core only through the phosphate residue closest to the core in each turn. This leads to the model diagramed in Figure 6-12. This model is also consistent with the observation of McGhee and Felsenfeld (1979) that almost all of the DNA residues in the core particle are accessible to a small reagent such as dimethyl sulfate. The DNA does not seem to be buried in the histone surface but, rather, contacted by it at certain rather widely spaced points.

The concept of such a "loose" connection between the histone core and the DNA supercoil is supported by a number of studies of DNA motions in the nucleosome. Researchers using a variety of techniques, in-

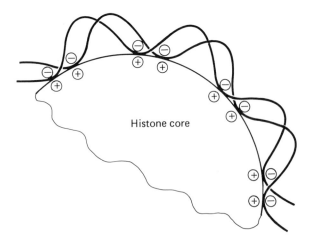

Fig. 6-12. A schematic view of the kind of histone–DNA interactions that could correspond to one electrostatic interaction of each DNA strand in each turn.

cluding ^{1}H NMR (Feigon and Kearns, 1979), triplet anisotropy decay (Wang et al., 1982), and fluorescence anisotropy (Ashikawa et al., 1983a, 1983b) all reached the same conclusion: The DNA in the core particle has nearly as much torsional freedom as does DNA free in solution! This result was wholly unexpected, and more recent studies suggest that it needs to be qualified. Schurr and Schurr (1985), in a reanalysis of the data of Wang et al., conclude that while the DNA retains considerable flexibility, it must be "clamped" at a number of points. Hilliard et al. (1986) have demonstrated by ^{13}C NMR that sugar motions are much inhibited in nucleosomal DNA. Finally, it should be noted that studies using fluorescent dyes can be compromised by effects of the dye binding on nucleosomal DNA. We have observed, for example, that ethidium bromide causes partial dissociation of core particles (McMurray and van Holde, 1986). Therefore, the conformational flexibility of DNA in chromatin should probably be regarded as still uncertain.

Mapping Histone–DNA Contacts

It has been argued above that the histones contact the DNA at a limited number of points and that the strength of interaction is nonuniform along the DNA. A more sophisticated question is: Which portions of the DNA are in contact with which histones? Can a "map" of histone–DNA interactions be constructed?

There are a number of ways of approaching this problem. One obvious method is to use nucleases to cleave chromatin or core particles into "subnucleosomes" and see which portions of the DNA remain associated with different histones in these fragments. The method involves the implicit assumption that rearrangement or redistribution does not occur following cleavage. This approach was pioneered by R. Rill and his collaborators (see Rill et al., 1975; Nelson et al., 1977; Rill and Nelson, 1977). A set of fairly well defined particles can be obtained. Bakayev and co-workers have expended considerable effort in the characterization of these particles (see Bakayev et al., 1977, 1981; Domanskii et al., 1982). The results are summarized in Table 6-6. The particles form an informative series. It may be presumed that SN1, SN2, and SN6 come from linker regions. SN4 and SN7 are an interesting pair, for they form a complementary set, each carrying a 5' end label and summing to 148 bp. One carries equimolar quantities of H2A and H2B, the other lacks only the H2A·H2B dimer. Very similar particles have been isolated in Rill's laboratory by either micrococcal nuclease or DNase I digestion (Rill et al., 1980; Nelson et al., 1982). In addition, these workers find a particle containing only H3 and H4, plus 70 to 80 bp of DNA. The implication of all of these experiments is that the H2A·H2B dimers lie near the ends of the DNA coil in the core particle.

A quite different approach has been employed by other workers to map histone–DNA interactions. Core particles carrying ^{32}P label on 5' DNA ends are reacted with dimethyl sulfate so as to methylate purine bases.

Table 6-6. Subnucleosomal Particles[a]

Particle	DNA size (bp)	Histones present
SN1	~20	None
SN2	~27	A basic nonhistone protein
SN3	~27	H4 plus the above
SN4	40[b]	H2A + H2B
SN5	~55	H2A + H2B + H3
SN6	~35	H1
SN7	108[b]	(H2A·H2B) (H3·H4)$_2$
SN8	~120	All core histones

[a]Data collated from Bakayev et al. (1977, 1981); Domanskii et al. (1982). For each particle, most recent data have been used. These particles have been obtained by digestion of mouse nuclei, hence H1 and nonhistone proteins are found as constituents.
[b]These DNA fragments have been shown to contain the 5' end (Bakayev et al., 1981).

The product is then depurinated; the aldehyde formed in this reaction produces a Schiff's base with a neighboring lysine, which is then stabilized by reduction with sodium borohydride (Levina and Mirzabekov, 1975). Extensive nuclease digestion will then leave, as the only histone linked to the 5' end, that which was closest to this end. One has only to isolate the histones and find which ones bear the radioactive label. The first attempts to apply this method to core particles seemed to indicate that both H4 and H3 lay near the ends of the core DNA (Simpson, 1976). However, these experiments were later shown to be compromised by the presence of a protein kinase in the core particle preparation. Repetition under conditions where this artifact was eliminated led to the observation of rather weak binding of the terminal phosphate to H3, and to no other histone (Simpson, 1978c). Thus, according to these experiments, H3 is close to both ends, a result seemingly in disagreement with the subnucleosome studies.

Reconciliation of these results, together with a much more detailed mapping of histone–DNA interactions, came through further development of Mirzabekov's technique (Mirzabekov et al., 1977, 1978; Belyavsky et al., 1980; Shick et al., 1980; Mirzabekov et al., 1982; Ebralidse and Mirzabekov, 1986). Essentially, the method is the same as described above for locating the histone at the ends of the DNA. End-labeled DNA is reacted with dimethyl sulfate, depurinated, and thus crosslinked at the depurination site with the adjacent histone. However, in the experiments we consider now, only low levels of reaction are allowed, so that a random distribution of DNA segments crosslinked to different histones is produced (see Fig. 6-13). The complex is then denatured and the histone–DNA adducts are separated. Cleavage of the DNA–histone bonds allows identi-

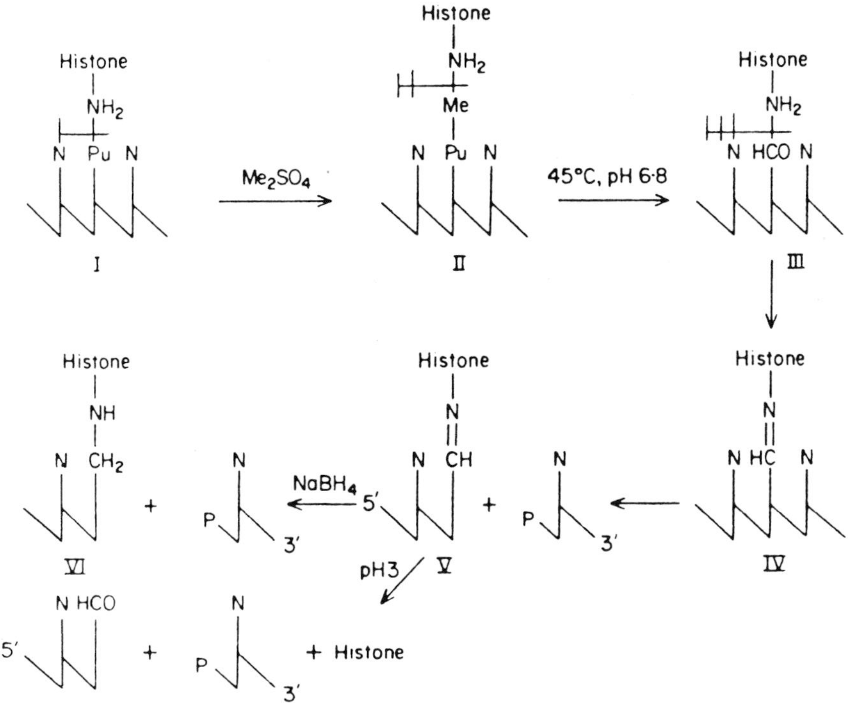

Fig. 6-13. A schematic of the technique developed by Mirzabekov and his collaborators to locate DNA–histone interactions in the nucleosome. This is the technique that produced the results shown in Figure 6-14.

fication of the histones and sizing of the DNA pieces. Autoradiography is used, so that only DNA fragments carrying an intact 5′ end are observed.

The experiment is complicated, and its interpretation has changed somewhat as techniques have been perfected. The major results are given here in Figure 6-14. Essentially, it is found that the histones bind to each strand in a symmetrical linear array. There appears to be some evidence for "crossing over"; that is, each histone has a preferred site, on a given strand, but can also interact less frequently with the opposite strand. In only a few instances do two histones interact with the same region of a given strand. About 20 residues on the 5′ end of each strand appear not to be bound to core histones, but the DNA duplex is definitely attached to the core in this region, for the 3′ ends are bound to both H3 and H2A. This result is consistent with Simpson's finding that the 5′ end could bind with low efficiency to H3 and can be reconciled with the subnucleosome studies if it is assumed that the interaction with H3 is a weak one, easily broken when nuclease cleaves off the SN4 particle. Further support comes from scanning transmission electron microscope studies of platinum-

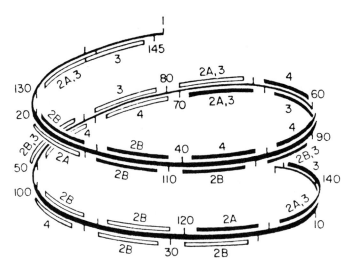

Fig. 6-14. Regions of histone–DNA contact in the core particle, as deduced by Mirzabekov and co-workers, using the technique described in Figure 6-13. From Shick et al. (1980). (Reprinted with permission from J. Mol. Biol. *139*, p. 491, Shick et al. Copyright 1980 Academic Press.)

labeled nucleosomes (Stoekert et al., 1984). These experiments locate H3 near the entry points of DNA into the nucleosome.

All of the above results indicate that the central region of the nucleosomal DNA is primarily in association with the (H3·H4)$_2$ tetramer. There exists a large body of experimental data to support the idea that the tetramer plays a major role in stabilizing the core particle structure. For example, H3 and H4 are the last to be dissociated from DNA as the salt concentration is raised (see Fig. 4-2). Taking advantage of this fact, Stockley and Thomas (1979) have depleted chromatin of all histones except H3 and H4 by gel filtration in 1 *M* salt. The chromatin was then digested with micrococcal nuclease and the products analyzed on sucrose gradients. Two components were obtained, a 6 S particle containing one (H3·H4)$_2$ tetramer and about 70 to 80 bp of DNA, and a 9.7S particle with *two* (H3·H4)$_2$ tetramers and DNA of the core particle length. It seems likely that the latter was formed by a disproportionation reaction, in which one tetramer is transferred to complex with another tetramer. Both types of particles could also be prepared by treating core particles with 1 *M* NaCl.

These experiments shed light on a series of studies that began with the demonstration by Camerini-Otero et al. (1976) that the pair of histones H3/H4, when reconstituted onto DNA by salt gradient dialysis, could protect a significant length of DNA from micrococcal nuclease. At low H3·H4/DNA ratios, the length protected was always about 70 to 80 bp. Further, it was shown that these two histones, especially at higher histone/DNA ratios, could effectively introduce supercoils into closed circular DNA (Camerini-Otero and Felsenfeld, 1977a) and generate particles that resemble nucleosomes in the electron microscope (Oudet et al., 1977a; Bina-Stein, 1978). A number of authors have demonstrated the formation of particles with $S_{20,w} \cong$ 6–7 S and 9–10 S upon reconstitution; higher histone/DNA ratios favor the 9–10 S particles (Camerini-Otero et al., 1977; Bina-Stein, 1978; Klevan et al., 1978). While there is some disagreement about details, the following is probably a fair summary of current understanding:

1. The (H3·H4)$_2$ tetramer is more strongly bound to the DNA than other histones; it is the first to associate as ionic strength is lowered and last to dissociate as ionic strength is increased.
2. A single tetramer of H3/H4 can bind to DNA, protecting about 70 to 80 bp in a 6–7 S particle. Long DNA to which single tetramers have been added is "primed" for the subsequent addition of H2A·H2B dimer. (Camerini-Otero et al., 1977).
3. At high ratios of (H3·H4)$_2$/DNA, a 9.7 S "psuedo core particle" is formed that contains an octamer of H3 and H4 and protects a core-length DNA from nuclease. This apparently produces full DNA supercoiling. The "psuedo core particle" can apparently also be formed by a disproportionation reaction (see above) at high salt concentration.

However, these observations must not be taken to imply that H2A and H2B do not play a significant role in core particle structure. It must be remembered that when DNA is reconstituted with a full, equimolar complement of the four core histones, two H2A·H2B dimers are incorporated in preference to another (H3·H4)$_2$ tetramer, and core particles indistinguishable from native particles are obtained. It would appear that contacts between H2A·H2B dimers and the (H3·H4)$_2$ tetramer are favored over (H3·H4)$_2$·(H3·H4)$_2$ interactions, a hypothesis supported by the fact that an H3/H4 octamer has never been reported free in solution. The (H3·H4)$_2$ tetramer seems to form a scaffold about which the DNA can initially be wrapped and into which H2A and H2B are inserted to complete the structure. However, the order of events need not be rigid, for it has been shown that DNA will fold readily about crosslinked octamer cores already containing all four histones (Stein et al., 1977). On the other hand, attempts to form core particles by *first* binding H2A and H2B are unsuccessful.

The weaker binding of H2A·H2B dimers in the core particle is attested to not only by salt-dissociation experiments but by a number of other techniques that allow its selective removal. Sibbet and Carpenter (1983) found that treatment with 0.25 M NaCl + 4 M urea quantitatively and reversibly removed these dimers. Jordano et al. (1984a, 1984b) reacted histone lysine residues with dimethylmaleic anhydride. This causes release of most of the H2A·H2B dimers. The residual particles exhibited modified circular dichroism and thermal denaturation and could protect only about 70 bp of DNA from nuclease digestion. Finally, as we shall see in Chapter 9, there is some evidence for the exchange of H2A·H2B dimers in vivo.

All of these results can be readily reconciled with the models presented above. Taken altogether, both chemical and physical studies yield a remarkably coherent picture of the overall organization of histones and DNA within the core particle. However, the picture has been *deduced* in rather indirect ways, and it lacks spatial detail. To proceed further, we must turn to physical studies of the particle conformation, with particular emphasis on the results from diffraction techniques.

A Model for Core Particle Structure

On the basis of the above information, together with scattering and diffraction data, it is now possible to present a fairly detailed model for the core particle. As indicated in Chapter 2, early hydrodynamic and electron microscopy studies indicated the particle to be compact, with a diameter of roughly 100 Å (see Table 2-1; also Olins et al., 1976). Low-angle neutron scattering from solutions led to considerable refinement; the data were best fitted by a disk-shaped object, about 50 Å thick, with the DNA wrapped in two coils or a helix of about 110 Å outer diameter, about a disk-like protein core of 60–70 Å diameter (Pardon et al., 1977a, 1977b; Suau et al., 1977; Braddock et al., 1981). A disk model of this kind was actually first suggested from images of unstained, unshadowed particles obtained with the scanning transmission electron microscope (Langmore and Wooley, 1975).

However, the most detailed information comes from diffraction studies with crystals of core particles. This work was initiated in the MRC laboratory at Cambridge, under the direction of A. Klug. The Nobel Prize awarded to Dr. Klug in 1983 was earned in part by these experiments. Since the first publication (Finch et al., 1977), the crystal quality and resolution have been steadily improved. A listing of the most important papers is given in Table 6-7. This program, which has combined X-ray diffraction, neutron diffraction, and electron microscope image resolution, has now reached the stage at which a quite detailed model of the core particle can be presented. I shall concentrate on the most recent results; readers interested in the development of the model should consult the earlier papers listed in Table 6-7.

In the X-ray study by Richmond et al. (1984), crystals of core particles from beef kidney were employed. Isomorphous replacement with two heavy atom derivatives was used for phase determination. The overall resolution is about 7 Å, close to the limiting resolution for these crystals. I shall refer to this as an "intermediate resolution" model. To obtain truly high resolution (~3 Å), particles homogeneous in both DNA and histones would probably be required. These experiments reveal a wealth of new detail about the core particle, which can best be summarized by considering first the overall particle and then the DNA and histone components.

1. *The particle:* The overall shape and dimensions of the core particle are very close to those deduced previously (see Fig. 6-15a). Symmetrically

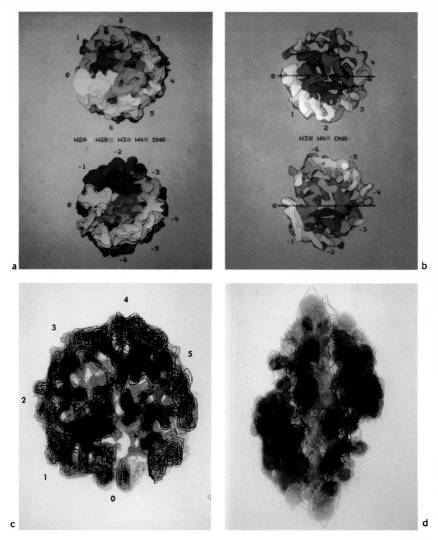

Fig. 6-15. Results from x-ray diffraction studies of nucleosomal core particles. (a) and (b) are from the laboratory of Dr. A. Klug. (Reprinted by permission from *Nature* vol. 311, pp. 532–537, Richmond et al. Copyright © 1984 Macmillan Magazines Ltd.) (a) shows "top" and "bottom" views of the whole core particle. Turns of DNA are numbered from $+7$ to -7, with the dyad axis parallel to the plane of the paper and passing through O. In (b), the structure has been opened up about the dyad axis, and only H3 and H4 histones are shown. Figures courtesy of Dr. A. Klug. (c) and (d) are from the studies of E. Uberbacher and G. Bunick. (See Uberbacher and Bunick, 1985a.) In (c) the "upper" half of the core particle is shown, with DNA running from bp 73 at position zero (dyad) to the end (past 5). H3 is shown in blue, H4 in green, H2A in violet, H2B in dark brown, and DNA in orange. (d) shows one of a series of cross-sections of the particle, looking down the dyad axis from near position 4. A portion of the DNA can be seen, and the central position occupied by H4; H3 is on the other side of the molecule. H2A·H2B dimers lie near the DNA ends. Courtesy of E. Uberbacher and G. Bunick.

Table 6-7. Diffraction Studies of Core Particles and Histone Octamers

Method	Resolution	Comments	Reference
X-ray diffraction	25 Å	Histone somewhat proteolyzed	Finch et al. (1977)
Image reconstitution from EM photographs of histone octamer fibers	22 Å	Gives information concerning shape of histone core	Klug et al. (1980)
Neutron diffraction	25 Å	Used protein and DNA matching to get low-resolution images of both portions	Finch et al. (1980)
X-ray diffraction	not stated	Used new crystal form; histones not proteolyzed	Finch et al. (1981)
Neutron diffraction	25 Å	Similar to Finch et al. (1980), but using new crystals described in Finch et al. (1981)	Bentley et al. (1981)
Neutron diffraction	16 Å	Resolves individual dimer domains	Bentley et al. (1984)
X-ray diffraction	7 Å	See text	Richmond et al. (1984)
X-ray diffraction	15 Å	See text	Uberbacher and Bunick
X-ray diffraction of histone octamer	3.3 Å	See text	Burlingame et al. (1985)

placed features revealed at this resolution argue strongly for a psuedo dyad axis passing through the center of the DNA molecule.

2. *DNA:* As expected, the DNA is wrapped about the histones in a left-hand double helix. There are about 7.6 turns of DNA per superhelical turn. Although the DNA structure is clearly of the B type, present resolution does not allow an exact determination of the number of residues per turn. Nor can the exact number of turns be determined, for the ends of the DNA are indistinct. This may be partly a consequence of the slight DNA length heterogeneity and partly a result of crystal packing effects; there is some interaction between DNA and histones in adjacent particles in the lattice. A most striking feature is that the DNA is not uniformly bent; there are regions of quite sharp bending at four symmetrically located positions (± 1, ± 4 DNA turns, taking the dyad axis position as zero). These sharp bends, which are not quite so extreme as previously postulated "kinks," can be clearly seen in Figure 6-16. Examination of the figure reveals, in fact, that the winding of the DNA is quite nonuniform, as might be expected from recent high-resolution studies of DNA structure (see Chapter 3). In particular, the widths of the major and minor grooves change quite dramatically along the helix. Interestingly, the positions of sharp bends (± 1, ± 4) correspond exactly to the positions observed for maximum protection of the DNA from DNase I cutting (see *DNA–Histone Interactions in the Core Particle,* above). Furthermore, positions ± 1 are those ob-

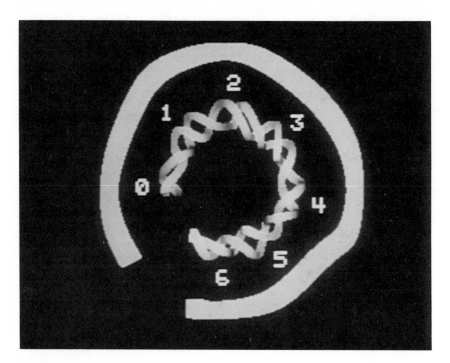

Fig. 6-16. Half of the DNA coil in the core particle, as depicted by Richmond et al. (1984). (Reprinted by permission from *Nature* vol. 311, p. 532–537. Copyright © 1984 Macmillan Magazines Ltd.) The white band has been drawn to emphasize the irregularity of DNA bending in the coil.

served by McGhee and Felsenfeld (1979) to be maximally accessible to dimethyl sulfate. Richmond et al. suggest that these differences in reactivity may be consequences of local DNA conformation, rather than histone propinquity as had been assumed.

3. *Histones:* At the present resolution, it is not possible to exactly delineate individual histone molecules. Neither can one tell if all of the histone mass is resolved; it is possible that nonstructured histone tails may be missing. Nevertheless, certain remarkable general features appear, and some of the data, together with the histone–histone and histone–DNA interaction studies described in the preceding section, allow a reasonably certain assignment of the locations of individual histone molecules in the structure.

A prominent feature in Figure 6-15a and 6-15b is the way in which the histone octamer contacts the DNA. There is no evidence in the model of Richmond et al. for histone mass extending beyond the DNA, nor do segments extend around the grooves. Rather, the core seems to make periodic close contacts with the inner DNA surface. As Richmond et al. point out, if every histone contact involved one phosphate residue, about 20% of the phosphates would be neutralized. This is in excellent agreement with the estimate of McGhee and Felsenfeld (1980), based on melting the DNA terminal regions.

Hypotheses as to the locations of particular histone molecules were put forward in an earlier paper (Klug et al., 1980), based in part on the DNA–histone crosslinking studies of Mirzabekov and co-workers (see *DNA–Histone Interactions in the Core Particle,* above). These assignments are reinforced by the intermediate-resolution study. For example, it had been proposed that the H3 dimer lay across the dyad axis near the point where this axis passes through the center of the DNA. This is confirmed by the fact that the heavy metal label on cysteine residue 110 of H3 lies precisely on the dyad axis and in the expected region. A detailed picture is shown in Figure 6-17. Adjacent to these H3 molecules are found two more regions of high electron density, which Richmond et al. tentatively identify as portions of H4 molecules. Since these are related by dyad symmetry, they can form, together with the H3 dimer, the kind of "helical ramp" postulated by Klug et al. (1980) as the organizing center of the nucleosome.

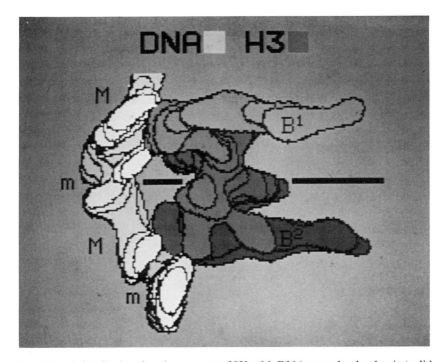

Fig. 6-17. A detail, showing the contact of H3 with DNA near the dyad axis (solid bar), according to Richmond et al. (1984). (Reprinted by permission from *Nature* vol. 311, pp. 532–537. Copyright © 1984 Macmillan Magazines Ltd.) Two H3 molecules are shown, making contacts with minor grooves of the DNA (m, minor groove; M, major groove). The position of the crosslinkable sulfhydryls lies near the center of this image.

The putative H4 molecules make substantial contacts with the DNA out to positions ±3.5, and minor contacts at slightly greater distances. This is in very good agreement with the observations (see above) that a single (H3·H4)$_2$ tetramer can protect about 70 to 80 bp of DNA.

This places the H2A·H2B dimers near the ends of the DNA, in accord with many other studies. The histone region identified by Richmond et al. as one of the dimers lies directly under the last 30 to 40 bp of DNA. The other dimer is oriented somewhat differently, as can be seen in Figure 6-15a. This is the sole clear exception to twofold symmetry. Both of the dimers make specific contacts with the DNA of adjacent nucleosomes, and in order to do so in the lattice they must be arranged asymmetrically. Richmond et al. point out that this facile displacement of an H2A·H2B dimer may have biological significance; it may represent an intermediate stage in nucleosome disassembly. It is also entirely consistent with the many observations cited in preceding sections that indicate relatively easy removal or replacement of H2A·H2B dimers. More difficult to reconcile with this model are the results of Ellison and Pulleyblank (1983a, 1983b, 1983c), who find the formation, upon reconstitution, of particles containing two H2A·H2B dimers, but only *one* H3·H4 dimer. These particles appear compact and can protect about 100 bp of DNA from digestion.

An independent X-ray diffraction study of the core particle has been reported by Uberbacher and Bunick (1985a). The crystal form used contains two core particles in the asymmetric unit and has been solved to a 15 Å resolution. The structure bears strong resemblance to that deduced by Richmond et al., at least insofar as *overall* dimensions, DNA conformation, and protein distribution are concerned. However, there are also some significant differences:

1. Uberbacher and Bunick see evidence for projection of some protein between the DNA gyres, particularly between positions $+4$ and -4, and near positions $+1$ and -1.
2. In the model of Richmond et al., there is evidence for protein density extending over the DNA near position zero, which has been suggested to bend or kink the DNA as it exits into the linker. Uberbacher and Bunick do not find this.
3. Whereas in the model of Richmond et al., the two H2A/H2B domains are different, leading to asymmetry in the particle, Uberbacher and Bunick find both particles in the asymmetric unit to be symmetrical and identical.

The model of Uberbacher and Bunick appears to more closely resemble, with respect to points (1) and (3), the results of a recent neutron diffraction study of the core particle (Bentley et al., 1984). These studies also reveal contacts between H3 and the ends of the DNA, in agreement with cross-linking results.

None' of the above-mentioned studies is consistent with the high-resolution model for the *histone core* presented by Burlingame et al. (1985). The discrepancies have generated considerable controversy (Moudrianakis et al., 1985a, 1985b; Klug et al., 1985; Uberbacher and Bunick, 1985b). As discussed in Chapter 4, it now seems likely that the discrepancy results either from differences in solution conditions under which crystallization occurred, conformational changes in the histone core upon DNA binding, or some combination of the above.

With three independent studies, using two different diffraction techniques, yielding generally consistent models, we may now believe that we have a reasonable, low-resolution picture of the core particle. While details must await higher resolution analysis, we have at this point a quite comprehensive and internally consistent model for this structure. There remain, however, certain areas in which the picture is incomplete. The role of the highly charged (and highly modifiable) histone tails is obscure. It is not known exactly how H1 and other lysine-rich histones interact with the core particle. Finally, certain nonhistone proteins have been shown to bind to the nucleosome. Do they change the structure, and if so, how? It is to these questions that we now turn.

The Sites of DNA–Histone Interaction

We still know relatively little concerning the actual sites on the histone core that interact with the DNA. Since elevated salt concentrations destabilize the core particle, electrostatic bonds presumably play a major role. But there are many (over 200) lysyl and arginyl residues in each histone core, and all of the evidence indicates that only a small fraction of these are actually involved in such bonding. The question, then, is whether these lie preferentially in certain portions of the histone sequences.

The N- and C-terminal tails of the histone molecules are especially rich in positively charged groups. An early hypothesis held that the "globular" regions of the histones were primarily involved in histone–histone interactions, and that the terminal tails, by wrapping around the DNA, secured it to the core (see, for example, van Holde et al., 1974a). Like many "obvious" hypotheses, this is probably at least partially incorrect. There are now several kinds of evidence that argue quite strongly against so simple a model.

First, the idea that the flexible tails wrap in DNA grooves is counterindicated by experiments which show that the grooves of the DNA must be quite uniformly accessible. McGhee and Felsenfeld (1979) have found that a small reagent such as dimethyl sulfate can react with virtually every base pair. Even stronger evidence against histone groove occupancy is the fact that glucosylated T4 DNA can form stable nucleosomes (McGhee and Felsenfeld, 1982).

Second, there are a whole series of experiments that indicate the terminal tails of the histones to be relatively unimportant in maintaining core particle structure. These are experiments in which these portions of the histones have been selectively excised by digestion with trypsin or other proteases. When histones free in solution are attacked by trypsin, digestion is more or less random and rapid because of the large number of lysyl and arginyl residues. But digestion of core particles with trypsin gives rise to a reproducible set of defined peptides, as first noted by Weintraub and van Lente (1974; see Fig. 6-18). A number of the sites of cleavage have been identified; in general it is found that the "unstructured" N- and C-terminal portions have been removed. The process of degradation has been studied in detail by L. Böhm and collaborators, and the results are summarized in Table 6-8. A number of research groups have examined the physical properties of particles modified in this way. The remarkable finding is that

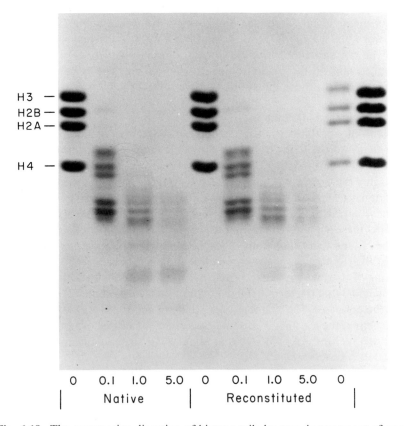

Fig. 6-18. The progressive digestion of histone tails by trypsin treatment of core particles. From left to right, chicken erythrocyte particles were digested with 0, 0.1, 1.0, and 5.0 μg/ml trypsin. Data from both native and reconstituted particles are shown. H3 is cleaved first, followed by H2A. The column headed "0.1" includes most of the defined products listed in Table 6-8. From the top, there are P1 and P1′ (superimposed here), P2 (migrating with H4), P3, and P4 and P5. From Tatchell (1978).

Table 6-8. Proteolytic Residues of Chicken
Erythrocyte Histones

Gel band	Histone	Residues	Reference
P1′	H3	21–135[a]	Böhm et al. (1981)
P1	H3	27–129	Böhm et al. (1981)
P2	H2A	12–118	Böhm et al. (1980a)
P2′	H2B	21–125[a]	Böhm et al. (1982)
P3	H2B	24–125[a]	Böhm et al. (1982)
P4	H4	18–102[a]	Böhm et al. (1981)
P5	H4	20–102[a]	Böhm et al. (1981)

[a]C-terminal residue.

tryptic cleavage of these regions, which altogether will remove about 40 to 50% of the positive charges of the histone core, seems to have relatively little effect on the core particle. For example, digestion to this limit of either chicken erythrocyte or HeLa core particles yields particles that still have high sedimentation coefficient (9–10S) and yield clear "ladder" patterns when digested with DNase I (Lilley and Tatchell, 1977; Whitlock and Simpson, 1977). Even more startling are the results of Whitlock and Stein (1978), who showed that histone cores from which the tails have been cleaved by trypsin were fully capable of refolding DNA into core particles in a standard reconstitution experiment. They have carried out calculations to show that the frictional coefficient of the particle has changed hardly at all, if the loss of histone mass is taken into account.

More recently, Rosenberg et al. (1986) have utilized chymotrypsin in a parallel fashion. The results are quite similar to the trypsin studies. Again, a "limit" digest is reached in which tails have been removed without major structural change.

It should not be assumed that these proteolyzed particles are *unchanged,* even though they remain intact. There is a significant increase in circular dichroism in the DNA region of the spectrum, and the thermal denaturation pattern is considerably modified; much more of the DNA now melts at lower temperatures. The DNase I pattern obtained with end-labeled DNA is distorted, particularly near the ends of the core DNA. The data seem to indicate that the residual histone core, stripped of its terminal segments, remains intact and can still bind DNA, albeit not as strongly and perhaps not in the same conformation as does the complete core.

NMR measurements have provided additional information. In an important study, Cary et al. (1978) reported that whereas the H2A and H2B tails exhibited partial flexibility even at low ionic strength, the H3 and H4 tails appear to be tightly bound and immobile up to at least 0.3 M salt. Rill and Oosterhof (1982) interpret somewhat differently results of studies of the accessibilities of arginine residues to an arginine-specific protease. They find that residues in the N-terminal region of H3, H4, and H2B are all accessible to cleavage in nucleosomes, but that H2A is highly resistant. Accessibility must mean more in this context than simply surface exposure, for arginines 132 and 134 in H3 are not cleaved, even though Muller et al. (1982) have shown by immunological studies that the C-terminal hexapeptide is exposed. Rill and Oosterhof also suggest that ^{13}C NMR measurements can be interpreted somewhat differently than the ^{1}H NMR studies of Cary et al. They propose that the N-terminal regions of H3 and H4 may be relatively free, whereas the N-terminal region of H2A and the C-terminal region of H2B are more constrained. In summary, it seems that there is evidence for both freedom and restraint in these tail regions, but some uncertainty as to specifics.

As was emphasized in Chapter 4, the N-terminal tails of the core histones are the prime targets for acetylation. Such modification of lysine residues reduces the positive charge in these regions. This reduction is not dramatic; even with hyperacetylation, only about 12 to 17 lysine residues out of 78 in the tail region have been neutralized. But such modification can have effects on the core particles comparable to that of removal of the entire tails by proteolytic cleavage (see Simpson, 1978a; Bode et al., 1980, 1983; Bertrand et al., 1984, for example). Results from different laboratories are not entirely consistent; this may be because the levels of acetylation are not equal in various studies. Recent experiments in our laboratory have utilized fractionated HeLa core particles having extremely high levels of acetylation (Ausio and van Holde, 1985; see Fig. 4-18). Such particles exhibit a lower sedimentation coefficient (about 9.5S) and an enhanced molar ellipticity (Fig. 6-19) as compared to HeLa core particles with low levels of acetylation. Furthermore, in agreement with Bode et al. (1980), the thermal denaturation curve shows a substantial increase in the amount of DNA melting in the lower transition, as well as decreases in T_m values (Fig. 6-20). It seems, then, that the effects of high levels of histone acetylation are similar to, though not so extreme as, the effects of tryptic cleavage of the histone tails. This is consistent with the fact that acetylation sites are clustered near the extreme N terminus in most histones (see Fig. 4-17, Chapter 4). Our data do not agree, however, with the conclusion of Bode et al. that *major* conformational changes follow upon acetylation.

From all of these data, we conclude that the globular portions of the histones interact sufficiently strongly with the DNA to wrap it about the core. At least some of the N-terminal regions must interact with the DNA as well, to confer additional stability. The effect of histone acetylation, at the core particle level, is not to disrupt but to somewhat destabilize the nucleosome. It is conceivable that acetylation has additional effects on higher-order structure in chromatin. This possibility will be discussed in Chapter 7.

If the major electrostatic interactions of the DNA phosphate are with the globular part of the histone core, what is the nature of the protein sites? There is mounting evidence to suggest that the majority may be

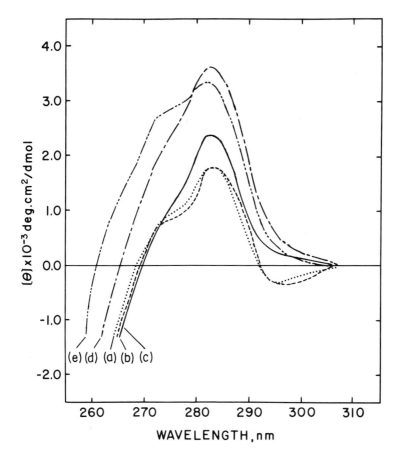

Fig. 6-19. The effect of hyperacetylation on the circular dichroic spectrum of core particles. Curves are (a) (• • • • •), chicken erythrocyte; (b) (------), nonacetylated HeLa; (c) (———), hyperacetylated HeLa particles; (d) (–•–), as in (b), but in 0.65 *M* NaCl, (e) (–• •–), as in (c), but in 0.65 *M* NaCl. (Reprinted with permission from *Biochemistry 25*, p. 1421, Ausio and van Holde. Copyright 1986 American Chemical Society.)

arginine residues. As early as 1976, Mansy et al. (1976) obtained results from Raman spectroscopy suggesting hydrogen bonding to N-7 of guanine both in chromatin and in poly-L-arginine–DNA complexes. More direct evidence has come from accessibility studies. Rill and Oosterhof (1982) found that an arginine-specific protease had "reasonable access" to at most eight arginines in the entire core particle, and that most of these lay in N- or C-terminal regions. The most convincing data, however, are those of Ichimura et al. (1982). Using low molecular weight reagents specific for lysine and arginine, they obtained the data shown in Table 6-9. If these data are interpreted literally, they indicate that the only residues in the

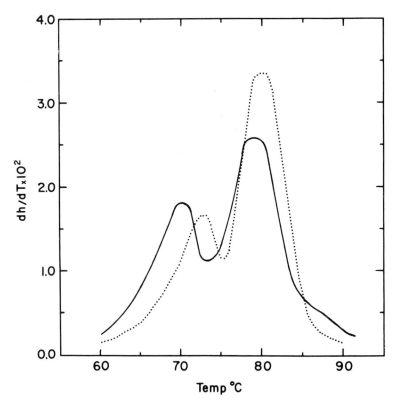

Fig. 6-20. Differential thermal denaturation graphs for (••••••) nonacetylated HeLa core particles; (———) hyperacetylated HeLa core particles. (Reprinted with permission from Biochemistry *25*, p. 1421, Ausio and van Holde. Copyright 1986 American Chemical Society.)

Table 6-9. Accessibility of Lysine and Arginine Residues in Core Particles[a]

	Lys	Arg
1. Total residues in particle	116	104
2. No. in N- and C-terminal regions	78	44
3. No. in globular regions	38	60
4. Accessible in 20 mM buffer	0	0
5. Accessible in 0.3–0.6 M NaCl	78	42
6. Accessible in 2.0 M NaCl	78	56
7. Made accessible by releasing histone tails [(5)–(4)]	78	42
8. Made accessible by removing DNA [(6)–(5)]	0	14

[a]Data of Ichimura et al. (1982). Accessibility is defined as reactivity of lysines with 2,4,6-trinitrobenzoic acid, reactivity of arginines with 2,3-butanediol.

globular core involved in histone–DNA interaction are 14 arginines. Most of the remaining 84 lysines and arginines in this region, Ichimura et al. suggest, are involved in salt bridges with the 74 aspartic and glutamic acid residues. Certainly this is an extreme view, and it is possible that the results are confused by rearrangements that occur when the DNA is released in high salt. The data are also in seeming conflict with the report by Cary et al. (1978) that some of the histone tails exhibit motional freedom even at low salt. Finally, one must ask why arginine residues should be so preferred over lysines for such interactions. Ichimura et al. suggest that this is because arginines can form both hydrogen-bonding and electrostatic interactions with phosphate residues:

$$- R - NH - C_{\oplus} \begin{array}{c} NH_2 \cdots O \\ \\ NH_2 \cdots O \end{array} \begin{array}{c} \\ \ominus P \\ \end{array} <$$

Indeed, there exist (unpublished) data of G. Felsenfeld to indicate that oligo-L-arginine binding to DNA is considerably stronger than oligo-L-lysine binding.

Other kinds of evidence concerning the histone core–DNA interaction are fragmentary and not wholly consistent. Taillandier et al. (1984) have examined the interactions of proteolytic fragments of histones with DNA by determining the length of DNA prevented by each from undergoing the B→A transition. They observe a correlation between the predicted α-helix content and the length of such "blocked" regions and argue therefore for interaction of helical regions of the protein with the DNA. The difficulty with this interpretation is that α-helices, when they bind to DNA in other DNA-binding proteins, seem invariably to fit into the major groove. Yet the reconstitution studies of McGhee and Felsenfeld (1982) using glucosylated T4 DNA indicate that the major groove must be open. Furthermore, Goodwin et al. (1979) argue from Raman spectroscopic data that DNA–histone interactions in nucleosomes are probably mostly through the minor groove.

Clearly, it is not yet possible to describe histone–DNA interactions in the core particle in any detail. The final answers will have to come primarily from higher resolution X-ray diffraction studies.

Organization of the Nucleosome Above the Core Particle Level

Chapter 2 presented evidence that the repeating unit of any chromatin—the nucleosome—was composed of two parts: a ubiquitous "core" particle and a "linker" region that varies between chromatins from different sources. In the preceding section, the structure of the core particle was described in considerable detail; now we examine the content and structure

of the linker region. We should first ask: What proteins are associated
with the linker, and how do they interact with the DNA and with the core
particle?

Interaction of H1: The Chromatosome
Almost from the first, it had been suggested that lysine-rich histones must
reside within the linker region (see, for example, van Holde et al., 1974a).
But the first *direct* experimental evidence to support this conjecture appears in 1976 (Varshavsky et al., 1976; Garrard et al., 1976). These workers
pioneered the use of gel electrophoresis for the analysis of mixtures of
nucleoprotein particles (as opposed to pure DNA or pure proteins). They
showed that nucleosome "monomers" as isolated from sucrose gradients
were in fact heterogeneous. At least two bands could be detected on such
gels; the slower had longer DNA and carried H1, the faster migrating had
shorter DNA and no H1. Subsequent studies by these groups and others
have refined this technique to yield highly discriminating analysis (see
Bakayev et al., 1977, 1978; Pospelov et al., 1977; Todd and Garrard, 1977,
1979; Albright et al., 1979, 1980). Several subclasses of mononucleosomal
particles can be electrophoretically resolved, differing in DNA length and
the nature and amount of proteins carried. A summary is provided in Table
6-10.

The major point of interest at the moment, however, is that the lysine-
rich histones are found *only* in particles with longer DNA ($\geqslant$ 160 bp).
Similar conclusions were drawn on the basis of nuclease digestion studies
by Whitlock and Simpson (1976b) and Noll and Kornberg (1977). These
workers provided evidence that digestion from about 160 bp to about 140
bp was accompanied by the loss of H1. The case was made conclusively
when Simpson (1978b) isolated and characterized a nucleoprotein particle
containing "about 160 bp" of DNA, a histone octamer, and one molecule
of H1. This particle he termed the *chromatosome*.

At the present time, it is generally accepted that the chromatosome
carries 166 to 168 bp of DNA, an amount that would correspond to two
full superhelical turns about the core, according to the model of Richmond
et al. (1984). From DNase I digestion experiments using end-labeled DNA,
Simpson concluded that about 10 bp have been added to each end of the
core DNA (see also Lawrence and Goeltz, 1981). The presence of the H1
molecule stabilizes the particle against thermal denaturation; not only is
the first transition suppressed, but the higher temperature transition is
shifted to a still higher T_m; thus, *all* of the DNA in the particle is stabilized
by the presence of H1 (Simpson, 1978b; Cowman and Fasman, 1980). On
removal of H1 from such particles, all of the DNA in excess of ~100 bp
melts in the first transition (Tatchell and van Holde, 1979; Cowman and
Fasman, 1980). It has also been demonstrated that the presence of H1
prevents the chromatosome from undergoing a conformational transition

Table 6-10. Composition of Electrophoretically Separated Nucleosome Components

Component	DNA size in bp			Proteins present (numerical values are moles per mole octamer)						
	Calf thymus[a]	Calf thymus[b]	Mouse mastocytoma[c]	Core histones, M1, M2[d]	uH2A[b,c,e]	HMG[b] 14,17	HMG[b] 1,2	H1[b,c,e]	M4[b] NF[d]	other[b] NHP
MV	190–220	185–220	200–220	+		+	+		+	+
MIV	185–205		190–210	+	0.14	+	+	0.72	+	+
MIII	160–210	160–210	160–200	+	0.12	+	+	0.82	+	+
MII	160–185	140–185	150–170	+	+	+	+	−	−	−
MI	140–175	140–175	140–160	+	<0.01	−	−	−	−	−

[a]Todd and Garrard (1977).
[b]Albright et al. (1980).
[c]Todd and Garrard (1979).
[d]Proteins M1, M2, M4, NF are minor histone variants (see Albright et al., 1980).
[e]Albright et al. (1979).

at low ionic strength (Burch and Martinson, 1980; see *Conformational Changes at Low Ionic Strength,* below).

It should be pointed out that while H1 stabilizes the chromatosome, as much as 168 bp of DNA can be resistant to micrococcal nuclease even in the absence of lysine-rich histones, if digestion conditions are chosen correctly. Weischet et al. (1979) have shown that nuclei depleted of H1 will yield 168 bp particles when the micrococcal nuclease digestion is carried out in salt concentration from 0.3 to 0.4 M. Similarly, Todd and Garrard (1977) and Albright et al. (1980) find an MII fraction containing only core histones but with DNA as large as 185 bp (see Table 6-10). Thus, there appears to be the possibility of contact between the ends of chromatosome length DNA and the histone core, under some conditions. From the model presented in *A Model for Core Particle Structure,* above, one might presume these contacts to be with H3 molecules. In support of the existence of such defined points of contact are the reconstitution experiments of Tatchell and van Holde (1979). It was found that whereas particles reconstituted with end-labeled DNA molecules smaller than or equal to 166 bp gave well-defined DNase I digestion patterns (suggesting defined "registry" of the DNA on the core), a reconstitute with 177 bp DNA yielded a "smear" on DNase I digestion, indicative of multiple alignments.

The sum of all of this evidence is to strongly suggest that the H1 molecule must interact with the extra 10 bp at each end of the chromatosome DNA, linking these together. In further support of this idea, Thoma et al. (1979) point out that extended fibers of H1-containing chromatin have a zigzag structure, as if the points of entry and exit of the DNA from the individual nucleosomes were close together; in contrast, H1-depleted chromatin fibers are more extended and look as if exit and entry points are far apart. These authors have suggested a specific model for the interaction of H1 with the ends of the DNA and its role in determining higher-order structure in chromatin. Such matters will be discussed in Chapter 7.

There is abundant evidence that the lysine-rich histones are also bound to the histone core of the nucleosome. Crosslinking studies using carbodiimides as contact site reagents yield H1-H2A adducts (Bonner and Stedman, 1979; Ring and Cole, 1979; Boulikas et al., 1980). A variety of other crosslinking agents have also been employed in such studies (see, for example, Ring and Cole, 1979; Glotov et al., 1978; Thomas and Khabaza, 1980). The partners most frequently found for H1 are H2A and H3.

In the study by Boulikas et al., the zero-length crosslink was found to form between the globular domains of both H1 and H2A. A number of laboratories have shown that the central, globular region of the lysine-rich histone is primarily responsible for the stabilization of the chromatosome (see, for example, Allan et al., 1980; Ishimi et al., 1981; Puigdomenech et al., 1983). Hence, we must conclude that this portion of H1 must be so positioned as to contact *both* H2A and the DNA ends. As Allan et al. point out, the globular domains of the lysine-rich histones are of about the right diameter (29 Å) to lie between the coils of DNA emerging from the core particle. This idea is supported by comparative studies of the interaction of H1 with superhelical and relaxed DNA molecules (Vogel and Singer, 1975a, 1975b; Singer and Singer, 1976, 1978; Triebel et al., 1984). Histone H1 has a marked preference for superhelical DNA. Furthermore, this preference is exhibited only by the central, globular region; isolated H1 "tails" bind without preference.

This raises a problem in stoichiometry and symmetry. How *many* lysine-rich histone molecules actually interact with each core particle? Dyad symmetry of the core particle would presumably generate two identical sites. However, the evidence is not in favor of two H1 molecules per nucleosome. First, Simpson finds only one H1 molecule on each chromatosome. The available data on the stoichiometry of lysine-rich histones in chromatin, while subject to numerous sources of error, is in general agreement that there is about one (or even less than one) H1 molecule per nucleosome in most chromatins (see Table 6-11 for a compilation of recent data). Although these data are consistent in pointing to an *overall* stoichiometry of one lysine-rich histone per core particle, it might be argued that some nucleosomes have two and some none. The experiments of Hayashi et al. (1978) indicate that this is not a general pattern. After carefully stripping nucleosomes of the *core* histones by gentle SDS treatment, they find only 140 bp DNA fragments with no H1, and 160 bp fragment with one molecule of H1. *No* fragments containing two are observed. The chromatins of avian erythrocytes, which have H5 as well as H1, are an evident exception to the above rule, but these certainly represent special cases. In experiments in which H1 is added back to stripped chromatin, Nelson et al. (1979) do find evidence for a second H1 binding site. However, this site is occupied only after each nucleosome carries an H1 in the first site. Evidently, the first site is much stronger. Furthermore, these workers show that occupation of the first site alone is sufficient to protect about 160 bp of DNA from micrococcal nuclease digestion.

It is not difficult to devise models in which two sites exist that are initially equivalent, but become nonequivalent upon binding of one ligand (see Nelson et al., 1979). For example, if we assume there to be two closely spaced H1 sites, occupation of one by the globular portion of an H1 molecule could exclude a second molecule from the companion site. Alternatively, a region of the nucleosomal DNA might be covered by the N-terminal or C-terminal tail of the first H1 molecule, preempting this binding from the second. Indeed, Belyavsky et al. (1980) present evidence to indicate that the tails of one H1 molecule can interact with much of the chromatosome DNA. If correct, this would quite effectively block binding by a second H1 molecule.

In any event, the association of a single, asymmetric H1 molecule with the DNA-core complex effectively destroys its symmetry, as was pointed out by Hayashi et al. (1978). This must be kept in mind in subsequent

Table 6-11. Lysine-Rich Histone Contents of Various Chromatins[a]

Organism	Tissue/cell	Moles/mole octamer		Reference
		H1	H5	
Neurospora crassa	—	0.8		Goff (1976)
Drosophila melanogaster[b]	Adult	1.1		Holmgren et al. (1976)
	Larva	1.1		
Chicken	Erythroblasts (3.5 day)	1.5	0.6	Urban et al. (1980)
Chicken	Erythroblasts (4 day)[c]	1.0	0.2	Weintraub (1975)
Chicken	Erythrocytes (adult)	0.86	1.0	Weintraub (1975)
		0.66	1.66	Urban et al. (1980)
		0.54	1.86	Olins et al. (1976)
		0.44	0.90	Bates and Thomas (1981)
Chicken	Liver	0.88		Bates and Thomas (1981)
Mouse	Liver	0.74		Bates and Thomas (1981)
Mouse	Thymus	1.08		Smith et al. (1980)
Mouse	Myeloma cells	1.0		Caron and Thomas (1981)
Mouse	Hepatoma cells	1.2		Paul and Duerksen (1977)
Mouse	L5178 cells	1.0		Hayashi et al. (1978)
Mouse	Mastocytoma cells	0.64		Albright et al. (1979)
Rat	Liver	0.93		Thomas (1977)
		0.79		Bates and Thomas (1981)
Rat	Spermatogonia	0.64		Chiu and Irvin (1983)
Chinese hamster	CHO cells	0.97		Rall et al. (1977)
Rabbit	Thymus	1.1		Goodwin et al. (1977)
Bovine	Lymphocytes	1.1		Renz et al. (1977)
Ox	Glia	1.07		Bates and Thomas (1981)
Average[d]		0.95		

[a]Data from 1975 and after.
[b]Data for other *Drosophila* species also given in this reference.
[c]Data for embryos at different ages also given.
[d]Excludes data for chicken erythroid cells.

discussions of higher-order structure: While one *could* devise symmetric structures involving an even number of identical lysine-rich histones per pair of nucleosomes, the prevailing stoichiometry and heterogeneity of these proteins argues for asymmetry in the chromatin chain.

Examining the distribution of repeat lengths reported for a wide variety of chromatins (Table 7-1), one is struck by the fact that the minimum DNA size coincides almost exactly with the chromatosome size of 160 to 170 bp. Indeed, I know of only three examples in which smaller repeats have been proposed and not later revised. These are the value of 154 ± 9 bp for *Aspergillus* chromatin (Morris, 1976a), 159 ± 1.2 bp for *Achlya ambisexualis* chromatin (Silver, 1979), and 153 ± 7 bp for logarithmically growing *Paramecium* (Prince et al., 1977). From the data given, the latter value appears dubious; in any event all could be included in the chromatosome framework within the error limits stated. Furthermore, there are now very few cases in which chromatins are claimed to lack H1 (see Chapter 4).

Thus, it would appear proper to postulate the chromatosome as a general, asymmetric element of chromatin structure. Indeed, there are certain cases (i.e., *Physarum,* Stalder and Braun, 1978; *Trypanosoma,* Rubio et al., 1980) in which a chromatosome-sized DNA accumulates during digestion, but no evidence for a stable core particle has been found. It should not be assumed, however, that *all* nucleosomes contain lysine-rich histones. There is considerable evidence that actively transcribed regions may be depleted in these proteins (see Chapter 8). Furthermore, nucleosomes have been isolated in which HMG 1 and HMG 2 substitute for H1 (see below).

It would be most naive, moreover, to assume that the chromatosome element (~168 bp + the octamer + H1) constituted the *whole* of the repeating unit. Most chromatins have average DNA repeat sizes appreciably larger than this and seem to exhibit internal heterogeneity in repeat. Speculations concerning the *reasons* for the variety of repeat sizes in different chromatins are intimately connected with ideas about higher-order structure. Therefore, further consideration of this topic should be delayed until Chapter 7.

Finally, lest we be lured into considering the chromatosome to be a static unit of chromatin structure, we should heed experiments which indicate that H1 molecules may be surprisingly mobile, even at ionic strengths below the physiological range. Caron and Thomas (1981) have convincingly demonstrated that H1 molecules can rapidly exchange with either intact or H1-depleted chromatin in 70 mM NaCl solution (see also Thomas and Rees, 1983). If this should also prove to be true in the more complex ionic milieu of the nucleus, many ideas about the regularities of chromatin structure may have to be reconsidered. However, in a more recent study, Huang and Cole (1984) find that 0.2 M salt allows separation of long chromatin oligomers into insoluble and soluble fractions. The soluble fraction is depleted in H1; the insoluble fraction is enriched in this histone. Furthermore, there does not appear to be H1 exchange between these fractions. Huang and Cole suggest that exchange, if it does occur at this ionic strength, must be only *within* the soluble portion.

Nonhistone Chromosomal Proteins in the Nucleosome

It is now very clear that many nonhistone chromosomal proteins can also be associated with the linker DNA. Some even bind to the core particle itself. In the first section of this chapter, it was pointed out that the protein uH2A substitutes for H2A in a fraction of nucleosomes. Since uH2A can be considered a covalent adduct of H2A and ubiquitin, this can be thought of as a binding of ubiquitin to the cores. It is of interest in this connection that Bonner and Stedman (1979) found that H1 can be crosslinked to protein uH2A, as well as to H2A via a contact-site crosslinker.

The first definite suggestion that individual nucleosomes might carry nonhistone proteins is found in the important early paper of Lacy and Axel (1975). These workers carefully isolated rat liver mononucleosomes and showed them to contain ~185 bp DNA (with a bit of a 160 bp component) and all five histones. But they also found that the total protein/DNA ratio (2:1) exceeded the histone/DNA ratio (1.2:1) and stated: "It is therefore probable that the subunits contain acidic proteins in amounts analogous to that of intact chromatin." In the same year, Gottesfeld et al. (1975) demonstrated the presence of nonhistone proteins in a "transcriptionally active" nucleosome fraction. Liew and Chan (1976) used two-dimensional gel electrophoresis to show that rat liver mononucleosomes carried many such proteins. Since then, a host of such reports have followed. This is not surprising, for it may be expected that virtually any chromatin-bound nonhistone protein will appear in nucleosomes if the digestion is sufficiently mild to preserve most of the linker region. Indeed, many of the proteins listed in Chapter 5 have been found to be associated with mononucleosomes. However, the reservations emphasized in that chapter apply even more strongly here. It is very difficult to exclude the possibility of protein exchange, especially when the preparation processes are carried to the point of isolating nucleosomes.

As might be expected, the best characterized of the nucleosomal nonhistone proteins are HMGs 1, 2, 14, and 17. There is a considerable body of evidence to indicate that these proteins are by no means randomly distributed among nucleosomes of a given tissue; indeed, their abundance seems to correlate with the transcriptional activity of the particular chromatin region. We shall examine this point in much greater detail in Chapter 8. For the moment, we shall be concerned instead with the question of the *localization* and *interactions* of nonhistone proteins within the nucleosome.

It seems likely that most of the larger, unidentified nonhistone proteins are associated with the linker region, rather than being bound to the core particle itself. This is suggested by several observations that such proteins are found primarily on those nucleosome components that carry a longer DNA, and hence more of the linker. These proteins tend to be lost when the DNA is digested down toward the core particle size (see, for example, Bakayev et al., 1978; Albright et al., 1980). However, Grebamier and Pogo (1981) have shown that small amounts of some quite large nonhistone proteins can be crosslinked to H3, via disulfide bonds, in situ, suggesting that some must at least make contact with the histone core.

The HMG proteins 1 and 2 also seem to be associated with the larger classes of nucleosomes (Jackson et al., 1979; Albright et al., 1980; Jackson and Rill, 1981). Jackson et al. were able to isolate, from very mild nuclease digests, 200 bp nucleosomes containing no H1, but 1 or 2 copies of HMG 1 or HMG 2. Schröter and Bode (1982) have found that nucleosomes containing 180 bp DNA could bind one molecule of HMG 2, whereas core particles could not. A significant study is that of Bernúes et al. (1983), who showed that HMG 1 could be crosslinked in vitro to H2A·H2B and (H3·H4)$_2$. It may also be of importance that HMG 1 and 2 have been shown to interact in highly specific ways with H1 fractions (Smerdon and Isenberg, 1976c; Yu and Spring, 1977; see Chapter 5 for details). Although these results imply that HMG 1/2 may contact core and linker, there is one report that does not fit at all well with this interpretation. Mita et al. (1981) have found that HMG 2 can apparently *substitute* for H2A in the reconstitution of core particles. However, there is, to my knowledge, no in vivo parallel to this observation.

It is the proteins HMG 14 and HMG 17 whose interactions with nucleosomes have been studied the most thoroughly. Two groups (Sandeen et al., 1980; Mardian et al., 1980) first provided evidence that these proteins can bind specifically to the core particle itself. Two strong-binding sites are available on each core particle. These sites appear to be approximately equivalent and independent in binding at low ionic strength, but to exhibit cooperative binding at higher ionic strength. In the presence of excess HMG 14/17, even more nonhistone proteins can be bound, albeit weakly (Zama et al., 1984). HMG 14 and 17 appear to be equivalent and interchangeable in binding.

There exists considerable evidence concerning the nature and location of the binding sites. They must be located near the two ends of the DNA coil, for DNase I digestion is markedly inhibited in these regions upon HMG 14/17 binding. Furthermore, Espel et al. (1985) have observed that HMG 14 can be crosslinked to H2A, H2B, and H1, all associated with the DNA "exit" region. (See also Espel et al., 1983). Neutron scattering studies (Uberbacher et al., 1982) show that the HMG 14/17 binding increases the protein radius of gyration by only about 0.9 Å, but increases the DNA radius of gyration by about 2.7 Å. This would imply that these proteins bind tightly to the histone core, but in such a manner as to slightly displace the ends of the DNA coil. All of these data suggest that HMG 14/17 binding might stabilize the core particle with respect to the first melting transition (see above), and this is precisely what Sandeen et al. have observed. Paton et al. (1983) find that the HMG-containing core particles are also stabilized against urea denaturation.

In a *functional* sense, this increased stability is somewhat surprising, for HMG 14 and 17 appear to be preferentially associated with transcriptionally active chromatin. It was expected by many that these proteins would therefore *destabilize* nucleosomes. The situation is obviously complicated, however, for Sandeen et al. . have found that in "titrating" a mixture of bulk chicken erythrocyte core particles with HMG 14/17, the nonhistone proteins preferentially attach themselves to particles carrying DNA from the β-globin gene. Thus, there is a preference for binding to "active" chromatin. What is the basis for this preference? Neither of the two major modifications associated with active chromatin seem to affect

HMG 14/17 binding: histone hyperacetylation (Brotherton and Ginder, 1982) or ubiquitination (Swerdlow and Varshavsky, 1983). The latter authors, however, have made one most interesting observation; nucleosomes that have even a few (3 to 5) base pairs of DNA more than the core particle limit show as much as 500-fold higher affinity for HMG 14/17. Perhaps a critical part of the HMG 14/17 binding site is this "extra-core" DNA, which may only be exposed in H1-depleted chromatin. This would predict a competition between H1 binding and HMG 14/17 binding. Furthermore, these observations raise the interesting possibility that binding to active chromatin regions may be quite different in its effects than the binding to bulk core particles that has been the subject of most studies.

A further indication that HMG 14 and 17 may bind near the ends of core particle DNA can be inferred from the observations of Schröter et al. (1985). They find that intercalating agents like ethidium bromide specifically release HMG 14/17 from chromatin at low ionic strength. McMurray (1987) has shown that ethidium bromide binding begins at the ends of nucleosomal DNA and facilitates release of H2A·H2B dimers (see also McMurray and van Holde, 1986).

The point to be emphasized is that there exist a variety of interactions with other proteins which can make nucleosomes quite specific and very complex structures. In addition, the existence of multiple histone variants, variants of nonhistone proteins, and a whole array of posttranslational modifications of both would seem to allow for almost unlimited variability in nucleosome structure. These multivariant nucleosomes form the "building blocks" of the chromatin strand. It should not be surprising to find that that strand can adopt a variety of conformations. These will be the subject of the next chapter.

The idea that different combinations of nonhistone proteins may modify chromatin conformation is strongly supported by the fact that specific fractionation and digestion techniques have been successful in separating nucleosomes and oligonucleosomes containing different subsets of these proteins (see, for examples, Levy-Wilson et al., 1977, 1979; Levy-Wilson and Dixon, 1978; Jackson et al., 1979; Mathew et al., 1980). That the nonhistone proteins, which are present overall in much smaller stoichiometric quantities than the core histones, can nevertheless be concentrated in certain regions is indicated, for example, by the observation of Jackson et al. that HMG 1 and 2 may approach the concentration of H1 (and apparently substitute for it) in particular regions.

Conformational Transitions of the Nucleosome

The preceding sections have outlined what we know about the structure of the core particle and the nucleosome. There has been a great deal of interest in studying the ways in which these conformations respond to changes in the solvent environment. It would seem that this interest is generated by two expectations. On the one hand, it is hoped that such responses can tell us something about the forces stabilizing the particles. Further, it is believed by some that such conformational changes can yield clues as to how the nucleosome might be rearranged in vivo to allow such processes as transcription or replication to proceed.

I have already discussed one kind of transition—the "melting" processes that occur when the temperature is increased. I turn now to those responses induced by changes in the ionic strength, or pH, or by the addition of denaturants such as urea.

Conformational Changes at Low Ionic Strength

Either very low or very high ionic strength has significant effects on nucleosome conformation. Gordon et al. (1978, 1979) were the first to systematically examine the behavior of nucleosomes in very dilute salt solutions. Using mainly hydrodynamic methods, they claimed evidence for *two* rather abrupt unfolding steps, one centered at about 7 mM salt, the other at about 1 mM. Over the next several years, this behavior was reexamined by a number of investigators, using a variety of hydrodynamic, optical, and chemical techniques.

Unfortunately, there has been a considerable amount of confusion and outright contradiction between these studies. To begin with, most investigators following Gordon et al. have not been able to detect two sharp transitions in purified core particles. Most agree on the existence of an abrupt change in a number of physical parameters in the vicinity of 1 mM salt, although there is considerable disagreement as to the exact ionic strength at which this is centered. Instead of a sharp transition at 7 mM, most of the more recent studies detect a gradual change in physical properties over the range between about 20 mM to 2 mM salt (see Burch and Martinson, 1980; Libertini and Small, 1980, 1982, for example). Figure 6-21 depicts data from the experiments of Libertini and Small, in which

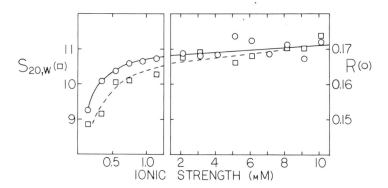

Fig. 6-21. The changes in sedimentation coefficient ($S_{20,w}$) and tyrosine fluorescence anisotropy (R) exhibited by chicken erythrocyte core particles at low ionic strength. (From Nucleic Acids Research, vol. 8, issue 16, p. 3530, Libertini and Small. Copyright 1980 IRL Press.)

highly purified core particles were used. Both the gradual decrease in S, and the abrupt change at 1 mM salt are evident. The latter, and to a much lesser extent the former, are paralleled by changes in the fluorescence anisotropy of histone tyrosines.

To avoid confusion, it is probably best to consider these transitions one at a time. The gradual change between 20 mM and 2 mM salt is characterized by (1) a loss of some but not all of the H2B–H4 contacts identifiable by contact-site crosslinkers (Martinson et al., 1979c), and (2) insensitivity to the presence of H1 in the nucleosome or mild crosslinking of the nucleosome core (Burch and Martinson, 1980). Changes in this range must be very subtle, however, for neither Mita et al. (1983) nor Uberbacher et al. (1983) detect any variation in radius of gyration in this range by low-angle neutron scattering. Perhaps the conformational behavior in the 20 mM to 2 mM range may best be characterized as a slight swelling of the particle, driven by the mounting electrostatic repulsion between the DNA coils as the ionic strength decreases. There is, however, evidence from electric dichroism measurements that something more dramatic happens to 175 bp particles (lacking H1) in this same ionic strength range (Schlessinger et al., 1982). Again, Uberbacher et al. could observe *no* corresponding change in the radius of gyration. Uberbacher et al. suggest that some of the confusion may result from the use of heterogeneous nucleosome samples. Dieterich et al. (1980) report that the "low salt" transition is displaced up to about 7 mM salt for 173 bp particles. Thus, a sample containing a mixture of 175 and 145 bp particles might show *two* transitions. A problem with this explanation is that the two transitions seen by Schlessinger et al. are clearly different.

Despite all of this controversy, all workers agree that there is an abrupt conformation change at about 1 mM salt. It must involve substantial unfolding, for Burch and Martinson find it to be completely inhibited by either the presence of H1 in the nucleosome or crosslinking of the histone core. The latter observation also distinguishes the low-ionic strength transition from the first stage of the thermal transition, for it has been shown that the latter is insensitive to protein crosslinking (Simpson, 1979). Martinson et al. (1979c) have provided evidence from crosslinking studies that H2B–H4 contacts are broken at very low ionic strength. Thus, it seems that the transition must involve some disruption of the histone core itself.

Electric dichroism studies of core particles at very low ionic strength demonstrate substantial increases in both the dipole moment and rotational relaxation time. The combination of changes is consistent with an opening of the particle into a shallow one-turn helix of about 150 Å diameter (H.M. Wu et al., 1979). In a later paper, the same laboratory investigated the low-salt unfolding of H1-depleted nucleosomes carrying 175 bp DNA. For these, they report evidence for an even more extended form at very low ionic strength (Schlessinger et al., 1982).

Fluorescence measurements have also been employed in such studies.

Dieterich et al. (1979) have followed the low salt transition using reconstituted core particles in which fluorescent dyes had been attached to the H3 sulfhydryls. From the decrease in energy transfer, they propose that the dye molecules move apart from an initial separation of ≤30 Å to a distance of about 48 Å at very low ionic strength. In later studies, Dieterich and Cantor (1981) have analyzed the kinetics of the unfolding process. They conclude that the process is monomolecular but kinetically complex. The data are consistent with (but do not require) a model in which two alternative intermediate states precede the final unfolding. Similarly, Ashikawa et al. (1982b) have concluded from fluorescent-labeling experiments with ethidium bromide that three distinguishable states of H3–H3 interaction can be observed. Unfortunately, one difficulty in evaluating all of these fluorescence studies resides in the possible effects of the bulky dye molecules on the stability of the particle in extreme circumstances.

The approach taken by Libertini and Small (1980, 1982) avoids this difficulty by using the *intrinsic* fluorescence of histone tyrosines as a structural probe. They find that the fluorescence change with ionic strength can be resolved into two steps and propose two possible models for the unfolding (Fig. 6-22). In either case, it is presumed that the driving force for the transition is the electrostatic repulsion between DNA coils, which increases as the ionic strength is lowered. Both models involve the breakage of contacts between H2A·H2B dimers and the (H3·H4)$_2$ tetramer, in agreement with the results of Martinson et al. (1979b).

Is there any basis for choosing between these (and other imaginable) models for the transition? There exist several experimental results that may bear on the question but yield ambiguous results. First, Eshaghpour et al. (1980) have measured singlet–singlet energy transfer between dye molecules attached to the ends of the DNA and dyes attached to the single SH group of each H3 molecule. They obtain a distance measure of 50 to 53 Å in the native core particle, which is reasonable if the SH groups lie near the center of the particle. Most interestingly, they find that this distance increases little if any in low salt; 50 to 58 Å is the estimate. This result is quite consistent with model (b), but difficult to reconcile with model (c). The problem is that a structure like model (b) has been rejected by H.M. Wu et al. (1979) as inconsistent with their dipole moment measurements, which point to a model like (c) with both ends extended. It is conceivable, of course, that the more expanded structures could be approached by the *sequential* release of the two DNA ends. A "half-opened" structure would relieve some of the electrostatic repulsion and at the same time keep one DNA end close enough to the H3 sulfhydryl position to allow efficient energy transfer. Such a two-step opening is, in fact, what Libertini and Small propose. Such a process might also account for the complex kinetics of the unfolding detected by Dieterich and Cantor (1981).

Another, though less compelling, argument for model (b) comes from the DNA–histone crosslinking studies of Zayetz et al. (1981). These work-

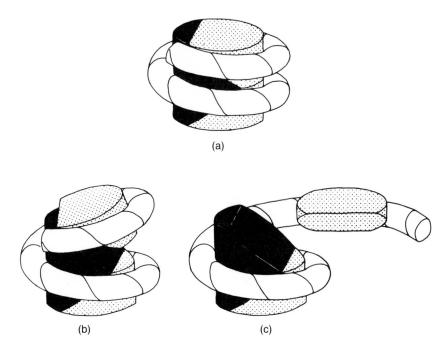

Fig. 6-22. Two possible ways in which the core particle (a) could unfold at low ionic strength. In model (b) (left) the DNA remains coiled, but at least one H2B · H2A–(H3 · H4)₂ contact is broken. In (c) (right) the DNA unfolds, taking at least one H2A · H2B dimer with it. Libertini and Small point out that (b) could be a prelude to (c) and, indeed, to further unfolding. (Reprinted with permission from Biochemistry *21*, p. 3327, Libertini and Small. Copyright 1982 American Chemical Society.)

ers have examined DNA–histone contacts under a number of denaturing conditions, including low salt. They find that at 0.5 mM salt most of the contacts seen in the intact core particles are maintained. In particular, the contact between the 3′ end of the DNA and histone H3 seems to remain intact. The simplest interpretation is model (b)—the nucleosome unfolds, but the DNA ends stay tied down. However, an alternative explanation is that H3 contacts this portion of the DNA through an extendable part of the chain (the N-terminal region, for example), and this segment is merely "stretched" during a type (b) unfolding. This is, in fact, the model suggested by Zayetz et al. Such an interpretation is perhaps supported by the observation of Grigoryev and Krasheninnikov (1982) that cleavage of histone tails by trypsin greatly facilitates the low-salt transition.

It might be hoped that low-angle neutron scattering experiments could resolve the matter, but experiments to date seem only to add to the confusion. In a very careful study, using both 146 bp and 175 bp particles,

Uberbacher et al. (1983) report no significant change in the protein radius of gyration, but a significant increase (greater for the long-DNA particles) in the R_G of the DNA. They propose a model in which the DNA tails are released from the histone core, in the fashion suggested for the first stage in thermal denaturation (*DNA–Histone Interactions in the Core Particle*, above). A very similar proposal has been made by Harrington (1982) on the basis of flow birefringence studies. Completely different results were reported from other neutron scattering experiments by Mita et al. (1983), who claim that the DNA radius of gyration actually *decreases* at low salt, while that of the protein *increases*. In fact, according to their measurements, the protein R_G becomes greater than the DNA R_G below 1 mM salt! There is no obvious way in which these neutron scattering results can be reconciled, nor does either set seem consistent with all of the other data. I find serious difficulties with interpretations of both sets of data. The model proposed by Uberbacher et al. requires breaking of histone–DNA bonds under precisely those conditions where it would be least expected—low ionic strength. The data of Mita et al., if accepted at face value, suggest that the nucleosome is literally turned inside out at low salt. It is very difficult to imagine how this could happen.

In search of further information concerning the behavior of nucleosomes at very low ionic strength, one might turn to numerous electron microscope studies of the unfolding of chromatin in low salt. While it must be kept in mind that even H1-depleted chromatin may behave differently than nucleosomes, and that interaction with the EM grid may play a role, the observations are suggestive. Oudet et al. (1977b) in a study of SV-40 minichromosomes, found that the apparent number of beads observed on each minichromosome doubled at low ionic strength. The new beads seemed smaller than the normal ones and were paired, as if each core particle had opened up in a "clamshell" fashion, in an extreme version of model (b), Figure 6-22. After long exposure to *very* low salt concentrations, Oudet et al. report that the SV-40 minichromosome took on an extended conformation, in which individual nucleosomes were no longer visible. However, histones were apparently still present, since the fibers were thicker than free DNA and an increase in salt concentration caused reformation of nucleosomes. Somewhat similar results were obtained by Thoma et al. (1979) in electron microscope studies of H1-depleted rat liver chromatin. While half particles were not reported in this case, unraveling of nucleosomes was observed in 1 mM buffer, containing 0.2 mM EDTA. Such behavior was not observed with chromatin containing H1. Thus, these electron microscope studies on chromatin parallel to some extent the observations with isolated nucleosomes.

Are these results consistent with the hydrodynamic studies of individual nucleosomes in low salt? To approach this question, I have calculated theoretical values for the sedimentation coefficient of the core particle in different conformations (Table 6-12). The lowest values of S that have

Table 6-12. Predicted
Sedimentation Coefficients for
Unfolded Nucleosome
Structures[a]

Structure	$s_{20,w}$ (Pred.)
Intact nucleosome[b]	11.4
Open circle	8.8
Linear chain	7.2
C-shaped hemisome	5.8
Linear hemisome	5.7

[a]All were calculated using the Kirk-
wood formalism (see van Holde,
1975), assuming that the nucleosome
is made of 8 spherical beads each
containing ⅛ of the total mass of the
core particle.
[b]Modeled as a helical ramp of beads,
with 4 beads per turn.

been *observed* at low ionic strength are about 9S (see Fig. 6-21, for example). This value is too high for a linearly extended core particle, but it
is close to the value expected for an opened circle (8.8S) or the similar
structure proposed by Wu et al. It is also about what one would expect
for a fully opened "clamshell" structure.

It may be that the highly extended forms observed in the electron microscope are favored by interaction with the EM grid and do not represent
likely solution conformations. On the other hand, it should be noted that
the chromatin is not usually seen as a linearly extended structure; rather,
the strands show considerable local curvature (see Thoma et al., for example). Finally, one should note that none of the hydrodynamic studies
of nucleosomes actually show a "plateau" in the parameters at low salt.
In fact, in an investigation of the low-salt unfolding of sea urchin sperm
core particles, Simpson and Bergman (1980) postulated a value of $s_{20,w} =$
8.2S for the low ionic strength limit. This would correspond to a structure
even more unfolded than the open circle of Table 6-12. Thus, contradictions
between the EM studies of chromatin and the hydrodynamic studies of
core particles may be more apparent than real.

In any event, we must conclude that the nucleosome can unfold to a
considerable extent at very low ionic strength. Perhaps some of the apparent contradictions in the literature result from the fact that there exists
a continuum of structures and that different methods of investigation are
biased to examine different aspects of this continuum.

Effects of High Ionic Strengths
In salt concentrations ranging between about 20 mM and 300 mM, most
physical methods indicate little change in nucleosome conformation. It

has already been mentioned that at very high ionic strengths (~800 mM or greater), stepwise removal of histones from the core particle is observed. In the intermediate range, from about 300 mM to 800 mM, there is strong evidence for another kind of nucleosome unfolding, which is often referred to as the "high salt transition."

It should be mentioned at the outset that studies of nucleosome conformation changes in high salt are invariably complicated by a certain amount of dissociation of the particles into free DNA and histones. This dissociation increases with increasing salt concentration and will slowly approach its equilibrium value at each ionic strength (Fig. 6-23). Since this is an association–dissociation equilibrium, the fraction of intact nucleosomes will decrease as the total nucleosome concentration is lowered. Thus, as Stacks and Schumaker (1979), Lilley et al. (1979), and Cotton and Hamkalo (1981) have observed, very dilute solutions can exhibit considerable dissociation even at moderate ionic strengths (see also Eisenberg and Felsenfeld, 1981; Yager and van Holde, 1984; Ausio et al., 1984a, 1984b). The slow dissociation reaction can be distinguished from a rapid conformational change observed at salt concentrations above 0.3 M by kinetic studies (see Fig. 6-24). This reversible conformational change, which reaches a plateau at about 0.7 M, is characterized by a lower sedimentation coefficient (about 9 S) and an increased molar ellipticity in the spectral region where DNA absorbs (Wilhelm and Wilhelm, 1980;

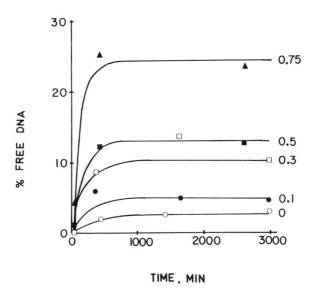

Fig. 6-23. The dissociation of core particles observed when diluted into NaCl concentrations of the indicated molarities. The dissociation is time-dependent but approaches an equilibrium value at each salt concentration. The amount dissociated was determined from sedimentation velocity experiments at different times after mixing.

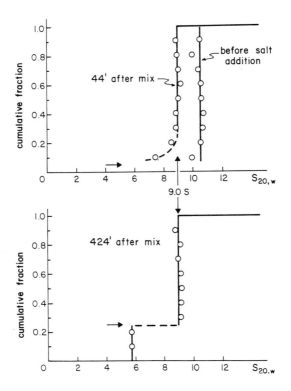

Fig. 6-24. Sedimentation coefficient distributions of chicken erythrocyte core particles at 0, 44, and 424 minutes after diluting into 0.75 M NaCl. The decrease in S of the particles to ~9.0 appears to be immediate; the dissociation to produce about 25% free DNA is a slower process.

Eisenberg and Felsenfeld, 1981; Ausio et al., 1984a, 1984b; see also Fig. 6-25).

In interpreting these changes, one should note that there are a number of other interesting things going on in or near this range of salt concentrations. For example, Cary et al. (1978) have interpreted high-resolution proton NMR measurements to indicate that the N-terminal tails of histones H3 and H4 become free to move in the range from 0.3 to 0.6 M salt. In parallel with this, Whitlock and Stein (1978) have found that core particles from which these tails had been excised by trypsinolysis exhibit marked dissociation at this ionic strength. Whitlock (1979) has noted changes in reactivity of nucleosomes with the diol-epoxide of benzo [a] pyrene in the same salt range, and Ashikawa et al. (1982a) see tyrosine fluorescence changes.

Perhaps of equal significance are the numerous experiments that demonstrate "sliding" of histone cores in H1-depleted chromatin under comparable salt conditions (see, for example, Steinmetz et al., 1978; Beard,

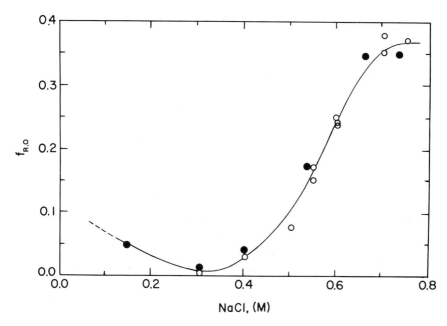

Fig. 6-25. The change in ellipticity at 282.5 nm observed when core particles are raised to various salt concentrations. The data are expressed as the fraction of the total change from the minimum ellipticity to that of free DNA. Data are for chicken erythrocyte core particles (○, data of T. Yager) and HeLa core particles (●). (Reprinted with permission from Biochemistry *25*, p. 1421, Ausio and van Holde. Copyright 1986 American Chemical Society.)

1978; Weischet, 1979; Spadafora et al., 1979; Glotov et al., 1982; van Holde and Yager, 1985).

All of this evidence points to a general "loosening" of the DNA–protein interactions in the ionic strength range between 0.3 and 0.8 *M*. This is emphasized by the fact that at only slightly higher ionic strength (0.8 to 1.0 *M*), histones H2A and H2B are released from the core particle (Burton et al., 1978). These observations point up the fundamental differences that must exist between the low-salt and high-salt transitions. At low ionic strength, binding of the DNA to the histones must be very firm, and the primary driving force for unfolding will be electrostatic repulsion between like groups: DNA–DNA and histone–histone. On the contrary, at higher ionic strength electrostatic repulsions are largely shielded, and it is the electrostatic interaction between unlike charges (DNA–histone) that is weakened.

Effects of pH on Nucleosome Stability
The response of nucleosomes to changes in pH has been mainly studied by two groups. The first systematic investigation appears to have been

initiated at the laboratory of D. Olins (Zama et al., 1978a; Gordon et al., 1979). Pronounced changes in sedimentation coefficient, diffusion coefficient, and circular dichroism were observed in the pH range between 5 and 6. Since the $s_{20,w}/D_{20,w}$ ratio remained unchanged, neither dissociation nor aggregation appear to be significant in this range. However, at pH 4.2 strong aggregation was detected. Major changes in physical parameters were observed *only* at low ionic strengths (below 10 mM). It would seem that this pH transition is closely linked to the low-salt transition described above. This relationship has been explored in detail by Libertini and Small (1982, 1984). Although their complete analysis is complex, the major point is the following: Hydrogen ions can substitute for other cations, thus suppressing the low-salt transition at pH $<$ 5.

Of greater potential interest is a small conformational transition that Libertini and Small have detected under more nearly physiological conditions. This conformational change, which is centered at pH 7, can be detected by either fluorescence anisotropy or circular dichroism (at 284 nm), but not by sedimentation velocity. It is inhibited by core crosslinking. Thus, it must correspond to some minor rearrangement, or loosening, of the core particle structure. Fluorescence data are shown in Figure 6-26. Although the effect is not large, the fact that it occurs in a pH and ionic strength range comparable to those found in vivo may mark it as of special significance. It is possible that this change is related to a change in dissociation pattern of the histone octamer noted over this same pH range by Kawashima and Imahori (1982).

Effects of Intercalators

The binding of intercalating agents such as ethidium bromide (EtBr) to nucleosomes has been the subject of some controversy. Erard et al. (1979) and Genest and Wahl (1981) observed cooperative binding, with initial binding much weaker than for free DNA. Electron microscope studies indicated the formation of "tailed" particles. Wu et al. (1980), in a fluorescence and electric dichroism study, obtained quite different results. The binding was reported to be noncooperative, and much *stronger* than to free DNA. Evidence for nucleosome unfolding was presented.

In a more recent examination of the problem, McMurray and van Holde (1986) found that *dissociation* of the core particle was produced above a low level of ethidium binding. When the core particle binding isotherm was obtained by correcting total binding for that from free DNA, cooperativity was evident (McMurray, 1987). The dissociation was found to be kinetically slow and to proceed through an intermediate that had lost an H2A·H2B dimer. Although these data are in general agreement with those of Erard et al., it seems to me that all earlier studies are compromised by the fact that the dissociation was not recognized. (See also Schmitz, 1982.)

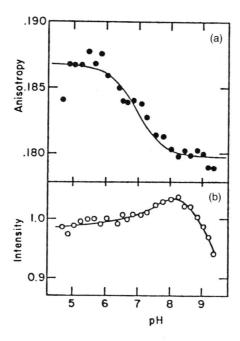

Fig. 6-26. The effect of pH on (a) the tyrosine fluorescence anisotropy, and (b) the fluorescence intensity of core particles. Panel (a) clearly demonstrates a small conformational change, with pK of 6.97. The intensity data are complicated by the fact that tyrosine fluorescence intensity is greatly quenched by tyrosine ionization at high pH. (From Nucleic Acids Research, vol. 12, p. 4351, Libertini and Small. Copyright 1984 IRL Press.)

Conformational Changes Induced by Urea
Urea is an agent that disrupts protein–protein interactions (at moderate concentrations) and the secondary and tertiary structure of proteins at high concentrations. In an early study, Shih and Lake (1972) had noted that chromatin structure was grossly disorganized in 5 M urea, but that the histones still seemed to be attached to the DNA. The first such studies with nucleosomes appear to be those of Whitlock and Simpson (1976a), who observed that the sedimentation coefficients of core particles were reduced to about 6.2 S in 6 M urea. More detailed studies in the Olinses' laboratory provided hydrodynamic evidence for a progressive unfolding of the particles with increasing urea concentrations (see Olins et al., 1977b; Zama et al., 1978b). The limiting $s_{20,w}$ in high salt was found to be about 6 S, slightly higher than the value for free core particle DNA (about 5 S) and close to that expected for a linear chain (Table 6-12). In the electron microscope, the particles were observed as extended nucleoprotein

strands. At lower concentrations of urea (3 to 4 M), intermediate, partially extended structures can be found (Woodcock and Frado, 1977).

Using contact-site crosslinkers, Martinson and True (1979a) noted that the H2A·H2B and H2B·H4 contacts were lost at urea concentrations above about 4 M. Thus, the data seem to suggest a two-stage process in urea denaturation. At concentrations below about $4M$, the particle is loosened but swollen. Above this urea concentration, histone–histone interactions are broken and the entire structure unravels. Such a two-stage process is consistent with the ^{31}P NMR data of Shindo et al. (1980), who find that the linewidth increases in a stepwise fashion, with a clear plateau in the region of 3 to 4 M urea. That histone–DNA contacts persist above this level is indicated by the data of Zayetz et al. (1981), who found a loss of specificity in DNA–histone crosslinking only at 5 to 6 M urea.

Overview: The Instability of the Core Particle

The core particle is a fragile object. Outside of a narrow environmental range, it is unstable, unfolding in a variety of ways. This should not be surprising, for the particle represents a precarious marriage of basically incompatible partners. The histone octamer is not itself stable at low or even moderate ionic strengths; in the absence of DNA, it will dissociate into dimers and tetramers. The DNA, on the other hand, has been bent to a degree inconsistent with its normal propensity. Only the strong interaction between the DNA and the histones compensates for these tendencies to unfold or dissociate.

In the *physiological* range of pH, ionic strength, and temperature the core particle is marginally stable. Stability appears to be increased by inclusion of H1 in the chromatosome and in the full chromatin structure. It seems reasonable that this incipient instability, which can be unlocked by removal of lysine-rich histones and by modification of the particles, is an essential feature of nucleosome function. An irreversibly condensed chromatin structure would be physiologically useless; what is seemingly required and provided is a structure that can exist in several levels of stability. The following chapters will be concerned with the ways in which the potential instability can be exploited.

7

Higher-Order Structure

The preceding chapter has described in considerable detail the repeating unit of chromatin structure, the nucleosome. Now we turn to the more difficult problem of how these subunits are arranged to form the complex entities that are chromosomes. There are three separable but interrelated aspects to this problem, which can best be expressed as questions:

1. How are the nucleosomes arranged along the DNA strand, and with respect to its sequence? This is essentially a problem of *linear* organization.
2. How is this linear sequence of nucleosomes coiled or wrapped so as to form the chromosome fibers observable in the electron microscope?
3. In both interphase and metaphase nuclei, there is evidence that the chromatin fibers are themselves further coiled and arranged in defined ways. What is the functional significance of such organization? Both (2) and (3) are clearly questions of *spatial* arrangement.

These questions are obviously interconnected. Certainly, the arrangement of nucleosomes along DNA will impose constraints upon the three-dimensional structures that can be generated. But from the point of view of biological function, the requirement for particular spatial organizations may be a selective force that dictates the linear arrangement. Since we do not as yet know what these requirements may be, it seems most sensible to simply describe what is now known about chromatin organization, beginning with the problem of linear arrangement.

Nucleosome Arrangement

One can imagine at least five different ways in which the nucleosomes might be arranged. These are depicted in Figure 7-1. Since the nomenclature in this field has been hopelessly inadequate (all kinds of regular

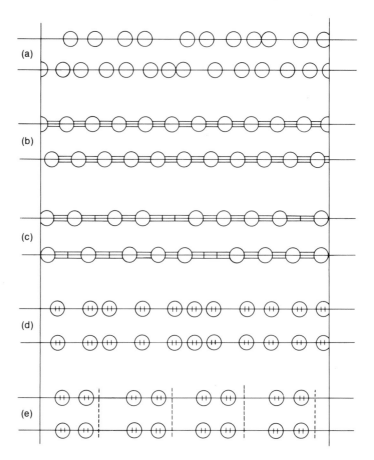

Fig. 7-1. Some possible kinds of linear arrangements of nucleosomes. In each case, two copies of the same region of the DNA are shown to indicate that some arrangements "copy" whereas others do not. See the text for explanation of each kind of arrangement. (a) Random. (b) Uniformly spaced. (c) Quantized spaced: blocks correspond to H1 or other proteins. (d) Positioned: marks indicate specific DNA sites. (e) Phased: dotted lines indicate DNA repeat.

arrangements are referred to as "phasing"), I shall propose a series of definitions, as illustrated by the figure.

1. A *random* arrangement (Fig. 7-1a) is one in which nucleosome placement bears *no* specific relationship either to the DNA sequence or to the position of other nucleosomes. Any two copies of the genome will be wholly different.
2. A *uniformly spaced* arrangement (Fig. 7-1b) is one in which adjacent members of a sequence of nucleosomes are separated by a uniform DNA linker length. The arrangement is *not* tied to the DNA sequence,

unless a specific nucleosome of the group is positioned on the DNA. Spacing uniformity could be provided by the presence of specific linker histones and/or nonhistone proteins.

3. Nucleosomes are said to show *quantized spacing* (Fig. 7-1c) if the spacer lengths are restricted to a few definite values. There need be no specific registry with the DNA sequence. A limited set of linker-associated proteins could provide such regularity.

4. Nucleosomes are *positioned* (Fig. 7-1d) if their individual locations on the DNA sequence are fixed. This might occur directly by core–DNA interactions or indirectly via interaction with sequence-specific non-histone proteins. A variant is "alternative positioning" in which a small number of positions is available to each nucleosome.

5. The word *phasing* (Fig. 7-1e) will be used only to denote the regular positioning of nucleosomes on a repeating DNA sequence. Again, there may be "alternative phases" with respect to a given sequence. If there is one nucleosome per repeat and only one phase, the arrangement will also be "uniformly spaced."

Note that of these arrangements, only (d) and (e) (as shown) must be identical in various copies of the genome, and there could exist variants of (e) that do not copy. However, (b) could show identity in different copies if an entire "string" of nucleosomes were "locked" with respect to sequence, either by positioning of one (or more) nucleosomes, or by one or more "positioning proteins." As has been pointed out by Kornberg (1981), it may be very difficult to distinguish cases in which a number of nucleosomes are packed between positioned nucleosomes or proteins from true positioning of the entire set.

It may well be that this listing does not encompass all possibilities. It is certainly unlikely that any whole genome will turn out to be completely described by any one of the arrangements listed above. There seems in fact to be experimental evidence for each of these kinds of organization. I turn now to consideration of that evidence. Additional details may be found in the comprehensive review by Eissenberg et al. (1985).

First, is there *general* evidence for regularity in nucleosome spacing? The simplest approach is to examine the data on nucleosome repeat lengths, of which much exists. A summary of such data is given in Table 7-1. Figure 7-2, abstracted from this table, suggests some regularities. The whole range of viral, fungal, protist, plant, and animal tissues seem to favor nucleosome repeats at three broad maxima, lying at about 175, 195, and 220 bp. Curiously, cells grown in culture show a single maximum at about 185 bp. What all of this means (if anything) is obscure, and there are several reasons for being cautious in interpretation. First, in at least some cases the data come from tissues in which there exist a variety of cell types. Even for pure cell lines, there is abundant evidence that the repeat is not uniform throughout the genome (see below). Thus, the value

Table 7-1A. Nucleosome Repeat Lengths—Viruses

Virus	Source	Repeat (bp)	Comment	Reference
Adenovirus	HeLa cells	165	Several hours post-infection; not found in mature virion	Tate and Philipson (1979)
Adeno-associated	HeLa cells	184	Apparently duplex DNA in cells	Marcus-Sekura and Carter (1983)
Epstein-Barr	Raji cells	196		Shaw et al. (1979)
Polyoma	Virion	191	Two populations: 70% @ 185 bp, 30% @ 205 bp	Ponder et al. (1978)
	Mouse embryo cells	195		Ponder et al. (1978)
SV-40	CV-1 African green monkey cells	196		Bellard et al. (1976)
		187 ± 11	If small oligonucleosome values neglected, get 191	Shelton et al. (1978)
		195 ± 9		Nedospasov et al. (1981)

Table 7-1B. Nucleosome Repeat Lengths—Protists

Species	Nuclear type	Growth phase	Repeat (bp)	Reference
Euglena gracilis	N.A.	S-phase	225 ± 13	Magnaval et al. (1980)
Olisthodiscus luteus	N.A.	Vegetative	220	Shupe et al. (1980)
Oxytrichia fallax	Macro	Vegetative	198 ± 5	Wada and Spear (1980)
	Macro	Starved	198 ± 5	Wada and Spear (1980)
Oxytrichia sp.	Macro	Starved	220 ± 5	Lawn et al. (1977)
Paramecia aurelia	Macro	Log	168 ± 8[a]	Prince et al. (1977)
	Macro	Stationary	178 ± 6	Prince et al. (1977)
Porphyridium aerigineum	N.A.	Vegetative	173	Barnes et al. (1982)
Stylonichia mytilis	Micro	Vegetative	202 ± 3	Lipps and Morris (1977)
	Macro	Vegetative	217 ± 4	Lipps and Morris (1977)
	Macro	Stationary	220 ± 3	Lipps and Morris (1977)
Tetrahymena pyriformis	Micro	Vegetative	175	Gorovsky et al. (1977)
	Macro	Vegetative	202	Gorovsky et al. (1977)
	Macro	Log	207 ± 10	Prince et al. (1977)
	Macro	Stationary	230 ± 10	Prince et al. (1977)
Trypanosoma cruzi	N.A.	Log	212 ± 6	Rubio et al. (1980)

[a]A value of 153 ± 7 is given in the paper. However, I believe that 168 bp, obtained from data at a shorter digestion time, is likely more representative.

of a "repeat length" for a given tissue or cell type is an average from samples of unknown dispersion.

There is a more fundamental reason for suspecting data on "average" nucleosome lengths. They may not, in fact, even represent true averages of a given genome. Since there is continual "trimming" of the ends of nucleosome oligomers during digestion, researchers tend to rely most heavily on sizes determined at early stages of the process. But the oligomers produced at this time may overrepresent a special subset of the whole, especially if the digestion is carried out in nuclei where both accessibility and the state of condensation of different portions of the genome may differ. Few studies have attempted to determine whether the average repeat is the same in different regions. In one such case (Gottesfeld and Melton, 1978), it is shown that in mouse liver bulk chromatin, transcribed regions, rDNA regions, and satellite regions all gave the same repeat length. On the other hand, it has been found that the repeat length for *Xenopus* erythrocyte bulk chromatin (187 bp) differed from that of the 5S genes contained therein (178 bp) (Humphries et al., 1979; Gottesfeld, 1980). Similarly, while rat liver chromatin has a "bulk" repeat of 198 bp, the satellite I chromatin shows a repeat of 185 bp, a value just one-half of the DNA repeat (Omori et al., 1980). The increasing frequency of such results, in the limited data available to us, questions the meaningfulness of global measurements like those tabulated in Table 7-1. We must turn to more definitive kinds of data.

Table 7-1C. Nucleosome Repeat Lengths—Fungi

Species	Common name	Phase or cell type	Repeat (bp)	Reference
Achlya ambisexualis		Mycelia	159 ± 1	Silver (1979)
Aspergillus nidulans		Mycelia	154 ± 9	Morris (1976a)
Dictyostelium discoideum	Cellular slime mold	Log	187 ± 8	Bakke et al. (1978)
Entomophthona aulicae		Log	197 ± 1	Ralph-Edwards and Silver (1983)
Neurospora crassa		Log	170 ± 5	Noll (1976)
Physarum polycephalum	Slime mold	Not given	171 ± 2	Compton et al. (1976)
		Plasmoidia	190	Johnson et al. (1976)
			181 ± 2	Stalder and Braun (1978)
		Amoeba	176 ± 8	Stalder and Braun (1978)
Saccharomyces cerevisiae	Yeast	Log	162 ± 6	Hörz and Zachau (1980)
			165 ± 5	Thomas and Furber (1976)
		Stationary	165[a]	Lohr and Ide (1979)
		(2 μm minichromosome from log phase)	165[a]	Lohr and Ide (1979)
			165	Nelson and Fangman (1979)

[a]Corrected by 3% for calibration error.

Table 71D. Nucleosome Repeat Lengths—Higher Plants

Species	Common name	Tissue	Repeat (bp)	Reference
Brassica pekinensis	Chinese cabbage	Seedlings, flower petals	175 ± 8	Leber and Hemleben (1979a)
Glycine max	Soybean	Roots, hypocotyls, cotyledons, leaves	175 ± 8	Leber and Hemleben (1979b)
Hordeum vulgare	Barley	Leaves	195 ± 6	Philipps and Gigot (1977)
Matthiole incana	Stock	Seedlings, flower petals	175 ± 8	Leber and Hemleben (1979b)
Nicotiana tabacum	Tobacco	Leaves	194 ± 6	Philipps and Gigot (1977)
Phaseolus aureus	Mung bean	Seedlings	175 ± 8	Leber and Hemleben (1979b)
Pisum sativum	Pea	Seed, ungerminated	195 ± 3	Grellet el al. (1980)
		Seedling, germinated	194 ± 7	Grellet et al. (1980)
Secale cereale	Rye	Embryos	200 ± 5	Cheah and Osborne (1977)

Table 7-1E. Nucleosome Repeat Lengths—Invertebrates

Species	Common name	Tissue, stage, or cell type	Repeat (bp)	Reference
Sipunculus nudus	Acorn worm	Erythrocyte	177	Mazen et al. (1978)
Holothuria tubulosa	Sea cucumber	Gonad	227 ± 6	Cornudella and Rocha (1979)
		Mature sperm	228 ± 5	Cornudella and Rocha (1979)
Arbatia lixula	Sea urchin	Gastrula	218	Spadafora et al. (1976)
		Sperm	242	Spadafora et al. (1976)
Arbatia punctulata		Hatching blastula	220 ± 22	Keichline and Wassarman (1977)
		Pluteus	220 ± 22	Keichline and Wassarman (1977)
		Sperm	260	Keichline and Wassarman (1977)
Lytichinus pictus		Morula, blastula	213 ± 3	Arceci and Gross (1980)
		Gastrula	217 ± 3	Arceci and Gross (1980)
		9–16 day embryos	230 ± 4[a]	Arceci and Gross (1980)
		Sperm	248 ± 3	Arceci and Gross (1980)
			239 ± 2[b]	Savic et al. (1981)
Strongylocentrotus intermedius		Sperm	237 ± 5	Zalenskaya et al. (1981)

Table 7-1E. *Continued*

Species	Common name	Tissue, stage, or cell type	Repeat (bp)	Reference
Strongylocentrotus purpuratus		Morula, blastula, pluteus	222 ± 10	Keichline and Wassarman (1979)
		Adult gut	222 ± 12	Keichline and Wassarman (1979)
		Sperm	250 ± 12	Keichline and Wassarman (1979)
			260	Simpson and Bergman (1980)
			243 ± 3[c]	Savic et al. (1981)
		2-cell embryo	225	Shaw et al. (1981)
		Blastula, gastrula	212 ± 5	Shaw et al. (1981)
Palaemon serratus	Shrimp	Hepatopancreas	189 ± 5	Sellas and van Wormhoudt (1979)
Aphelasteria japonica	Starfish	Sperm	224 ± 6	Zalenskaya et al. (1981)
Spisula solidissima	Surf clam	Germinal vesicle	200	Boothby et al. (1977)
		Embryo	220	Boothby et al. (1977)
		Sperm	230	Boothby et al. (1977)
Drosophila melanogaster	Fruit fly	Salivary chromosomes	~200	Hill et al. (1982)
		1.688 satellite	~190	Hill et al. (1982)

[a]Average of values for 9, 11, 16 day. See reference.
[b]Approximate data also given for 4, 16 day embryos, blastula.
[c]Approximate data also given for several developmental stages.

Table 7-1F. Nucleosome Repeat Lengths—Vertebrates

Species name	Common name	Tissue or cell type	Repeat (bp)	Reference
Bos taurus	Calf	Cerebral cortical neurons	168	Allan et al. (1983)
		Kidney	193 ± 5	Weber and Cole (1982)
		Thymus	191 ± 8[a]	Todd and Garrard (1977)
		Satellite	186 ± 7	Weber and Cole (1982)
Gallus domesticus	Chicken	Erythroblasts[b]	190–205	Weintraub (1978)
		Erythroblasts, blastophilic[c]	205 ± 2	Schlegel et al. (1980a)
		3-day erythroid cells	202 ± 5	Wilhelm et al. (1977)
		12–16 day erythrocytes	207 ± 4	Weintraub (1978)
		Anemic peripheral blood	218 ± 3	Schlegel et al. (1980a)
		Erythrocytes (adult)	212 ± 3	Schlegel et al. (1980a)
			207 ± 2	Compton et al. (1976)
			216 ± 5	Wilhelm et al. (1977)
			212	Mazen et al. (1978)
			212 ± 5	Morris (1976b)
			210 ± 3	Hörz and Zachau (1980)
		Liver	202 ± 6	Morris (1976b)
			200 ± 5	Wilhelm et al. (1977)
		Oviduct	196 ± 1	Compton et al. (1976)
(not specified)	Duck	Reticulocytes	185	Sollner-Webb and Felsenfeld (1975)
Xenopus laevis	Frog	Erythrocyte	184 ± 4	Gottesfeld (1980)
			189 ± 4	Humphries et al. (1979)
		Erythrocyte, 5S genes	175 ± 5	Humphries et al. (1979)
			176 ± 2	Gottesfeld (1980)
		Kidney	178 ± 2	Gottesfeld (1980)
		Liver	178 ± 4	Gottesfeld (1980)
			178 ± 5	Humphries et al. (1979)

Table 7-1F. *Continued*

Species name	Common name	Tissue or cell type	Repeat (bp)	Reference
		Liver, 5S genes	175 ± 5	Humphries et al. (1979)
			178 ± 5	Gottesfeld (1980)
Mesocricetus auratus	Syrian hamster	Kidney	196 ± 1	Compton et al. (1976)
		Liver	196 ± 1	Compton et al. (1976)
Mus musculus	Mouse	Brain	207 ± 6	Gaubatz et al. (1979)
		Bone marrow	209 ± 5	Dean et al. (1985)
		Heart	208 ± 6	Gaubatz et al. (1979)
		Ascites tumor	197 ± 4	Dean et al. (1985)
		Liver	204 ± 7	Gaubatz et al. (1979)
			199 ± 6	Keichline and Wassarman (1979)
			195 ± 5	Gottesfeld and Melton (1978)
		Liver: transcribed ribosomal and nontranscribed satellite	195 ± 5	Gottesfeld and Melton (1978)
Bos taurus	Ox	Glial cells	162	Pearson et al. (1984)
		Cerebral neuron	201	Pearson et al. (1984)
Oryctolagus cuniculus	Rabbit	Cerebellar neuron	197 ± 6	Thomas and Thompson (1977)
		Cerebral cortex neuron	162 ± 6	Thomas and Thompson (1977)
			162	Brown (1978)
		Glial cells	200 ± 6	Thomas and Thompson (1977)
			198	Brown (1978)
		Kidney	198	Brown (1978)
			200 ± 4	Silver (1979)
		Liver	200 ± 5	Thomas and Thompson (1977)
Rattus norwegicus	Rat	Bone marrow	192 ± 1	Compton et al. (1976)
		Cerebellar neurons[d]	201 ± 5	Whatley et al. (1981)
			218 ± 5	Jaeger and Kuenzle (1982)
		Cerebellar neurons (fetal)[d]	165 ± 5	Jaeger and Kuenzle (1982)
		Cerebellar glial cells	207 ± 4	Whatley et al. (1981)
			200 ± 5	Jaeger and Kuenzle (1982)

Table 7-1F. *Continued*

Species name	Common name	Tissue or cell type	Repeat (bp)	Reference
		Cortical neurons[a]	174 ± 3	Ermini and Kuenzle (1978)
			175 ± 11	Whatley et al. (1981)
			170 ± 4	Jaeger and Kuenzle (1982)
		Cortical neurons (fetal)[d]	195 ± 8	Ermini and Kuenzle (1978)
			195 ± 11	Whatley et al. (1981)
			200 ± 5	Jaeger and Kuenzle (1982)
		Cortical glial cells	207 ± 8	Whatley et al. (1981)
		Kidney	196 ± 1	Compton et al. (1976)
		Liver	196 ± 1	Compton et al. (1976)
			200 ± 5	Morris (1976b)
			195 ± 6	Thomas and Thompson (1977)
			207 ± 4	Ermini and Kuenzle (1978)
			195	Omori et al. (1980)
		Liver (fetal)	193 ± 2	Compton et al. (1976)
			185 ± 3	Ermini and Kuenzle (1978)
		Liver (satellite DNA)	185	Omori et al. (1980)
Salmo trutto	Trout	Liver	206 ± 6	Bailey et al. (1980)

[a] Actually, 3 other minor repeats reported in tissue: 203 ± 9, 170 ± 8, 142 ± 3.
[b] 3–6-day chicks; repeat increases with age.
[c] Bone marrow cells from anemic adult chicken.
[d] More data are given in Whatley et al. and Jaeger and Kuenzle for changes during prenatal and postnatal development. See Berkowitz et al. (1983) for further changes in aging.

Table 7-1G. Nucleosome Repeat Lengths—Cells in Culture

Source	Cell line designation	Repeat (bp)	Reference
African green monkey	CV-1	189 ± 2	Compton et al. (1976)
		182 ± 6	Shelton et al. (1978)
		188	Fittler and Zachau (1979)
Chicken	Friend	183 ± 5	Schlegel et al. (1980b)
Drosophila melanogaster	Kc	185	Samal et al. (1981)
	Kc, HSP genes	180 ± 4	Levy and Noll (1980)
Hamster	BHK	190 ± 2	Compton et al. (1976)
	B8	186	Sperling and Weiss (1980)
	3460-3B	189	Sperling and Weiss (1980)
	CHO	178 ± 1	Compton et al. (1976)
	CHO	178 ± 5	Ferrer et al. (1980)
	CHO-genes	178 ± 5	Ferrer et al. (1980)
Human	HeLa	188 ± 1	Compton et al. (1976)
		195	Tate and Philipson (1979)
Mouse	BW1J	192	Sperling and Weiss (1980)
	LC	194	Sperling and Weiss (1980)
	P815	188 ± 2	Compton et al. (1976)
	F9.22	196 ± 6	Oshima et al. (1980)
	PSA1	195 ± 5	Oshima et al. (1980)
	PSA1 Endo	205	Oshima et al. (1980)
	P4S2	185 ± 3	Oshima et al. (1980)
	PFHR9	187 ± 3	Oshima et al. (1980)
	L1210-ascites in culture	186 ± 6	Dean et al. (1985)
Rat	C6	198 ± 2	Compton et al. (1976)
	FaO	189	Sperling and Weiss (1980)
	FaZa 967	189	Sperling and Weiss (1980)
	Fu5-5	189	Sperling and Weiss (1980)
	FaoflC2	189	Sperling and Weiss (1980)
	H5	184	Sperling and Weiss (1980)
	HF1	184	Sperling and Weiss (1980)
	HF1-5	189	Sperling and Weiss (1980)
	P4	192	Sperling and Weiss (1980)
	RG6A	191	Sperling et al. (1980)
	Kidney, primary culture	191 ± 2	Compton et al. (1976)
	Hepatoma	188 ± 2	Compton et al. (1976)
	Teretoma	188 ± 2	Compton et al. (1976)
	Myoblast	189 ± 2	Compton et al. (1976)
	Myotubule	193 ± 2	Compton et al. (1976)
Xenopus laevis	Kidney cells	169 ± 2	Gottesfeld (1980)
	Kidney cells, 5S genes	165 ± 2	Gottesfeld (1980)

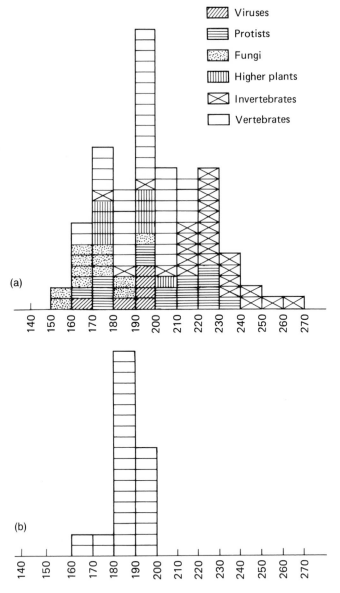

Fig. 7-2. (a) Overall distribution of nucleosomal repeat lengths found in viruses, protists, fungi, higher plants, and animals. (b) Distribution of repeat lengths for cells in tissue culture.

One indication that there may be a certain "spacing" regularity for at least some sets of nucleosomes comes from studies of the digestion of nuclei with DNase I. In early experiments, Noll (1974b) argued that the ladder of bands observed on a denaturing gel extended to "at least 200 bp." This was generally regarded with skepticism, for it clearly exceeded the core DNA length. However, in later experiments, using gels of high resolution, Lohr et al. (1977b) and Lohr and van Holde (1979) showed that the ladder extended even further. Figure 7-3 shows the results of electrophoretic analysis of such digestions for yeast and chicken erythrocyte chromatin. The important point is that the bands extend not only

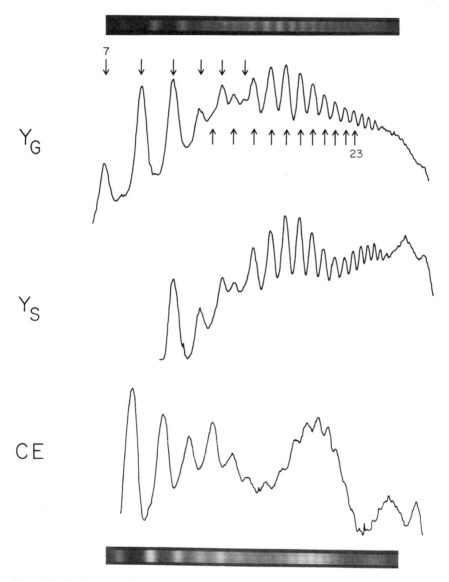

Fig. 7-3. Single-strand DNA fragments obtained by DNase I digestion of growing yeast nuclei (top), stationary phase yeast nuclei (middle), and chicken erythrocyte nuclei (bottom). Bands 1–6 have been run off the gel to resolve the larger bands. he arrows indicate the two interleaving series of bands in yeast (see Fig. 7-4). (From Lohr and van Holde, 1979.)

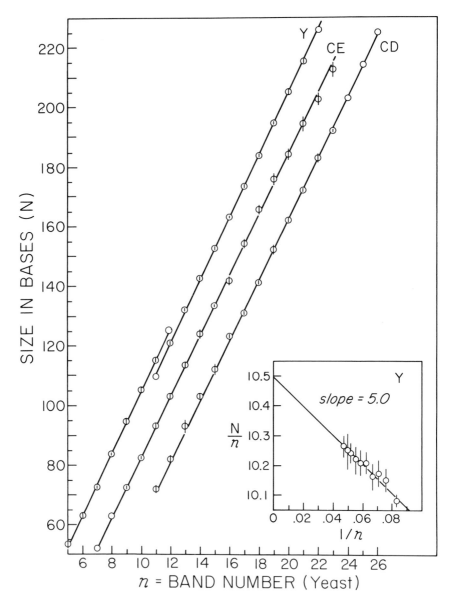

Fig. 7-4. The sizes of DNA fragments like those shown in Figure 7-3, graphed versus band number. Y, yeast nuclei; CE, chicken erythrocyte nuclei; CD, compact dimers from chicken erythrocyte. The band number (n) is for yeast; the lines for CE and CD have been displaced for clarity. The displacement of the two series of yeast fragments is obvious. The insert shows a method for quantitating the band spacing and displacement, according to the formula

$$\frac{N}{n} = r - \frac{d}{n}$$

where r is the number of bases between nicking sites and d is the displacement between series. The line gives $r = 10.5$ bp, $d = 5$ bp. (From Lohr and van Holde, 1979.)

beyond the core DNA size, but even beyond the repeat size. Since the highest bands can only come from cuts in adjacent nucleosomes, the data would seem to require, as a *minimal* explanation, quantized spacing of a substantial portion of the nucleosomes. The yeast patterns, which extend to the greatest distance beyond repeat size, reveal another intriguing feature. The upper set of bands, as shown in Figure 7-4, are offset by 5 bp with respect to the lower set. This implies that the permitted linker lengths are not 10 n base pairs (n integer) but 10 $n \pm 5$ base pairs. Although the phenomenon was first detected in yeast, it has now been observed in other cell types as well. By extending the chicken erythrocyte ladder to larger sizes, D. Lohr (private communication) has been able to observe a similar shift. Using rather different techniques, Karpov et al. (1982) and Strauss and Prunell (1983) have demonstrated the dominance of 10 $n \pm 5$ bp linkers in *Drosophila* and rat liver chromatin, respectively.

Lest these results be overinterpreted, certain reservations must be noted: (1) The bands in question (i.e., Fig. 7-3) are superimposed on a broad diffuse background. It is difficult to tell *how much* of the chromatin is so organized, although it cannot be a minor fraction. (2) Most, if not all, of these data could be explained by regular spacing between nucleosome *pairs*, for it is difficult to resolve bands beyond two repeat lengths in any case. Indeed, Burgoyne and Skinner (1981) claim evidence for a two-nucleosome unit as the repeating entity in rat liver and chicken erythrocyte chromatins. (3) It should be emphasized that neither the "extended ladder" data nor the experiments of Karpov et al. or Strauss and Prunell show that the nucleosomes contributing to the pattern are *uniformly* spaced. It simply shows that the spacings must be given by (10 $n \pm 5$) where n is an integer (see Fig. 7-3). The regions involved *could* be uniformly spaced, but these data do not establish it.

A linker of (10 $n + 5$) bp imposes interesting topological constraints on the orientation of DNA strands in adjacent nucleosomes. As Figure 7-5 (from Strauss and Prunell) shows, the resulting effect depends strongly on whether each DNA turn in the nucleosome core particle involves 10.0 or 10.5 bp. The implication is that the (10 $n + 5$) rule may be a *consequence* of the way in which nucleosomes are packed in the chromatin fiber.

Whatever the reasons for preferences for certain linker sizes may be in vivo, there is now evidence that lysine-rich histones alone are sufficient to establish regular spacing under certain circumstances in vitro. Stein and Bina (1984) have reconstituted "chromatin" using poly (dA·dT)·poly(dA·dT), chicken erythrocyte histone cores, and the erythrocyte-specific lysine-rich histone H5. The histone cores were first assembled on the polynucleotide by salt-gradient dialysis (see Chapter 6) and the H5 then added at 0.15 M NaCl, along with polyglutamic acid. When H5 was provided at about 1 molecule per nucleosome, a regular repeat of about 210 bp was generated. Such an experiment is a specific test for the ability of the lysine-rich histone to promote the normal spacing;

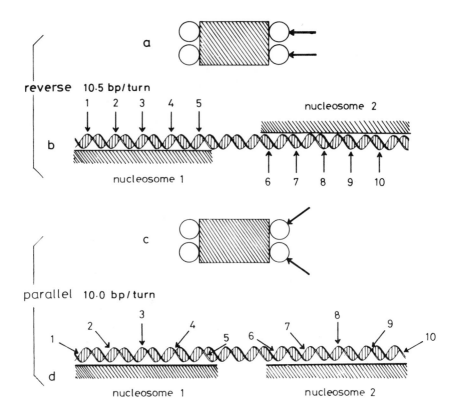

Fig. 7-5. Two possible interpretations of the (10 n + 5) rule, based on the directions of accessibility of nucleosomal DNA to DNase I or DNase II cleavage. In the top model (a, b) the nucleosomal DNA is assumed to have 10.5 bp/turn. In this case, the results of nuclease nicking experiments (see Chapter 6) can be explained only if the direction of maximum accessibility is perpendicular to the core axis (a). The 10 n + 5 rule then requires that alternate nucleosomes lie on the DNA in a "reverse" sense, as shown in (b). Alternatively, if the nucleosomal DNA has 10.0 bp/turn, then the nicking result can be explained by maximum accessibility being oriented as in (c). If there are 10.0 bp/turn, there are 14.5 turns in a nucleosome, and adjacent nucleosomes will be oriented as in (d). (From The EMBO Journal, vol. 2, issue 1, p. 55, Strauss and Prunell. Copyright 1983 IRL Press.)

it says nothing about the relative importance of positioning effects, which are necessarily absent with this uniform polynucleotide substrate.

In sum, the above data indicate that regular *spacing* of nucleosomes can be accomplished in vitro, and probably occurs in some form in vivo. But a more intriguing question is this: Does specific *positioning* and/or *phasing* also occur in vivo? Until fairly recently, there was little evidence to indicate that it does, and a number of experimental studies suggested that it does not (see, for example, Steinmetz et al., 1975; Cremisi et al., 1976; Baer and Kornberg, 1979; Prunell and Kornberg, 1982). For some time, it was generally believed that interactions between the histone core and DNA were wholly nonspecific, and that if nucleosomes were arrayed in any particular register with the polynucleotide sequence, this must be a consequence of sequence-specific phase-setting nonhistone proteins, together with protein-mediated regular spacing. There are now, however, a number of solid experiments which show that nucleosomes can indeed recognize, or avoid, certain specific sequences in in vitro reconstitution experiments. The pioneering studies of this kind were performed by Chao et al. (1979, 1980a, 1980b), who reconstituted histone cores onto restriction fragments containing the *E. coli* lac operator. A limited number of distinct positions were observed. In somewhat similar experiments utilizing a 145 bp *E. coli* DNA fragment, Ramsay et al. (1984) found a single preferred position. Remarkably, this is such that one end of the DNA projects by 17 bp from the core. In further studies, Ramsay (1986) has made deletions in this sequence, seeking those regions that dictate histone core placement. It was found that only a central region, approximately 40 bp in length, was necessary to maintain positioning. Near the edges of this region lies a symmetrically placed TGG sequence. Ramsay cautions, however, that the determinants of positioning are probably more complex than this.

Perhaps most appropriate to eukaryotic chromatin structure are the experiments of Simpson and Stafford (1983), in which reconstitution was carried out using a 260 bp cloned fragment of sea urchin DNA carrying a 5S RNA gene. Particles formed in which the histone core occupied residues 20 to 165. In further studies, Fitzgerald and Simpson (1985) have constructed mutants of this sequence. Some of these position nucleosomes, others do not. Although no specific "positioning" sequences were recognized, the region 20 to 30 bp either side of the core center appeared important in accord with the results of Ramsay (1986). Although details remain to be resolved, it is clear that experiments of this kind establish that the histone octamer *can* exhibit selectivity toward certain DNA sequences, without the aid of additional nonhistone proteins or "assembly factors."

What is not clear at the present is how general this positioning phenomenon may be, or how histone octamers recognize certain sequences. Comparison of preferred sequences has not revealed any obvious common features. It is possible that recognition may involve quite subtle aspects of DNA structure, such as "bendability," as suggested by Trifonov and his collaborators (see, for example, Trifonov, 1980, 1985; Mengeritsky and Trifonov, 1984; Zhurkin, 1981; Drew and Travers, 1986). Such a concept gains credibility with the observations from X-ray diffraction studies that the path of the DNA on the nucleosome is not uniformly curved, but rather is sharply bent at certain points (Richmond et al., 1984).

Given that sequence-specific recognition can occur in in vitro reconstitution experiments, the question follows as to whether it *does* occur in vivo. This topic has been the subject of intense interest, because of its obvious significance to the question of the physiological role of nucleosomal structure in chromatin. Although there have been a number of approaches to this problem, two methods have emerged that appear to give the most reliable results. The first of these is the *indirect end-labeling* method, developed independently by Nedospasov and Georgiev (1980) and Wu (1980). The essentials of the technique are described in Figure 7-6A; it can, in principle, be applied to analysis of chromatin structure in any region for which a suitable probe is available. To map accurately over extended regions, multiple probes are required. A different approach that is particularly applicable to repetitive DNA sequences has been developed by Zhang et al. (1983). This method, shown in Figure 7-6B, is capable of very high precision in some cases; accuracies of ±1 bp have been claimed. However, there are potential pitfalls in both techniques, especially if micrococcal nuclease is used for the initial cleavage. This enzyme exhibits considerable sequence specificity, preferring to cut in A–T-rich regions (see, for example, Dingwall et al., 1981). Free DNA, when lightly cleaved by micrococcal nuclease, will yield a distinctive band pattern. Furthermore, especially sensitive sites will be cut even if they lie within nucleosomes. Thus, the band pattern obtained from indirect end-labeling experiments on chromatin must be carefully compared with that found for a naked DNA control. If chromatin is *extensively* digested with this enzyme, the preference to cut at A–T sites may bias the distribution of remaining nucleosomes, again leading to possible artifacts (McGhee and Felsenfeld, 1983). For these reasons, a number of workers have turned to other cleavage agents, seeking ones that exhibit less sequence preference. These include some nonenzymatic reagents. An attractive candidate is "Dervan's reagent," methidiumpropyl·EDTA·Fe(II). Cartwright et al. (1983) find that this compound exhibits little sequence specificity and cleaves very selectively in chromatin linker regions (see also Cartwright and Elgin, 1984). In contrast, another nonenzymatic reagent for DNA cleavage, the $(1,10\text{-phenanthroline})_2\cdot Cu(I)$ complex, has been found to exhibit sequence preferences much like those of micrococcal nuclease (Jessee et al., 1982; Cartwright and Elgin, 1982a).

Table 7-2 lists a number of cases in which positioning or phasing in vivo has been demonstrated with reasonable certainty. The table excludes numerous older reports, in which methods of questionable reliability were

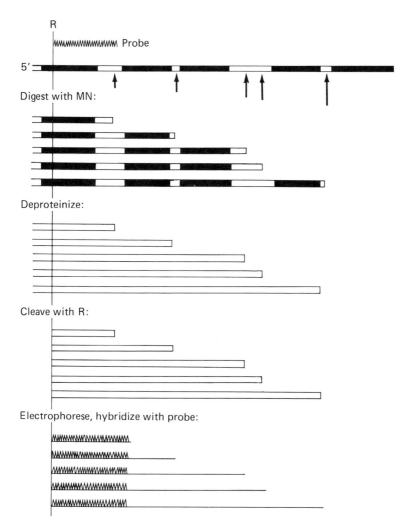

Fig. 7-6A. The indirect end-label method of Nedospasov and Georgiev (1980) and Wu (1980) for locating positioned nucleosomes. Use is made of a probe that abuts a particular restriction site to identify fragments that begin at the site and extend to nuclease-sensitive (in this case internucleosomal) sites.

employed, or in which adequate controls were not provided. It also excludes the many studies in which specific genomic regions have been reported to contain nucleosomes, but definite information as to their linear organization is lacking. Finally, the table is not intended to be a comprehensive survey of examples; it is a sampling. For additional information, the reader is referred to the excellent reviews by Reeves (1984) and by Eissenberg et al. (1985).

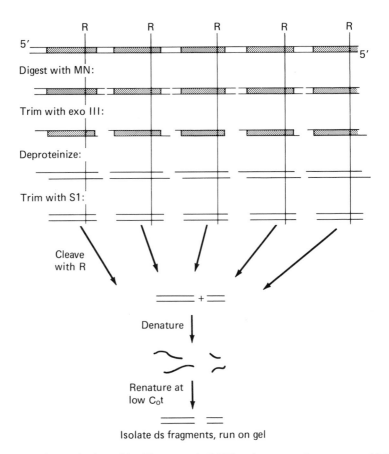

Fig. 7-6B. The method used by Zhang et al. (1983) to locate nucleosomes on highly repetitive DNA. A particularly simple case is chosen as an example, where there is one nucleosome in one phase on each repeat. There is a restriction site (R) within the nucleosomal DNA. Core particle DNA is isolated, end-labeled, and cleaved with R. The highly repetitive sequences are separated from the bulk nucleosomal DNA by denaturation and renaturation to low C_0t. In practice, several restriction enzymes will usually be employed.

The several categories of genomic regions listed in Table 7-2 require separate comment, with particular attention to certain important examples.

Satellite Chromatin

The satellite DNAs of many eukaryotic organisms consist of multiple repetitions of identical or nearly identical sequences. Interestingly, many of these seem to have sequence repeat lengths that are approximately equal to typical nucleosome repeat lengths or simple multiples thereof. It would seem reasonable that such a periodic structure might exhibit phasing of

Table 7-2. Examples of Defined Nucleosome Arrangements

Organism	Genomic region	Arrangement	Methods	Reference
I. Satellite DNA				
Rat	Satellite I	Phased; 2 major, 16 minor modes	Restriction digest, trimming, second digest	Igo-Kemenes et al. (1980); Bock et al. (1984)
Mouse	Satellite	Phased; 16 approximately equivalent modes	As above	Zhang and Hörz (1984)
African green monkey	α-Satellite	Phased; 1 major, 7 minor modes	As above	Zhang et al. (1983) (see text for earlier references)
II. RNA genes				
Chick embryo	tRNA genes	Phased; 1 mode	Isolation of nucleosome tetramers, cloning of tRNA genes, location of gene, on tetramer	Wittig and Wittig (1979)
Xenopus laevis	tRNA genes (inactive)	Phased; 1 mode	Indirect end-label	Bryan et al. (1983)
Xenopus laevis	5S RNA genes	Phased; 4 modes	Variant of indirect end-label method	Gottesfeld and Bloomer (1980)
Dictyostelium discoideum	rRNA (nontranscribed terminal region of minichromosome)	Spaced or positioned	Indirect end-label	Edwards and Firtel (1984)
Tetrahymena thermophila and *T. pyriformis*	rRNA (nontranscribed central region of minichromosome)	Spaced or positioned	Indirect end-label	Palen and Cech (1984)
III. Structural genes, regions				
Saccharomyces cerevisiae	Acid phosphatase gene (repressed)	Positioned	Indirect end-label	Bergman and Kramer (1983)

Table 7-2. *Continued*

Organism	Genomic region	Arrangement	Methods	Reference
Drosophila melanogaster	Heat-shock locus 87A7 (intergenic spacer)	Positioned	Indirect end-label	Udvardy and Schedl (1984)
Drosophila melanogaster	Histone locus (intergenic spacer)	Positioned	Indirect end-label	Samal et al. (1981); Worcel et al. (1983)
IV. Other Structures				
Oxytrichia sp.	Single gene chromosomes	Regularly spaced or positioned	Indirect end-label	Gottschling and Cech (1984)
Saccharomyces cerevisiae	Sequences flanking centromere	Positioned	Indirect end-label	Bloom and Carbon (1982)
Saccharomyces cerevisiae	TRP1/ARS1 minichromosome	Positioned	Indirect end-label	Thoma et al. (1984); Thoma and Simpson (1985)

nucleosomes, in the strict sense of the word. If there is *any* preference of nucleosomes for particular local base sequences, then in scanning across a repeat one should find one site of minimum free energy. Since this will recur at regular intervals commensurate with allowable nucleosome spacing, a phased relationship would seem a logical consequence. Such a very regular arrangement might be well poised for condensing the chromatin fiber into a solenoidal helix (see *Higher-Order Structures: Chromatin Fibers,* below). Indeed, there is now evidence that phasing does occur in satellite chromatins, although the details turn out to be more subtle than might have been anticipated.

The history of studies of the African green monkey α-satellite chromatin is instructive, both in illustrating the experimental difficulties in the field and in the complexity of the final result. As early as 1977, Musich and co-workers suggested that nucleosomes were phased on this chromatin (Musich et al., 1977b). Their conclusion was based primarily on the fact that micrococcal nuclease cleaved the satellite chromatin into ~172 bp fragments, a size exactly equivalent to the DNA repeat. From accessibility to restriction nucleases, they suggested that the linker lay near position 30. These conclusions were rather strongly attacked. Fittler and Zachau (1979) showed that the *free* satellite DNA was also cleaved into 172 bp fragments by micrococcal nuclease, suggesting that the results of Musich et al. were an artifact of enzyme preference. Singer (1979) obtained hybridization results that cast further doubt on phasing. In further studies, Musich et al. (1982) reiterated the claim for a unique nucleosomal position, but now suggested that the linker lay near site 126 (Fig. 7-7). In rebuttal, Hörz et al. (1983) carried out quantitative studies on the cleavage of free DNA and showed that 50% of all early micrococcal cuts occurred at sites 123 and 132. Wu et al. (1983), using EcoRI to solubilize a particular fraction of the α-satellite chromatin, suggested that the structure must involve both phased and random arrangements (see Smith and Lieberman, 1984, for a similar conclusion).

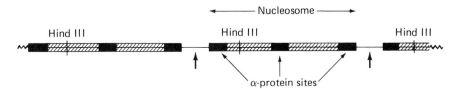

Fig. 7-7. Nucleosome phasing on the African green monkey α-satellite DNA. Two copies of the 172 bp repeat are shown. Only the major nucleosome phase of the eight detected by Zhang et al. (1983) is illustrated. The arrows indicate the positions (near 126) where Musich et al. (1982) found major nuclease cleavage and proposed as the linker regions. The black boxes represent the binding sites on the DNA sequence identified by Strauss and Varshavsky (1984) for the HMG-like, α-satellite specific protein.

The most detailed study to date, however, is that of Zhang et al. (1983). Using the method of micrococcal digestion, followed by end-trimming and subsequent restriction nuclease cleavage of the DNA, they find eight different phases. Amusingly, the major phase, which comprises about 35% of the material, corresponds quite closely to that proposed by Musich et al. (1982).

This observation of multiple phases seems to be the rule, rather than the exception in satellite chromatins. As Table 7-2 shows, similar results have been obtained for the rat satellite I and the mouse satellite chromatin. Why should such multiple phases exist? In some cases, it is possible that they reflect subtle variances in the repeating DNA sequence. Another and likely possibility is that phasing is set over extended regions by the intervention of specific DNA-binding proteins. Musich et al. (1977a) were the first to point out that there are specific nonhistone proteins associated with α-satellite chromatin. Recently, Strauss and Varshavsky (1984) have isolated an HMG-like protein that binds specifically to three sites on the α-satellite DNA. Remarkably, two of these sites correspond to the ends of the core DNA in the major frame (see Fig. 7-7). The third lies near the center of the core particle. Such "phase setting" proteins need not be present in stoichiometric quantities to exert a profound effect. The presence of the α-satellite protein on an occasional nucleosome would seemingly be capable of setting the major phase over a large region. It is of interest that the locations of two of the sites (near the ends of the core DNA) are very similar to the localization of HMG 14 and 17 in core particles (see Chapter 6).

RNA Genes
The genetic arrangements of the genes for tRNAs, 5S RNAs, and the larger ribosomal rRNAs are frequently unusual. The smaller RNA genes are often arrayed in tandem repeats, reminiscent of the satellite sequences. In a number of lower eukaryotes, the ribosomal RNA genes are carried on autonomous, highly duplicated minichromosomes. These features have made the chromatin organizations of these genes favored objects of study. Although the data for the small RNA genes are not as precise as those presently available for some of the satellite chromatins, there is considerable evidence in support of phasing, as the examples in Table 7-2 illustrate. Even the general pattern of multiple phases seems to be present. There appear to be, in at least some cases, significant differences in nucleosome positioning between active and inactive genes. This is a topic more appropriately considered in Chapter 8.

The ribosomal RNA genes of the protist *Tetrahymena* and the slime mold *Dictyostelium* are in both cases carried on palindromic minichromosomes. Investigation of *nontranscribed* regions has given indication of very regular nucleosomal structure. Edwards and Firtel (1984) have found that the terminal, nontranscribed 10 kb of the *Dictyostelium discoideum* minichromosome contains an array of fixed nucleosomes. Since the spacing is not regular, positioning of the cores on the DNA sequence is suspected. In *Tetrahymena thermophila,* the nontranscribed central 1400 bp of the palindromic rDNA chromosome contains precisely seven nucleosomes, the fourth being centered on the axis (Palen and Cech, 1984). The closely related *T. pyriformis* has only 1000 bp in this region and a corresponding five nucleosomes. In these cases, it is not clear whether the nucleosomes are actually positioned or simply packed between boundaries. In both cases, nucleosome-free regions are found near the 5' ends of the symmetrically placed genes, and there is no evidence for regular nucleosome structure in the coding regions. Again, discussion of such features will be deferred until the following chapter.

Structural Genes

Most of the genes coding for proteins exist as one or a few copies in the eukaryotic genome. The general features of nucleosome structure in and around such genes will be considered in Chapter 8; at this point, we are concerned only with the question of whether positioned or regularly spaced core particles are found in such regions.

Only with the advent of the indirect end-labeling technique (see above) has progress been possible with such systems. There is, moreover, frequent difficulty in establishing solid evidence for *positioning,* as opposed to simple packing of nucleosomes between or adjacent to defined boundaries. This is because evidence for highly regular structure in the neighborhood of such genes is usually confined to short intergenic or flanking domains. As will be detailed in Chapter 8, there is accumulating evidence for the binding of specific nonhistone proteins near the 5' ends of genes. Such fixed points could serve to delimit short strings of nucleosomes, producing the appearance of positioning. It is often argued (see below) that irregular spacing of nucleosomes in defined locations is evidence for positioning. However, light micrococcal nuclease digestion into linker regions might well be influenced by DNA sequence preference in such a way as to give a false appearance of irregular spacing. Thus, it is probably wise to regard all claims made to date for nucleosome positioning in the vicinity of structural genes with some degree of skepticism.

A number of examples are given in Table 7-2. For details of these, the original literature should be consulted; additional, less definitive studies are described in the review by Eissenberg et al. (1985). Further examples are discussed in Chapter 8. I shall treat only one case as a model; the careful study by Udvardy and Schedl (1984) of the 87A7 locus of heat-shock genes in *Drosophila.* At this locus, two copies of the *hsp*70 gene are transcribed in opposite directions (Fig. 7-8). The indirect end label was employed to investigate two cell lines that differ in genome structure at this locus. The major conclusions are shown in Figure 7-8. Regions 3' to the genes, as well as the transcribable sequences, exhibit micrococcal

proximal   distal

ORS

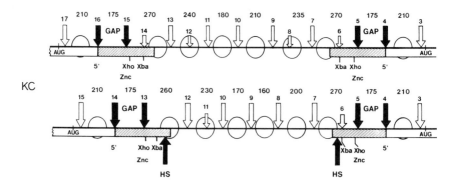

Fig. 7-8. Chromatin organization near the 5' end of the *Drosophila hsp*70 heat-shock genes and in the intergenic spacer. The 5' ends of the two gene copies are shown by boxes; transcription is both to the left and right. The arrangements for two different cell lines (OR5 and KC) are shown. Arrows are above preferred sites for micrococcal digestion, with distances (bp) between major sites noted. Especially strong are the sites in the GAP region near the 5' end of either gene. Upon heat shock, two more sites appear at the 5' ends in KC cells (arrows marked HS). (From Journal of Molecular Biology *172*, p. 385, Udvardy and Schedl. Copyright 1984 Academic Press.)

digestion patterns similar to naked DNA, although other data indicate that nucleosomes are present. Therefore, it is concluded that in these areas no regular, reproducible pattern exists, although occasional positioned nucleosomes may be present. There does seem to be evidence, in the repressed state, for one or two nucleosomes positioned beyond the 3' end and one within the 5' portion of the gene itself. The nontranscribed intergenic region contains a sequence of very well defined micrococcal cutting sites, clearly distinguishable from those in the free DNA. The authors believe the nucleosomes to be positioned here, since a wide range of spacings is observed (see Fig. 7-8). This kind of pattern—a lack of defined nucleosome organization within transcribed or transcribable regions, together with positioned or at least highly restricted nucleosomes in some flanking regions—appears to be fairly common. I shall return to consideration of its possible significance in Chapter 8.

It should not be assumed, however, that all structural genes are surrounded by positioned or regularly spaced nucleosomes. As one example, consider the inactive κ-immunoglobin genes in mouse liver. A careful study by Weischet et al. (1983a, 1983b) rules out positioning, multiframe uniform-repeat spacing, or a truly random distribution. There is evidence for some kind of order, but its precise nature cannot presently be specified.

Special Genomic Structures

Some of the clearest evidence for nucleosome positioning comes from studies of yeast centromeres and plasmids. Bloom and Carbon (1982) have examined the chromatin structure surrounding the centromeres of yeast chromosomes 3 and 11 (see also Bloom et al., 1984). They find evidence for specific placement of at least 12 nucleosomes flanking a special, 220 to 250 bp central region. That these are positioned, rather than representing regularly spaced arrays built about this center, is demonstrated by the fact that plasmids lacking the centromeric region, but containing the flanking sequences in question, showed exactly the same arrangement in the yeast DNA.

A somewhat similar study has been reported from the laboratory of R.T. Simpson. Thoma et al. (1984) prepared a yeast plasmid containing the autonomous replicating sequence (ARS) and a *trp*1 gene. Evidence for specific location of three nucleosomes was obtained by the indirect end-labeling method. In further experiments, Thoma and Simpson (1985) proved that these nucleosomes were positioned on the DNA sequence by inserting additional DNA into the linker regions (see Fig. 7-9). Small additions had no effect on the position of adjacent histone cores. Additions of 155 bp or more resulted in the incorporation of another nucleosome. Further studies (Thoma, 1986) show that the positions of nucleosomes on the *trp*1 gene are influenced by flanking sequences. Thus, at least *two* mechanisms for nucleosome arrangement are functioning here.

From these experiments with yeast, we find the definitive proof that nucleosome positioning does occur in vivo. On the other hand, these and other experiments described above show definitively that not all nucleosomes are so ordered. There is good circumstantial evidence to suggest that each of the kinds of linear organization postulated at the beginning of this chapter may be found in one region or another of the eukaryotic genome. We are only beginning to understand the significance such variegated organization may have on the three-dimensional structure of chromatin.

Higher-Order Structures: Chromatin Fibers

A typical eukaryotic genome, if displayed as a single extended DNA molecule, would be of the order of 3 m in length. Yet all of this DNA is compacted into a nucleus a few micrometers in diameter. Obviously, the chromatin must be highly coiled or folded within the nucleus. Wrapping DNA about the nucleosome core produces only about a fivefold compaction at best, so it is clear that the linear array of nucleosomes must itself be further folded. The full elucidation of the manner of folding, both in the interphase nucleus and in chromosomes at metaphase, is a problem that has attracted intense interest. Many feel that chromatin compaction is of importance not only to the packing problem, but also to the mech-

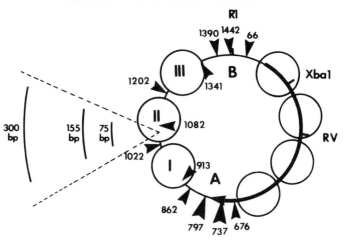

Fig. 7-9. Chromatin structure of the TRP1ARS1 plasmid. The open reading frame of the *trp*1 gene is indicated by the heavy arrow. The putative origin of replication lies in nucleosome-free area A, and the *trp*1 promoter in nucleosome-free area B. The dark arrows show restriction sites; those on the inside of the circle are protected in the chromatin, those on the outside are accessible. Three positioned nucleosomes are shown. The four nucleosomes on the *trp*1 are imprecisely positioned or unstable. To the left is indicated the position at which inserts of 75 bp, 155 bp, and 300 bp were made. Insertion of 75 bp merely lengthened the spacer between nucleosomes I and II; the three nucleosomes retained their positions on the sequence. Insertion of 155 bp or 300 bp allowed the addition of a fourth positioned nucleosome. (Reprinted by permission from *Nature*, vol. 315, p. 250. Copyright © 1985 Macmillan Journals Limited.)

anisms of gene repression. The field has been beset with controversy. Part of the difficulty lies in the fact that it is very hard to obtain information concerning the structures that exist in nuclei. Most studies of "higher-order structure" carried out to date have utilized chromatin fragments isolated from lysed nuclei. There is always a danger of artifact when such preparations are made, and many arguments have revolved around the adequacy of preparative methods. In this section, I shall concentrate primarily on such in vitro studies. The much more limited attempts to discuss structure in intact cells or nuclei will be considered in the next section. Both topics are also dealt with in considerable detail in recent reviews (Butler, 1983; Crane-Robinson et al., 1984).

. Many techniques have been employed in the in vitro studies of chromatin fibers. X-ray, neutron and light scattering, circular dichroism, electric field dichroism, and hydrodynamic methods have all added important information. But the generative technique has been electron microscopy. Although sectioning and whole-mount methods have become increasingly important, most of our basic notions about higher-order structure have emerged from transmission electron microscopy (EM) studies of samples spread on grids.

There are a number of steps involved in the preparation of macromolecular samples for conventional transmission electron microscopy. Since the danger of artifact is always present, it is appropriate to examine these steps in some detail.

1. The macromolecules or fibers must be isolated and suspended in solution. The method of isolation and choice of solution medium may profoundly modify the structure. For example, some early (1974–1975) EM studies employed shearing to yield manageable chromatin; it is now known (Noll et al., 1975; DeMurcia et al., 1978) that this has drastic effects on chromatin structure. Most workers now use a mild nuclease digestion, followed by nuclear lysis. The ionic strength and the nature of the ions present during the preparation can play a major role in determining structure (see below).

2. The sample may, or may not, be fixed by a crosslinking reagent such as glutaraldehyde. With some methods of sample handling, fixation is essential to preserve structure (see Thoma et al., 1979; DeMurcia and Koller, 1981, for discussion). On the other hand, some have expressed concern that fixation may *distort* structure (i.e., Fulmer and Bloomfield, 1982; Wijns et al., 1982).

3. The electron microscope grid may be an uncoated carbon film or may be treated in various ways to enhance absorptivity. DeMurcia and Koller (1981) have demonstrated that the absorptivity of the film may be of critical importance. Fibers deposited on weakly absorbing films tend to become stretched, presumably during the subsequent drying procedure, and may present a distorted appearance.

4. The grid must be *dried;* this can be accomplished by air drying, freeze drying, drying in ethanol, or critical point drying. Again, the nature of the process and the substances present may produce artifacts. For example, ethanol drying of DNA can cause compaction into supercoiled forms (Eickbusch and Moudrianakis, 1978a).

5. For all techniques except scanning transmission electron microscopy (STEM), it is necessary to treat the grid in some manner so as to make the particles visible. This can be done by positive staining, negative staining, or shadowing. While these techniques probably do not cause gross distortion of structures, they can give rise to different apparent sizes. Thus, the diameters reported for positively or negatively stained nucleosomes usually range from 80 to 100 Å, whereas particles visualized by shadowing often appear to be about 120 to 130 Å across, even after correction for the thickness of the metal deposit.

Thoughtful investigations and discussions of these problems have been presented by Olins (1979), Thoma et al. (1979), DeMurcia and Koller (1981), and most recently, by Tsanev and Tsaneva (1986). As these authors emphasize, electron microscopy, though an invaluable tool in ultrastructure research, is capable of yielding misleading results.

In an interesting sense, this may have been true of the earliest EM studies to reveal nucleosomal structure. These experiments, which are described in Chapter 2, utilized the Miller and Bakken (1972) spreading technique. In this method, the chromatin fibers are exposed to very low ionic strengths and tend to be stretched on the electron microscope grid during drying. The extended "beaded strings" produced in this way are ideal for demonstrating the existence of a particulate subunit structure in chromatin fibers, but they almost certainly do not reveal the in vivo structure (see Fig. 7-10). These comments are not intended to denigrate the importance of these pioneering experiments. They were, in fact, the only kind of experiments that could provide convincing initial evidence for nucleosomes. Another important early study (Langmore and Wooley, 1975) used dark-field microscopy in the STEM microscope (see also Wooley and Langmore, 1977). This work is significant on two accounts; first, it demonstrated nucleosomes in *unfixed* chromatin without staining. Thus, a number of potential sources of artifact were eliminated. Second, it provided the first evidence that the core particles are disk-like and gave a quite accurate estimate of their thickness (50 Å).

It is now clear that the structure adopted by the chromatin fiber is critically dependent on both ionic strength and the nature of the ions present. Furthermore, removal of lysine-rich histones produces profound effects. There have been many EM studies of such conformational changes, but the most complete in my opinion is that carried out by Thoma and collaborators (Thoma and Koller, 1977, 1981; Thoma et al., 1979). In these experiments, rat liver nuclei were lightly digested with micrococcal nuclease. Nuclei were lysed and the chromatin fixed in glutaraldehyde in buffers of varying ionic strength. In some experiments, H1 was removed by incubation of chromatin with an ion exchange resin in 50 mM Na phosphate, 100 mM NaCl.

The conclusions, which are in general agreement with the observations of other recent studies, can be summarized as follows:

1. In very low salt (0.2 mM EDTA, 1 mM triethanolamine chloride [TEAC1]), the chromatin appears as zigzag fibers of nucleosomes (see Fig. 7-11a). The DNA seems to enter and leave each particle at nearly the same point. Such structures have been observed by many others (i.e., Worcel et al., 1978; Olins et al., 1980; and Woodcock et al., 1984). Chromatin depleted of H1 appears to have lost most nucleosomal structure under these same conditions; irregular thick fibers are often observed (see Thoma et al., 1979). This apparent "unfolding" of H1-

(a)

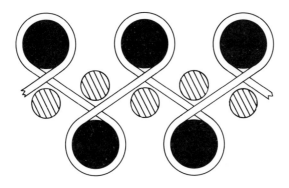

(b)

Fig. 7-10. The effects of H1 removal and stretching on the appearance of the chromatin fiber. In (a) the DNA takes about 2 turns around each core, giving a zigzag fiber. If H1 is removed and the fiber is stretched, the DNA contact with the core is reduced, and the beaded-string appearance is produced.

depleted nucleosomes at very low ionic strengths is in accord with the similar observations with isolated core particles (see Chapter 6). It should be noted that under such conditions *chromatosomes* containing H1 do *not* unfold (Martinson and True, 1979a). The zigzag appearance of H1-containing fibers at low ionic strength is consistent with the model of the chromatosome described in Chapter 6. When such fibers are prepared under tension, they can be stretched to produce almost rectilinear strings of nucleosomes, the "100 Å fibers" referred to frequently in the earlier literature.

2. In slightly higher ionic strength (5 mM TEACl, 0.2 mM EDTA), native chromatin assumes the form of a flat ribbon about 250 Å wide. The zigzag arrangement is usually still present, but the nucleosomes are

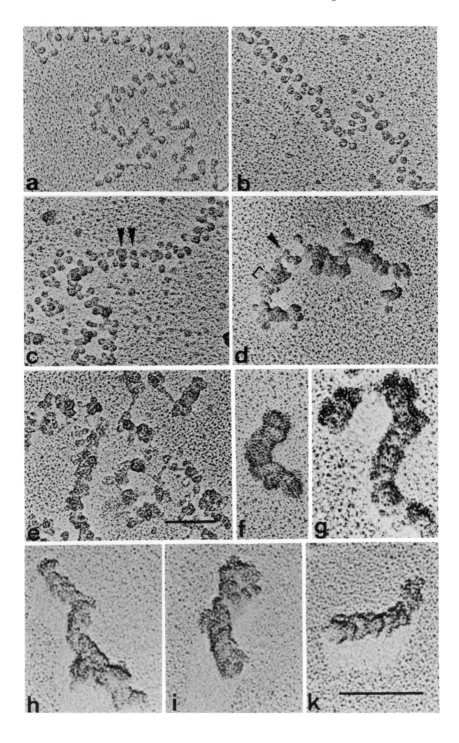

now more tightly packed, and connecting strands of linker DNA are seen only infrequently (Fig. 7-11b,c). In contrast, H1-depleted chromatin under these conditions exhibits a "beaded-string" appearance without a regular zigzag arrangement (Fig. 7-10). Linker DNA is clearly seen and does not appear to enter and leave most nucleosomes at the same points.

To this level of organization, the appearance of chromatin fibers is essentially what would be expected from current knowledge of the core particle and its interaction with H1.

3. At still higher ionic strengths ($\sim$20 mM to 100 mM), chromatin undergoes a progressive condensation to form irregular, roughly rod-like entities with a diameter of 250 to 350 Å (see Fig. 7-11d–k). Such structure is observed *only* in the presence of lysine-rich histones. Chromatin depleted of H1 aggregates under such conditions but only forms irregular clumps. Divalent ions are much more effective than monovalents in producing this regular condensation; only about 1 Mg^{2+} per DNA phosphate (corresponding to a few millimolar under the usual conditions) is required (Finch and Klug, 1976; Thoma et al., 1979; McGhee et al., 1983a).

This "300 Å fiber" has been the object of a great deal of interest, for its dimensions and general appearance are very similar to fibers that have been observed in sectioned nuclei (Davies, 1968; Davies et al., 1974; Walmsley and Davies, 1975), in freeze-fractured nuclei (Haggis and Bond, 1979), and in whole metaphase chromosomes (Ris and Kubai, 1970; Labhart et al., 1982). Thus, the fiber is believed to represent one order of chromatin folding in vivo.

The *structure* of the 300 Å fiber has been the subject of intense debate. There exist two major classes of models, each of which has its articulate proponents. One group, which is at the moment the majority, holds that the 300 Å fiber is a helical winding of the string of nucleosomes. The opposition contends that the fiber is in fact a linear aggregation of "superbeads," each of which is an aggregate of a number of nucleosomes.

The "superbead" hypothesis had its origins in apparently independent publications by Franke et al. (1976) and Kiryanov et al. (1976). Both had noted the irregular appearance of the 300 Å fiber and the fact that electron

◁―――

Fig. 7-11. High magnification electron micrographs of rat liver chromatin fibers fixed at different ionic strength. Fixation was performed in the following solution at pH 7.0: (a) 1 mM TEAC1, 0.2 mM EDTA; (b,c) 5 mM TEAC1, 0.2 mM EDTA; (d) 40 mM NaC1, 5 mM TEAC1, 0.2 mM EDTA; (e) same as (d) but freeze dried after absorption, then washed on water, dehydrated in ethanol, and air dried; (f–k) 100 mM NaC1, 5 mM TEAC1, 0.2 mM EDTA. The bars are 1000 Å. See references for further details. (Reproduced from *The Journal of Cell Biology*, 1979, 83, 403–427, Thoma et al. by copyright permission of the Rockefeller University Press.)

micrographs often suggested a segmented structure (see Fig. 7-11d–k, for example). Kiryanov et al. also showed that digestion of rat liver chromatin with the endogenous Ca^{2+}–Mg^{2+}-dependent nuclease produced roughly spherical particles of the same diameter as the fiber. In the following years, a number of papers, from several research groups, appeared in support of this model (see, for examples, Hozier et al., 1977; Strätling et al., 1978; Butt et al., 1979; Renz, 1979; Zentgraf et al., 1980a, 1980b; Scheer et al., 1980; Pruitt and Grainger, 1980). It was found that the particles could be released by mild nuclease digestion and that they could be reconstituted either in vitro or in vivo (on plasmids in an amphibian oocyte system; Scheer et al., 1980). However, none of these workers were able to provide a detailed model for the internal packing of the superbeads, and the number of nucleosomes contained in each was suggested in various papers to be as small as 6 or as great as 48. The sedimentation coefficient of the monomeric unit was observed to be about 40S (Hozier et al., 1977) or 33S (Strätling et al., 1978). The latter workers also reported putative dimeric and trimeric particles of about 60S and 90S, respectively. The model received a serious blow when Muyldermans et al. (1980b), repeating the conditions used by Strätling et al., found that the peaks in the sucrose gradient profile were most likely due to ribonucleoprotein particles. In a parallel study, Webb-Walker et al. (1980) reported that 40S ribonucleoprotein particles (hnRNP) could be obtained from brief micrococcal nuclease digestion of HeLa nuclei. Perhaps too much attention was paid these results, however, for researchers as early as Kiryanov et al. (1976) had demonstrated that the particles they obtained had a normal chromatin composition.

Other workers have suggested that the "superbeaded" structure so often seen in electron microscope studies of chromatin may be a consequence of partial breakdown of an originally smooth fiber during exposure to suboptimal salt concentration during specimen preparation, or by interaction of the fiber with the EM grid (Butler and Thomas, 1980; Bates et al., 1981; see Azorin et al., 1982, for significant experimental studies). It is perhaps significant that Derenzini et al. (1984) report that EM studies of their frozen sections of nuclei reveal *smooth* fibers. Because of these and other criticisms (see below) the "superbead" model has lost favor with many investigators. A thorough summation of the negative view is given in the review by Butler (1983). (See also Felsenfeld and McGhee, 1986.) Nevertheless, the debate continues. In a more recent publication, Zentgraf and Franke (1984) have endeavored to meet some of the criticisms. Carrying out nuclease digestions of several cell types at relatively high salt (100 mM), they have prepared well-defined particles without exposure to low salt conditions (see Fig. 7-12). The paper demonstrates quite convincingly that the particles contain nucleosomes and a full complement of histones. They are definitely *not* hnRNP particles. Interestingly, the superbeads seem to vary in size (and number of nucleosomes contained) depending

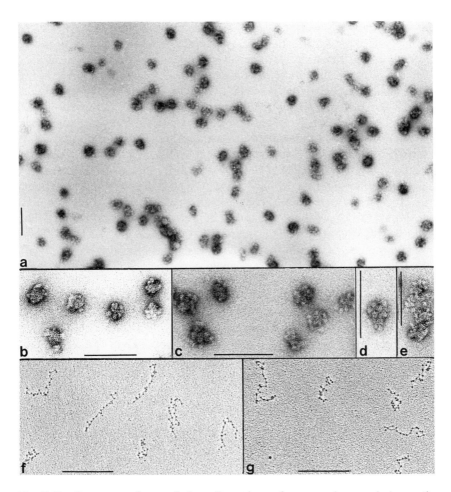

Fig. 7-12. Compact and extended configurations of supranucleosomal chromatin particles from chicken erythrocytes. (a) Particles from a sucrose gradient fraction spread with aldehyde fixation in 100 mM NaC1. (b–e) Particles showing various degrees of unfolding and internal structure. (f,g) Extended arrays obtained from the compact particles by dialysis against low-salt buffer. These contain an average of 20 nucleosomes each. Bars for scale are (a) 1.0 μm; (b–e) 0.1 μm; (f,g) 0.5 μm. (Reproduced from *The Journal of Cell Biology*, 1984, 99, 272–286, Zentgraf and Franke by copyright permission of The Rockefeller University Press.)

on the chromatin source, ranging from about 8 nucleosomes (chicken and rat liver) to 48 nucleosomes (sea urchin sperm). Finally, it should be noted that although the concept of superbeads has been criticized because of the heterogeneity of some preparations (i.e., Butler, 1983), a method developed by Wijns et al. (1982) produces remarkably homogeneous particles.

I shall not attempt, at this point, to judge whether the "superbead" model is "correct" or "incorrect," for I believe the question to be a more subtle one. In any event, before any overall assessment can be made, it is necessary to consider the alternative model.

The "solenoid" model was proposed by Finch and Klug (1976) in the same year as the "superbead" model. Curiously, the electron micrographs cited as evidence for a solenoidal structure look to the untutored eye remarkably like those that had led others to propose superbeads. But Finch and Klug were also considering the earlier X-ray diffraction studies of chromatin (Chapter 2). These had yielded a set of diffraction maxima, at about 110, 55, 37, 27, and 21 Å. In particular, the 110 Å reflection was observed only under those ionic conditions that favor the 300 Å fiber; thus, it was suggested that this corresponds to the pitch of the solenoid (see Fig. 7-13). Given this pitch and an approximate diameter for the fiber, a number of roughly 6 nucleosomes per turn is calculated. The general model has received support from a wide variety of physical studies (Table 7-3), which are more or less in agreement concerning the basic parameters. A particularly important paper is that of Thoma et al. (1979), in which electron microscopy was used to study samples under a wide variety of conditions. Some of their micrographs are reproduced in Figure 7-11; frames f–k depict the fully condensed chromatin fiber. As Thoma et al. point out, cross-striations can *occasionally* be seen in the fibers (see

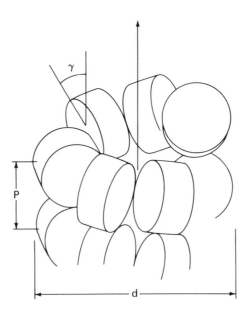

Fig. 7-13. The parameters that will be used in describing a solenoid of nucleosomes. The solenoid has n nucleosomes per turn (not necessarily integral), a pitch P, and a diameter d. The nucleosome faces are tilted at an angle γ with respect to the solenoid axis.

Table 7-3. Parameters Reported for the 300 Å Solenoid

Source of chromatin (treatment)	Ionic conditions[a]	Method[b]	Diameter (Å)	Pitch (Å)	Nucleosomes per turn	Orientation angle	Reference
Rat liver	> 0.2 mM MgCl₂	EM, XS	300–500	120–150	~6–7	ND	Finch and Klug (1976); Sperling and Klug (1977)
Chicken erythrocyte	10 mM Tris, 0.7 mM EDTA, 60 mM NaCl	LS	300[c]	120[c]	6[c]	ND	Campbell et al. (1978)
Rat liver (glutaraldehyde fixed)	5 mM TEACl, 0.2 mM EDTA, 60 mM NaCl or 0.3 mM MgCl₂	EM	200–300	110	~6	ND	Thoma et al. (1979)
Chicken erythrocyte	10 mM Tris, 2 mM MgCl₂	NS	340	120	7 ± 1	ND	Suau et al. (1979)
Chicken erythrocyte	0.2 mM Tris-cacodylate 0.1 mM MgCl₂	ED	300[c]	110[c]	6[c]	20°–25°	McGhee et al. (1980)
Calf thymus (crosslinked)	0.1 mM Tris-HCl 0.1 mM MgCl₂ for most expts.	LS, DLS	330 ± 30	112 ± 8	7.5 ± 0.05	39°	Lee et al. (1981)
Rat liver (kept at ~100 mM salt during preparation)	20 mM Tris, 5 mM citric acid, 100 mM NaCl	XS	320–340	ND	ND	ND	Brust and Harbers (1981)
Chicken erythrocyte	10 mM Tris-HCl, 1 mM EDTA, 80 or 100 mM NaCl	LS, DLS, S	480	253	11.5	ND	Fulmer and Bloomfield (1982)

Table 7-3. *Continued*

Source of chromatin (treatment)	Ionic conditions[a]	Method[b]	Diameter (Å)	Pitch (Å)	Nucleosomes per turn	Orientation angle	Reference
Calf thymus: a. crosslinked	0.1 mM Tris-HCl, 0.1 mM MgCl$_2$	ED	ND	ND	ND	38°	Yabuki et al. (1982)
b. not crosslinked	10 mM Tris-HCl, 0.1 mM EDTA, 0.5 mM MgCl$_2$	ED	ND	ND	ND	30°	
Rat liver, chicken erythrocytes, sea urchin sperm, CHO cells, HeLa cells	0.13 mM cacodylate, 0.12 mM NaOH, 0.003 mM EDTA + MgCl$_2$ to condense	ED	300[c]	110[c]	6[c]	21°–33°	McGhee et al. (1983a)
Calf thymus	0.1 mM Tris-HCl, 0.01 mM EDTA, 0.5 mM MgCl$_2$	ED, PD	ND	ND	ND	33° ± 3°	Mitra et al. (1984)
Calf cerebral cortex neurons (fixed for EM, otherwise not)	EM: fixed in 80 mM NaCl ED: 0.22 mM cacodylate + MgCl$_2$ to condense	EM, ED	200–300	110[c]	5.5	30°	Allan et al. (1983)

Table 7-3. *Continued*

Source of chromatin (treatment)	Ionic conditions[a]	Method[b]	Diameter (Å)	Pitch (Å)	Nucleosomes per turn	Orientation angle	Reference
Chicken erythrocyte (formaldehyde fixed)	10 mM TEACl, 1 mM EDTA, 0.4 mM PMSF	a. 20 mM NaCl EM b. 100 mM NaCl EM	320 400	120 ND	3.1 8.2	ND ND	Woodcock et al. (1984)
Chicken erythrocyte	Not given	XS	340	110	ND	ND	Widom and Klug (1985)
Sea urchin sperm, chicken erythrocyte	Several used; see ref.	XS	[d]	[d]	ND	ND	Widom et al. (1985)
Sea cucumber sperm, chicken erythrocyte, (glutaraldehyde fixed)	Several used, see ref.	EM	300	100	5–6	ND	Subirana et al. (1985)
Several, with different repeat lengths; see ref.	Several used, see ref.	EM, XS	Varies with repeat	260[e]	Varies with repeat	ND	Williams et al. (1986)

[a] In some cases, other conditions were employed in addition to those listed here. The table lists only those that are known to stabilize the 300 Å fiber. Abbreviations for compounds: TEACl, triethanolamine chloride; PMSF, phenyl methyl sulfonyl flouride.

[b] Abbreviations for methods: DLS, dynamic light scattering; ED, electric field dichroism; EM. electron microscope: LS. static light scattering; NS. neutron scattering; PD, photochemical dichroism; S, sedimentation; XS. X-ray scattering.

[c] Values not necessarily unique, but consistent with data. See reference for details.

[d] Structure of sea urchin sperm chromatin stated to be virtually identical to that of chicken erythrocyte.

[e] For two-start helix.

h–k), with an apparent pitch of 110 to 150 Å. Curiously, there seems to be no consistent left- or right-handedness, both cases being roughly equally represented (see also Finch and Klug, 1976). Nor is the fiber of uniform diameter; even the best specimens show irregular bumps and indentations.

A question not readily resolved by examination of micrographs is the orientation of nucleosomes with respect to the fiber axis. Indeed, it is exceedingly difficult to even recognize individual nucleosomes in such photographs. (But see Subirana et al., 1985, for recent advances.) Electric field dichroism measurements have been helpful with respect to this problem (see McGhee et al., 1980, 1983a; Lee et al., 1981; Yabuki et al., 1982; Allan et al., 1983; Mitra et al., 1984; Sen and Crothers, 1986). There are complications with this technique, arising primarily from the fact that very low ionic strengths are required. Thus, the 300 Å fiber must be stabilized either by crosslinking or by the use of low concentrations of divalent ions in place of the more "physiological" 100 to 200 mM NaCl. Probably as a consequence of the different ways in which this problem has been surmounted, there are minor disagreements between the results of the different studies listed above. A further complication in interpretation arises from assumptions made concerning the contribution of the linker DNA to the dichroism. However, Mitra et al. (1984) have developed a technique of "photochemical dichroism" that provides direct information on this matter (see also Sen et al., 1986). Despite these problems, the studies are in agreement as to the major point: The core particles must lie with their flat faces more or less parallel to the axis of the 300 Å fiber. As Table 7-3 shows, the recent studies center on an orientation angle between 20 and 40°. Considering the apparent irregularity of the fibers observed in the microscope, it would seem inappropriate to argue about small differences.

The most important consequence of these results is seen in Figure 7-14. Of three reasonable orientations for nucleosomes in a solenoid, the electric dichroism measurements definitely rule out C. A distinction between A and B can then be made on the basis of neutron scattering data. The results of Suau et al. (1979) are inconsistent with B, which would require too large a cross-section radius of gyration and a significant hole in the center of the fiber. A structure like A, with the core particles somewhat tilted, appears to be the only one consistent with all of the physical data.

X-ray diffraction studies of chromatin fibers date back to the early work of Wilkins, Pardon, and their collaborators (see Chapter 1). Although these early studies demonstrated reflections at 110 Å, 55 Å, 37 Å, 27 Å, and 21 Å, there was considerable difficulty in assigning these to specific structural features because of the poor orientation of the samples. Recently, Widom and Klug (1985) have found that chromatin solutions can be quite effectively oriented in capillary tubes. With such samples, the 110 Å reflection is shown to meridional and is attributed by Widom and Klug to the packing of nucleosomes with their long dimensions roughly parallel

Fig. 7-14. Possible idealized orientations of nucleosomes in a solenoidal fiber, as viewed down the fiber axis. Model C is ruled out by electric dichroism, and model B is inconsistent with neutron scattering results. (See the text.)

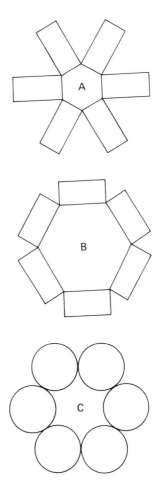

to the solenoid axis. The remaining reflections are equatorial and presumably arise from the side-by-side packing of the nucleosomes in the fiber. In addition, a 300 to 400 Å equatorial reflection is now revealed, which is reasonably interpreted as due to side-by-side packing of solenoids.

In a continuation of these studies, Widom et al. (1985) have studied the long-repeat chromatin from sea urchin sperm. They find a structure indistinguishable from that observed with chicken erythrocyte chromatin and conclude that the solenoidal structure is independent of linker length. This conclusion is disputed by Williams et al. (1986), who report solenoidal diameters strongly dependent on nucleosome repeat. The issue is presently unresolved.

Thus, a wide variety of experimental results are consistent with a solenoidal structure with a pitch of about 110 Å, a diameter of approximately 300 Å, and 6 to 8 nucleosomes per turn. The only set of data that seems to be in marked disagreement with this is that of Fulmer and Bloomfield (1982). From light scattering and hydrodynamic studies, they arrive at a quite different picture: an open helix with much greater pitch and diameter (see Table 7-3). Furthermore, Fulmer and Bloomfield argue that the sedimentation data of Butler and Thomas (1980) are also inconsistent with the 300 Å solenoid. A somewhat similar conclusion was reached by Osipova (1980) and by Ausio et al. (1984b), who noted that the frictional coefficients observed for supposed solenoids are too high. Fulmer and Bloomfield suggest that the fixation procedures used in EM studies shrink the fiber. However, this explanation cannot apply to a number of the experiments listed in Table 7-3, in which no fixation was employed. Nor does it seem consistent with the observations of Langmore and Paulson (1983), who carried out low-angle X-ray scattering studies on whole (living) cells, isolated nuclei, and chromatin. In cells and nuclei, they observed a 300 to 400 Å reflection, which they attribute to packing of 300 to 400 Å fibers. Finally, it is not evident how the 110 Å and 55 Å (or 60 Å) reflections, observed by Langmore and Paulson as well as by investigators in many other scattering studies, can be explained by the Fulmer–Bloomfield model. Such reflections are not seen with extended chromatin fibers at low ionic strength. Perhaps an easier explanation for the hydrodynamic data can be provided by the idea that the solenoids, as studied in solution, are periodically broken or exhibit frayed ends (see Ausio et al., 1984b).

There have also been a series of systematic studies of the effects of ionic strength and nucleosome chain length on the sedimentation coefficients of oligonucleosomes (Butler and Thomas, 1980; Thomas and Butler, 1980; Osipova et al., 1980; Bates et al., 1981; Pearson et al., 1983). In general, these experiments reveal marked changes in sedimentation behavior at two particular sizes—at about 5 to 6 nucleosomes and at about 50 to 60 nucleosomes. The former has been attributed to the completion of a single turn of the solenoid, the latter to a critical size for stability of the solenoid. However, interpretation of these data has tended to be qualitative rather than quantitative, and they do not, therefore, constitute additional proof for the solenoidal model. The only studies in which a detailed analysis of the hydrodynamics has been carried out are those of Osipova (1980), Marion et al. (1981), and Fulmer and Bloomfield (1982). Unfortunately, the first two include data only for small oligonucleosomes; the latter does not agree with the solenoid model. One aspect of the sedimentation experiments requires special comment. As Figure 7-15 shows, data from chromatins ranging in repeat from 162 bp to 210 bp nearly superimpose (Pearson et al., 1983; Butler, 1984). Like the electric dichroism results of McGhee et al. (1983a), these experiments suggest a uniform higher-order structure for all chromatins.

Is it possible, on the basis of all of these data, to reconcile the superbead and solenoid structure for the 300 Å fiber? In my opinion, the answer is yes, if certain assumptions are entertained. If superbeads have any reality,

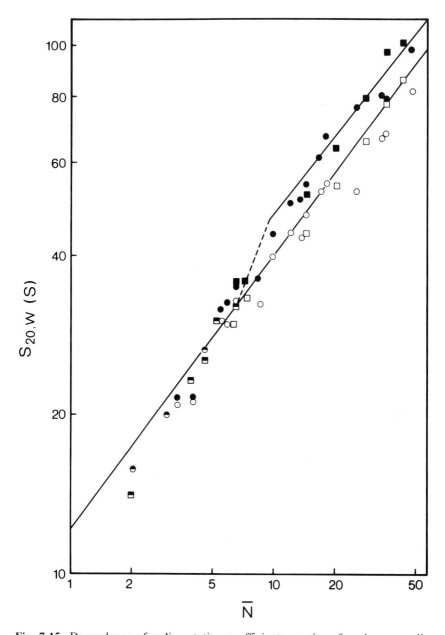

Fig. 7-15. Dependence of sedimentation coefficients on size of nucleosome oligomers (N̄) at different ionic strengths, using chromatin of different repeat lengths. The open symbols and lower line are for 25 m*M* ionic strength, the filled symbols and upper line correspond to 65 m*M*. The two *lines* represent regression fits to data for the 162 bp repeat ox brain chromatin. Points are: ○ and ●, 196 bp repeat rat liver chromatin; □ and ■, 210 bp repeat chicken erythrocyte chromatin. (From The EMBO Journal, vol. 3, issue 11, p. 2600, Butler. Copyright 1984 IRL Press.)

they must be subject to numerous constraints. Any superbead structure must be able to explain the diffraction data, the electric dichroism results, and the overall consistency demonstrated in Table 7-3. The easiest explanation would be that superbeads are the products of periodic or quasiperiodic interruptions in a helical 300 Å fiber. This is precisely the position of Subirana et al. (1985). But why should such interruptions exist? Several possibilities bear consideration:

1. It has been noted (Finch and Klug, 1976; Thoma et al., 1979) that the handedness of the wrapping of the nucleosome chain seems to vary along the fiber. If there were, in fact, changes between right- and left-handed segments, discontinuities should exist at these points.
2. Data on the lysine-rich histone content of chromatin indicate about one (or even less than one) molecules per nucleosome (see Chapter 6). Such a stoichiometry, as pointed out in Chapter 6, can lead to an asymmetry in the chromatin chain. If the direction of that asymmetry were to periodically reverse, interruptions could be generated.
3. There is evidence that the binding of H1 to depleted chromatin at physiological ionic strength is a highly cooperative process (Watanabe, 1984). This could lead to a clustering of lysine-rich histone, with intervening regions unsaturated and susceptible to both unfolding and nuclease digestion. Immunological studies by Mazen et al. (1982) have in fact indicated such an irregular, clustered distribution of H5 in chicken erythrocyte chromatin. Furthermore, a number of experiments demonstrate a preference of lysine-rich histones for the larger oligonucleosomes and chromatin fragments (see, for example, Ruiz-Carillo et al., 1980; Thomas and Rees, 1983; Huang and Cole, 1984).

These possible explanations are by no means mutually exclusive. In fact, random nucleation of H1-rich domains with alternate symmetries during chromatin assembly could result in a mosaic structure. None of these suggestions provides an obvious mechanism for a highly regular periodicity, but this may not be a serious objection, for most superbead preparations are markedly heterogeneous.

It is conceivable that periodic breaks in a solenoidal structure could be induced by a "vernier" effect. Even if the 6-nucleosome-per-turn structure may be universally favored by packing considerations (as is suggested by the comparative studies of McGhee et al., 1983a), different nucleosome spacing, as defined by linker size, might give rise to accumulated strain, eventually causing a break in the solenoid. In this connection, see the recent studies by Williams et al. (1986), which suggest a fiber diameter dependent on linker. Such a mechanism might explain the apparent dependence of superbead size on linker length observed by Zentgraf and Franke (1984).

There has also been considerable debate over how the linker (and therefore H1 molecules) are fitted into the solenoid. Intimately tied to this problem is the question of how the chain of nucleosomes is folded into the fiber; for example, is it a continuous solenoidal wrapping or a two-start helix made from a zigzag chain? The folding of the nucleosome chain, including the linker, also has bearing on the "linking number paradox." It will be recalled from Chapter 6 that the DNA makes about 1.75 left-hand turns about the core particle and probably 2 turns about the chromatosome. Thus, each nucleosome should decrease the linking number by $\Delta L_k = -2$. Yet data on closed circular minichromosomes indicate a value closer to $\Delta L_k = -1$ per nucleosome. There are two obvious ways in which this paradox might be resolved. First, as mentioned in Chapter 6, the DNA might be wound differently in the nucleosome than is free B-DNA in solution, and there is now strong evidence to support this. Alternatively, particular kinds of coiling of the linker might compensate for part of the left-hand superhelical coiling; this could allow the DNA twist on the nucleosomes to be the same as in solution. (See Strogatz, 1983, for a theoretical analysis.)

A number of models for the 300 Å fiber are listed in Table 7-4, and some of these are depicted in Figure 7-16. Not all proposals have been included. For example, Staynov (1983) has proposed an ingenious "non-sequential" model, but since this is applicable over only a relatively narrow range of repeat lengths, it cannot be generally applicable. The various structures described in Table 7-4 and Figure 7-16 differ in two fundamental respects. Some are "one-start" helices, formed from a continuous chain of nucleosomes (Figs. 7-16B,D), others are "two-start" helices, formed by the winding of zigzag chains (Figs. 7-16C,E), and one is a twisted ribbon formed from such a zigzag structure (Fig. 7-16A). The other fundamental differences lie in the way in which the linker DNA is arranged, which has a major effect on the linking number per nucleosome (see Table 7-4).

Despite statements to the contrary from some proponents, it would probably be premature to rule out any of the above models on the basis of current evidence. There are, however, certain experimentally accessible questions that may be of aid in evaluating models for the solenoid. These include:

1. *The accessibility of lysine-rich histones and the linker DNA:* If, as is proposed in several models, the linker DNA and accompanying proteins lie exclusively within the core of the solenoid, they should be at best only marginally accessible to large enzymes. It is frequently argued that micrococcal nuclease ($M = 16,800$) should readily penetrate the helix, and that the preferential susceptibility of linker DNA to this enzyme is therefore not an argument against internal linker. However, even this small enzyme has a diameter of roughly 35 Å, and most of the solenoidal models do not show openings of this size. It could be argued that digestion proceeds from the ends of solenoidal regions, which should result in a nonrandom distribution of oligonucleosomes

Table 7-4. Some General Models for the 300 Å Solenoid

Figure	Distinguishing features	Location of lysine-rich histones	Range of repeat lengths	$\left(\dfrac{\Delta L_k}{N}\right)^a$	Reference
—	Continuous coiled chain of nucleosomes, variable number per turn	Inside of solenoid	Not defined, since linker conformation not specified	Not defined	Thoma et al. (1979)
7-16A	Twisted ribbon of zigzag chain, with "crossover" of linkers	Close to solenoid surface	Above ~ 170 bp	−1	Worcel et al. (1981)
7-16B	Continuous coil (as Thoma et al.) with linker coiled as continuation of nucleosome DNA coil	Variable; distribution depends on linker length	At least 186–243 bp	Variable	McGhee et al. (1980, 1983a)
7-16C	Helix of zigzag chain; no linker "crossover"	Close to solenoid surface	Above 176; smaller possible with modification	−2	Woodcock et al. (1984)
7-16D	Continuous coil, with linker as reverse loop inside	Inside of solenoid	Above 176; a different ~ 165 bp structure is proposed	−1	Butler (1984)
—	Layered zigzag structure	Some inside, some external	Not defined	Not defined	Subirana et al. (1985)
—	Helix (left-handed) of zigzag chain (2-start)	Inside of solenoid	No obvious restriction	−1 or −2	Williams et al. (1986)

[a] Change in linking number per nucleosome in a long helix.

Fig. 7-16A. The 300 Å fiber model of Worcel, Strogatz, and Riley. A ribbon of nucleosomes is twisted to form the fiber. From Worcel et al. (1981). For details of the structure, consult the original paper.

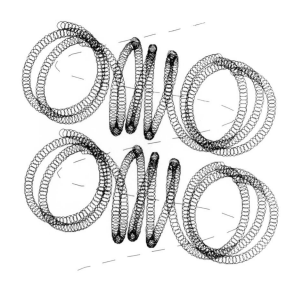

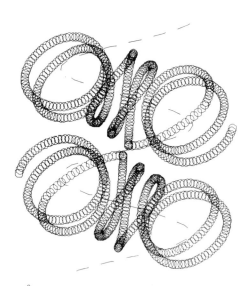

Fig. 7-16B. The 300 Å fiber model of McGhee, Nickol, Felsenfeld, and Rau. The chromatosomes are proposed to lie radially with flat faces tilted $26° \pm 6°$ from the solenoid axis. The linker DNA between nucleosomes is supercoiled about the helix (dashed line) that passes through the chromatosome centers. For clarity, only the three nucleosomes on the front surface of the fiber are seen. The histone cores are also omitted. The two models shown are:

	Chromatin	Linker length	Turn in linker	Tilt angle
Top	Sea urchin sperm	77 bp	1	20°
Bottom	HeLa	20	¼	30°

From McGhee et al. (1983a).

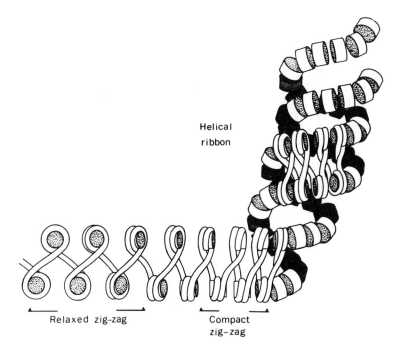

Helical
ribbon

Relaxed zig-zag   Compact
                         zig-zag

Fig. 7-16C. The 300 Å fiber model of Woodcock, Frado, and Rattner. The zigzag fiber is compacted and then wound into a two-start helix. For clarity, the DNA is shown only in places, and nucleosomes are shaded in various ways. Note that the linkers are alternately to the outside and inside of the helix. (Reproduced from *The Journal of Cell Biology*, 1984, vol. 99, p. 42, Woodcock et al., by copyright permission of The Rockefeller Univ. Press.)

following digestion. In a test of this idea, LaFond et al. (1981) did find the distribution to be nonrandom but observed that it was independent of the condensation state of the chromatin. It seems surprising, if the linker DNA is indeed buried inside the structure, that the multitudinous exposed *intra*nucleosomal sites on the surface are not cleaved preferentially.

Similar considerations apply to the digestion of H1. Lysine-rich histones are almost invariably the first to be proteolyzed. Furthermore, Marion et al. (1983) have shown that trypsin immobilized on collagen membranes yields the same result; H1 is efficiently cleaved and is the first histone to be degraded. It would seem that evidence of this kind at least favors those models in which the linker DNA and its accompanying protein lie near the surface of the 300 Å fiber.

2. *Evidence for dinucleosome cleavage:* A number of experiments suggest that cleavage of linker DNA, especially by bulky enzymes, leads to a preferential cleavage at every other nucleosome (see, for example, Ar-

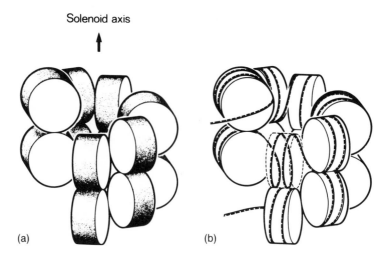

Solenoid axis

(a)   (b)

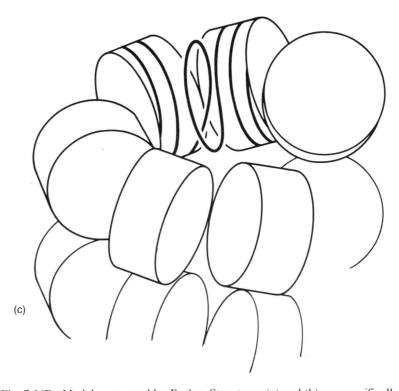

(c)

Fig. 7-16D. Models proposed by Butler. Structures (a) and (b) are specifically for short-repeat chromatin (~165 bp). Note that DNA passes directly from one chromatosome to the next. This is a right-hand helix. Model (c) is applicable to chromatins with repeats of greater than 176 bp. It is a left-hand helix, in which the linker DNA makes internal reverse loops. In both structures, the lysine-rich histones would be confined to the inside of the helix. (From The EMBO Journal, vol. 3, issue 11, Butler. Copyright 1984 IRL Press.)

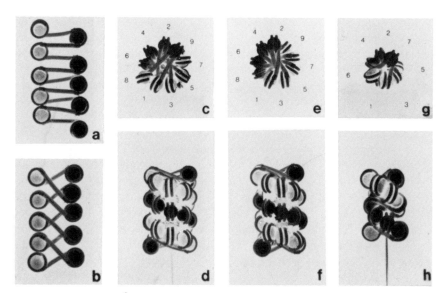

Fig. 7-16E. The 300 Å fiber model of Williams et al. This is a crosslinker model with a two-start helix. Odd and even nucleosomes are shaded and white, respectively. (a–f) show models with a 48 bp linker. (a) and (b) are extended ribbons with $L = -1$ and -2. (c) and (d) show a solenoid with $L = -1$. (e) and (f) with $L = -2$. (g) and (h) show a $L = -1$ structure with a short linker (22 bp). (Reproduced from the Biophysical Journal, 1986, 49, 233–248, Williams et al. by copyright permission of the Biophysical Society.

ceci and Gross, 1980; Burgoyne and Skinner, 1981, 1982; Pospelov and Svetlikova, 1982; Burgoyne, 1985; and the discussion in Staynov et al., 1983). This suggests that alternate linkers are not equivalent, a feature difficult (though not impossible) to explain by continuous helix models. Either of the zigzag structures listed in Table 7-4 or the model of Staynov (1983) leads easily to such a result.

3. *Linker orientation:* If the average orientation of the linker could be accurately measured, certain models could be eliminated. The first results (Mitra et al., 1984; Sen et al., 1986) allow a number of possibilities, depending upon whether the linker is assumed to be straight or curved in some fashion. These researchers assert that models such as that of McGhee et al. (1983a), in which the linker follows the same kind of path as the core DNA, are ruled out by their experiments. It would seem that the model of Staynov (1983), in which the linker lies nearly perpendicular to the solenoid axis, is also disfavored. However, zigzag models such as those of Worcel et al. or Woodcock et al. do not seem to be inconsistent with these data, if somewhat modified.

What maintains the integrity of the 300 Å fiber? One factor has been repeatedly emphasized in the preceding discussion: Either a high concentration of monovalent cations, or a much lower concentration of divalent cations is essential. It seems likely that these serve mainly to reduce electrostatic repulsion between DNA segments. Indeed, as has been described in Chapter 6, histone cores have a strong tendency to aggregate into fibrous structures. Dubochet and Noll (1978) have shown that even core particles will associate into long helices if low levels of spermine are present. Thus, the particles seem to be intrinsically sticky, and aggregation is favored whenever charges are sufficiently screened.

The formation of a *regular* solenoidal fiber from chromatin seems to require, in addition to a sufficient cation concentration, the presence of lysine-rich histones. (See Renz et al., 1977; Strätling, 1979; Thoma et al., 1979; Osipova et al., 1980; Allan et al., 1981; Makarov et al., 1984; and Watanabe, 1984, for a variety of demonstrations of this fact.) In the absence of H1 and/or H5, the fiber of nucleosomes will still condense in high salt, but only irregular clumps will be formed (Thoma et al., 1979). The essential requirement for lysine-rich histones led to the initial idea that such protein molecules might so interact as to form a helical ladder, to stabilize the solenoid. The concept received support by early reports that H1 molecules in chromatin could be crosslinked to another (Chalkley and Hunter, 1975). However, Thomas and Khabaza (1980) showed that such oligomers were readily formed even at *low* ionic strength, conditions under which the 300 Å fiber is unstable (see also Thomas, 1984). Other experimenters have expressed skepticism about the importance of lysine-rich histone homotypic interaction in fiber stabilization. Russo et al. (1983) find no evidence for specific interactions between the globular domains of these molecules and conclude that H1 is not important in chromatin condensation. This seems hardly compelling evidence for so broad a conclusion, especially since Ring and Cole (1983) have shown that H1–H1 crosslinks are between tail regions. The experiments of Marion et al. (1983) make a stronger case. They observed that proteolytic digestion of H1 to the point where no intact molecules remained had negligible effect on the circular dichroism or electric birefringence of rat liver oligonucleosomes. It is conceivable that the major role of lysine-rich histones is to help set regular spacing of nucleosomes (cf. Stein and Bina, 1984), which will in itself aid in the formation of a regular higher-order structure and perhaps also to help fold the linker DNA in some specific way. It should be noted that the data of Marion et al. do not specify the degree of destruction of the H1 molecules, and substantial portions might have remained in contact with the DNA.

Other experiments have provided evidence that the tails of the *core* histones may be involved in stabilizing the 300 Å fiber. In the course of their digestion studies, Marion et al. (1983) observed that significant changes in the physical properties of the chromatin first appeared only as H3 began to be appreciably cleaved. These results are entirely in accord with earlier experiments by Allan et al. (1982a). By first removing H1 and H5 from chicken erythrocyte chromatin, carefully digesting the depleted chromatin with trypsin, and then adding back H1 or H5, Allan et al. were able to assess directly the effect of the core histone tails. The experiments demonstrate quite elegantly that whereas addition of lysine-rich histones to nontrypsinized chromatin allowed refolding to 300 Å fibers, the same was not effective when the core histones had been proteolyzed. In a related study, Harborne and Allan (1983) found that the order of susceptibility of histone tails to proteolysis was drastically changed as chromatin was exposed to higher salt concentrations. Finally, Jordano et al. (1984c) have shown that removal of H2A and H2B induces solenoid relaxation even when H1 and H5 remain.

These experiments suggest an answer to a question raised and left unanswered in Chapter 6: What is the role of the histone tails? It would appear that they play some significant part in the stabilization of higher-order structure.

Higher-Order Folding of the Chromatin Fiber: The Chromosome

Each chromosome of a eukaryotic organism contains a single DNA molecule (Kavenoff and Zimm, 1973). Such molecules will be several centimeters in length, and even compaction into a 300 Å chromatin fiber will reduce that length at most 50-fold. Thus, the 300 Å fiber must itself be further folded or coiled in order to pack this chromatin into a metaphase chromosome, or within the confines of an interphase nucleus. While this necessity has long been recognized, it is only recently that the nature of that folding has become clearer. The remainder of this chapter will be devoted to discussing, separately, the higher-order folding in metaphase and interphase. First, however, mention should be made of some important early observations that lie behind much of the current thinking on this question.

Benyajati and Worcel (1976) very gently isolated intact interphase chromosomes from *Drosophila* cells and subjected them to mild digestion with DNase I, so as to produce single-strand nicks. They found that the sedimentation coefficients of the chromosomes decreased gradually and then reached plateau values. Such behavior is reminiscent of the response of the *E. coli* nucleoid in similar experiments (Worcel and Burgi, 1972) and suggests a similar explanation: The chromosomal DNA is arranged in a large number of independently supercoiled *domains* or *loops,* each of which can be relaxed by a single nick (see Fig. 7-17). From analysis of the effects of single-strand cleavage, Benyajati and Worcel estimated the average loop size to be about 85,000 bp. Independent confirmation of this concept came from studies by Igo-Kemenes and Zachau (1977) of the digestion of rat liver nuclei by micrococcal nuclease and restriction nucleases. Again employing very mild digestion conditions, they found that in no case were fragments larger than about 75,000 bp obtained. A math-

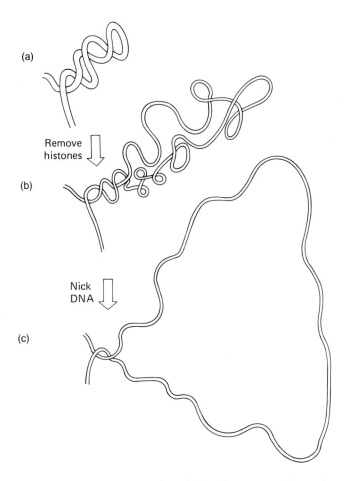

Fig. 7-17. A schematic representation of domain structure in a eukaryotic chromosome. (a) The intact loops as folded 300 Å fibers. (b) Upon removal of histones, supercoiled loops are formed. (c) Nicking of loops leads to their individual relaxation and further expansion.

ematical analysis of the progress of the digestion indicated an average domain size of about 35,000 bp.

There have been numerous other efforts to estimate chromatin domain size, using a variety of techniques. Some of these are presented in Table 7-5. In assessing these values, it should be noted that the numbers represent, at best, estimates of average size in very heterogeneous populations. Some methods will yield a number average domain length, but with others the kind of average obtained is unclear. Probably the most that can be said at the moment is that the number average domain size is usually between 40 and 80 Kbp, with much smaller and much larger domains present in any given genome. A further reason for caution in the inter-

Table 7-5. Average Domain Lengths in Chromatins

Cell source or type	Average length (Kbp)	Method	Reference
Yeast	< 250	Sedimentation, estimation of fragment size	Pinon and Salts (1977)
Drosophila	85	Nicking, observation of sedimentation coefficient change	Benyajati and Worcel (1976)
Chicken erythrocyte	45	Nuclease digestion, analyzed distribution of fragments[a]	Ganguly et al. (1983)
Chicken erythrocyte	~ 40	Sizes from nuclease digestion	Hyde (1982)
Mouse 3T3	90	Dimensions of "halo" in nucleoid	Vogelstein et al. (1980)
Mouse P815	53	EM on nucleoids	Hancock and Hughes (1982)
Mouse L	62	Digestion, sedimentation	Razin et al. (1979)
Rat liver	35	Nuclease digestion, analyzed distribution of fragments	Igo-Kemenes and Zachau (1977)
Rat liver	80	Estimate of no. of loops per genome	Berezney and Burchholtz (1981)
HeLa	42	EM on histone-stripped metaphase chromosomes	Paulson and Laemmli (1977)
HeLa	220	Effect of γ-irradiation on ethidium binding	Cook and Brazell (1978)
HeLa	83	EM of histone-stripped metaphase chromosomes	Earnshaw and Laemmli (1983)

[a] Is increased to ~ 150 Kbp by incubation in 0.135 M salt (Ganguly and Bagchi, 1984).

pretation of such data is the observation by Ganguly and Bagchi (1984) that the average domain size in chicken erythrocyte chromatin can be greatly increased by mild salt treatment.

The concept that the eukaryotic genome is folded into a large number of topologically independent domains is satisfying in a number of respects. First, it indicates a structural continuity with prokaryotes. Second, the possibility that different domains may contain different sets of functionally linked genes suggests elegant methods for the control of gene expression. More will be said of these aspects later; for the moment, the primary issue is that such domain structure imposes limitations on possible chromosome models.

I shall begin with the mitotic chromosome, because its structure is more easily visualized and is in fact probably simpler than that of the more dispersed chromosome at interphase. For an overview of recent work on both kinds of structure, the reader is referred to the *Journal of Cell Science*, Suppl. 1, a collection of papers edited by Cook and Lasky (1984), and reviews by Butler (1983) and Eissenberg et al. (1985). Broad overviews of the eukaryotic nucleus are provided by Bostock and Sumner (1978), Hancock and Hughes (1982), and Hancock and Boulikas (1982).

Metaphase Chromosomes

Modern ultrastructure research on mitotic chromosomes began with the development by Wray and Stubblefield (1970) of a new and gentle method for their isolation. This technique avoids the osmotic shock of earlier water-spreading methods and retains the histone composition typical of chromatin (Wray et al., 1980). Using chromosomes prepared in this way, many workers, employing a wide variety of electron microscopic techniques, see essentially the same picture. The intact chromosome at high ionic strength exhibits thick (~500 Å) fibers that appear to extend radially from

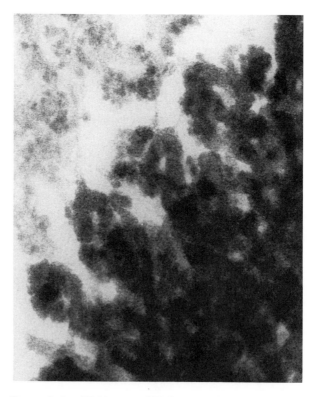

Fig. 7-18A. Transmission EM image of HeLa metaphase chromosomes. The micrograph shows the edge of a chromosome prepared in hexylene glycol buffer. The fibers are approximately 500 Å thick and are believed to consist of back-folded or twisted 300 Å fibers, projecting as loops. (From Cell *17*, p. 849, Marsden and Laemmli. Copyright 1979 M.I.T. Press.)

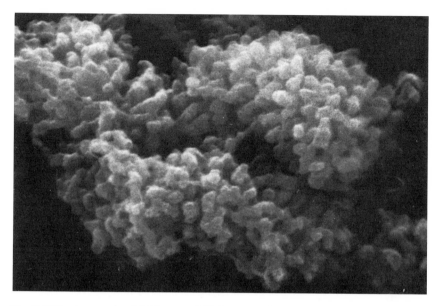

Fig. 7-18B. A scanning EM image of HeLa metaphase chromosome preparation. The knobby projections correspond to the thick fibers. (From Cell *17*, p. 849, Marsden and Laemmli. Copyright 1979 M.I.T. Press.)

a central core (see Mace et al., 1977; Marsden and Laemmli, 1979; Adolph, 1981, for examples). Representative views are shown in Figure 7-18A and B. These thick fibers often give the appearance of back-folded or coiled 300 Å fibers. At lower ionic strength, the structure unfolds to reveal loops of 300 Å fibers, and if divalent ions are removed, extended loops of the "100 A fiber" are found.

To what are these loops attached? In a very early study, Stubblefield and Wray (1971) claimed the existence of a "core ribbon" at the axis of sheared metaphase chromosomes. However, the nature of this structure was unclear, and it was not until Paulson and Laemmli (1977) examined histone-depleted chromosomes that a more detailed picture could be presented. They noted enormous loops of DNA emerging from a "scaffold" structure (see Fig. 7-19; also Wray et al., 1977; Hadlaczky et al., 1981a). The size of these loops corresponds nicely to estimates of domain size. Where individual loops can be traced, they appear to leave and enter the scaffold at nearly the same point. Subsequent studies (Laemmli et al., 1977; Jeppesen et al., 1978) provided strong evidence that the scaffold structure was in fact a complex of nonhistone chromosomal proteins. The fact that this structure was seen only after histones had been extracted (by salt, polyelectrolytes, or low pH) led some to wonder if the scaffold might be an artifact produced by precipitation of nonhistone proteins under extraction conditions (Okada and Comings, 1980; Hadlaczky et al., 1981b; Labhart et al., 1982). However, the facts that the same structure can be

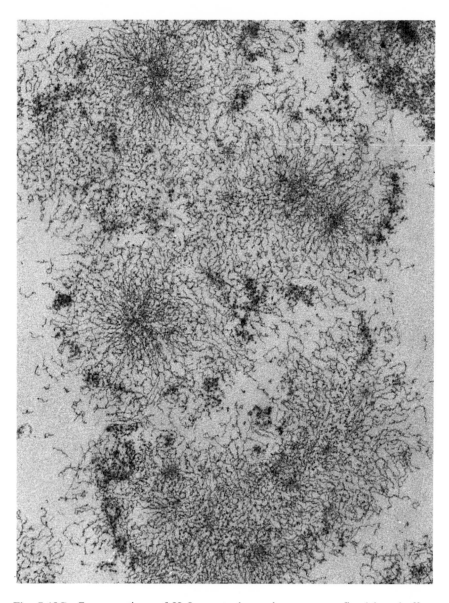

Fig. 7-18C. Cross sections of HeLa metaphase chromosomes fixed in a buffer containing 1 mM MgCl$_2$. This preserves the 300 Å fibers, but the 500 Å folding is lost. Note dense center and radial loops. (From *Cell 17*, p. 849, Marsden and Laemmli. Copyright 1979 M.I.T. Press.)

Fig. 7-19. The metaphase HeLa chromosome scaffold and surrounding DNA. Histones and most nonhistone proteins have been removed from HeLa chromosomes by heparin-dextran sulfate treatment. The extremities of some of the DNA loops can be seen at the top of the picture. (From Cell *12*, p. 817, Paulson and Laemmli. Copyright 1977 M.I.T. Press.)

obtained by many techniques (Lewis and Laemmli, 1982) and that a protein-rich core can be observed by silver staining of chromosomes (Howell and Hsu, 1979; Earnshaw and Laemmli, 1984) argue for its reality (but see Zheng and Burkholder, 1982 for a different opinion).

In more recent studies, Laemmli and his co-workers have concentrated on the composition and stability of the scaffold. Two major proteins are found, SC1 with $M_r = 170,000$ and SC2 with $M_r = 135,000$ (Lewis and Laemmli, 1982). Interestingly, the former is now believed to correspond to the eukaryotic topoisomerase II (Earnshaw et al., 1985; Earnshaw and Heck, 1985; Gasser et al., 1986). If the enzyme is active, it may play a role in regulating supercoiling in individual domains. Between them, SC1 and SC2 comprise at least 40% of the total metaphase protein scaffold (Earnshaw and Laemmli, 1983). Certain metal ions are essential for stability of the scaffold; Cu^{2+} and Ca^{2+} have proved efficacious. The requirement is quite specific, for Mn^{2+}, Co^{2+}, Zn^{2+}, or Hg^{2+} will not substitute (Lewis and Laemmli, 1982). The picture that emerges is one of far greater simplicity than might have been anticipated. An excellent overview, contrasting the metaphase chromosome scaffold with the "interphase scaffold" (see below), has been provided by Lewis et al. (1984).

Chromosomes in Interphase

Many years ago, D.E. Comings (1968) published a remarkably prescient model for the organization of chromatin in the interphase nucleus. Drawing on early EM studies, examples from prokaryotes, and pulse-labeling studies of replicating DNA in nuclei, he proposed that the interphase chromosomes were periodically connected to the nuclear envelope and projected as loops into the interior. However, it was nearly ten years before evidence began to accumulate to support a "chromatin loop" model for the interphase nucleus. The pioneering studies of Benyajati and Worcel (1976) and Cook and co-workers (Cook and Brazell, 1976; Cook et al., 1976) were soon followed by many others in which eukaryotic "nucleoids" were prepared by removal of the detergent-soluble nuclear membrane and extraction of most nuclear proteins (see Cook and Brazell, 1978; McCready et al., 1979; Vogelstein et al., 1980; Hancock and Hughes, 1982; Lebkowski and Laemmli, 1982a, 1982b; Jackson et al., 1984, for examples). The kind of structure most frequently observed is depicted in Figure 7-20; an extended "halo" of DNA loops extends from a complex cage-like proteinaceous structure, referred to by Laemmli and co-workers as the "nuclear scaffold." As will be shown below, it is probable that this nuclear scaffold contains elements of both the nuclear lamina and the nuclear matrix. Lebkowski and Laemmli (1982a) have noted two stages of DNA folding in such structures. If nuclei are extracted with digitonin plus either 2 M NaCl or a heparin-dextran sulfate mixture, a "type I" structure is obtained; if thiols are present during the extraction, a more expanded, "type II" structure results. In further studies, the protein compositions of the two

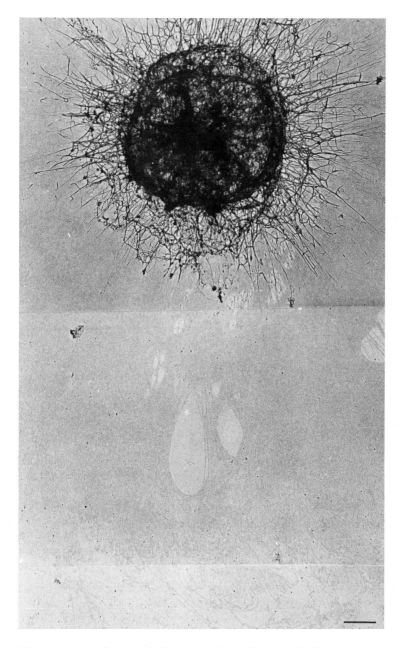

Fig. 7-20. A portion of a spread, histone-depleted HeLa nucleoid. A tangled mass of superhelical fibers extends to the edge of the field. The ''cage'' structure (top) is about 15 μm in diameter. (From J. Cell Sci. (Suppl. 1), pp., 59–79, copyright 1984 The Company of Biologists Limited.)

Table 7-6. Proteins Identified in Chromosome
Scaffolds[a]

Protein	Metaphase scaffold	Interphase scaffold I	Interphase scaffold II
SC1	+	+	−
SC2	+	?	−
Lamins A, B, C	−	+	+
Polymerase II	−	+	−

[a]Data from Lewis et al. (1984).

kinds of scaffolds have been investigated, with results summarized in Table
7-6 (Lewis et al., 1984). Both kinds of interphase scaffold contain the
lamin proteins A, B, C, which are major constituents of the inner nuclear
lamina (see Gerace and Blobel, 1980; and Gerace et al., 1984, for properties). These proteins are the major constituents of type II structures (see
also Hancock and Hughes, 1982). Type I scaffolds also contain SC1, a
major component of the *metaphase* scaffold, plus a number of other proteins. Electron micrographs of nuclear scaffolds from which most of the
DNA has been removed by nuclease digestion reveal further differences.
Type I scaffolds strongly resemble "nuclear matrix" preparations, showing
a fine filamentous structure within the residual nucleus, whereas type II
scaffolds show only the peripheral, laminar layer. This has led to the proposal that the DNA in interphase nuclei is attached to *two* kinds of sites—
some on the peripheral laminar structures and others on the matrix fibers
within the nucleus. The results provide a somewhat new perspective for
the numerous studies of the nuclear matrix (see Chapter 5).

These observations can form the basis for a possible synthesis of models
for the interphase and metaphase chromosomes. It is known that the nuclear envelope disintegrates during metaphase. Lamins A and C become
cytoplasmic, whereas B is associated with residual vesicles of the nuclear
membrane (Gerace and Blobel, 1980). DNA contacts with the membrane
are apparently lost in prophase; therefore, the metaphase scaffold retains
only the matrix contacts (or a subset) including the SC proteins. At tel-
ophase, lamins can once again be seen to be associated with the swelling
chromosomes in anticipation of the reformation of the nuclear envelope
(Jost and Johnson, 1981; McKeon et al., 1984).

A possible intermediate in the DNA–lamin contacts has been discovered
by McKeon et al. (1984). A 33,000 dalton protein that they term *perichromin* is observed to be associated with the nuclear envelope during
interphase, but to remain attached to the chromosomes during their metaphase condensation (see Fig. 7-21). Perichromin resembles in many ways
the protein XO, noted by Laemmli and collaborators to be a major constituent of metaphase chromosomes. XO is not found in isolated *metaphase*

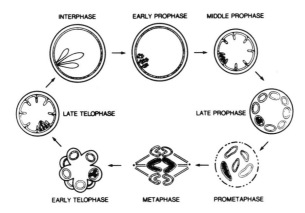

Fig. 7-21. A schematic representation of mitosis, emphasizing the redistribution of lamin (dark line) and perichromin (broken line). (From Cell *36*, p. 83, McKeon et al. Copyright 1984 M.I.T. Press.)

scaffolds, a result that should be expected if it is attached to peripheral regions of the DNA.

As will be discussed in the following chapters, there has been considerable evidence to suggest that the nuclear matrix plays some role in such processes as transcription and replication. It is noteworthy that Lewis et al. (1984) find polymerase II to be associated with the type I scaffold. The idea that the interphase chromatin strands connect to both the peripheral lamina and the internal matrix structure may allow the reconciliation of many puzzling observations.

What is the nature of these contacts? Do they involve specific DNA sequences for functional regions? There have been many attempts to answer this question by stripping most proteins from chromatin, digesting the DNA with nucleases, and examining the residual DNA fragments attached to the scaffold (see, for earlier examples, Jeppesen and Bankier, 1979; Razin et al., 1979; Cook and Brazell, 1980; Bowen, 1981; Kuo, 1982; Robinson et al., 1982; Ciejek et al., 1983). The results have been discordant, some finding evidence for specific kinds of DNA at attachment sites, others not.

A common feature of most of these studies was the use of high salt concentrations to remove histones. Since it is well documented that DNA slides on nucleosomes at even moderate salt concentrations, Mirkovitch et al. (1984) suggested that milder techniques might be more appropriate. They found that extraction with lithium diiodosalicylate efficiently removed histones at low salt. Use of this technique produced a remarkable result. As shown in Figure 7-22, the *Drosophila* histone gene clusters were found to be periodically attached to the type I scaffold via A–T-rich sequences lying in the H1–H3 spacers. Similar sequences were observed to

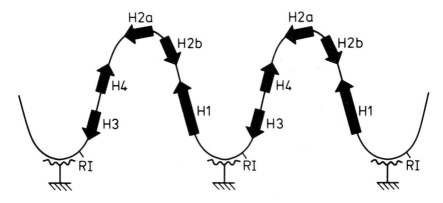

Fig. 7-22. A representation of the organization of the *Drosophila* histone genes and their mode of attachment to the interphase scaffold. (From Cell *39*, p. 223, Mirkovitch et al. Copyright 1984 M.I.T. Press.)

serve as attachment sites for heat-shock gene clusters. Further characterization of these sites in the *Drosophila* histone gene cluster has shown them to be protein-associated and to lie close to topoisomerase II sites (Gasser and Laemmli, 1986). Similar regions have been demonstrated close to the enhancer element in the κ-immunoglobulin gene region (Cockerill and Garrard, 1986). While these represent as yet only a small sampling, and the histone gene loops are unusually small (only about 5000 bp), these observations suggest exciting models for interphase chromatin organization.

Although very much remains to be done, the first hints of a coherent model for chromosome organization are beginning to appear. The experiments described above are providing information at the molecular level. Insofar as the overall morphology of the nucleus is concerned, much is to be expected from a new approach instigated by J. Sedat and his colleagues. Using a novel combination of fluorescence microscopy and digital image processing, they have mapped the positions of polytene chromosomes in individual *Drosophila* salivary gland nuclei (Sedat and Manelidas, 1978; Agard and Sedat, 1983; Mathog et al., 1984; Gruenbaum et al., 1984). They find certain general features: The centromeres are always grouped together at one region of the nuclear envelope, telomeres are attached at the opposite end of the nucleus, and each chromosome has its own nuclear domain; it is not entangled with other chromosomes. While this approach is as yet restricted to the (perhaps special) study of polytene chromosomes, future developments in this direction may be expected to nicely complement the biochemical and molecular studies.

8

Chromatin Structure and Transcription

Transcription of eukaryotic genes appears to proceed in a quite different manner and to obey very different rules than the corresponding process in prokaryotes. In a typical prokaryotic organism, the genome is small, and much of it is transcribed. Polycistronic messages are common, but intervening sequences are not. A single RNA polymerase is used for all genes. In eukaryotes, much of this has been changed. In a typical differentiated cell of a higher organism, the genome is immense, and only a small fraction (~5 to 10%; Davidson, 1976) is transcribed. Polycistronic messages are rare, but intervening sequences are common, and many of the transcripts must be spliced to produce messages. Three different RNA polymerases are employed, each specific for certain kinds of genes.

Perhaps most striking is the complexity of gene control in eukaryotes. In each differentiated cell type there are expressed both "housekeeping" genes, common to many cell types, and those differentiation-specific genes that provide the products for specialized cell functions. Consider, as an example, the ovalbumin gene in chickens. Although carried by all chicken cells, it is expressed only in certain cells of the oviduct tissue and even there only in response to hormonal stimulation. The same gene in other tissues is unresponsive to the hormone. A hierarchy of control systems must be involved. Perhaps one of the reasons for the complexity of chromatin organization in higher organisms is to allow such multilevel control of gene expression.

It is now clear that the specificity of gene expression in differentiated eukaryotic cells resides at the transcriptional level and that each cell type selects the appropriate genes to transcribe from a complete genomic library. The history of this "variable gene activity theory" of differentiation and the abundant evidence in favor of it are succinctly outlined in the first chapter of *Gene Activity in Early Development* (Davidson, 1976). The reader interested in these topics, or in details concerning changes in gene expression during embryogenesis, is referred to this volume.

I shall say little concerning the basic machinery of eukaryotic transcription—the polymerases themselves. Their properties and structures are described in a number of excellent reviews (see, for example, Chambon, 1975; Roeder, 1976; Lewis and Burgess, 1982). There are three recognized types: Polymerase I is specific for the transcription of the major ribosomal RNAs; polymerase III is used for the small tRNAs and 5S RNA, and polymerase II is employed for transcription of the genes for all structural proteins. As we shall see below, there are hints that they may interact with the chromatin structure in quite different ways, but very little is yet known concerning the details of those interactions. Before such problems can even be approached, it will be necessary to understand the basic differences between transcribed regions of chromatin from those not transcribed. That problem constitutes the principal subject matter of this chapter. The subject is so vast, and research has been so extensive, that it is impossible to cover it in its entirety. For more detailed expositions of the state of the subject at various times, recourse to some of the many reviews is required. Particularly useful are those by Mathis et al. (1980), Cartwright et al. (1982), von Beroldingen et al. (1984), Reeves (1984), and Eissenberg et al. (1985).

The Nature of the Problem

To begin, we need to be mindful of the difficulties in defining the relevant chromatin states. Certainly, a chromatin region that is being transcribed at a given moment is active, and one which is not is inactive. But the complexities of eukaryotic development suggest that some other distinctions need to be made. Are chromatin domains that are destined to be transcribed at some future time in the development of a particular cell line somehow marked or prepared for transcription? Do sequences that have once been transcribed continue to carry some structural imprint of that event? Some genes are transcribed at a very high rate throughout much of the cell cycle; others seem to be read only rarely. Do these categories possess distinguishing features? Surely "active" and "inactive" must be too simple a lexicon to describe the subtleties of eukaryotic gene behavior.

On the other hand, it will serve no purpose to erect some complex system of classification based on ideas of what *ought* to happen. We are attempting to understand a game whose rules are unknown to us, and we can only deduce them from observations of how the pieces are moved. At the present time, the experimental methods that will be described in the next part of this chapter suggest a few distinct categories of transcribability. Further experimentation may reveal a more complex pattern, but at the moment we seem obliged to make at least the kinds of distinctions outlined in

Table 8-1. Although some readers may object to the way in which words are used, the categories described here seem real, and *some* terminology is required for the purposes of exposition.

A few examples of the use of these terms may be helpful. Consider the adult β-globin gene in chickens. This is certainly a *functional* gene, since it is transcribed under some circumstances. In erythroid cell lines, it is also *competent,* whereas in liver cells (for example) it is *incompetent.* In immature erythrocytes, it may be found in an *active* state, whereas in mature erythrocytes it is *inactive.* It should be emphasized that these distinctions have been made because there are at least some hints that structural corollaries exist. As we shall see, there is evidence that the β^A-globin chromatin in mature chicken erythrocytes is somehow differently conformed than is the corresponding region in chicken liver nuclei, even though it is not transcribed (and can be said to be "inactive," in the more specific sense) in either case. Although it may be useful, we should not take such a classification scheme too seriously. It should not be a procrustan bed in which ideas must be chopped to fit. It is only a framework for discourse.

Ever since the discovery of nucleosomal structure in chromatin, a basic question has been: What relationship does this structure bear (if any) to the process of transcription? A more specific question is: What *happens* to nucleosomes during this event? For *if* nucleosomes are present in regions about to be transcribed, something quite drastic must occur. Both strands of the DNA are bound periodically to the histone core, over a length of

Table 8-1. Classification of Chromatin Regions

	Nonfunctional	vs.	Functional
Organism level	DNA never transcribed in any cell line	↓	DNA can, under some circumstances, be transcribed in some cell lines
	Incompetent	vs.	Competent
Tissue level	Not transcribed in a particular cell line	↓	Transcribed at some developmental or cell cycle stage in a particular cell line
	Inactive	vs.	Active
Temporal level	Not being transcribed at time of observation		Being transcribed at time of observation

at least 146 bp. There is no way in which the canonical nucleosome structure could be maintained during the read-through of one of these strands. Indeed, a number of experimental studies (see Wasylyk and Chambon, 1979, for example) have shown that addition of nucleosome cores to DNA severely inhibits transcription in vitro. This would argue that in regions to be transcribed, some special provision must be made. Three possibilities can be imagined: (1) The histone cores are detached from active regions during transcription; (2) the cores slide out of the way of the polymerase, or are passed around it; or (3) the nucleosomes unfold in a gross sense, so that each DNA strand remains attached to only a portion of the histones, and certain histone–histone contacts are broken (see Fig. 8-1). Certainly there may be other possible models, but all proposed to date seem to fall into one of these categories.

Evidence that nucleosomes may play an active part in the transcription process comes from several reports of polymerase–nucleosome complexes (Baer and Rhodes, 1983; Sakuma et al., 1984; Sargan and Butterworth, 1985). The study by Baer and Rhodes is of particular interest in that the complexes were found to contain only about half the normal complement of H2A/H2B.

It is probably not necessary that there be specific *chromatin* features that distinguish functional and nonfunctional regions of the genome. Nonfunctionality is more apt to reside at the DNA sequence level, in the simple absence of a promoter, for example, although there may be differences at the chromatin level as well. But competence and incompetence, activity and inactivity exist on the same DNA sequences, in different tissues and at different times. Unless DNA methylation plays a more direct role than is presently believed, we are forced to the conclusion that something is different in the *chromatin* structure of different regions. What are these differences, and how are they established and regulated? This question lies at the heart of the problem of cell differentiation.

Thus, the basic questions can be stated succinctly: What are the structural differences between competent and incompetent, active and inactive genes? What happens to chromatin structure when a polymerase molecule passes through?

Experimental Approaches

The fundamental difficulty is this: Most genes in eukaryotic cells are transcribed only rarely, and even then are embedded in a mass of untranscribed material. There exists no general, wholly adequate method for the separation of chromatin into active/inactive or competent/incompetent fractions, although some progress has been made toward this goal (see below). To be sure, there are some exceptions to the above statement: For example, the ribosomal genes in actively growing cells are highly amplified, transcribed at a high rate, and concentrated in the nucleolus. In some organisms, they are carried on separate minichromosomes (see Chapter

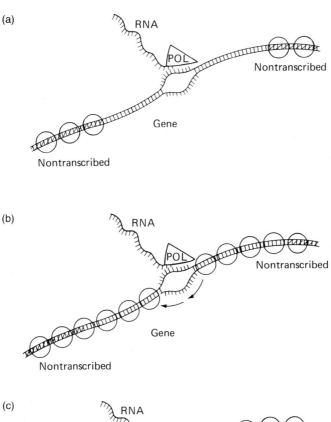

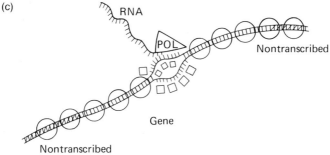

Fig. 8-1. Some conceivable mechanisms for transcription in eukaryotes. (a) Histone cores are removed before transcription occurs. (b) The cores are removed as the polymerase passes, or are passed around the enzyme. (c) Nucleosomes unfold in such a way as to allow DNA strand separation to occur.

7). For this reason, these genes can be readily isolated as nucleoprotein complexes. In a few eukaryotes, such as yeast, the major part of the genome appears to be at least transcriptionally competent, and much is probably active at any time. But the chromatin structure in these and similar exceptional cases may not be very representative of that in those single-copy genes which are specifically expressed in differentiated cell lines.

As a consequence of these difficulties, there has been a tendency to seize upon whatever techniques will reveal *any* differences in chromatin structure between various regions of the genome. As we shall see, this has all too often resulted in observations of unclear meaning and uncertain relatedness. Before considering the results of studies of different gene types, it will be helpful to review some of the techniques that have proved most useful.

Electron Microscopy

Obviously, the most direct way to tell what a region of chromatin is like is to look and see. Unfortunately, it is not so simple. In the first place, it is in most cases impossible to identify (in the genetic sense) a particular region of the genome. There are, to be sure, exceptions; the tandemly repeated ribosomal genes, when actively transcribing, are readily identifiable in spread preparations, and it has been possible to identify some very special genes, such as that of silk fibroin (McKnight et al., 1976) or very large genes (Olins et al., 1984, 1986). But these are the exceptions rather than the rule. A more fundamental problem lies in the methods that must be used to prepare specimens for electron microscope study. While sectioning of nuclei and whole-mount chromosome preparation have been useful in examining overall chromatin structure in situ (see Chapter 7), the study of actively transcribing genes has to date mainly depended on variations of the spreading technique of Miller and Beatty (1969). In the versions of this method most used at the present time, nuclei are lysed in a low ionic strength medium (typically 1 mM or less) and centrifuged onto the EM grids through a sucrose cushion containing a fixative (i.e., glutaraldehyde). The grids are then ethanol or air dried. Labhart and Koller (1982a, 1982b) claim that the low ionic strength conditions can cause the adventitious attachment of proteins to the fibers and suggest a salt-washing procedure to eliminate this. The low ionic strength and drying procedures also have the potential for extending and stretching the chromatin fibers. Thus, it is quite difficult to be certain that a structure observed in the electron microscope actually corresponds to the in situ conformation. As we shall see later, there has been considerable controversy concerning the interpretation of electron microscope results. It seems probable that much of this arises from differences in sample preparation.

Nuclease Digestion

An altogether different and complementary approach to the analysis of eukaryotic gene structure is provided by nuclease digestion experiments. Two such experiments, carried out in the first years following the discovery of nucleosomes, were of seminal importance. First Noll (1974a) showed that a large fraction of the rat liver genome was organized into nucleosomes; this hinted that these particles probably would have to be reckoned with in transcription. A much more direct demonstration was performed by Garel and Axel (1977), who prepared mono- and oligonucleosomes by micrococcal nuclease digestion of chicken oviduct nuclei (see also Garel et al., 1977). They showed that DNA from these particles hybridized to ovalbumin cDNA. Similar results were obtained for ribosomal DNA by Reeves (1977, 1978) and Mathis and Gorovsky (1977). In a particularly careful study, Bellard et al. (1977, 1980) not only confirmed these results but also demonstrated that the ovalbumin gene was being actively transcribed; an estimate of "at least 5" polymerase molecules per gene was made.

As will be discussed below, the idea that nucleosomes are actually present during transcription has now been questioned, at least for some kinds of genes. Nevertheless, these early studies were important in that they pointed out another feature of micrococcal nuclease digestion of active genes; such genes were found to be much more readily digested by the enzymes than was bulk chromatin. Many subsequent studies have employed micrococcal nuclease digestion in this way. With the development of the "Southern" blotting technique (Southern, 1975), it became possible to test directly on gel electrophorograms for the presence of particular DNA sequences in a "ladder" pattern. A number of such experiments will be described in *The Chromatin Structure of Some Specific Genes*.

At about the same time, it was found that other nucleases (DNase II, Gottesfeld et al., 1975; DNase I, Berkowitz and Doty, 1975; Weintraub and Groudine, 1976) also preferentially digested active genes. However, Weintraub and Groudine noted a subtle difference between DNase I and micrococcal nuclease selectivities. The enhanced sensitivity to DNase I was also expressed by the globin genes in *mature* chicken erythrocytes. These genes are surely not active but can be classed as *competent* (see above) in these cells. This was a first indication that nuclease digestion might be sensitive to subtle, residual changes in chromatin regions that had once been transcribed.

A special feature of DNase I digestion was revealed when C. Wu et al. (1979) subjected to gel electrophoresis the DNA products of a very mild digestion with this enzyme. As Figure 8-2 shows, such digestion preferentially cleaves the chromatin at certain "hypersensitive" regions. It is possible to map the locations of these points (or the points of preferential cutting in a mild micrococcal digestion) by a very clever technique devised by Wu (1980) and Nedospasov and Georgiev (1980). This "indirect end-labeling" method has been described in Chapter 7. Extensive use of the technique has allowed detailed studies of the chromatin structure of competent or active gene regions, and many examples will be discussed in later sections. The surprising discovery by Larsen and Weintraub (1982) that the single-strand specific nuclease, S1, can sometimes cut in hypersensitive sites has sparked even more interest in these regions. Examples

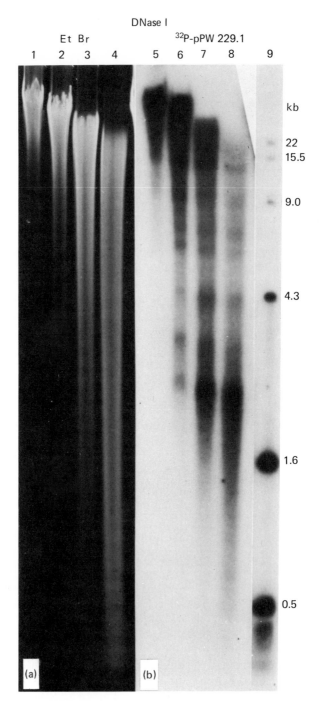

of and the significance of this observation will also be discussed in later sections of this chapter.

In evaluating the results of nuclease digestion studies of the chromatin structure of genes, certain limitations and potential pitfalls in these methods must be kept in mind:

1. All such techniques at least partially destroy the region being investigated. The consequences of this may be even more important than would appear at first glance. Any tension on the DNA resulting from supercoiling is relaxed at the first cut, and any structural consequences of this may be lost. If digestion is extensive, lysine-rich histones and some nonhistone proteins may be displaced from their original sites and could conceivably rebind in other regions.
2. The nucleases usually employed have sequence preferences on naked DNA. As a consequence, it is always necessary to compare the digestion pattern obtained in nuclei or on chromatin with that observed with the corresponding purified DNA. Only differences between the digestion patterns obtained with chromatin and purified DNA have significance.
3. McGhee and Felsenfeld (1983) have pointed out another potential source of ambiguity in micrococcal nuclease studies of phasing of nucleosomes with respect to genes. Since this enzyme has a preference for A–T-rich sequences and *can* cut within nucleosomes, it is possible to produce a biased population by extensive digestion. Those nucleosomes positioned with respect to the DNA sequence so as to have internal A–T segments exposed on the surface will be preferentially destroyed.

It is evident, therefore, that nuclease digestion, like electron microscopy, has certain limitations in the elucidation of active chromatin structure. Furthermore, neither technique is particularly well suited to address a very important question: What differences are there in protein composition and/or protein modifications between active and inactive regions of chromatin? For this purpose, some kind of chromatin fractionation procedure is needed. This is by no means an easy task, since active genes represent only a small fraction of the chromatin in differentiated cells and are widely scattered and interspersed with inactive regions. There are, to be sure,

⊲————————————————————————————

Fig. 8-2. The first evidence for DNase I "hypersensitive" sites in chromatin. Nuclei from *Drosophila* tissue culture cells were digested for 3 minutes with increasing (left to right, each panel) concentrations of DNase I, and the products electrophoresed on 0.8% agarose gels. In the left panel (a), the gels were stained with ethidium bromide to reveal the total digestion pattern. On the right (b), the gel was Southern blotted and probed with a specific, radiolabeled *Drosophila* probe. The autoradiograph is shown. A number of fairly discrete "hypersensitive" sites are revealed. Channel 9 shows marker DNAs. (From Cell *16*, p. 797, Wu et al. Copyright 1979 M.I.T. Press.)

exceptional cases—the nucleolar genes for ribosomal DNA, certain extrachromosomal elements, and animal cell viruses such as SV-40, which have a nucleosomal structure. But these represent a small (and possibly atypical) sampling.

Chromatin Fractionation

A great deal of effort has been expended over many years in attempts to devise methods to separate "active" from "inactive" chromatin. While a wide variety of techniques have been proposed, the more successful seem to have been the following.

Fractionation on ECTHAM-Cellulose Columns

This method, developed by Smith and Billett (1982a, 1982b), uses a cellulose to which Tris has been coupled with epichlorohydrin. Mechanically sheared (or, more recently, nuclease-cleaved; see Smith and Billett, 1982b) chromatin is eluted with a salt gradient, and the "active" chromatin fractions preferentially elute at 80 to 100 mM salt.

Fractionation by Partition in a Dextran/Poly(Ethylene Glycol) Two-Phase Mixture

This is the procedure of Hancock and co-workers (Turner and Hancock, 1974; Faber et al., 1981; Gabrielli et al., 1981). A fraction is obtained from mouse cells comprising only 16% of the total DNA but 63% of the pulse-labeled RNA.

Fractionation Based on DNase II Digestion Followed by Differential Solubilization in 2 mM MgCl$_2$

Developed by Gottesfeld et al. (1975), this method has been quite widely employed (see, for example, Gottesfeld and Butler, 1977; Gottesfeld and Partington, 1977; Hendrick et al., 1977; Sarkander and Dulce, 1979; Pashev et al., 1980; see Fig. 8-3a). Smith and Billett (1982b) find that the Mg^{2+}-soluble "active" fraction behaves consistently on their ECTHAM-cellulose columns.

Methods Based on the Selective Digestion of "Active" Chromatin by Micrococcal Nuclease

A host of techniques seem to rely on the fact that micrococcal nuclease preferentially cleaves in active (and competent?) regions of the genome. In addition, many of these methods exploit solubility differences (or perhaps differences in diffusibility through the nuclear envelope) of the "active" fragments. Perhaps the most widely used variant is that of Sanders (1978), as diagrammed in Figure 8.3b. The fragments eluted from the nuclei at lowest ionic strength are found to be enriched in active genes (see, for example, Rocha et al., 1984). Similar methods, based on limited micrococcal nuclease digestion, have been employed by Jackson et al. (1979),

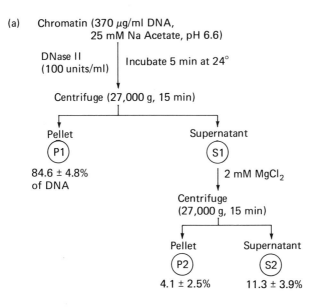

(a)   Chromatin (370 μg/ml DNA,
                      25 mM Na Acetate, pH 6.6)

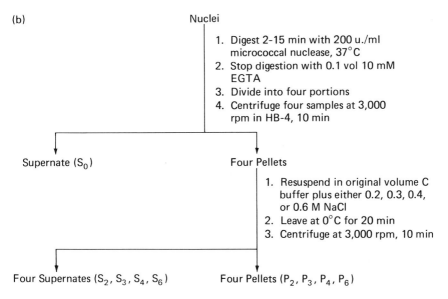

(b)

Fig. 8-3. Two widely used methods for the separation of chromatin into transcriptionally "active" and "inactive" fractions. (a) The method of Gottesfeld et al. (1975; see also Gottesfeld and Partington, 1977). The S2 fraction is enriched in active gene sequences. (b) The method of Sanders (1978). Fractions extracted by low salt are enriched in active, DNase I sensitive sequences. (Reproduced from *The Journal of Cell Biology*, 1978, 79, 97–109, Sanders by copyright permission of The Rockefeller University Press.)

Levy-Wilson and Dixon (1979), Dimitriadis and Tata (1980), Aström and von der Decken (1980), and Kitzis et al. (1982). These techniques, as usually employed, yield an "active" fraction that consists mostly of monosomes and small oligonucleosomes. Komaiko and Felsenfeld (1985) point out that the enhanced salt solubility of these particles may be a consequence of loss of H1 through redistribution. They find that the solubility differences can be obliterated by addition of H1. Thus, the mechanism behind the fractionation remains obscure.

A most interesting variant of this method has been developed by Wurtz and Fakan (1983). By carrying out a *very* mild micrococcal nuclease digestion on nuclei, followed by a long (15 hr) dialysis against digestion buffer, they were able to obtain a soluble fraction comprising only 2 to 10% of the DNA, but containing 50 to 70% of the pulse-labeled RNA and 90% of the chromatin-engaged polymerase II. Most remarkably, this material was composed mostly of long (20 to >100) polynucleosome chains, many of which carried nascent RNA transcripts. The technique is a potentially interesting one, for the size of these fragments is reminiscent of the "domain" sizes mentioned in Chapter 7.

It is very difficult to compare the efficiency of these various techniques, for quite different criteria (presence of nascent RNA, enrichment in active genes, presence of polymerase, etc.) have been employed by different researchers. What can be said in general is that the results are not overwhelmingly impressive. Enrichment factors of about 5 for active genes are frequently claimed. At first glance this might seem good, but when it is noted that only a few percent of the total genome may be active, it is clear that even the best preparations are still grossly contaminated with "inactive" chromatin. None of these techniques gives any hope for purification of a specific gene present in small copy number. Nor is it by any means clear that the various techniques select for the same level of activation. Gottesfeld and Partington (1977) suggest that their method may be selecting only those genes that are actively transcribing. On the other hand, it is obvious that the long segments isolated by Wurtz and Fakan (1983) *must* include substantial nontranscribed regions. Different levels of gene activation may have their correlates in different modifications of the chromatin structure, and the various "fractionation" procedures may each respond to one or another of these perturbations. The unhappy conclusion is that none of the above chromatin fractionation procedures is very efficient in separation, or clearly defined in its selectivity. There has recently appeared, however, a novel technique that holds considerable promise for the isolation of particular chromatin regions in high purity.

Nucleoprotein Hybridization
Workman and Langmore (1985a, 1985b) have developed a way to use the high selectivity of nucleic acid hybridization for chromatin fractionation. A DNA probe is prepared that abuts the end of the chromatin region of

interest. Mercury is covalently bound to the cloned probe, and a few hundred bases are exposed at its 5' end by exo III digestion. The chromatin is cut with the same restriction enzyme used to prepare the probe, and a region of 3' end is exposed by digestion with λ-exonuclease. Probe and chromatin are then mixed, hybridized under mild conditions (so as to not disturb the proteins), and passed over a sulfhydryl column. Only those chromatin fragments carrying 3' terminal regions complementary to the probe will be retained, and then may be selectively eluted from the column. Although the method has, at this writing, received only limited testing, the preliminary results are impressive. SV-40 minichromosomes could be picked out of a fifteenfold excess of sea urchin chromatin at 88% purity and 63% yield. This is the kind of purification that will be needed to answer questions concerning the composition and chromatin structure associated with genes of low copy number.

The Chromatin Structure of Some Specific Genes

In this section, I will summarize present knowledge of the chromatin structure of a number of kinds of genes. The section begins with the smallest and simplest genes, those for tRNA and 5S RNA, proceeds to the ribosomal genes, and concludes with some examples of well-studied structural genes.

Transfer RNA Genes

The chromatin organization of tRNA genes in a few vertebrates has been studied extensively and might be hoped to provide specific information concerning the relationship between the structure and the transcriptional process. These are small genes (~75 bp) and are always present in multiple copies. In *Xenopus laevis,* for example, blocks of 8 tRNA genes have been found to occupy, with interspersed spaces, regions 3.18 Kbp in length. This sequence is repeated about $100 \times$ per haploid genome. Hofstetter et al. (1981) found that particular short sequences were essential for transcription of these genes. Remarkably, the required sequences lie *within* the structural genes. Similar internal control sequences have been found in 5S RNA genes, also transcribed by polymerase III (see below, and von Beroldingen et al., 1984, for a review of the control of these genes). Such genes should be ideal candidates for definitive study. Unfortunately, the results to date have been confusing.

Wittig and Wittig (1979) were the first to report evidence for nucleosome phasing with respect to the tRNA genes. Using chicken embryos, tetranucleosomes were fractionated from a micrococcal nuclease digest and the DNA prepared from these. By hybridization, the tRNA genes in these fragments were found to lie in four locations, each of which would lie within a nucleosome if the nuclease cutting were, as expected, in linker regions (see Fig. 8-4). Such locations would appear, at first glance, hardly

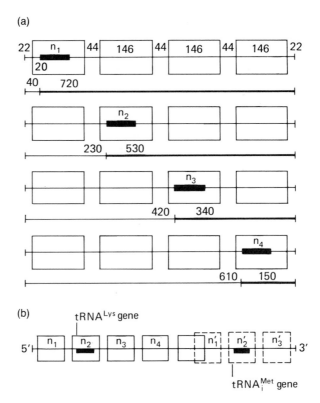

Fig. 8-4. The relationship of tRNA genes to nucleosome positions in chicken embryo chromatin, according to Wittig and Wittig (1979). In the top portion (a) are shown the four possible locations (n_1–n_4) of the genes with respect to the 760 bp repeat. Putative nucleosome positions are shown as open boxes. Part (b) depicts a possible phase shift between two kinds of tRNA genes. (Adapted from Wittig and Wittig, Copyright 1979 M.I.T. Press.)

auspicious for transcription. Nevertheless, in a later publication, Wittig and Wittig (1982) provided evidence that in vitro reconstituted tRNA chromatin fragments are maximally transcribable with just this phasing. In fact, the favored location for transcribability is claimed to be that in which one of the critical sequences detected by Hofstetter et al. (TTCGA) lies exactly on the nucleosome dyad axis. Furthermore, this phasing corresponds closely to that observed earlier in the nuclease digestion experiment. The Wittig experiments are complex, involving the reconstruction of the repeat with an insertion of a 30 bp palindromic fragment and the construction of a single-strand region to define the phasing. The cellular extracts used in the reassembly and transcription are not well defined, and the potential effects of the inserted and single-strand regions are difficult to evaluate. Nevertheless, the results make a certain amount of mechanistic sense. An effector molecule (or the polymerase itself), binding at the dyad axis, might serve to open the nucleosome for transcription. As we shall see below, there is evidence for just such an unfolding process in both 5S RNA and in rRNA transcription.

A rather different picture has emerged from a very careful study of *Xenopus* tDNA carried out in M. Birnstiel's laboratory (Bryan et al., 1981, 1983). They have compared the nucleosome arrangement in the mature erythrocytes, where the tDNA genes are inactive, with that in cultured liver and kidney cells, where a high level of activity is observed. Micrococcal digestion of either, followed by probing with a tDNA probe, yields a highly regular oligonucleosome ladder. However, there are striking differences. The repeat in erythrocytes is 198 ± 5 bp; in the cultured cells, it is about 185 bp. Furthermore, whereas the indirect end-labeling technique demonstrates a single, precise nucleosome phasing in the inactive erythrocyte tDNA, no evidence for a preferred phase could be found in the transcriptionally active cells. The nucleosomes here appear to be regularly spaced, but with random phases with respect to the DNA sequence. This result seems, at first inspection, to be wholly inconsistent with the model proposed by Wittig and Wittig. However, the situation is not fully understood, and a number of complexities exist. Wittig and Wittig (1979), for example, found evidence that the phase was frequently "reset" in the genome. Bryan et al. have pointed out the possibility that the nuclease digestion experiments of Wittig and Wittig might have selected for a subset of particularly nuclease-resistant tetramers.

If the phasing reported by Wittig and Wittig is critical for transcription in vivo, then the altered repeat length observed by Bryan et al. for inactive erythrocyte chromatin would provide an explanation for the inactivation; such a change would necessarily result in all but a very few genes having the critical sequence incorrectly placed in the nucleosome. Alternatively, Bryan et al. have suggested that the lengthened and regular repeat in inactive tDNA may promote the formation of a higher-order structure incompatible with transcription. In either event, we are left with the intriguing question of *how* the repeat is changed.

More recent studies by DeLotto and Schedl (1984) add still a third view of tRNA chromatin. These researchers have investigated tRNA gene clusters in *Drosophila* and find nuclease-sensitive sites in two regions: upstream from the start sequence, and within the genes, close to or overlapping the control regions. The latter are also sensitive to S1, the single-strand nuclease. Since the distance between these two regions (60 to 80 bp) is too small to contain a nucleosome but too large for a linker, they suggest two phases, one of which would presumably correspond to the active gene structure.

It is very difficult to compare these experiments. In the first place, three different organisms have been used, and in at least two cases—chicken embryos (Wittig and Wittig) and *Drosophila* (DeLotto and Schedl)—the extent to which the genes are actively transcribed is unknown. The observation of Bryan et al. that phasing disappears in activation is reflected in a number of studies of polymerase II genes (see below). It is discouraging to find that even in this simple transcription system, the structure of active chromatin remains obscure.

5S RNA Genes

Genes for the 5S RNA of eukaryotes, unlike those of prokaryotes, are located apart from the remainder of the ribosomal RNA genes. They are present in multiple copies, ranging in number from about 150 per haploid genome in yeast to as many as 300,000 in some amphibians. Each repeat contains a copy of the gene, nontranscribed spacer, and in some cases a pseudogene as well. Like the tRNA genes, 5S RNA genes are transcribed by polymerase III. Extensive information concerning these genes and their products has been summarized in Chapters 1 through 4 of Vol. 11 of *The Cell Nucleus* (H. Bush and L. Rothblum, eds., Academic Press, 1982; their control is reviewed by von Beroldingen et al., 1984). In this section, I shall be concerned mainly with what is known concerning the relationship between chromatin structure and transcription of 5S DNA.

Like the tRNA gene, the 5S gene has been shown to possess an internal transcription control sequence (Sakonju et al., 1980; Bogenhagen et al., 1980; Sakonju and Brown, 1982). This lies in the region between residues 45 and 95, near the center of this 120 bp gene. Furthermore, protein factors that stimulate transcription have been isolated, and one (now called TFIIIA) has been shown, by nuclease protection experiments, to bind to the same region (Engelke et al., 1980; Smith et al., 1984). The interaction between TFIIIA and the control region is critical for 5S gene activation (Bieker et al., 1985).

Xenopus laevis proves to be an exceptionally interesting organism in which to study 5S genes. Not only are the genes highly reiterated, but there are two distinguishable types, oocyte-specific and somatic genes. The former are transcribed only in the oocyte; the somatic genes are active in either milieu. However, in mature erythrocytes the somatic genes are also inactive. Thus, comparisons of the same gene in different environments and related but distinguishable genes in the same environment are possible. Furthermore, as first demonstrated by Brown and Gurdon (1977, 1978), the 5S DNA can be correctly transcribed if injected into the germinal vesicles of oocytes, where it is assembled into a chromatin structure, presumably by the nucleosome assembly system discovered by Lasky et al. (1977; see also Chapter 9). This observation is of especial interest because Parker and Roeder (1977) have shown that the naked 5S genes do not transcribe correctly in vitro.

Initial studies of the 5S chromatin (Gottesfeld, 1980) provided indications for a nucleosomal structure. In further experiments, Gottesfeld and co-workers obtained evidence for phasing of the nucleosomes with respect to the 5S genes (see Gottesfeld and Bloomer, 1980; Reynolds et al., 1982). Four arrangements, differing slightly in phase, were reported. In each of these, the 5S gene lies mostly within one or two nucleosomes. Various salient features are exposed in linker regions in different phases—the origin and termination sites and the control site. Young and Carroll (1983) examined the oocyte-type 5S genes in *Xenopus* erythrocytes. In these inactive cells, they find evidence for a single, regular phasing of the nucleosomes with respect to the 5S gene.

A most interesting in vitro study of a ternary complex between a 5S RNA gene, TFIIIA, and a nucleosome has been carried out by Rhodes (1985). Using a cloned DNA fragment containing the *Xenopus borealis* 5S gene and flanker regions, she has demonstrated that a histone core will reconstitute onto this with almost exactly the same positioning as on the sea urchin 5S gene studied by Simpson and Stafford (1983), covering about 80 bp of the 5' end of the gene. Even though only about 15 bp of the TFIIIA site extends beyond the 3' end of the nucleosomal region, it is still possible to bind this factor to form a ternary complex. Rhodes points out that this must perturb at least 20 bp of the nucleosomal DNA. It is suggested that the "finger" binding domains first make attachment at the 3' end of the domain and then invade the nucleosomal region (see Miller et al., 1985, for details of TFIIIA structure).

The results also have possible significance with respect to the interaction of this complex with polymerase, for the coiling of the nucleosomal region brings one end of TFIIIA in close contact with the putative entry site for the polymerase. Finally, Rhodes points out that the presence of H1 adjacent to such nucleosomes would effectively block the TFIIIA site, and suggests that this may account for the observation of Schlissel and Brown (1984) that H1 inhibits 5S gene transcription.

A very different approach to the problem is exemplified by the microinjection studies of Worcel and co-workers (Ryoji and Worcel, 1984; Gargiulo and Worcel, 1983; Gargiulo et al., 1984). They have observed that when oocyte-type and somatic-type 5S genes, inserted into plasmids, are injected into oocytes, the somatic genes seem to compete preferentially for an activating factor or factors, presumably including TFIIIA. Furthermore, they report two types of chromatin structure on these plasmids. One form, which they term "dynamic chromatin," is apparently under torsional strain, for it can be relaxed by topoisomerase I. The other, which they call "static chromatin," does not exhibit this behavior. Since relaxation also shuts off 5S RNA transcription, it is suggested that the torsionally strained "dynamic" structure is in fact the transcriptionally active form. In later studies, Glikin et al. (1984) present evidence that ATP is required for the assembly of "active" chromatin. Further details are given in *DNA Torsion and Transcription*, below.

Although the *Xenopus* system has allowed the most sophisticated experiments, studies of 5S chromatin structure in other organisms seem to give generally comparable results. For example, an analysis of the chromatin organization of *Drosophila melanogaster* 5S genes has been reported by Louis et al. (1980). Dinucleosomes were prepared from *Drosophila* nuclei and the DNA from these digested with restriction endonucleases. Simple fragment patterns were obtained when the gels were blotted against a 5S DNA probe. The results were consistent with a two-phase arrangement, as shown in Figure 8-5. Both phases have interesting properties. In A, part of the "control" sequence lies in linker, as in the studies of Rhodes (see above). In B the 5' end of the gene is exposed. As in the *Xenopus* system, much of the gene appears to be buried in nucleosomes.

Although the picture is far from complete, a common *motif* appears in both the tRNA and 5S-RNA gene systems: the appearance of multiple but regular phases, often related to the internal control site. With such short genes, smaller than the DNA contained in a single nucleosome, a highly localized mode of action might be expected. Perhaps polymerase III is an enzyme especially suited for this kind of interaction. In terms of transcriptional mechanism and regulation, we should perhaps not expect much in common with polymerase I and polymerase II, which transcribe much longer regions.

The Major Ribosomal RNA Genes

In addition to the 5S RNA, eukaryotic ribosomes contain three other RNA molecules, which in most species have sedimentation coefficients of about 18S, 5.8S, and 28S. These are transcribed as a ~40S precursor (35S in

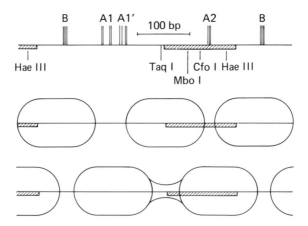

Fig. 8-5. Two arrangements of nucleosomes on the 5S gene repeating structure in *Drosophila* chromatin, as deduced by Louis et al. (1980). The diagram at the top shows the position of the gene within the repeat, together with strong sites of micrococcal nuclease cleavage. The bottom diagrams represent the two possible nucleosome phases consistent with these results. See reference for details. (From Cell *22*, p. 387, Louis et al. Copyright 1980 M.I.T. Press.)

yeast), which is then processed into the final products (see Fig. 8-6). Transcription is by polymerase I. The gene organization and transcription of ribosomal genes is reviewed in exhaustive detail in Volumes 10–12 of *The Cell Nucleus* (Busch and Rothblum, 1982). As elsewhere herein, I shall emphasize what is known concerning the chromatin structure of these genes and how it may relate to their transcription. The reader interested in other details should consult the above-mentioned volumes.

One aspect of ribosomal gene organization should be mentioned for clarification of what is to follow. In most higher eukaryotes, the ribosomal genes are arranged in long tandem arrays, separated by lengths of nontranscribed spacer (see Fig. 8-6b). However, in some of the more primitive eukaryotes (i.e., *Tetrahymena, Physarum*), most of the rDNA is carried as extrachromosomal elements, which are found to be giant palindromes, with central and terminal nontranscribed regions (Fig. 8-6a). In any event, rDNA is almost entirely confined to the nucleolus and represents virtually all of the DNA contained therein. This has made isolation of relatively pure ribosomal chromatin a comparatively easy task, and the distinctive arrangement of these genes has facilitated their identification in electron micrographs. With the spreading technique developed by O.L. Miller and co-workers (Miller and Beatty, 1969; Miller and Bakken, 1972), active ribosomal genes of higher eukaryotes appear as highly characteristic, tandemly arranged transcription units (Fig. 8-7). Since the DNA length of these genes is quite accurately known, it is possible to deduce the packing

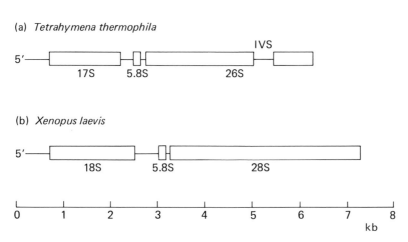

Fig. 8-6. The two major kinds of organization of ribosomal genes in eukaryotes. (a) *Tetrahymena thermophila,* showing half of the palindromic minichromosome. The center of symmetry is at the left. Note that in *thermophila* there is an intervening sequence (IVS) within the 26S region. (b) *Xenopus laevis,* showing one of the tandemly repeating units. In both examples, transcription is from left to right. (From The Cell Nucleus, vol. 10, pp. 171–204, Cech et al. Copyright 1982 Academic Press.)

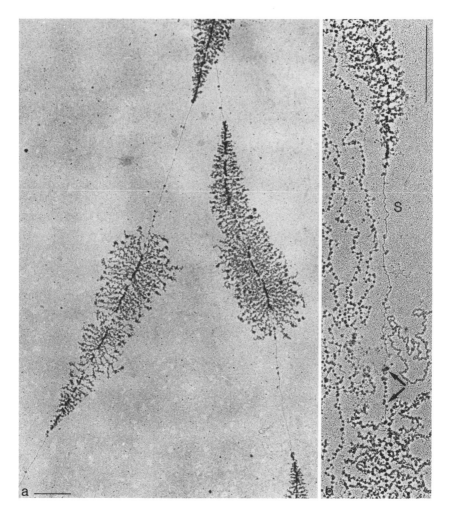

Fig. 8-7. An electron micrograph of tandemly repeating ribosomal genes in oocytes of *Triturus*. Note that in this case, even the intervening nontranscribed spacers present a smooth, nonbeaded appearance. (From Eur. J. Cell Biol. *23*, 189, Scheer. Copyright 1980 Wissenach. Verlag.)

ratio of the chromatin. In parallel studies from a number of laboratories, it was shown that the packing ratio is very close to unity; therefore, active ribosomal genes cannot be organized into a folded nucleosomal structure, for which a packing ratio of at least 2 would be expected. Furthermore, transcriptional units that were lightly transcribed, so that the underlying chromatin strand could be observed, were found to exhibit a nonbeaded structure (Woodcock et al., 1976; Laird et al., 1976; Franke et al., 1976; Foe et al., 1976). In their careful studies of the activation of rDNA in *Oncypeltus*, Foe and co-workers distinguished two levels of organization (see Foe et al., 1976; Foe, 1977). The first, which was called "nu chromatin" was found in nonribosomal portions of the genome and in the nontranscribed spacers of ribosomal genes; it had the characteristic beaded morphology associated with nucleosomes, and a packing ratio of about 2. A second form, designated "rho chromatin," was observed in the ribosomal genes themselves, first appearing at a stage in embryogenesis *before* transcription has begun. This structure appeared as a smooth fiber, with a packing ratio of 1.0 to 1.2. From staining characteristics and fiber thickness, it was judged to be protein-covered. Actively transcribed genes, covered with nascent RNA transcripts, were also found to have a packing ratio close to unity; in instances where the underlying structure could be seen between transcripts, it seemed to resemble rho chromatin. In the following years, a multitude of studies, using rDNA from a wide variety of organisms, have confirmed that actively transcribed ribosomal chromatin is fully extended (see, for examples and further references, Franke et al., 1980; Scheer et al., 1981; Labhart and Koller, 1982a, 1982b; Ness et al., 1983; Derenzini et al., 1985). Furthermore, there seems to be a general consensus that *wholly* inactive ribosomal chromatin is observed in the electron microscope as compacted into a nucleosomal structure (Foe, 1977; Franke et al., 1977, 1980, 1984). Foe's developmental studies, together with similar observations by other workers, indicate that the extended, rho chromatin structure is not simply a consequence of transcription per se. Rather, it appears to represent a transcriptionally competent state, established in some way before transcription itself commences.

There is, however, considerable disagreement among electron microscopists concerning both the state of the nontranscribed spacers and the nature of rho chromatin. Whereas Foe and a number of other researchers (e.g., Angelier and Lacroix, 1975; Woodcock et al., 1976; Franke et al., 1976; Grainger and Ogle, 1977) have observed nucleosome-like structures in these regions, it has been more recently argued that these may be artifacts of specimen preparation, corresponding either to polymerase molecules or other proteins (see Franke et al., 1977; Scheer, 1980; Borkhardt and Nielsen, 1981; Labhart and Koller, 1982a, 1982b). In fact, Labhart and Koller, who have used a step-gradient centrifugation method to salt-wash the chromatin as it is applied to EM grids, suggest that *all* of the evidence for protein on active ribosomal chromatin may be a consequence of nonspecific binding during preparation. They cannot distinguish, in their preparation, any differences between either nontranscribed spacers or intragenic chromatin, and naked DNA! This is surely an extreme view, which must be considered in the light of other kinds of evidence described below. Furthermore, in a more recent study of *Dictyostelium* ribosomal chromatin, the same group acknowledges the presence of nucleosomes in nontranscribed spacers, while maintaining that the transcribed regions are nucleosome-free (Ness et al., 1983).

Nuclease digestion experiments have provided an alternative and entirely independent method for analyzing rDNA structure. By the use of Southern blotting against specific probes, it is possible to examine the structures of both transcribed and nontranscribed regions. Furthermore, since the experiments can be carried out with intact nuclei, or isolated nucleoli, potential preparation artifacts that plague the electron microscopists can be averted. Some of the earlier experiments seemed to indicate a preponderance of nucleosomal structure in rDNA (Mathis and Gorovsky, 1977; Butler et al., 1978). Studies by Lohr (1983a) on yeast ribosomal genes give a similar result, but Lohr points out the uncertainty as to the extent of transcription. The work of Reeves (1977) and a number of subsequent studies have pointed up the complexity of the situation. Using *Xenopus* oocytes at different developmental stages, where different rates of rRNA synthesis are occurring, Reeves found that whereas the rDNA of Stage I oocytes, which have a low transcriptional activity, seemed to be 85% nucleosomal, that of Stage II oocytes which are maximally active was only about 37% in this form. Borchensius et al. (1981), probing different regions of the palindromic *Tetrahymena* rDNA molecules, found evidence for a defined nucleosomal structure in the nontranscribed spacer but no such structure in the gene itself. Similar results have been obtained by Gottschling et al. (1983), who also observed differences in the rDNA from logarithmically growing and starved *Tetrahymena*. Perhaps the most detailed study of this kind is that of Davis et al. (1983), who have compared the rDNA chromatin in two kinds of mouse cells: liver cells, which are transcriptionally inactive, and a very active cultured cell line. Fractionating the chromatin by nuclease digestion and salt extraction, they observed that the ribosomal genes were distributed very differently for the two cell types (see Table 8-2). The "pellet" material, which represented a large fraction of the rDNA in the transcriptionally active P815 cells, did not yield a nucleosome ladder upon digestion. This result applies to the nontranscribed spacers as well as to the intragenic regions. In liver cells, most of the ribosomal chromatin ended up in a soluble fraction, which appeared to have a nucleosomal structure. Quite similar results have been obtained by Ness et al. (1983) and Parish et al. (1986) with *Dictyostelium*, by Udvardy et al. (1984) using *Drosophila*, and by Gottschling et al. (1983) with *Tetrahymena*. In these cases, the transcribing regions do not give evidence for well-defined nucleosome structure, but the spacer regions do.

Support for this model comes from psoralen-crosslinking studies by Sogo et al. (1984). They observe a regular (approximately 200 bp) spacing of crosslinks in the nontranscribed spacers of *Dictyostelium* r-chromatin, but no such regularity in the transcribed regions. It seems possible to reconcile the digestion studies with most of the electron microscope investigations. It would appear that the rDNA repeats are in a "conventional" nucleosomal structure when wholly inactive, but lose this structure, at least in the transcribed regions, when activated for transcription. It is possible,

Table 8-2. Distribution of Ribosomal Gene Sequences Among
Different Chromatin Fractions from Mouse Cells[a]

Cell type	Fraction[b]	Sequence distribution (% of total)				
		Bulk DNA	NTS[c]	ETS[c]	28S[c]	Satellite[c]
	S1	7	< 1	< 1	< 1	< 1
P815[d]	S2	76	7	9	6	74
	P	17	93	91	94	26
	S1	3	< 1	< 1	< 1	< 1
Liver	S2	93	80	78	82	95
	P	4	20	22	18	5

[a]Data from Davis et al. (1983). Only results from a digestion yielding 5% acid-soluble DNA are given here. The paper gives in addition quite similar data from a 15% digestion.
[b]S1 and S2 are fractions soluble after nuclear digestion, P is the pellet. See Davis et al. for details.
[c]Abbreviations: NTS, 5' nontranscribed spacer; ETS, external transcribed spacer (at 5' end of 45S transcript); 28S, transcribed region for 28S product; satellite, mouse satellite DNA.
[d]An exponentially growing cell line. These cells are very actively transcribing ribosomal genes, whereas the level of transcription in liver cells is much lower.

and even likely, that the process is not of an all-or-none nature. During the process of activation, some regions may undergo the conformational shift before others, which might explain some of the seemingly contradictory results. It is now clear that both the electron microscopy and nuclease digestion approaches have limitations. In the former case, there is a natural tendency to focus on those genes that are active. Nuclease digestion of whole ribosomal chromatin, on the other hand, may emphasize nucleosome ladders that result from only a small inactive fraction of the rDNA copies.

These results, however, still leave unclear the nature of the chromatin associated with the active genes. Is it really, as Labhart and Koller suggest, devoid of histones? Or are the nucleosomes present, but in some kind of unfolded conformation? It would seem that a clear answer could be obtained from a protein analysis of nucleolar chromatin in actively transcribing cells. But even here there are inconsistent results. For example, Jones (1978) reports that the ribosomal chromatin of logarithmically growing *Tetrahymena* cells contains only 40% of the normal histone complement. In contradiction, Colavito-Shepanski and Gorovsky (1983) find a histone/DNA ratio almost the same as in bulk chromatin.

If nucleosomes are still present in active ribosomal genes, they must exist in a modified form. A model for an unfolded structure has emerged from studies in V. Allfrey's laboratory. Micrococcal nuclease digestion of *Physarum* chromatin led to the observation of a peculiar mononucleosome, with sedimentation coefficient of about 5S, containing about 170 bp of DNA (Johnson et al., 1978). Although the initial reports suggested a depletion of some histones in this particle, more recent studies have indicated the normal histone stoichiometry but an unfolded conformation. Furthermore, the particle is now said to contain 144 bp of DNA (Johnson et al., 1979; Prior et al., 1983). The presence of two specific nonhistone proteins of molecular weights 30,000 and 32,000 was also reported. Electron micrographs suggest a bipartite structure, so Prior et al. have proposed the "lexosome" structure depicted in Figure 8-8.

Such a model provides the basis for a reasonable (though still hypothetical) scenerio for ribosomal gene activation. Specific nucleolar nonhistone proteins bind near the nucleosome dyad axis, opening the structures into lexosomes. These are presumed to allow reading of the rRNA genes and to be the basis for the rho chromatin structure.

Although the lexosome model is an attractive one, it should be noted that it has met with serious objections. A reinvestigation of peak A by Stone et al. (1985) finds that peak A and core particle DNA run together on polyacrylamide gels and that the peak A band does not stain for protein. Furthermore, its thermal denaturation behavior is that of free DNA, and peak A particles do not give a DNase I ladder. Finally, it should be pointed out that the sedimentation coefficient reported for peak A (5S) is equal to that of mononucleosomal DNA. From arguments given in Chapter 6, it is difficult to see why a particle with normal histone complement and two "extra" proteins should sediment so slowly, even if unfolded.

While there are still some knotty problems (and undoubtedly some surprises) remaining with respect to ribosomal gene transcription, it would appear that parts of a consistant picture are beginning to emerge. It is not

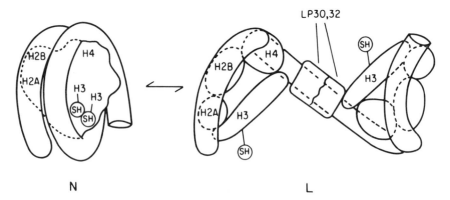

Fig. 8-8. The model postulated for the "lexosome" or unfolded nucleosome in transcriptionally active chromatin. It is suggested that two nonhistone proteins bind to the "bridge" region between the nucleosome halves. (From Cell *34*, p. 1033, Prior et al. Copyright 1983 M.I.T. Press.)

necessarily to be expected, however, that ribosomal genes will suffice as models for the understanding of the many genes selectively transcribed by polymerase II. Here another factor enters; polymerase II genes include not only those that are transcribed in every cell, but also those that are transcribed in certain cell lines at certain stages of differentiation.

Structural Genes Transcribed by Polymerase II

There have been very many studies of the chromatin organization in polymerase II genes and their flanking regions. Unfortunately, most have been fragmentary, consisting in the main of observations of nuclease sensitivity and nuclease hypersensitive sites. Rather than discuss each separately, which would be tedious, repetitious, and confusing, I shall concentrate on a few systems that have either received intensive study or reveal what

Table 8-3. Studies of Chromatin Structure of Some Polymerase II Genes

Gene	Source	Observations	Reference
Albumin	Rat	DNase I sensitive in liver, not kidney	Nahon et al. (1984)
Alcohol dehydrogenase-1	Maize	5′ DH[a] sites respond reversibly to derepression	Ferl (1985)
Alcohol dehydrogenase	Yeast	DNase I sensitivity, DH sites	Sledziewski and Young (1982)
Amylase	Rat	Active genes in MNase-sensitive conformation	Levy-Wilson (1983b)
Apo-very low density lipoprotein	Chicken	Extra DH sites upon hormone induction, lost on withdrawal	Kok et al. (1985)
c-myc	Human	Translocation has little effect on DH sites	Dyson and Rabbitts (1985); Dyson et al. (1985)
α-collagen	Human	Very large DNase I sensitive domains; DH sites display tissue specificity	Barsh et al. (1984)
Cuticle gene	*Drosophila*	Introducing transposable element has little effect on chromatin organization	Eissenberg et al. (1984)
Dihydrofolate reductase	Human	DH sites in 5′ flank; undermethylated	Shimada and Nienhuis (1984)
Dihydrofolate reductase	Mouse	Nucleosomal structure	Barsoum et al. (1982)
		Variant nucleosomes at 5′ end	Barsoum and Varshavsky, (1985)

Table 8-3. *Continued*

Gene	Source	Observations	Reference
α-fetoprotein	Rat	DNase I sensitivity in liver, not kidney	Nahon et al. (1984)
GAL-gene cluster	Yeast	Chromatin organization in intergene region	Lohr (1984b)
Galactokinase	Yeast	Changes in nucleosome arrangement on expression	Lohr (1983b, 1984a)
Glue protein	*Drosophila*	Change in DH sites upon activation	Shermoen and Bekendorf (1982)
Glyceraldehyde-3-phosphate-dehydrogenase	Mouse	DH sites upstream	Sen et al. (1985)
Histone gene cluster	*Drosophila*	Nucleosome arrangement, DH, MNH sites	Samal et al. (1981)
Histone gene cluster	*Drosophila*	Different nucleosome arrangement in spacer, genes	Worcel et al. (1983)
Histone gene cluster	Sea urchin (*Psammechinus*)	MN, DNase I sensitivity changes upon activation	Bryan et al. (1983)
Histone gene cluster	Sea urchin (*Paracentrotus*)	MN sensitivity changes during development	Spinelli et al. (1982)
Histone H4	*Physarum*	preferential sensitivity to DNase I throughout cell cycle	Wilhelm et al. (1982)
Histone H4	*Xenopus*	DH, MNH sites flanking gene	Gargiulo et al. (1985)
Histone H5	Chicken	S1 sites Nucleosome arrangement	Ruiz-Carrillo (1984) Renaud and Ruiz-Carrillo (1986)
Immunoglobulin (μ-heavy)	Human	Rearrangement yields new DH sites; one in J_H-C_μ intron	Mills et al. (1983)
Immunoglobulin (κ-light)	Mouse	New DH sites in active genes; one in J_κ-C_κ intron	Parslow and Granner (1982, 1983)
Immunoglobulin (κ-light)	Mouse	Compares nucleosome arrangement, DNase I sensitivity in active and inactive genes	Weischet et al. (1982, 1983a, 1983b)

Table 8-3. *Continued*

Gene	Source	Observations	Reference
Immunoglobulin (κ-light)	Mouse	DH sites differ in rearranged and nonrearranged; site in intron	Chung et al. (1983)
Interferon-β	Mouse	Induction yields increased DNase I sensitivity and new DH sites	Higashi (1985)
Lysozyme	Chicken	Flanking DH sites	Fritton et al. (1983)
Phaseolin	French bean	Comparison of DNase I sensitivity in cotelydon, leaf	Murray and Kennard (1984)
Phosphatase (acid)	Yeast	DNase I, MN sensitivity; DH sites in repressed, derepressed, nucleosome positions	Bergman and Kramer (1983); Bergman, (1986); Bergman et al. (1986)
Phosphoenol-pyruvate carboxy-kinase	Rat	Methylation difference in fetal, adult tissue	Benvenisty et al. (1985)
Preproinsulin	Rat	5' DH compared in different cell types	Wu and Gilbert (1981)
Ribosomal protein 49	*Drosophila*	5'DH sites	Y.C. Wong et al. (1981)
Secretory protein	*Chironomus*	Chromatin structure depends on activity	Widmer et al. (1984)
Thymidine kinase	Herpes-into mouse cells	Nucleosome arrangement	Camerini-Otero and Zasloff (1980)
Thymidine kinase	Herpes-into mouse cells	5' flank hypersensitive to restriction nucleases	Sweet et al. (1982)
Vitellogenin	Chicken	Response of DH sites to hormone	Burch and Weintraub (1983)
Vitellogenin	*Xenopus*	DNase I sensitivity changes, but methylation does not in response to hormone	Folger et al. (1983)
SV-40		Nuclease-sensitive regions	Choder et al. (1984); Weiss et al. (1986)

[a]DH site, DNase I hypersensitive site; MNH, micrococcal nuclease hypersensitive site.

seem to be especially important features of active chromatin structure. A representative listing of other polymerase II genes is presented in Table 8-3. References to some of these studies will be made in the final section of this chapter, in which I shall attempt to summarize current ideas about the organization of chromatin in structural genes.

Hemoglobin Genes in Vertebrates

The relationship between chromatin structure and transcription of the vertebrate globin genes has been the object of intensive study. The system is a particularly attractive one, for these genes exhibit well-defined "switching" during development. In all higher organisms, both the α- and β-globin genes exist in multiple forms, expressed at different times in embryonic, fetal, and adult life. The variants of the α genes and of the β genes are each arrayed within a defined domain. Figure 8-9 shows some typical arrangements.

The globin genes, although apparently present in all tissues of each vertebrate, are expressed only in *erythroid* cell lines. The erythroid cells arise from a class of precursor "stem" cells, which differentiate to produce a number of different cell types. Concise descriptions of this process are given by Till (1982) and Nathan (1983). For a broader view, the reader should consult Volume 134 of *Progress in Clinical and Biological Research* (Stamatoyannopoulos and Nienhuis, eds., 1983).

The chick embryo has been a favorite object for studies of globin gene switching, for it provides a system in which various stages of development can be clearly identified and investigated. Background and concise descriptions of the phenomenon are given by Weintraub (1975) and by Stalder et al. (1980a, 1980b). In one-day-old chick embryos, no globin genes are expressed. Hemoglobin is first found at about 35 to 40 hours, following the development of a "primitive" line of erythroid cells. These express exclusively the embryonic genes (Fig. 8-9). At about six days into development, a second cell line (the "definitive" erythroid cells) arises. This lineage expresses adult-type genes of the "hatching" class (e.g., β^H, Fig. 8-9) and matures into transcriptionally inactive mature erythrocytes in about one week. Finally, at about 14 days of development a second line of definitive cells appears. These express only adult globin genes and also mature into inactive erythrocytes. It should be noted that each lineage derives independently from precursor cells; one type does not mature into another.

Changes in chromatin structure accompanying these developmental stages have now been well documented. The seminal study in this field is that of Weintraub and Groudine (1976), who showed that the globin genes have a nucleosomal structure, but that the sensitivity of that structure to DNase I changes during development. In subsequent studies, the general sensitivity to DNase I, the presence and locations of DNase I hypersensitive sites, and the level of DNA methylation have all been cor-

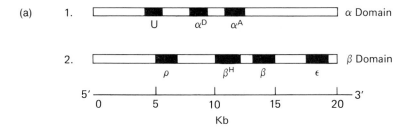

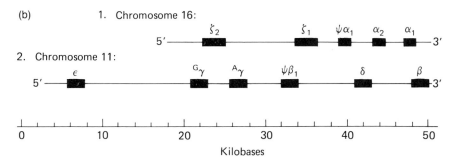

Fig. 8-9. Examples of globin gene arrangements in vertebrates. In all cases, transcription is left to right.

(a) Chicken
1. α-globin domain: The U gene is embryonic, the α^D and α^A genes are expressed only in adults.
2. β-globin domain: ρ and ε are embryonic genes, β^H is the "hatching" gene, and β the adult gene.

(b) Human
1. α-globin domain: ζ_2 and ζ_1 code for embronic α-chains, $\psi_{\alpha1}$ is a nonexpressed pseudogene, and α_2, α_1 are copies of the adult α chain.
2. β-globin domain: ε is embryonic, the two γ varieties are fetal, β is the adult gene (accompanied by the minor variant δ), and $\psi_{\beta1}$ is a pseudogene.

related with developmental stages. Results can be briefly summarized as follows:

1. *Day 1 (before the onset of globin synthesis):* All of the globin genes are insensitive to DNase I, and the DNA is highly methylated (Weintraub and Groudine, 1976).
2. *Days 2 through 6 (embryonic globin synthesis from the primitive cell line):* DNase I sensitivity is now observed over large domains surrounding the α and β gene clusters. Sensitivity is especially pronounced in the embryonic gene coding regions. In addition, there appear sites hypersensitive to DNase I. (See Stalder et al., 1980a, 1980b; Weintraub et al., 1981b.) Some of these sites also exhibit S1 sensitivity (Larsen and Weintraub, 1982). As Figure 8-10 shows, strong DNase I sensitive

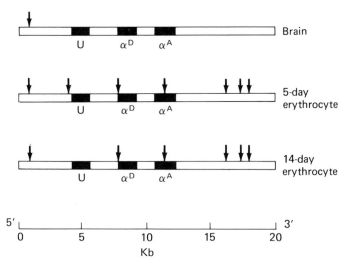

Fig. 8-10. Changes in DNase I hypersensitive sites during development in the chicken. The α-globin domain is shown, including the embryonic U gene and the adult α^D and α^A genes. In brain nuclei, only one site is detected. Six new sites appear in erythrocytes from 5-day embryos. In 14-day embryos, when only the adult genes are expressed, the site 5' to the U gene has disappeared. Adapted from Weintraub et al. (1981)

sites are present near the 5' termini of the embryonic genes, but also in the vicinity of the adult α^D gene. The latter sites do not reflect activity; these adult genes do not even initiate transcription in these cells.

3. *14 Days (synthesis of adult globins from the definitive cell line):* While DNase I sensitivity is maintained over the globin domains, the most sensitive regions are now found to reside in the neighborhood of the *adult* globin genes. Sites near the 5' ends of the embryonic genes, which were undermethylated in earlier stages, are now observed to be methylated. DNase I hypersensitive sites also shift; those immediately 5' to the embryonic genes have disappeared, and new sites are found in the 5' flanking regions of the adult genes (Stalder et al., 1980a, 1980b; Weintraub et al. (1981a, 1981b); Wood and Felsenfeld, 1982). Sensitivity to the single-strand specific nuclease, S1, is noted in the same regions (Larsen and Weintraub, 1982).

In more detailed studies, the DNase I "site" near the adult β gene has been shown to be a *region*, extending between about $-60 \rightarrow -260$ bp (Fig. 8-11). The DNA in this region is also accessible to restriction nucleases, and Msp I can be used to excise a 115 bp fragment ($-109 \rightarrow -224$). At least a portion of these fragments behave like protein-free DNA (McGhee et al., 1981). The importance of this 5'-flanking region is emphasized by the observations of Dierks et al. (1978) that the sequence

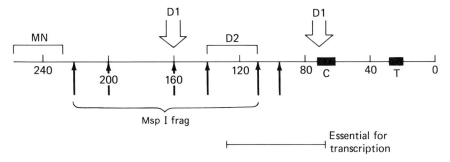

Fig. 8-11. The 5'-flanking region of the chicken adult β-globin gene. The region between about −60 and −260 bp is accessible to a number of nucleases. The brackets designated MN and D2 show areas especially sensitive to micrococcal nuclease and DNase II, respectively. DNase I sensitivity, while high over the whole region, is especially pronounced in the regions indicated by the broad arrows (D1). A number of restriction nucleases readily cleave, at points indicated by the thin arrows. The 115 bp fragment selectively removed by Msp I is indicated by the bracket. This overlaps a region shown to be essential for correct transcription. The conserved CCATT and TATA sequence are indicated by C and T, respectively. See the text for further details and references.

between −70 and −110 is essential for correct transcription of the adult β genes.

All of the evidence points to major changes in chromatin conformation in the 5'-flanking region accompanying adult gene activation. Emerson and Felsenfeld (1984) suggested that a protein factor or factors might be involved. In their experiments, plasmids carrying the $β^A$ gene and its 5'-flanking region were reconstituted with histones and tested for DNase I hypersensitivity. It was found that a factor from chick embryos was required to induce the hypersensitive site. This appears to be a 60 Kd protein. More precise mapping (Jackson and Felsenfeld, 1985) showed the presence of *two* protein-protected sites, one lying between −136 and −163, the other between −175 and −200. The latter contains a stretch of 16 G residues, which is also an S1-sensitive region (Nickol and Felsenfeld, 1983; Wang and Hogan, 1985). Finally, more recent studies by Plumb et al. (1986) reveal at least three proteins that were shown to bind in this 5'-flanking region. We shall see evidence for similar protein factors in other polymerase II systems and discuss their properties in more detail later.

It is of interest to attempt interpretation of these chromatin changes during erythropoiesis in the chicken in terms of the earlier discussion of hypothetical gene states. In this vocabulary, the globin genes in precursor cells are functional, but at this stage, incompetent. At some early time in the development of each of the erythroid cell lines, a first level of competence is conferred, via an event that somehow marks a very large region. Observationally, this change is characterized by the establishment of an intermediate level of DNase I sensitivity over the whole domain. Since this is observed in all erythroid cell lineages, it is tempting to suggest that this event occurs in stem cells from which the various cell types arise. Each lineage, however, is characterized by a further increase in the level of DNase I sensitivity in the *particular* genes to be transcribed and in their intervening and surrounding regions. In addition, particular DNase I and S1 hypersensitive sites emerge. Plumb et al. suggest that some of these steps may involve successive binding of protein factors. That these features reflect *competence*, rather than *activity*, is indicated by the fact that they are observed *before* the genes are actually expressed (Weintraub et al., 1981b). Furthermore, it has been shown that once these sites have been established, they can be propagated through many cell generations (Groudine and Weintraub, 1982).

Similar studies have been carried out using the globin genes from other vertebrates. Although the data are less complete, investigations of the nuclease sensitivities of human fetal and adult hemoglobin gene chromatin has led to a picture rather similar to that in the chick (Cockerill and Goodwin, 1983; Groudine et al., 1984; Shen, 1983; Zhu et al., 1984; Forrester et al., 1986).

A question raised by all of these studies is: What *structural* changes can account for the differences in nuclease sensitivities that occur during gene switching? Before attempting to answer this, it is worthwhile to consider some very interesting results that have emerged from studies of another globin system, the murine erythroleukemia cells. These cell lines, sometimes referred to as "Friend" cells, do not transcribe the globin genes at a significant level but can be induced to do so by the application of agents such as dimethyl sulfoxide (DMSO) or hexamethylene bisacetamide (HMBA). Stimulation of transcription is variously reported as ten- to thirtyfold (Sheffery et al., 1982; Salditt-Georgieff et al., 1984). Thus, they provide an excellent system for the study of the competent → active transition. Early experiments by Wallace et al. (1977) showed that the globin genes could be isolated in the "active" fraction (after DNase II digestion) from *either* induced or noninduced Friend cells. Interestingly, in another line, which had lost the capacity for induction, the globin genes were found in the "inactive" fraction. In contrast, Nose et al. (1981) found that the globin genes in both inducible and noninducible cells were DNase I sensitive. Clearly, it is necessary to take care in assessing "competence." The globin genes in uninducible cells would, by a functional definition, be incompetent, yet they have been marked in some manner during the history of these cells. It also seems likely, from the above results, that DNase II and DNase I may examine different features of chromatin structure.

A number of significant changes in the globin chromatin structure of erythroleukemia cells have been observed following administration of inducing agents:

1. An overall increase in DNase I sensitivity (Hofer et al., 1982; Shefferey et al., 1982, 1984; Smith et al., 1984a).
2. The induction of new DNase I hypersensitive sites in the vicinity of the genes (Hofer et al., 1982; Shefferey et al., 1982, 1984; Smith and Yu, 1984).
3. An increase in sensitivity to micrococcal nuclease (Smith and Yu, 1984; Yu and Smith, 1985).

The major advantage of the inducible erythroleukemia cells is that they allow following the time course of these changes after induction. Such studies have been carried out by Salditt-Georgieff et al. (1984), Smith and Yu (1984), and Yu and Smith (1985). It is found that the increase in DNase I sensitivity and the appearance of new hypersensitive sites appear between 24 and 36 hours after induction. At least one round of cell division is required. Increased sensitivity to *micrococcal* nuclease develops somewhat later, becoming fully expressed only after about 48 hours, by which time transcription of the globin genes is fully activated. The reagent dexamethasome, which inhibits differentiation, blocks *both* of these changes. On the other hand, the addition of imidazole, which inhibits globin synthesis but not differentiation, blocks the development of micrococcal nuclease sensitivity, but *not* the increase in sensitivity to DNase I. However, *ongoing* transcription does not appear to be responsible for the micrococcal nuclease sensitivity, for actinomyosin D does not abolish the effect.

The picture emerges, then, of several successive stages in the activation of the globin genes in erythroleukemia cells. First, there is a low level of DNase I sensitivity, which distinguishes these cells from nonerythroid cells. Upon stimulation, DNase I sensitivity increases, and specific hypersensitive sites are generated. As with the chicken globin genes, some of these sites are also S1 sensitive (Sheffery et al., 1984). Following these steps, and apparently accompanying the onset of transcription itself, there is a further structural change that confers increased sensitivity to micrococcal nuclease.

Interestingly, there appear to be *no* changes in DNA methylation during this gene activation process (Sheffery et al., 1982). This may signify that any role played by demethylation occurs very early in the series of steps leading to expression. It would be of the greatest interest to compare the methylation state in uninducible mutants of the Friend cells.

There have been a number of attempts to discern the chromatin structural features that might account for different levels of nuclease sensitivity within the globin domains. Electric dichroism studies of the chicken adult β-globin gene were taken to indicate that this chromatin remained in a solenoidal structure (McGhee et al., 1983b). However, more recent investigations have indicated differences between expressed and nonexpressed genes in erythroid cells. Kimura et al. (1983) have used hybridization techniques to probe the sucrose gradient centrifugation profiles for various genomic regions. In each case, the sedimentation coefficient range observed was compared with that expected for the probed fragments on the basis of DNA fragment size; thus, differences in digestibility were presumably corrected for. The data show that the β-globin genes, in contrast to nonexpressed genes, sediment more slowly, as would be expected for unfolded chromatin. Support for this idea comes from observations on murine erythroleukemic cells by Smith et al. (1984b). They find that the difference in DNase I sensitivity observed between the transcriptionally active adult globin genes and the inactive embryonic genes is abolished at low ionic strength. *All* regions become sensitive under these conditions where the higher-order structure has been relaxed (see Chapter 7). Furthermore, increasing the ionic strength restores the differential sensitivity. Removal of H1 has a similar effect to that of low ionic strength. All of these experiments strongly suggest that *one* feature of gene activation may be a relaxation of higher-order structure. Possible causes for such a relaxation and its relation to other aspects of chromatin structure will be discussed in a later section.

There is now evidence concerning the arrangement of nucleosomes in the β-globin gene neighborhood. Benezra et al. (1986) have used MPE-Fe(II) cleavage to compare nucleosome structure in mouse L cells and erythroleukemia cells. In the former, they find a positioned array from −3000 bp to +1500 bp, relative to the start of the β^A gene. In erythroleukemia cells, regular positioning cannot be observed between −200 to +500. Since the same result is obtained with either induced or uninduced cells, the disruption of structure must precede induction. Sun et al. (1986) describe nuclease digestion studies that indicate unusual nucleosome structure in chicken β^A-globin genes. An anomalous "ladder," suggesting modified nucleosome structure is found in the neighborhood of the transcribed region.

Heat-Shock Genes

Many cells and organisms respond to elevated temperatures by major modifications in transcription and protein synthesis. Transcription of normally active genes decreases, and a special class of "heat-shock" genes is activated. Such a controllable "switching on" of a particular set of genes provides an excellent model system for the investigation of chromatin changes in gene activation. The phenomenon has been studied in considerable detail in *Drosophila* (see Ashburner and Bonner, 1979, for a succinct review of the background to these studies). The heat-shock response can be observed at the cytological level in *Drosophila* polytene chromosomes. Nine new "puff sites" are observed when the temperature is raised from the normal value of about 25°C to 35–40°C. These occur at loci 33B, 63BC, 64F, 67B, 70A, 87A, 87C, 93D, and 95D. Protein products from a number of these loci have been identified. They are characterized by their molecular weights in kilodaltons: *hsp*82 (formerly called *hsp*83) from 63BC; *hsp22*, *hsp23*, *hsp26*, and *hsp28* from 67B; *hsp70* (several genes) from 87A, 87C; and *hsp68* from 95D. The function of these heat-shock proteins is unknown. There is one report that H2B synthesis is specifically elevated in *Drosophila* heat shock (Sanders, 1981), but H2B is clearly not one of the protein products listed above. A study by Levinger and Varshavsky (1982) indicates that most of the heat-shock proteins are nuclear but not associated with nucleosomes. They seem to be proteins that are released from the chromatin only at high ionic strength.

Many of the recent studies have centered on the 87A and 87C loci, which code for a total of five copies of the 70,000 dalton *hsp70* protein (see Mirault et al., 1979, for a map). In pioneering experiments, Wu et al. (1979a, 1979b) showed that DNase I sensitivity of the *hsp70* (and *hsp83*) chromatin increased markedly on heat-shock activation. Even more interesting was the fact that the sharp nucleosomal pattern displayed by the uninduced genes upon micrococcal nuclease digestion was smeared to the point of invisibility after about 30 minutes of heat shock. After 3 hours recovery from heat shock, the "normal" pattern was regained. In complementary studies, Levy and Noll (1980, 1981) examined the micrococcal nuclease digestion in considerable detail. They noted that in the uninduced state, a region somewhat larger than a structural gene showed *greater* resistance to micrococcal nuclease than bulk *Drosophila* chromatin. Within the coding region, the nucleosome arrangement was quite regular, perhaps corresponding to a small number of discrete sets of positionings. Yet when the genes were induced by heat shock, the micrococcal digestion pattern became blurred, as observed by Wu et al. Levy and Noll suggest that before heat shock the repressed genes may be compacted into a higher-order structure, which is then relaxed, and the regular nucleosomal arrangement disrupted.

In parallel studies, DNase I hypersensitive sites were observed near the 5' ends of *hsp70* and *hsp82* genes (Wu, 1980) and the four *hsp* genes at the 67B locus (Keene et al., 1981). The latter workers note that these sites are evident even in the uninduced genes and suggest that they may represent a necessary condition for transcription. Cartwright and Elgin (1982b) also report enhanced sensitivity to micrococcal nuclease and the nonenzymatic reagent 1-10-phenanthroline·Cu in the same regions.

Using the indirect end-labeling technique, Wu (1982) has mapped these "hypersensitive" sites in noninduced cells with some precision. As in other cases, they turn out to be *regions,* rather than specific sites (see Fig. 8-12). The hypersensitive regions, which extend from -215 to -38 and -8 to $+100$, with a maximum sensitivity at -93, bear interesting relationship to other known features in the 5'-flanking regions of the heat-shock genes. The "TATA box," for example, lies in a relatively resistant region between -8 and -38 bp (Fig. 8-12). Using deletion mutations, Pelham (1982) has demonstrated that the region from -47 to -66 bp is necessary for heat-shock induction of *hsp70*. The reasons for DNase I

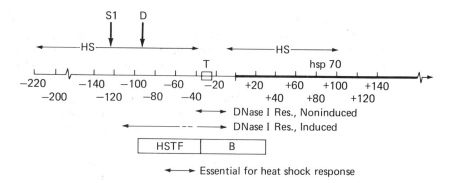

Fig. 8-12. Fine structure near the origin of transcription of the *hsp*70 gene. Code: HS, DNase I hypersensitive region in noninduced cells; D, most sensitive site; S1, nuclease S1 cutting site; T, TATA box; HSTF, B, heat-shock transcription factor and B-protein binding site, respectively. For appropriate references, see the text.

hypersensitivity are unclear. Costlow et al. (1985) have shown that deletion for a significant distance (~20 bp) *past* the very hypersensitive region at −93 does not destroy the site; it still exists even on "foreign" DNA. Thus, it cannot be a consequence of the DNA sequence at this point, but must reflect an effect of downstream sequences, starting to the right of −74. Deletions extending to −43 or −23 *did* destroy the very sensitive site at −93. The DNase I hypersensitive region also contains potential S1-sensitive sites. Using supercoiled plasmids containing *hsp*70 flanking sequences, Mace et al. (1983) demonstrated torsion-dependent S1 sensitivity at −124. They suggest that this may arise from "slippage structures" involving short direct repeats in this region. S1-sensitive sites have also been mapped near the *hsp*82 genes by Han et al. (1985b).

There has for some years existed indirect evidence that some specific "factors" might be involved in stimulation of the heat-shock response (see, for example, Compton and Bonner, 1977). More recent studies indicate that these factors are proteins and that they bind within the 5'-flanking regions of the heat-shock genes. Craine and Kornberg (1981) showed that a factor which could activate *Drosophila* heat-shock genes in vitro was heat-labile and protease-sensitive (see also Jack et al., 1981). More recently, Parker and Topol (1984a, 1984b) and Wu (1984a, 1984b) have obtained more direct evidence for protein binding. Although the techniques employed are quite different, with Parker and Topol using footprinting and Wu a new exonuclease-protection method, the results are generally compatible and can be summarized as follows: Two kinds of protein factors have been identified. One, which binds to the region covering the TATA box, and into the 5' ends of the genes themselves

(Fig. 8-12), appears to be similar to, but distinguishable from, proteins that bind to this region in other *Drosophila* genes (Parker and Topol, 1984a). In addition, there appear to be specific proteins that bind to the regions upstream from the TATA box to activate the heat-shock genes. Thus, at least one factor in the gene switching that occurs in the heat-shock response may involve activation and deactivation of particular DNA binding proteins. (It is argued that protein synthesis and degradation are not likely involved, since the response is so rapid.) The binding of these proteins may in turn perturb the chromatin structure in the genes and their 5'-flanking regions.

That such perturbation occurs is suggested not only by the several references cited above, but also by the studies of Karpov et al. (1984). Using a protein–DNA crosslinking technique, these researchers have followed changes in histone interactions with coding-region DNA following heat shock. They observe first the loss of H1 and, with increased transcription, a diminution in core histone content as well. Such changes could account for the progressive "smearing" of the micrococcal nuclease digestion pattern described above.

Studies of heat-shock genes may also provide information concerning boundaries of functional domains in chromatin. In an extended study of the regions surrounding the *hsp*70 genes, Udvardy et al. (1985) have discovered peculiar chromatin structures many kilobases downstream from each of these divergently transcribed genes. In each case, a 350 bp nuclease-resistant region is flanked on both sides by unusually spaced micrococcal nuclease-sensitive sites. Udvardy et al. suggest that these specialized chromatin structures (SCS) may serve as domain boundaries and play some role in condensation–decondensation processes.

Although much of the work on heat-shock genes has utilized the *Drosophila* system, there is also a very nice study of yeast heat-shock genes that adds to the above picture. Szent-Gyorgyi et al. (1987) have shown that while sequences bordering these genes have precisely positioned nucleosomes, the structure within the transcribed and 5'-flanking regions is more complex. In fact, they find evidence for a half-nucleosome ladder in a portion of the transcribed region.

In summary: Although the function of heat-shock proteins remains unknown, the chromatin structure of these genes has become one of the most thoroughly investigated of all polymerase II systems. In many respects, the changes observed during heat-shock activation resemble those seen in the developmental activation of the globin genes.

The Ovalbumin Gene Cluster

The chicken ovalbumin gene cluster is a most promising system for the study of gene activation. These genes, which include the ovalbumin gene itself and two related genes (X and Y; see Fig. 8-13), are tandemly arranged

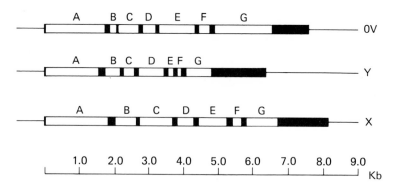

Fig. 8-13. Structure of the X, Y, and ovalbumin (OV) genes in the ovalbumin cluster in chickens. Exons are dark, and the corresponding intron regions are designated by letters. These genes are ordered in the arrangement X·Y·OV in the domain (Fig. 8-14) and transcribed left to right. (Reprinted with permission from Biochemistry *19*, p. 5586, Colbert et al. Copyright 1980 American Chemical Society.)

in a ~100 Kbp domain. The X and Y genes exhibit some homology to the ovalbumin gene and have, in fact, been termed *pseudogenes* (see Colbert et al., 1980, for a description of these genes and references to earlier work). The function of the X and Y gene products is unknown. Although present in all chicken tissues examined, the ovalbumin cluster is expressed only in the tubular cells of the oviduct and then only in response to steroid hormone stimulation. Transcription of the stimulated genes is by no means equal; ratios of OV:Y:X of roughly 100:10:1 are reported (Colbert et al., 1980). Following withdrawal of hormone, transcription of these genes ceases.

Some of the earliest attempts to investigate the structure of transcriptionally active chromatin utilized the ovalbumin gene. Garel et al. (1977) showed that the ovalbumin gene, in oviduct nuclei, was more sensitive to DNase I than bulk chromatin. Garel and Axel (1977) and Bellard et al. (1977) demonstrated that the ovalbumin gene had at least partially nucleosomal structure, even under conditions where transcription was occurring; however, the active genes were found to be more sensitive to micrococcal nuclease than the inactive globin genes in the same nuclei. Bloom and Anderson (1979) showed that this micrococcal nuclease sensitivity decreased upon hormone withdrawal, in parallel with the decrease in transcription.

A more detailed picture of the nuclease sensitivity has emerged from a series of studies in the laboratory of B.W. O'Malley. Lawson et al. (1980, 1982) have shown that the DNase I sensitivity extends over a very large domain, which includes the X, Y, and ovalbumin genes and their intervening and flanking sequences (Fig. 8-14) (see also Bellard et al., 1980, and Stumph et al., 1982). This sensitivity was not observed in a number of cell lines in which these genes are inactive. Upon withdrawal of hormone, the region remained DNase I sensitive, in contrast to the loss of micrococcal nuclease sensitivity.

This distinction between responses to DNase I and micrococcal nuclease mimics that observed in other polymerase II systems (see above). Preferential DNase I sensitivity appears to reflect a *competent* state, which can exist before, and persist after, the actual events of transcription. Micrococcal nuclease, on the other hand, seems to respond to a more dramatic change in chromatin structure that occurs when genes are actually prepared for transcription. It has been shown in other examples (see above) that the micrococcal nuclease ladder itself becomes smeared under these circumstances. Just this effect has been observed in the ovalbumin gene by Bloom and Anderson (1982). They note, however, that *ongoing* tran-

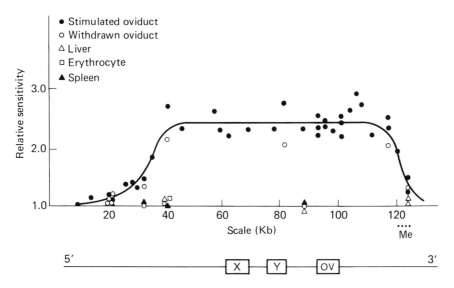

Fig. 8-14. Features of the domain containing the ovalbumin gene cluster. The graph at the top depicts relative DNase I sensitivity in nuclei from hen oviduct (●), estrogen-withdrawn chick oviduct (○), hen liver (△), chick spleen (▲), and hen erythrocytes (□). In the schematic drawing below, the X, Y, and OV genes are shown, and the region where DNA methylation is tissue specific (dotted line). (Adapted from Stumph et al., 1982, and Lawson et al., 1982. Stumph et al. from Gene Regulation, pp. 87–104, O'Malley and Fox, eds. Copyright 1982 Academic Press.)

scription may not be necessary for this chromatin state. Parallel studies by Bellard et al. (1982) utilized the indirect end-labeling technique to examine nucleosome structure in the 5'-flanking region and first (leader) exon of the ovalbumin gene. In inactive cells (erythrocytes), the region displayed a normal nucleosome pattern, with nucleosomes randomly placed. This pattern is disrupted in active oviduct tissue. However, new, specific sites for micrococcal nuclease cleavage are now found.

All of these results must be considered in the light of the most interesting experiments of Trendelenburg et al. (1980). These researchers inserted an 11.5 Kbp fragment, including the ovalbumin gene, into pBR322 and microinjected this plasmid into *Xenopus* oocyte nuclei. Most of the resulting plasmids were transcriptionally inactive and exhibited a nucleosomal structure with a compaction ratio of about 2:1. However, a small fraction showed active transcription, with closely packed polymerase molecules and a "Christmas tree" structure of transcripts. These showed a compaction ratio of only 1.1 to 1.2:1 in the transcribed regions. If it is accepted that these represent valid ovalbumin transcription, the transcribed region must either be nucleosome-free or contain wholly unfolded nucleosomes. Unfortunately, the small number of active plasmids makes critical evaluation difficult.

In further studies from O'Malley's laboratory, Anderson et al. (1983) have compared the micrococcal nuclease sensitivity of the very actively transcribed ovalbumin gene with that of the more sluggish X and Y genes. The latter, like transcriptionally inert sequences in the same domain, did not show enhanced susceptibility to this enzyme. In further experiments, sensitivity to an endogenous nuclease was investigated. This nuclease preferentially attacked the whole ovalbumin coding sequence and demonstrated "hypersensitive" sites in the 5'-flanking region. (Unlike many other genes, the ovalbumin gene has *not* been shown to exhibit DNase I hypersensitive sites near its 5' end.) These endogenous nuclease-sensitive sites persist after hormone withdrawal.

How do hormones stimulate the expression of the ovalbumin gene in competent oviduct cells? Chambon et al. (1984) present evidence for the existence of a repressor, associated with sequences upstream from the gene (see also Kaye et al., 1986). Conceivably, steroid hormone receptors interact with this repressor or with adjacent DNA regions to relieve the repression. This model raises the interesting prospect that the oviduct tubular cells are made competent for ovalbumin gene cluster expression by some event in cellular differentiation, but are then specifically repressed until hormone is present. Such a model could also easily accommodate the simultaneous but differential activation of the X and Y genes, and the loss of transcription following hormone withdrawal.

The behavior of the ovalbumin gene cluster suggests a tentative model for stepwise gene activation. The very large size of the DNase I sensitive region (~100 Kb) is strongly reminiscent of the "structural domains" discussed in Chapter 7. If so, it might be expected that the competent domain would be associated with nuclear matrix elements. This is exactly the result obtained by Robinson et al. (1982). Furthermore, they do not observe such association in chicken liver cells, nor do the globin genes behave this way in oviduct cells. In this connection, it may be of interest that members of a repetitive family of DNA sequences seem to be located near the ends of the DNase I sensitive domain. Stumph et al. (1982) suggest that they may mark the boundaries of the domain. If so, they might also be associated with matrix attachment.

In terms of such a "domain" model (Fig. 8-15), segregation of a group of genes during cell differentiation could select them for *potential* expression. Transcription itself would occur only through the intervention of cell-specific factors (hormone receptors, other proteins) that could in turn exhibit a higher level of selectivity on different genes within a given domain. In its broadest form, such a model is consistent with each of the three systems described in this section and with many other examples as well (see also Lawson et al., 1982).

The same recurring themes that have been noted in the specific examples described above are found to repeat in the chromatin structure of many other structural genes. As examination of papers listed in Table 8-3 shows, DNase I sensitivity and the existence of hypersensitive sites (often, but not always, upstream from transcription starts) are common features of

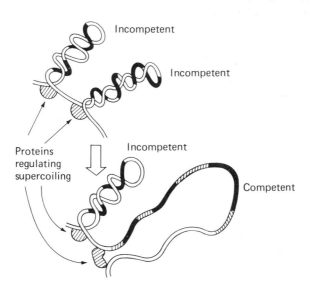

Fig. 8-15. A domain model for gene activation. Groups of genes that are to be expressed in a given cell line are segregated into individual loop domains. A primary act in differentiation is postulated to be some modulation of the chromatin structure of a loop that confers transcriptional competence on particular domains.

competent or active genes. In addition, very active genes seem in many cases to have lost their "regular" nucleosomal arrangement. To the extent that such observations are becoming common, we may claim to know *something* about active chromatin structure.

But in a certain sense, all of this is merely symptomatic. To take a specific example, one might ask: What are the *causes* of the observed differences in nuclease sensitivity between the ovalbumin gene chromatin in a hormonally stimulated oviduct cell, an unstimulated oviduct cell, or a liver cell? The underlying DNA sequence is the same, with one exception: The differentiated tubular oviduct cell DNA is hypomethylated at a cluster of sites near the 3' end of the domain (Stumph et al., 1982). Can this be the explanation for the complex behavior described above?

A Role for DNA Methylation?

Methylation of cytidine residues, especially in CpG sequences, is common in eukaryotes. This kind of modification, the only one known to exist at the DNA level in these organisms, has attracted much interest and is the subject of recent reviews (Felsenfeld and McGhee, 1982; Doerfler, 1983) and several books (Razin et al., 1984; Trautner, 1984; Taylor, 1984). The chemistry and distribution of DNA methylation has been briefly summarized in Chapter 3, *Composition and Modification of Eukaryotic DNA.* Of interest at this point is its potential importance in gene regulation. The fact that mechanisms exist for perpetuating DNA methylation patterns through somatic cell divisions makes such modification an attractive candidate for differentiation-dependent "marking" of particular sites in the genome.

The general observation is that transcriptionally active (or competent) regions seem to be undermethylated in comparison to their inactive counterparts in other cells. Differences are easy to detect, for the sequence CCGG will be cleaved by restriction endonuclease Msp I, but not by Hpa II if the internal C is methylated. Using this technique, together with DNase I treatment and nick-translation to mark active genes, Naveh-Many and Cedar (1981) found that about 70% of all such sequences were methylated in bulk animal DNA, but only about 30 to 40% in DNase I sensitive regions. Weisbrod (1982) noted that individual nucleosomes from active genes were hypomethylated (see also Hatayama et al., 1984). Many studies of individual genes have demonstrated the existence of particular hypomethylated sites in either competent or active genes (see the preceding section and Table 8-3 for examples). A particularly impressive study is that of Keshet et al. (1986). Insertion of unmethylated and highly methylated DNA sequences into mouse L-cells gave very different results. The former gave DNase I sensitivity, the latter were insensitive over the hypermethylated region.

One difficulty in interpreting the many observations of specific hypomethylated sites is that no general *pattern* emerges. Such sites are found sometimes within genes, sometimes in flanking regions, and sometimes far removed from the transcribed genes themselves. For example, the ovalbumin gene cluster in tubular oviduct cells exhibits a group of hypomethylated sites near the 3' boundary of the large, DNase I sensitive domain many kilobases from the genes themselves (Stumph et al., 1982). In contrast, the active human dihydrofolate reductase gene is hypomethylated in the 5'-flanking region, whereas the phosphoenolpyruvate decarboxylase gene in rats shows tissue-dependent hypomethylation in both the 5' flank and within the gene (see Table 8-3 for references).

Perhaps most disturbing are the observations that some organisms, in which gene regulation clearly occurs, exhibit either undetectable or *exceedingly* low levels of methylation. *No* DNA methylation has been detected in yeast (Proffitt et al., 1984), and *Drosophila* DNA contains only about one methylated cytosine in 12,500 nucleotides (Achwal et al., 1984). It is difficult to see how a theory that accorded to hypomethylation a major role in gene regulation could accommodate these cases. Of course, they may be exceptions, but it would seem strange that a general, fundamental mechanism could admit such exceptions.

Even if changes in DNA methylation play a role in gene regulation at some level, this surely cannot be the whole story. There are a number of examples in which dramatic changes in transcriptional activity occur without accompanying changes in methylation. (See, for example, the study by Sheffery et al., 1982, on the activation of globin transcription in erythroleukemia cells.) In consequence, it is necessary to examine other aspects of chromatin structure to seek the sources of transcriptional activation and the accompanying appearance of nuclease sensitivity. A natural place to look is in the protein complement of chromatin, which has been shown (Chapters 4 and 5) to exhibit enormous variability in composition and modification.

Chromosomal Proteins and Gene Activity

At the present time, our knowledge concerning the specific *composition* of active genes can best be described as chaotic. The problem, of course, is that such information depends, in the end, on efficient fractionation procedures, and these do not as yet exist. Therefore, many attempts at studying this problem have been essentially *correlative*. The dangers inherent in such an approach are obvious, and they are well illustrated in Chapter 4, where it has been noted that virtually every histone modification has been claimed by *somebody* to be associated with transcriptionally active chromatin. Similar roles have been ascribed to a number of nonhistone proteins (Chapter 5). In what follows, I shall attempt to summarize the arguments, pro and con, for a "special" protein composition in active chromatin.

Histone Stoichiometry

It has been reported by Baer and Rhodes (1983) that a complex between calf thymus polymerase II and core particles contains only half the normal complement of H2A and H2B. Either the 15% of calf thymus core particles that bind polymerase II are already so deficient, or the H2A and H2B are displaced upon binding. The fact that even a large excess of polymerase does not increase binding supports the former explanation but does not exclude the possibility that in a subset of the nucleosomes H2A and H2B are especially labile. If H2A·H2B depletion represents the in vivo state of nucleosomes that can interact with polymerase II, one would expect to find a deficit in H2A and H2B in "active" chromatin fractions. The evidence is ambiguous. Hutcheson et al. (1980) report one subfraction of mononucleosomes from trout testes, enriched for active genes, which contains only half the normal complement of H2A and H2B, in agreement with these expectations. On the other hand, Gabrielli et al. (1981) found *no* departure from normal histone stoichiometry in a transcriptionally enriched fraction of mouse cell chromatin (see also Faber et al., 1981). The problem of assessing the significance of such data lies, of course, in the inefficiency of fractionation. If only a subfraction of that small portion of chromatin that is transcriptionally active is so modified, its detection will be very difficult and may well depend upon some special feature of the fractionation technique. Conversely, fractionation methods that depend upon unknown principles may always be suspected of artifact.

Finally, it could be that loss of H2A·H2B is a dynamic process and that the polymerase II column of Baer and Rhodes has selected for that set of nucleosomes *capable* of this interchange. In this connection, it should be noted that at least two groups have reported evidence for exchange of histones H2A and H2B in chromatin (Louters and Chalkley, 1985; Schwager et al., 1985). Such a dynamic exchange would be entirely consistent with the idea that during active transcription these histones are *transiently* released and that it is only to such nucleosomes that the polymerase can attach.

Although these results are suggestive, it seems to me that the question as to whether the very interesting observations of Baer and Rhodes are germane to transcription in vivo must be left open, until more conclusive compositional evidence has been accumulated.

Claims that transcriptionally active chromatin is depleted in H1 have been much more frequent. (See, for example, Sarkander and Dulce, 1979; Davie and Candido, 1980; Davie and Saunders, 1981; Gabrielli et al., 1981; Czupryn and Toczko, 1985.) Again, caution is required in interpretation. It should be noted that a number of these studies have utilized quite extensive digestion by DNase I or micrococcal nuclease, treatments that preferentially attack linker regions. Furthermore, the well-documented propensity of H1 to redistribute at moderate ionic strengths leaves open the possibility of artifact. (See Chapter 7.) Indeed, there are experiments that can be interpreted to support the presence of H1 in active regions. For examples: Using a photoaffinity labeling method, Nielsen (1981) found H1 *preferentially* labeled in nuclease-sensitive regions, and Berent and Sevall (1984) report a strong H1 binding site at the 5' end of the rat albumin gene.

Whether or not it is ultimately decided that transcriptionally active chromatin is depleted in H1, there are other indications that the linkers in such regions may be unusual. For example, Smith et al. (1983) report that two actively transcribed genes (β-globin and IgM) exhibit distinctly longer chromatin repeats than do inactive genes in the same cells. (But see Berkowitz and Riggs, 1981 for an opposing example.) Weintraub (1984) has examined large chromatin fragments released from nuclei digested lightly with micrococcal nuclease. He finds that these "particles" from inactive regions remain intact, even though periodically cleaved by the enzyme. "Active" chromatin does not behave in this way, even though these experiments indicate that it still contains H1. Weintraub argues that the difference lies not in the presence or absence of H1, but in the way in which this protein interacts with adjacent nucleosomes (see also Weintraub, 1985). Such a model is consistent with the numerous suggestions that transcriptionally active or competent chromatin may exhibit a more "relaxed" higher-order structure. In this view, it is the role of H1 to help maintain the condensed structure of inactive chromatin. The experiments of Goodwin et al. (1985), who found that removal of H1 did not produce major changes in DNase I sensitive regions, are not inconsistent with this proposal. However, more recent work by Caplan et al. (1987) demonstrates that a chromatin fragment containing the chicken β^A globin gene sediments more slowly than comparable bulk chromatin fragments, and that addition of H1 + H5 does not remove this difference. From these studies, Caplan et al. conclude that active (or competent?) genes have a different, perhaps disjointed higher order structure because of the presence of unusually long linker regions.

If any such model is to be considered, the question that naturally arises is: What makes the difference? Does the relaxation of interactions derive from modification of the histones, from the presence of specific nonhistone proteins, or some other cause?

The same question can be addressed to those who contend that "active" nucleosomes are deficient in H2A and H2B, or are unfolded (see above). Thus, it is important to consider the data that correlate histone modification and/or nonhistone protein content with transcriptional activity.

Histone Modification

There is, in fact, a considerable body of experimental evidence correlating transcriptional activity with certain kinds of histone posttranslational modification. Primary among these is *acetylation,* which has been described in some detail in Chapter 4. In a very early paper, Allfrey et al.

(1964) suggested that histone acetylation might promote transcription. In more recent years, a number of laboratories have produced evidence to support this contention. Many of the experiments carried out prior to 1982 can be categorized into two groups: those that demonstrate enrichment of highly acetylated histones in "active" chromatin fractions (i.e., Davie and Candido, 1978; Levy-Wilson et al., 1979; Perry and Chalkley, 1981) and those that correlate increased sensitivity to nucleases with hyperacetylation (i.e., Nelson et al., 1978; Sealy and Chalkley, 1978b; Davie and Candido, 1980; Perry and Chalkley, 1981). This earlier work has been reviewed by Doenecke and Gallwitz (1982). Although such studies make a strong circumstantial case, they are all, in my opinion, to some extent compromised by the uncertainties of chromatin fractionation procedures and by our limited understanding of the underlying causes of nuclease sensitivity. There have, in fact, been a few contrary claims (see, for example, Loidl et al., 1983; Yukioka et al., 1983). Unfortunately, these papers do not provide any convincing explanations as to why the earlier studies might have been in error.

More recent experiments, approaching the problem from somewhat different directions, provide additional insight. For example, Vavra et al. (1982) have compared histone acetylation in the transcriptionally active macronucleus and inactive micronucleus of *Tetrahymena*. The former shows extensive modification, the latter little or none. Leiter et al. (1984) observe that butyrate treatment (see Chapter 4) increases hyperacetylation in both inducible and noninducible strains of Friend cells. Upon induction, a further increase occurs. From these data, they argue that histone hyperacetylation *alone* is not sufficient for transcribability but that it may act in concert with other factors. A somewhat similar note is sounded in the report of Malik et al. (1984). Utilizing an immuno-affinity column to isolate oligonucleosomes containing the nonhistone protein HMG 17, they find that over 90% of the histone acetyl groups are associated with this fraction. This suggests a local correlation of hyperacetylation with HMG 17, a protein that has been believed by some to be associated with actively transcribed regions (see below). Using *Physarum*, in which accurate cell cycle synchrony can be induced, Waterborg and Matthews (1983) provide evidence for two overlapping patterns of histone acetylation: one that they associate with DNA replication, the other with transcription (see Chapter 9).

It has long been believed that hyperacetylation, by weakening electrostatic interactions between the histones and DNA, might play a role in "opening" chromatin structure for transcription. Indeed, Bode et al. (1983) have reported electrophoretic evidence for unfolding of hyperacetylated core particles, and Bertrand et al. (1984) show EM results in support of this hypothesis. However, careful hydrodynamic studies (Simpson, 1978a; Ansio and van Holde, 1986) show no evidence for unfolding, nor do the neutron scattering experiments of Imai et al. (1986). Some of these studies *do* report, however, changes in thermal denaturation, DNase sensitivity, and CD accompanying hyperacetylation.

At the present, it appears more likely that hyperacetylation modifies higher-order structure in chromatin (see Vidali et al., 1978; Simpson, 1978a; Allan et al., 1982a). This could account for its association with transcription, if active genes must be relaxed. Supporting evidence of a kind comes from a number of studies of cells that are very *inactive* in transcription: maturing spermatids in which histones are being replaced by protamines. Analyses of spermatogenesis in rainbow trout (Christensen et al., 1982, 1984), roosters (Oliva and Mezquita, 1982), and rats (Grimes and Henderson, 1984a) all yield the same result: Histone acetylation is maximal at a stage where spermatids have virtually ceased transcription and are remodeling their chromatin.

Clearly, there is no necessary contradiction between these observations and the notion that acetylation may also loosen chromatin structure for transcription. The process occurring in developing spermatids is a massive one, involving all or most of the genome, whereas changes accompanying transcription should be local and perhaps transitory. Indeed, Perry and Chalkley (1982) have proposed that "waves" of acetylation pass through the interphase nucleus, presumably accompanied by local relaxation. Such a model is attractive, for it provides a role for histone acetylation commensurate with many other studies, yet requires that this modification be only *one* of a number of factors required for transcription.

It has been suggested from time to time that certain other kinds of histone modification (i.e., ubiquitination and ADP-ribosylation) or certain histone variants may be associated with gene activity. However, the evidence is still scanty in these cases. Such proposals have been discussed in Chapter 4 and will not be further considered here.

Nonhistone Chromosomal Proteins

There must be many nonhistone chromosomal proteins that play quite specific roles in transcription. Obvious candidates are the polymerases themselves, "transcription factors," and hormone receptors. These are discussed in Chapter 5, along with a listing of many other minor chromosomal proteins (see Table 5-2). As pointed out earlier in this chapter, evidence is mounting for the existence of many specific protein factors that bind in 5'-flanking regions and may be involved in the regulation of the transcription of particular genes (see Dynan and Tijan, 1985, for a recent review). With the exception of the polymerases, all of these are specialized proteins, present in only small quantities, and often showing cell specificity. Are there any nonhistone proteins that play a *general* role in the transcription process?

The primary candidates are the high mobility group (HMG) proteins 14 and 17. Aspects of the primary and secondary structures of these proteins have been described in Chapter 5, and their interaction with nucleosomes in Chapter 6. These are among the most abundant of the nonhistone proteins, amounting in some cells to several percent of the histone mass, or roughly one molecule for every ten histone cores (Goodwin and Mathew, 1982; Seale et al., 1983). Thus, they are at least *potential* candidates for proteins present in stoichiometric amounts in transcriptionally active nucleosomes. As has been pointed out in Chapter 5, there is some evidence that HMGs may be even more abundant than this estimate.

In 1979–1981, a number of papers appeared indicating that "active" chromatin fractions were enriched in HMG 14 and 17. This work has been reviewed in detail by Goodwin and Mathew (1982) and will not be reiterated here. It should be emphasized, however, that the enrichments observed were not striking and that HMG 14 and 17 have been also detected in definitely nonactive chromatin (i.e., satellite chromatin; see Goodwin and Mathew, 1982). As Goodwin and Mathew point out, it is difficult to evaluate these experiments, considering the generally low efficiency of chromatin fractionation and the possibility that the nonhistone proteins may redistribute during the fractionation procedures. However, it should be noted that Gabrielli et al. (1981) report that HMG 14 was found *only* in a transcriptionally enriched fraction of mouse cell chromatin. HMG 17 was not so uniquely distributed.

These results would probably not have attracted major attention had they not been accompanied by some rather striking experiments by Weisbrod and Weintraub (1979; see also Weisbrod et al., 1980). These investigators found that extraction of embryonic chicken erythrocyte chromatin with 0.35 *M* NaCl (which extracts, along with other proteins, HMG 14 and 17) led to a loss in the preferential DNase I sensitivity of the globin genes. More remarkably, reconstitution with the extract or with partially purified HMG proteins restored the sensitivity. Essentially equivalent results were obtained independently by Gazit et al. (1980). In subsequent experiments, Weisbrod and Weintraub (1981) prepared chromatographic columns to which the HMG proteins were bound and used these to fractionate nucleosomes. They reported that the nucleosomes binding to such columns were enriched in sequences corresponding to the coding regions of active genes. Such nucleosomes were found to be remarkably similar to "bulk" nucleosomes in DNA size and histone composition.

As has been pointed out in Chapter 6, *all* nucleosomes can bind HMG 14 and 17, so that the column must be selecting for a particular class with exceptionally strong binding sites. In further experiments, Weisbrod (1982) characterized in greater detail the nucleosomes retained on the HMG column. He found the particles to contain undermethylated DNA and a higher level of histone acetylation, as compared to bulk nucleosomes. However, less than one hyperacetylated histone per bound nucleosome was detected, so it was concluded that hyperacetylation alone could not account for the selection of these particles. A relevant observation is that of Malik et al. (1983), who passed *oligo* nucleosomes over an immunoaffinity column carrying antibodies to HMG 17. They found that virtually all of the radiolabeled acetate was retained on this column, suggesting that even if hyperacetylation and strong HMG binding did not always occur on the *same* nucleosome, they must occur in closely adjacent particles.

All of these results led to a strong conviction that these HMG proteins were somehow intimately connected with chromatin activation. The question that naturally arises is: Precisely what do they *do* to nucleosome or chromatin structure? Effects seem hard to find. Insofar as the core particle is concerned, the results of many experiments described in Chapter 6 seem to indicate that, if anything, binding of HMG 14 or 17 *stabilizes* the particle! This is hardly the result expected. It can, of course, be argued that experiments on bulk particles may not reveal the effects seen in that small class of nucleosome which have strong HMG binding sites. But the appropriate experiments do not seem to have been done.

If the effect is not at the core particle level, might it not be at the level of extended chromatin? Could HMG binding modulate higher-order structure? Attempts to answer this question by sedimentation and electric dichroism studies gave wholly negative results (McGhee et al., 1982; see also McGhee et al., 1983b). No evidence could be found to indicate that HMG 14/17 caused unfolding of either bulk chromatin or the active β-globin gene in 14-day chick erythrocytes (see Fig. 8-16; see also Caplan et al., 1987). It should be pointed out that these experiments, which are of necessity carried out on isolated, linear polynucleosomal fragments, differ in a fundamental respect from the studies of Gazit et al. (1980), which utilized a nuclear pellet, and those of Weisbrod and Weintraub (1979), which were conducted with pelleted, undigested chromatin from lysed cells. It is conceivable that active domains in intact chromatin are under torsional strain, which might in turn alter their response to HMG binding.

The confusion over the role of HMG 14 and 17 in transcription has been compounded by more recent studies that appear to be in direct conflict with some of the earlier observations. Seale et al. (1983) report little enrichment of HMG 14 and 17 in fractions enriched sixfold in active sequences and little correlation between the content of these proteins and transcriptional activity in stimulated Friend cells. Goodwin et al. (1985) find no change in differential gene sensitivity to DNase I in erythroid or fibroblast nuclei upon removal of HMG proteins by 0.35 M NaCl.

Given all of these conflicting results, it is extremely difficult to tell, at this point, precisely what role (if any) HMG 14 and 17 play in chromatin transcription. As Goodwin and Mathew (1982) point out, there is barely enough of these proteins to provide one molecule per nucleosome in the transcribed sequences in a typical cell. This means that either the reports of significant quantities of HMG 14 and 17 in inert regions are the consequence of redistribution during isolation, or that these proteins are present only on *some* nucleosomes in an active gene, or are present only part

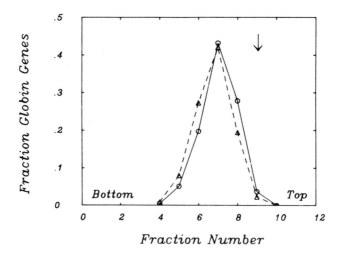

Fig. 8-16. A demonstration that HMG binding does not alter the higher-order structure of polynucleosomes from chick embryos carrying the active β-globin gene. The polynucleosomes (~20-mers) have been sedimented in an isokinetic sucrose gradient and those containing β-globin genes detected by dot-blot hybridization. (O), polynucleosomes stripped of endogenous HMG 14/17. (△), polynucleosomes to which had been added two moles HMG 14/17 per nucleosome. The arrow denotes the position where the peak would be expected if higher-order structure were unfolded. (From Nucleic Acids Research, vol. 10, issue 6, p. 2014, McGhee et al. Copyright 1982 IRL Press.)

of the time. The situation appears reminiscent of that encountered with acetylation. Probably the safest statement is that transcription in eukaryotes will turn out to be a considerably more complex process than has been so far imagined.

The situation is not clarified by reports that HMG 1 and HMG 2 may also be associated with transcriptional activity. Using a mouse myeloma chromatin preparation that supported ongoing transcription, Stoute and Marzluff (1982) noted that extraction with 0.35 M NaCl abolished this activity. Surprisingly, the extract was found to contain HMG 1 and HMG 2, but no significant amount of HMG 14/17. Addition of either the 0.35 M extract or purified HMG 1/2 resulted in renewed transcription. It is perhaps noteworthy, in this connection, that Jackson et al. (1979) had earlier reported the isolation of a class of nucleosomes and oligonucleosomes in which H1 was replaced by HMG 1/2. A most disturbing aspect to the Stoute and Marzluff study is the finding of these two proteins rather than the expected HMG 14 and 17. Had they not analyzed their extract, their results might have appeared to be a confirmation of the Weisbrod and Weintraub experiments! Clearly, the greatest care is required in this field.

An enormous amount of experimental effort has gone into attempts to define the special protein composition associated with transcribable chromatin. The results are not very gratifying, for the problem is still beset with ambiguities and seeming contradictions. One gains the impression that some important factor or factors have been overlooked. A suspicion is developing that one such may be the torsional state of chromatin in vivo.

DNA Torsion and Transcription

It has been known for a number of years that DNA supercoiling has a major effect on transcription in prokaryotes: In general, genes that are supercoiled are much more efficiently transcribed (reviewed by Smith, 1981; see also Brahms et al., 1985). Yet for several reasons, there has been reluctance to extrapolate this idea to eukaryotes. First, although eukaryotic DNA is *de facto* supercoiled by being wrapped into nucleosomes, there has been little evidence that most chromatin domains are under torsional stress. Indeed, psoralen binding experiments by Sinden et al. (1980), which demonstrated torsion in prokaryotic DNA, did not show such evidence for the bulk of eukaryotic DNA. This does not seem, in retrospect, to be a particularly compelling argument, since most eukaryotic DNA is not transcribed anyway. Finally, although eukaryotic analogs to both the topoisomerase I and topoisomerase II are known, the latter has not as yet been directly shown to exhibit the gyrase activity of the prokaryotic enzyme (Gellert, 1981). However, *indirect* evidence for a gyrase-like activity in *Xenopus* oocytes has been presented by Glikin et al. (1984).

At the same time, data have been accumulating to suggest that torsional stress may indeed be a major factor in the control of chromatin transcription (see Weintraub, 1985, for a recent review). Several groups have shown that eukaryotic RNA polymerases prefer supercoiled templates in vitro (i.e., Perdone et al., 1982). Second, as described in preceding sections of this chapter, the DNase I hypersensitive regions associated with active genes often show susceptibility to single-strand specific reagents, an observation most easily explained by the existence of torsional stress.

A number of recent studies, utilizing topoisomerase inhibitors, have produced evidence linking topoisomerase II activity to DNase I sensitivity. Villeponteau et al. (1984) find that the DNase I sensitivity of the active β-globin genes in chick embryo erythrocytes is rapidly lost following administration of novobiocin. In a following publication (Villeponteau et al., 1986), they report a highly specific cleavage of active genes in intact cells, *after* the loss of DNase I sensitivity. Han et al. (1985a) find that novobiocin can block the heat-shock response in *Drosophila*.

Hypersensitive *sites* have also been shown to be sensitive to the presence of topoisomerase II inhibitors. Using the reagent 4'-(9-acrydinylamino)-methanesulfon-m-anisidide (m-AMSA), Yang et al. (1985) observed cleavage of SV-40 DNA in infected monkey cells. A major site of cleavage was mapped into the DNase I hypersensitive region near the origin of late transcription (see also Luchnik et al., 1982). Similarly, Villeponteau and Martinson (private communication from H. Martinson) found that novobiocin administration caused relaxation of DNase I sensitivity in the *Drosophila* histone gene domain and that cleavage at some of the hypersensitive sites could be induced by SDS treatment of nuclei. Since $SDS + Mg^{2+}$ produces topoisomerase II induced cleavages in vitro, they argue that these are probably sites of interaction of the topoisomerase with the chromatin.

In addition, a number of studies indicate that topoisomerase I may be preferentially associated within active chromatin regions (see, for example, Weisbrod, 1982; Javaherian and Liu, 1983; Fleischmann et al., 1984). As has been described in Chapter 7, there is also strong evidence that topoisomerases are present at points where chromatin domains interact with the nuclear matrix. However, this seems to be a type II enzyme, as contrasted to the topoisomerase I present in active genes. Thus, a quite convincing argument can now be made that one and perhaps two kinds of topoisomerase activity is essential for gene transcription in eukaryotes. (See North, 1985, for a review.)

There are a number of lines of experimental evidence to indicate that one role of topoisomerases may be the *production* of superhelical torsion. Much of this evidence comes from studies in which plasmids have been microinjected into *Xenopus* oocytes. Harland et al. (1983) demonstrated that transcription of a plasmid-carried thymidine kinase gene was substantially reduced when the plasmid was cleaved in situ by injection of restriction endonucleases. Ryoji and Worcel (1984) provide evidence that both transcriptionally inactive ("static") and active ("dynamic") minichromosomes are assembled when plasmids are injected into *Xenopus* oocytes. The "dynamic" minichromosomes can be relaxed in vivo by injection of topoisomerase 1 or novobiocin. In later work, Ryoji and Worcel (1985) have been able to separate two types of minichromosomes on the basis of solubility in 85 m*M* KCl. The soluble material gives a normal digestion pattern with micrococcal nuclease, whereas the insoluble minichromosomes are more sensitive to both DNase I and micrococcal nu-

---

Fig. 8-17. Hypothetical steps in the activation of eukaryotic genes for transcription. (a) folded, torsionally relaxed domain with genes in inactive, incompetent state. (b) Upon specific signals, torsion is modified by topoisomerase at loop base. A region of the domain (containing the gene set) is distorted and becomes DNase I sensitive. (c) Within the distorted region, specific modifications and interactions with nonhistone proteins are favored, thereby maintaining the "open" state. (d) Specific protein signals can now interact with 5'-flanking regions of the appropriate subset of genes within the set. This produces further distortion of the nucleosome structure within the adjacent genes. (e) Transcription occurs in these genes.

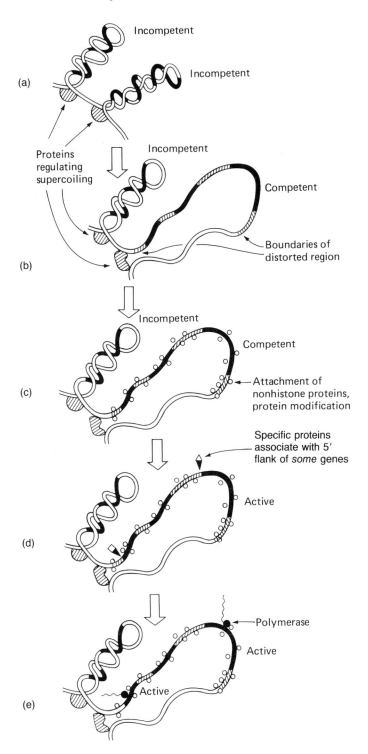

clease and yield a smeared digestion pattern, with evidence for "half-nucleosome" production. It is suggested, but not proved, that the insoluble portion corresponds to the "dynamic" transcriptionally active minichromosomes. The role of supercoiling has been further explored by this group (see Kmiec et al., 1986a, 1986b). Using 5S DNA plasmids, plus a *Xenopus* oocyte extract plus TFIIIA, they report that transcription increases with increased supercoiling.

Although the exact role of "gyration" in the transcription process is still partially unclear, and to some extent controversial, there seems little doubt that it must be an important factor in vivo.

A Tentative Model for Gene Activation

On the basis of all of the data presented in this chapter, a reasonable (though highly hypothetical) scenario can be constructed to describe gene activation in eukaryotes (see Fig. 8-17). Through the influence of a topoisomerase II (perhaps located at a loop origin on the matrix), supercoiling is changed in a particular chromatin domain. The torsional stress thus induced distorts the chromatin structure and imposes DNase I sensitivity on a *portion* of that domain. The boundaries of that portion may be marked by specific sequences and/or DNA methylation. In the sensitized portion, relaxation of chromatin higher-order structure may occur, and protein interactions (i.e., H1, HMG 14, 17) may be modified. These changes have produced transcriptional competence but are not in themselves sufficient to induce transcription. At this point, the chromatin will still produce a "normal" micrococcal nuclease ladder.

Activation of a particular gene or genes within the sensitized region of the domain to actually begin transcription may involve the binding of specific "transcription factor" proteins. Presumably, the binding sites were not accessible prior to sensitization. It seems likely that a topoisomerase I activity is at this point required in transcribing regions, to allow a controlled, local relaxation during transcription. Such final steps lead to a gross distortion of the nucleosome structure and the act of transcription itself.

Such a model is, of course, nothing more than an educated guess. Much has been left out—in particular, the roles of histone modification and HMG proteins. It seems possible to me that the confusion that has plagued those problems may result in part from the fact that many studies of the effect of such factors on chromatin structure have been carried out in vitro. In many such experiments, the torsional constraints on chromatin have necessarily been relaxed. Histone acetylation or HMG binding may have very different effects on "dynamic" chromatin.

9

Replication of Chromatin

The replication of the eukaryotic chromosome must necessarily be much more complicated than the corresponding process in prokaryotes. Not only must a vast amount of DNA be duplicated, but the specific structure of the whole nucleoprotein complex has to be reconstructed as well. In the preceding chapters, evidence has been presented that the arrangement and local modifications of at least a fraction of the nucleosomes on the genome play roles in determining the transcriptional capabilities of specific cell lines. Then we must conclude that such features are accurately copied in somatic cell division. Indeed, there is excellent evidence that such peculiarities as DNase "hypersensitive" sites are transmitted from parent to daughter chromatids (see Chapter 8). Furthermore, since there is evidence that differentiation of cell lines usually occurs at the point of cell division, this nucleoprotein structure must be mutable in some controlled fashion.

The basic *mechanism* of chromatin replication also raises fascinating questions. How does a polymerase proceed through a nucleosome-covered DNA? What happens to the parental nucleosomes? How are new nucleosomes assembled, and how are they apportioned between daughter strands? Attempts to answer such questions have engaged the attention of many researchers. Given the inherent complexity of the process, it is not surprising to find that the history of this research is riddled with seeming contradictions and intense controversy. As we shall see, many issues are still far from settled, although a crude picture of the process is finally beginning to emerge.

In describing our present knowledge of chromatin replication, it will be useful to divide the whole process into a number of stages, each with its associated questions:

1. *DNA replication:* How is it initiated, and how does it proceed through the nucleosome structure?
2. *Histone synthesis:* How is it controlled with respect to DNA replication? How are histones transported into the nucleus? Are there histone pools?
3. *Nucleosome assembly and histone distribution:* Is it conservative or nonconservative; that is, do old and new histones intermix in nucleosomes? How are old and new histones distributed on old and new DNA?
4. *Nucleosome segregation:* How are new and old nucleosomes apportioned between the daughter DNA duplexes generated at the replication fork?
5. *Nascent chromatin and its maturation:* What is the structure of newly formed chromatin? How does it acquire the nucleosome positioning, DNA and histone modifications, and complement of nonhistone proteins characteristic of the mature nucleoprotein complex?

This list corresponds to different but interrelated *processes* that must occur during chromatin replication. In addition to considering the *mechanisms* of these processes, we must consider the question of their spatial localization. Which are nuclear and which are cytoplasmic? Are the majority that are nuclear localized in special regions of the nucleus? In the following sections, I shall consider these processes in turn and attempt to interleave descriptions of their mechanisms with what information is available concerning spatial localization. For an excellent, detailed review of research before 1980, the reader is referred to DePamphilis and Wassarman (1980). More recent advances are given by DePamphilis et al. (1983).

DNA Replication

Insofar as the replication of the DNA itself is concerned, the process in eukaryotes bears marked similarity to that in prokaryotes. A single enzyme, polymerase α, appears to be involved; the other eukaryotic DNA polymerases (β,γ) are primarily associated with repair mechanisms and mitochondrial DNA replication. (See DePamphilis and Wassarman, 1980, for details.) Replication proceeds bidirectionally from discrete origins; the synthesis is continuous at one arm of each fork (leading strand) and discontinuous on the other (retrograde) strand (Perlman and Huberman, 1977; Kaufman et al., 1978). Initiation, as in prokaryotes, involves an RNA primer (see for example, Waqar and Huberman, 1975; Kowalski and Denhardt, 1979; Tseng et al., 1979; Kaufman, 1981). However, the Okazaki fragments in eukaryotes appear to be considerably shorter than those in prokaryotes. In SV-40, Perlman and Huberman (1977) found a weight average size of 145 b; Anderson and DePamphilis (1979) report a size distribution from 40 to 290 b.

There is a fundamental difference in the *number* of replication origins observed in eukaryotic and prokaryotic genomes. Whereas prokaryotes typically exhibit a single origin, very large numbers are observed in eukaryotic cells. For example, Anachkova and Russev (1983) report $\sim 10^4$ origins per haploid genome in ascites tumor cells, and Blumenthal et al. (1973) have estimated an average spacing of only 7.9 Kbp between origins in *Drosophila* embryos, although the average spacing in most eukaryotes appears to be larger (Sheinin and Humbert, 1978). It seems likely that these large numbers of initiation sites exist mainly to facilitate the rapid replication of the enormous eukaryotic genome, rather than being required by some peculiarity of chromatin replication. *Small* eukaryotic chromosomes, such as the SV-40 minichromosome, yeast plasmids, and the amplified ribosomal minichromosomes of *Physarum,* typically contain a single origin of replication. The replication origins of SV-40 and other papovaviruses have been studied in great detail; they include a long G–C-rich stretch with dyad symmetry, flanked by conserved sequences (Subramanian and Shenk, 1978; Soeda et al., 1979; DePamphilis and Wassarman, 1980; Innis and Scott, 1984). In SV-40, this region has a high probability of being nucleosome free and contains the binding site for the T-antigen protein required for replication. However, as Lasky and Harland (1981) point out, this may not be a good model for most eukaryotic replication origins, for papovavirus replication is not subject to cellular control. In fact, this structure may represent a mechanism for avoiding such regulation.

Much less is known concerning "true" eukaryotic origins. The autonomously replicating sequences (ARS sequences) in the yeast genome are the most thoroughly studied (Struhl et al., 1979; Hsiao and Carbon, 1979). These were first identified in yeast plasmids, but at least one subclass is present in 50 to 100 copies in the yeast genome (Chan and Tye, 1983).

Identification of replication origins in higher eukaryotic genomes has been difficult. However, Botchan and Dayton (1982) report evidence for such sites in the nontranscribed spacers of the tandemly repeated rDNA genes of a sea urchin.

There is mounting evidence that DNA replication in eukaryotes occurs in close association with some elements of the nuclear matrix and/or the nuclear lamina (see, for example, Berezney and Coffey, 1975; Pardoll et al., 1980). Support comes from studies in which it has been shown that a number of nonhistone proteins resembling major matrix or lamina proteins associate preferentially with putative origins of replication (Anachkova and Russev, 1983). Reddy and Pardee (1980) postulate that a multienzyme complex involved in nucleotide metabolism forms a portion of this matrix apparatus. Finally, it has been hypothesized that the same attachments of chromatin loops to the matrix of nuclear lamina that may define "domains" (see Chapter 7) may serve as the sites of initiation of replication (see Laskey and Harland, 1981). Since the experiments of Earnshaw et al. (1985) indicate that topoisomerase II is a component of the attachment sites, it may be of significance that Mattern and Painter (1979) have suggested that the inhibition of DNA synthesis by novobiocin is a consequence of this compound's inhibition of gyrase-induced supercoiling. However, at this writing, no gyrase-like activity of a eukaryotic topoisomerase has been directly demonstrated.

If nuclear matrix sites are involved, unraveling the precise sequence of events that occur at the replication fork may be extremely difficult, for the use of simplified in vitro systems may be uninformative or even misleading. The possibility of such complexity should be kept in mind throughout the following discussion. We turn now to the synthesis of the second major class of participants in chromatin replication: the histones.

Histone Synthesis

Like all other proteins, histones are synthesized in the cytoplasm. This was first clearly demonstrated by Robbins and Borun (1967), who noted that until then it had generally been assumed that the synthesis took place in the nucleus. There is evidence (Ruiz-Carillo et al., 1975; and see Chapter 4) that modification of histone H4 also occurs in the cytoplasm, but this modification is swiftly changed to the "mature" pattern after transport into the nucleus (Jackson et al., 1976; Wu and Bonner, 1981; Cousens and Alberts, 1982). In contrast, histone H3 appears to be initially unmodified, becoming acetylated only after incorporation into chromatin. H2A and H2B seem to be modified immediately and not to undergo subsequent change in state.

There has been much interest in the question of how tightly histone synthesis is coupled to DNA synthesis. Such coupling might be expected, since large and stoichiometric amounts of histones are required at each round of replication, whereas most somatic cells have very small histone pools (see below). The actual situation appears to be complex. Some eggs, oocytes, and embryos appear to have built up large histone pools and thus to have undergone extensive histone synthesis in the absence of replication. For example, the *Xenopus* oocyte accumulates an enormous excess of histones (Adamson and Woodland, 1974; Woodland and Adamson, 1977). Earnshaw et al. (1982) report 20,000 diploid equivalents per egg, stored in octamer-like structures. Smaller pools have been detected in sea urchin eggs (Cognetti et al., 1977; Salik et al., 1981) and in mouse oocytes (Wassarman and Mrozak, 1981). Embryos of both rice (Ahmed and Padayatty, 1982) and sea urchin (Arceci and Gross, 1977) also contain sizeable histone pools. In the latter case, it has been shown that histone synthesis continues even in enucleated cells. All of the above can be considered as special cases, where large amounts of histones are to be required for the rapid rounds of cell division that occur in early embryogenesis.

In somatic cells, there does not seem to be general evidence for the existence of *large* histone pools. However, there have been conflicting reports as to how tightly DNA and histone synthesis are coupled in such cells. Whereas many early studies (i.e., Prescott, 1966; Robbins and Borun, 1967) and some more recent ones (i.e., Hereford et al., 1982; Kelly et al., 1983; Plumb et al., 1983) have reported tight coupling, there have been quite a number of claims to the contrary. Nadeau et al. (1978) observed that blocking DNA synthesis in HTC cells by hydroxyurea did not cause a comparable decline in histone synthesis. Although DNA replication was reduced to 3 to 4% of normal S-phase levels, histone production remained at about 40%. Groppi and Coffino (1980) describe experiments with mouse lymphocytes and CHO cells that show substantial synthesis throughout G_1 as well as S phase. Most of the histones synthesized in G_1 were not associated with chromatin but were stored in a cytoplasmic pool and incorporated into chromatin in the next S phase.

Between these extremes lie the results of a number of careful studies that seem to indicate that there is a low basal rate of histone synthesis throughout the cell cycle *in addition* to a burst of synthesis during S phase. (See, for example, Spalding et al., 1966; Gurley et al., 1972; Seale, 1981; Wu and Bonner, 1981; Djondjurov et al., 1983; Graves and Marzluff, 1984). The studies of Wu and Bonner are of special importance, for they have shown that a different distribution of histone variants is synthesized in S phase than produced in G_1 and G_2 phases. This definitively eliminates the possibility that the observed G_1/G_2 synthesis might be an artifact from contaminating S-phase cells, a criticism that can be leveled against many other studies. Second, their results suggest that each kind of synthesis may play a particular role in organizing and modifying chromatin. The results are summarized in Table 9-1; for the postulated functional roles of these variants, see Chapter 4. The data in Table 9-1 also give an indication of the relative level of G_1 histone synthesis observed by a number of investigators.

An elegant series of studies has been carried out by L. Hereford's group at Brandeis, using the yeast *Saccharomyces cerevisiae* (Hereford et al., 1981, 1982; Osley and Hereford, 1981). Yeast possesses a number of unique advantages for such studies; there are only two copies of each histone gene, and there exist temperature-sensitive cell cycle mutants in which

Table 9-1. Relative Histone Synthesis in S and G_1 Phase by CHO Cells[a]

Histone	S/G_1
H3.2	68
H2A.1/.2	58
H3.3	6.2
H2A.Z	3.0
H2A.X	6.8
H4 (total)	35
H2A (total)	32
H2B (total)	29
H3 (total)	23

[a]Data from Wu and Bonner (1981).

DNA replication can easily be controlled. Hereford and co-workers were able to show that H2A and H2B messenger RNA synthesis is very low in early G_1 and increases dramatically in late G_1 near the onset of DNA replication. The mRNA levels peak *before* the DNA synthesis reaches its maximum rate (Fig. 9-1). Most interestingly, studies of cells in which DNA replication had been blocked by temperature shift showed a rapid decrease in histone mRNA levels. It appears that there exist mechanisms for the posttranscriptional degradation of histone mRNA in yeast. The precision and specificity of this control is illustrated by experiments in which an extra pair of H2A and H2B genes were inserted. Although transcription was increased, the steady-state levels of the corresponding mRNAs remained unchanged. The half-life of H2B transcripts was found to decrease from 15 min to 7 min. Furthermore, neither transcription nor stability of other histone mRNAs was influenced! Nurse (1983) presents a very nice, succinct review of yeast studies.

The overall picture, at least for somatic cells, appears to be one in which a low basal level histone synthesis throughout the cell cycle is augmented by a massive increase during, or just before, DNA replication. It may well be that different somatic cell types vary considerably in this respect, which would explain some of the seeming disparities in the literature. Certainly, some oocytes and embryonic cells behave differently. We know very little as yet concerning the mechanisms involved in this regulation, and these should constitute a fruitful ground for future study.

Once histones are synthesized in the cytoplasm, they must enter the nucleus. Does this involve active or passive transport? In a seminal study, Bonner (1975) investigated the fate of ^{125}I-labeled proteins injected into the cytoplasm of *Xenopus* oocytes. He observed that molecules of weight less than 20,000 daltons equilibrated between the cytoplasm and nucleus within 24 hours, whereas proteins of molecular weight greater than 65,000 were strongly hindered from passing the nuclear membrane. On this basis, histones would be expected to readily enter, and indeed they did so. Bonner also observed that free histones were concentrated in the nucleus, and the concentration factor was different for different histones. At equilibrium, H4 was found to be fortyfold more concentrated in the nucleus, whereas H2A, H2B, and H1 were only five- to sixfold more concentrated. There are two possible mechanisms that could account for such an unequal distribution between two compartments: active transport or the existence of a nontransferable binding substance. The latter seems to be indicated (although not exclusively) by the fact that Bonner observed competition between different histones. As we shall see later, there is now strong evidence for the existence of histone-binding proteins in oocyte nuclei.

This explanation may not be the whole story, however, for there is emerging evidence for an active (or at least facilitated) transport of some proteins across the nuclear envelope. Dingwall et al. (1982) have shown

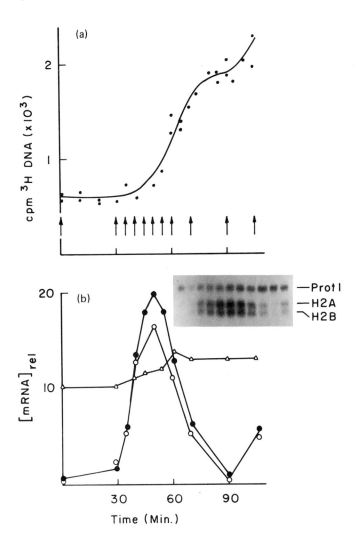

Fig. 9-1. Coupling of histone mRNA synthesis to DNA replication in yeast. Panel (a) shows DNA synthesis in the synchronized cells. In panel (b) are shown the levels of H2A mRNA (●) and H2B mRNA (○) at the times indicated by arrows in panel (a). The mRNAs were detected by hybridization against a probe carrying H2A and H2B genes, as well as the structural gene for an unrelated protein (protein 1). The Northern blots are shown in the insert. Note that whereas H2A mRNA and H2B mRNA syntheses are closely correlated, the synthesis of protein 1 (△) is largely independent. Also note that the histone mRNAs peak at the *beginning* of DNA replication (From Cell *24*, p. 367, Hereford et al. Copyright 1981 M.I.T. Press.)

that the concentration of protein nucleoplasmin (see below) in oocyte nuclei is blocked if a 10,000 dalton tail is removed from the 33,000 dalton polypeptide chain. Nucleoplasmin exists as a pentamer, and only one tail per pentamer is sufficient to allow transport. In a related and more directly relevant study, Dingwall and Allan (1984) have found that H1, *or* its C-terminal domain, will accumulate in *Xenopus* nuclei. The excised N-terminal and globular portions of the molecule do not enter the nucleus but are proteolyzed in the cytoplasm. Thus, the C-terminal tail of H1 appears to play a role similar to that of the C-terminal tail of nucleoplasmin. To what extent similar phenomena may be observed with other histones is an intriguing question.

Nucleosome Assembly and Histone Distribution

Once histones are synthesized and transported into (or diffuse into) the nucleus, how are they assembled into nucleosomes? It would be extremely naive to assume that histone cores spontaneously form and are then attached to DNA. In the first place, the histone octamer is not stable at physiological salt concentrations; one would expect at most the spontaneous assembly of $(H3 \cdot H4)_2$ tetramers and H2A·H2B dimers. Furthermore, it has been demonstrated that H3 and H4 tend to form very large aggregates, if present in physiological salt solution in appreciable concentration. This fact might not pose a problem in somatic cell nuclei where histone pools are small but demands consideration in oocytes, where as much as 50 ng of histones can accumulate per cell. Interestingly, mechanisms seem to have been evolved to deal with just this problem.

The first suggestion of this came with the discovery by Laskey et al. (1977, 1978) that a component in *Xenopus* oocytes could facilitate the assembly of histones into chromatin. This factor was subsequently identified as a highly negatively charged protein, *nucleoplasmin,* which is especially abundant in oocytes but also found in a wide variety of nuclei of actively growing cells. (See Laskey and Earnshaw, 1980, for a review.) The protein contains about 30% Asx plus Glx and is phosphorylated as well. The functional form is a pentamer of ~30,000 dalton chains (Lasky and Earnshaw, 1980). The acidic nature of the protein is apparently important, for other acidic polypeptides, including polyglutamic acid (see Chapter 6) and HMG 1 (Bonne-Andrea et al., 1984) can also facilitate nucleosome reconstitution in vitro.

It was initially assumed that most or all of the pooled histones in oocyte nuclei were associated with nucleoplasmin. However, subsequent experiments have cast considerable doubt on this idea. Krohne and Franke (1980) prepared antibodies to an abundant oocyte nuclear protein, presumably nucleoplasmin. Precipitation with these antibodies did *not* cause coordinate precipitation of histone. In further studies, Kleinschmidt and Franke (1982) showed that H3 and H4 (but not H2A or H2B) in *Xenopus* oocyte nuclear extracts were associated with another, acidic protein of 110,000 daltons. A smaller amount of all of the histones electrofocused with nucleoplasmin. Thus, it seems likely that the major *storage* function is provided by the Kleinschmidt–Franke protein, but this does not exclude the possibility that nucleoplasmin acts as a *transfer* agent. (See Annunziato and Seale, 1983a, for a careful discussion.) However, there still must be some mechanism for sequestering H2A and H2B. The reason for this is that H2A and H2B, in the absence of other histones, will bind strongly and nonspecifically to DNA to form nonnucleosomal structures. If the fertilized egg were to commence DNA replication in the presence of high concentrations of free H2A and H2B, these proteins might be expected to bind randomly. As will be shown below, there is now evidence that the $(H3 \cdot H4)_2$ tetramer may be added first in nucleosome assembly, a concept entirely consistent with the idea of separate "handling" of the two classes of histones.

All of the data cited above apply to *Xenopus* oocytes. Of much wider interest would be the demonstration of such "assembly" or "storage" proteins in other cells. Ishimi et al. (1983, 1984) have purified a 53,000 dalton protein from HeLa and mouse FM3a cells that appears to serve these functions. The protein forms a 12S complex with the four core histones, with one copy of each histone per copy of the binding protein. From the sedimentation coefficient and estimated molecular weight (~270,000), it seems likely that the complex may have two 53,000 dalton chains and an octamer of the histones. Like nucleoplasmin, this protein will assemble nucleosomes in vitro under physiological conditions. Its relationship (if any) to nucleoplasmin is obscure; antibodies to it do not cross-react. A possibly similar protein has been detected by Saffer and Coleman (1980) in calf thymus.

Although the data are still incomplete, and even to a certain extent confusing, it is becoming clear that complicated mechanisms exist for storing histones (when necessary) and incorporating these into nucleosomes.

When DNA replication occurs, two DNA duplexes are generated. Given the semiconservative nature of DNA replication, each of these contains one parental strand and one newly synthesized strand. These daughter duplexes are commonly called "new" DNA, to distinguish them from the "old" (as yet unreplicated) portions. In analyzing *chromatin* replication, we must also consider the way in which new and old histones are apportioned. The first question, of course, is whether old histones are utilized at all; it is at least conceivable that all old nucleosomes might be degraded at each cell division. Many experiments have shown this not to be the case (Seale, 1975). The most direct and convincing evidence is the fact that labeled histones have been shown to persist in cells through several generations (cf. Prior et al., 1980; Wu and Bonner, 1981; Leffak, 1984). Another kind of evidence comes from experiments that have utilized cycloheximide blocking of protein synthesis in cells undergoing DNA replication; it is found that the new DNA (which can be identified by radioactivity-labeling) contains nucleosomes but is about twice as sensitive to nuclease digestion as that in nonreplicating chromatin (Weintraub, 1976; Seale, 1976). The fact that even partial protection is preserved means that nucleosomes must be present. Since nucleosomes on the new DNA could be formed in these circumstances only from old histones, preservation is clearly established.

Given that old histones persist and that new histones are synthesized in each cycle, the next question is: Do old and new mix to create "hybrid" nucleosomes, or is the histone core a conserved entity? The problem has been controversial. The first study to seriously address this question (Leffak et al., 1977) used histones labeled in vivo with dense $[^{13}C, ^{15}N]$-amino acids; $[^{3}H]$-lysine was used as a marker for these new histones (see Fig. 9-2). The octamers were crosslinked and then sedimented to equilibrium in density gradients. The density observed for the radiolabeled proteins coincided with that found for "control" dense octamers; therefore, it was concluded that old and new histones did not intermix. These experiments were criticized by Annunziato and Seale (1983a), who pointed out that the use of a lysine label strongly biases the data to reflect the distribution

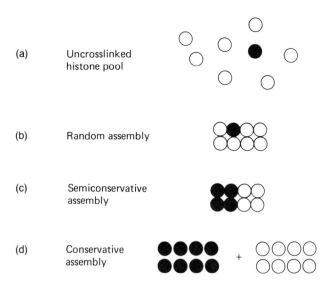

Fig. 9-2. Basis of the experiments of Leffak et al. (1977). New histones are density-labeled and constitute about 1 in 8 of all histones (a). If these are assembled randomly with light histones, the average density will be close to the "light" value (b). If they assemble semiconservatively, using half old histones and half new histones, the new octamers will have a density intermediate between the light and heavy values (c). But if they assemble conservatively, all new histones going together, the new octamer will be heavy and the old octamers light (d).

of H1-histones and (perhaps) some nonhistone proteins. The banding of H1 polymers (see Chapters 4 and 7) and nonhistone proteins at the dense locus might bias the results. As we shall see below, there is now strong evidence in support of this idea. However, in further studies Leffak (1983, 1984) has carried out electrophoretic analysis of the individual density gradient fractions and still reports octamers primarily in the dense position. Furthermore, experiments in which the fate of these octamers has been followed over several generations indicate very little (although perceptible) mixing. Leffak argues that even this apparent mixing is probably an experimental artifact and contends that the entire histone core is assembled *de novo* during replication.

Few researchers in the field would adopt so extreme a view at the present time. Even Kumar and Leffak (1986) have noted that the assembly of *active* chromatin involves mixing of old and new histones. There is, moreover, extensive evidence to indicate that the assembly of new nucleosomes may in general be a more complex process than simply wrapping DNA about new or old histone *octamers*. A number of research groups, utilizing a variety of techniques to distinguish "nascent" from unreplicated chromatin, have concluded that not all of the new histones are deposited simultaneously during replication. (See, for example, Worcel et al., 1978; Senshu et al., 1978; Cremisi et al., 1978; Cremisi and Yaniv, 1980; Russev and Hancock, 1981; Jackson and Chalkley, 1981a, 1981b, 1985a, 1985b; Jackson et al., 1981; Seale, 1981; Annunziato et al., 1982.) While the details of these experiments are complicated, all point to the same conclusion: A new H3 and H4 appear first and almost exclusively on new DNA, while new H2A and H2B are at least partially distributed between new and unreplicated DNA. In a number of cases where the deposition of new H1 was followed, it was observed to be either uniformly distributed between old and new DNA, or even preferentially associated with unreplicated parts of the genome (Worcel et al., 1978; Jackson and Chalkley, 1981a, 1981b).

The most convincing evidence for nonconservative deposition of new histones comes from recent studies by Jackson (1987a). Using formaldehyde crosslinking and density labeling, he finds two classes of "new" octamers: One contains a new $(H3 \cdot H4)_2$ tetramer and two old H2A·H2B dimers, the second contains one new H2A·H2B dimer associated with a hexamer of old histones. The conclusion is that new $(H3 \cdot H4)_2$ tetramers are laid down first and then are "filled in" with predominantly old H2A·H2B. The H2A·H2B lost from old octamers seem to be mostly replaced with new histones. In an accompanying paper (Jackson, 1987b), the crosslinking technique used by Leffak and co-workers is critically analyzed. It is claimed that nonhistone protein contamination leads to artifactual results.

Although the bulk of evidence now seems to support the existence of nonconservative assembly of new octamers, one should probably be cautious in making firm conclusions. Many of the careful studies of Leffak and co-workers cannot be easily dismissed, and the recent observations that transcriptionally active regions may behave differently than inactive regions (Kumar and Leffak, 1986) hints at possible further complexities.

Nucleosome Segregation

Even though there may be mixing of old and new histones during chromatin replication, it is clear that we may still distinguish between what are primarily old or preexisting octamers, and those that have been newly synthesized. The next question, then, is: How are these apportioned between the two daughter duplexes generated at the replication fork? Several imaginable modes of segregation are depicted in Figure 9-3. It will be noted

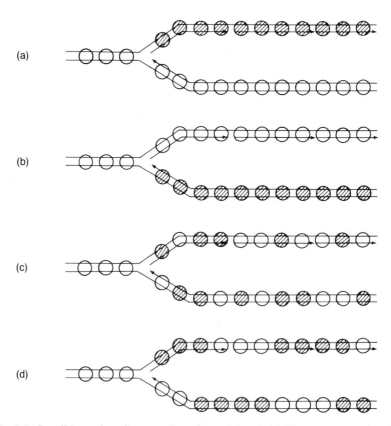

Fig. 9-3. Possible modes of segregation of new (●) and old (○) octamers on daughter strands. (a) Conservative segregation of old octamers on the leading strand. All new octamers go on the retrograde strand. (b) Conservative segregation of old octamers on the retrograde strand. All new octamers go on the leading strand. (c) Random segregation. (d) Random but clustered segregation on the two strands.

that at each fork there are three distinguishable duplexes: the *unreplicated duplex,* the *leading-strand duplex,* and the *retrograde-strand duplex.* The old or parental nucleosomes must be transferred to one or both of the daughter duplexes. If they are transferred to both, this might occur in a wholly random fashion, or with some tendency to clustering if the mechanism were such as to transfer nucleosomes in groups, alternatively to one daughter duplex and to the other. In any event, old nucleosomes can account for only half the complement needed for the replicated DNA, and new histone synthesis must provide the remainder to fill in whatever kind of gaps are left. There have been, and continue to be, intensive efforts to determine the mode of nucleosome segregation. Unfortunately, the subject is, if anything, even more confused by conflicting reports than is the problem of histone mixing.

At first it seemed that the question was to be quickly and easily resolved. Early experiments (Weintraub, 1973, 1976; Seale, 1976) were generally interpreted to indicate that parental nucleosomes segregated exclusively to one of the daughter duplexes. These experiments utilized cycloheximide to block protein synthesis during DNA replication. It was found that although the newly synthesized DNA was more susceptible to micrococcal nuclease cleavage than bulk DNA, it would still yield nucleosome "ladders" on DNA gels. As pointed out by Weintraub (1976), this could be explained in two different ways; either the new nucleosomes were on one strand, or they were "clustered" on both strands (see Fig. 9-4). The issue seemed to be settled by experiments of Riley and Weintraub (1979). They presented electron micrographs of chromatin replicated during cycloheximide block. A few long lengths of nucleosome-free DNA were observed, and a number of replication forks in which one strand seemed devoid of nucleosomes could be identified. However, it should be noted that no evidence for nucleosome-free DNA could be produced in the digestion experiments of Seale (1976). (See also Seale and Simpson, 1975.) A different kind of evidence for strand preference was presented by Seidman et al. (1979). Using SV-40 minichromosomes replicated under cycloheximide block, they digested these with micrococcal nuclease and then hybridized the nuclease-resistant DNA (e.g., nucleosomal DNA) to probes from both strands of SV-40. The hybridization experiments showed a clear preference for the *leading strand,* indicating that the parental nucleosomes had been preferentially associated with this strand.

Unfortunately, this seemingly clear picture has been clouded by subsequent studies. For example: The data of Cusick et al. (1981) indicate that nucleosomes are present on *both* sides of the SV-40 replication fork. To add to the confusion, Roufa and Marchionni (1982), studying SV-40 integrated into the cellular genome, find hybridization to only one strand— but to the *opposite* one to that reported by Seidman et al.! This result may not necessarily be contradictory, for it is probable that a different replication origin is employed for the integrated virus.

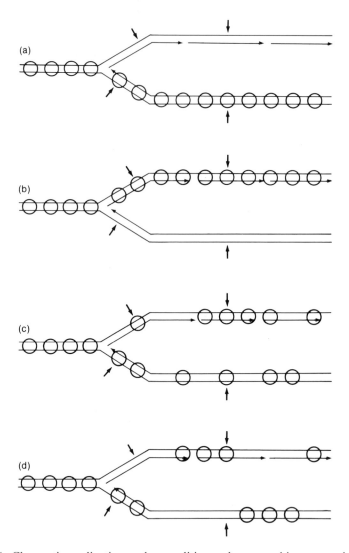

Fig. 9-4. Chromatin replication under conditions where new histone synthesis is blocked. Models (a)–(d) are as in Figure 9-3, except that new octamers are not added. Note that long oligonucleosomes can only be generated in (a), (b), or possibly (d). In models (a) and (b), long stretches of free DNA will be left, which should yield large DNA fragments if cleaved by an infrequently cutting restriction nuclease. (Restriction sites are indicated by short arrows.)

However, most studies of the SV-40 system tend to support the idea that new and old histones are distributed to both sides of the replication fork. Tack et al. (1981) studied the accessibility of newly replicated sequences to six different restriction endonucleases and found no difference in the two daughter strands. In a more detailed study from the same laboratory, Cusick et al. (1984) essentially repeated the studies of Seidman et al., with the very different result that dispersion of old nucleosome cores to *both* strands was observed.

Using a very different method, Sogo et al. (1986) provide strong evidence for dispersive segregation in SV-40 replication. Psoralen crosslinking was employed to "mark" the positions of nucleosomes, and the replicating DNA was then examined in the electron microscope. Clear evidence was found for nucleosomes within 300 bp of the fork on *each* daughter arm. These experiments are especially important because they do not require the use of cycloheximide or other blocks of protein synthesis. There has always been some concern that such agents might cause anomalies in replication. When Sogo et al. did add cycloheximide, they still observed nucleosomes on both strands, but these exhibited a tendency to cluster.

Recent studies using cellular (rather than SV-40) replication also cast more and more doubt on the conservative segregation model. Fowler et al. (1982), utilizing 5-bromo deoxyuridine density-labeling of new DNA and [^{3}H]-lysine labeling of histones, found no evidence to support conservative segregation in CHO cells. Pospelov et al. (1982a) treated ascites cell and L-cell chromatin replicated under cycloheximide block with the restriction nuclease BSp I. The expected result, according to the conservative model, would include the generation of long fragments of free DNA (see Fig. 9-4). This was not observed, even in short digestions. Furthermore, most (~90%) of the new nucleoprotein banded midway between the positions expected for complete chromatin and free DNA on metrizamide gradients, a result consistent with equal segregation to the two arms of the fork. Similar evidence has recently been reported by Annunziato and Seale (1984). They find that if HeLa chromatin is replicated under cycloheximide block and the nuclei digested either with micrococcal nuclease or Hae III, fractions insoluble and soluble at low ionic strength are produced. The insoluble portion is enriched in newly replicated chromatin, perhaps by virtue of its association with nuclear matrix. (See also Pospelov et al., 1982b.) DNA from this insoluble fraction does not yield a clear nucleosomal ladder after the initial digestion by micrococcal nuclease or Hae III, but redigestion of the nucleoprotein with micrococcal nuclease reveals mono- and small oligonucleosomes. This is precisely the kind of pattern that would be expected if parental nucleosomes were scattered onto both daughter duplexes. Annunziato and Seale argue that previous observations of oligonucleosome tracts on new DNA could be explained by a tendency of the nucleosomes to cluster.

A model that seems consistent with most of the presently available data can be described as follows: Old nucleosome cores (or possibly only portions thereof) are in some unknown fashion passed across the replication fork. Both daughter strands receive them, but there appears to be some clustering, at least under conditions where new nucleosome formation is blocked. Gaps in the array are filled by new histone cores, or more likely first by new $(H3 \cdot H4)_2$ tetramers. This new or "nascent" chromatin must then adopt its final conformation by some kind of maturation process.

Nascent Chromatin and Its Maturation

It has been clear for a number of years that the structure of the most newly synthesized chromatin ("nascent chromatin") is somehow different from that of "mature" chromatin. It is possible to experimentally distinguish very new chromatin from nonreplicated chromatin by the use of pulse labeling of the DNA with, for example, $[^3H]$-thymidine. In some experiments, labeling "pulses" as short as 7 seconds have been employed. Estimates of the rate of fork motion in eukaryotic replication suggest values of the order of 50 bp/second (Annunziato and Seale, 1983a). From this we can judge that such time periods allow examination of nucleosomal structure within a few hundred base pairs of the replication fork. Subsequent changes in chromatin structure can be observed by following the radiolabel pulse with "chase" periods, utilizing nonlabeled DNA precursors (see Fig. 9-5). The earlier researches in this area have been admirably summarized in the review by DePamphilis and Wassarman (1980).

Using techniques of this kind, many investigators have observed that nascent chromatin exhibits a greater sensitivity to nuclease digestion than does nonreplicating chromatin. (See, for examples, Seale, 1975; Klempnauer et al., 1980; Cusick et al., 1981, 1983; Schlaeger et al., 1983; Annunziato and Seale, 1983b.) This hypersensitivity is exhibited toward a number of kinds of nucleases. With micrococcal nuclease, it is expressed by a more rapid release of both small oligonucleosomes and acid-soluble DNA. Pulse-chase experiments show that the nascent chromatin passes through a "maturation" process during which the nuclease hypersensitivity is rapidly lost. Periods of the order of 10 to 20 minutes are frequently reported as sufficient for the recovery of normal sensitivity to nuclease digestion. Considering the rate of fork elongation mentioned above, even such a short period means that a chromatin region must be many kilobases behind the moving fork before its nuclease susceptibility is "normal." The overall picture of nuclease hypersensitivity and its maturation seems to be agreed upon by all who have conducted such experiments, a rare instance of unanimity in this field!

Such concurrence does not exist with respect to some other properties reported for nascent chromatin. Levy and Jakob (1978) measured the sizes of pulse-labeled DNA fragments produced by micrococcal nuclease digestion of sea urchin nuclei. The data indicated a nucleosome repeat for the new chromatin of about 150 bp, much shorter than that of nonreplicating DNA in the same embryos (~200 bp). Furthermore, like nuclease hypersensitivity, the anomaly disappeared within a short period in pulse-chase experiments (see Fig. 9-5). Similar results have been obtained by a number of research groups, using various cell types. (See, for example, Murphy et al., 1978, 1980; Yakura and Tanifuji, 1980; Annunziato and

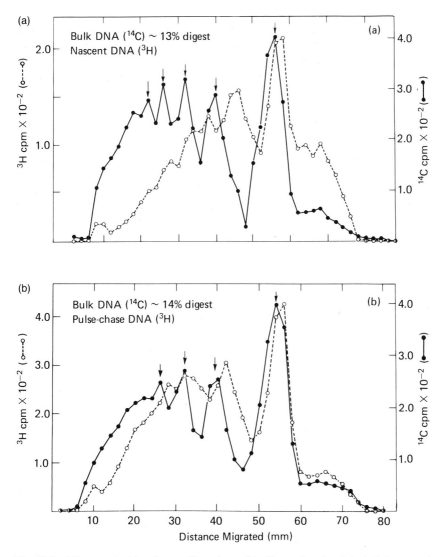

Fig. 9-5. Micrococcal nuclease digestion of bulk- and pulse-labeled (nascent) chromatin. (a) Sea urchin embryos were grown on [14C]-thymidine for 16 hr and then subjected to a very brief (7-second) pulse of [3H]-thymidine. Nuclei were digested with micrococcal nuclease, DNA extracted, and run on a 2.5% polyacrylamide gel. The gel was sliced and counted. The solid circles (●) show the bulk DNA digestion pattern, with the usual series of oligomers. The open circles (○) show the DNA from new chromatin. Note that the dimer and trimer bands are significantly shorter than in bulk chromatin. (b) As in (a), except that after the [3H]-thymidine pulse, the medium was replaced by one containing nonradioactive thymidine and chased for 2 min. Note that the oligomer DNA peaks of pulse-labeled material have already begun to shift toward the "normal" sizes. With longer chases, they become indistinguishable. (From Cell *14*, p. 259, Levy and Jakob. Copyright 1978 M.I.T. Press.)

Seale, 1982.) Nevertheless, the *significance* of the observation has been questioned. Jackson et al. (1981) have pointed out that during digestions at 37°C (the condition most commonly employed), the apparent repeat length shortens with increased digestion. They suggest that the results of the above authors, while real, do not necessarily indicate that a shortened nucleosome repeat exists in nascent chromatin. Rather, the results might be explained by sliding of the nucleosomes during digestion. There are indeed reports that nuclease digestion can, under some circumstances, induce sliding (i.e., Weischet and Van Holde, 1980). In an attempt to circumvent this potential difficulty, Annunziato and Seale (1982) carried out the digestions at 0°C. In these experiments, *no* change in apparent repeat with extent of digestion was observed, yet the nascent HeLa chromatin still appeared to have close-packed nucleosomes. Therefore, Annunziato and Seale argue that the phenomenon is real.

The debate has been continued by Vaury et al. (1983). Using CHO cells, they were able to repeat the earlier results, finding a spacing of 157 bp for nascent chromatin, as compared with 195 bp for nonreplicating chromatin. But when the nucleosomes in the nuclei were fixed with formaldehyde before nuclease digestion, the difference vanished—both new and old chromatin gave a value of about 195 bp! This observation has been confirmed by Smith et al. (1984) using HTC cells. In response, Jakob et al. (1984) have repeated their experiments on nascent sea urchin chromatin, with the inclusion of a formaldehyde fixation step. They still claim to see a shorter repeat in the newly replicated material. Thus, the issue remains in conflict.

Even if the apparent short repeat of nascent chromatin is a consequence of digestion-induced sliding, we must still ask why newly replicated and nonreplicating materials behave differently in this respect. Somehow, the positions of nucleosomes that have just been laid down on new DNA must be less firmly fixed. There are a number of possible explanations for this. First, as suggested by Smith et al. (1984), it may be that newly assembled histone cores are not as tightly associated with the DNA as are those of mature nucleosomes; there exists some evidence to support this contention (see Jackson et al., 1981; Seale, 1981). The reason for a weaker association is unclear, although Worcel et al. (1978) and Seale (1981) suggest that nascent nucleosomes may be nonstoichiometric in core histones (*see Nucleosome Assembly and Histone Distribution,* above).

A second possibility is that histones H1 and the concomitant development of higher-order structure are necessary for chromatin maturation. Worcel et al. (1978) have suggested that H1 is not deposited until 10 to 20 minutes after replication. (See also Annuziato and Seale, 1982; Schlaeger et al., 1983; Cusick et al., 1983.) However, the lability of H1 is such that it is very difficult to obtain definitive evidence concerning its in vivo location; the possibility of redistribution during preparation must always be considered. In a study that may have some bearing on the possible role of H1, D'Anna and Prentice (1983) investigated CHO cells blocked in S phase by hydroxyurea. Under these conditions, DNA replication is initiated but elongation is very slow. They found that in 10 hours 30% of the H1 had been lost, and the new DNA had a short repeat. By 24 hours, 70 to 80% of the H1 had disappeared, and both new and old chromatin exhibited a short repeat.

There is mounting evidence that the spacing of newly deposited nu-

(a)

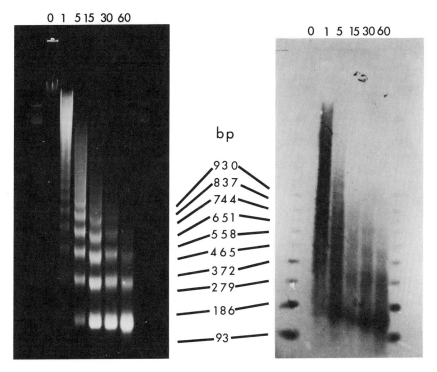

Fig. 9-6A. Newly replicated chromatin has a diffuse organization, which matures to yield a normal nucleosomal structure. Chromatin from HTC cells that have been pulsed with [³H]-thymidine for 2 minutes. The nuclei were digested with micrococcal nuclease, DNA extracted, and run on an agarose gel. To the left is the ethidium bromide-stained pattern, which corresponds almost entirely to bulk chromatin. To the right is a fluorogram, showing the diffuse pattern of the nascent chromatin. (Reprinted with permission from Biochemistry *23*, p. 1576, Smith et al. Copyright 1984 American Chemical Society.)

cleosomes, if not necessarily close packed, is at least less regular than that in mature chromatin. Smith et al. (1984) report that nascent chromatin exhibits a "smeared" pattern in DNA electrophoresis, even when fixed with formaldehyde; this matures in 5 to 30 minutes to yield a normal nucleosome ladder (see Fig. 9-6B). Similar results with cycloheximide-blocked chromatin have been described by Annunziato and Seale (1984). Such a pattern is to be expected if the nucleosomes are irregularly distributed with a wide variation in spacer length.

These results may suggest a model that reconciles a number of seemingly contradictory results. If we assume, as in Figure 9-7, that histone cores

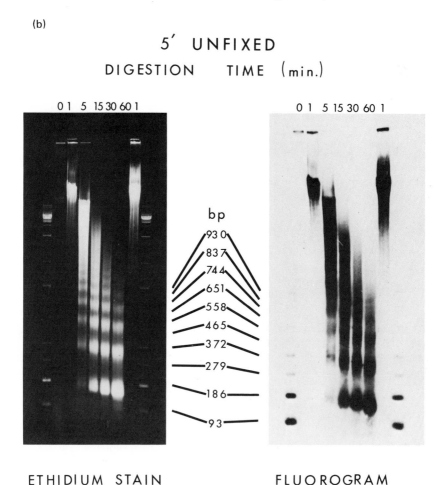

(b)

5′ UNFIXED

DIGESTION   TIME (min.)

ETHIDIUM STAIN   FLUOROGRAM

Fig. 9-6B. As A, except that the pulse has been chased for 5 minutes with cold thymidine. Now a nucleosome ladder can be detected in the fluorogram. (Reprinted with permission from Biochemistry *23*, p. 1576, Smith et al. Copyright 1984 American Chemical Society.)

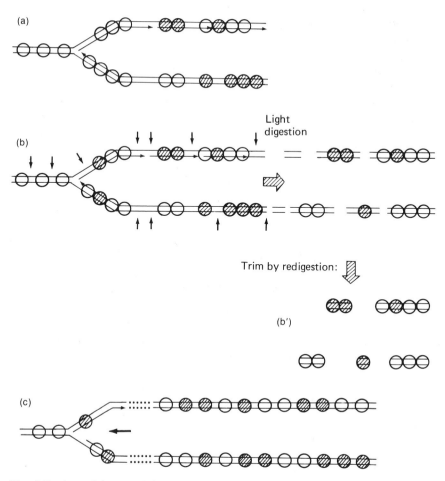

Fig. 9-7. A model to explain chromatin maturation: (a) Nucleosomes assemble on both arms of the replication fork, with some tendency to cluster. The spacings initially are irregular, so that light nuclease cutting (b) yields a smeared DNA gel pattern. Redigestion of these fragments produces trimming and well-defined small oligomers (b'). Upon maturation, gaps are filled and nucleosomes rearrange to yield the kind of regular pattern found in bulk chromatin (c).

first associate *randomly* on new DNA, an irregular pattern will result. Indeed, some gaps approaching 145 bp in size might be expected. These nucleosomes *must* rearrange to produce the mature, more regular spacing, and we may presume that this occurs by sliding. If new nucleosomes exhibit such a propensity to slide, it should not be surprising to observe nuclease-dependent sliding, with the production of compact oligomers and an apparent short-spacing. Whether the establishment of a final spacing is dictated by the cores finding preferred sites on the DNA, the interposition of H1, or some maturation process in the cores themselves is impossible to judge at this point.

Certainly, the maturation process is apt to be a complex one. Annunziato and Seale (1983b) report that butyrate treatment, which inhibits the deacetylation of histones (see Chapters 4 and 6), also delays some aspects of chromatin maturation. Interestingly, Cousens and Alberts (1982) have observed that nascent chromatin shows a heightened accessibility to the deacetylase. Movement away from the replication fork may be essential for maturation; Schlaeger et al. (1983) have found that inhibition of DNA synthesis prevents development of the "mature" response to micrococcal nuclease.

How close to the replication fork can nucleosomes be found? Electron microscopic studies (McKnight and Miller, 1977) show what appear to be core particles to within at least a few hundred base pairs of the fork, but cannot resolve the structure at shorter distances. The problem has been most closely studied by DePamphilis and Wassarman and their co-workers, using SV-40 as a model system (see DePamphilis and Wassarman, 1980; Herman et al., 1981; Cusick et al., 1981, 1983). Exonuclease digestion studies indicate that the closest distance from the fork to the first nucleosome on the leading strand is of the order of 125 bp, and on the retrograde strand, about 300 bp. These values are in general agreement with the psoralen crosslinking studies of Sogo et al. (1986) and are consistent with the observation that unligated Okazaki fragments are nucleosome-free (Herman et al., 1981). The overall picture suggested by these investigators is presented in Figure 9-8. The model should probably not be taken too literally, for much necessary machinery for replication is not illustrated, and considerable variation in local structure is doubtless possible. Nevertheless, it provides a framework for further analysis. The authors of this model distinguish three kinds of DNA regions behind the fork: prenucleosomal, immature nucleosomal, and mature nucleosomal. The first two probably correspond to what others have referred to as "nascent" chromatin. Cusick et al. (1983) point out that according to their data, the nuclease sensitivity of nascent chromatin cannot be explained by the prenucleosomal DNA alone. The immature nucleosomes must also exhibit increased susceptibility. These authors suggest that the immature region probably lacks H1 and that maturation occurs only between about 5 Kb and 40 Kb from the fork.

There are many aspects of chromatin replication that remain largely unexplored. For example, while much attention has been focused on the events *following* the passage of the fork, very little is known about what happens immediately ahead of the moving replication point. In prokaryotes, rather elaborate mechanisms for unwinding the DNA have been detected. In eukaryotes, a whole complicated chromatin structure must somehow be undone, and an even more complex machinery might be expected. Virtually nothing is known of this, although Sogo et al. (1986)

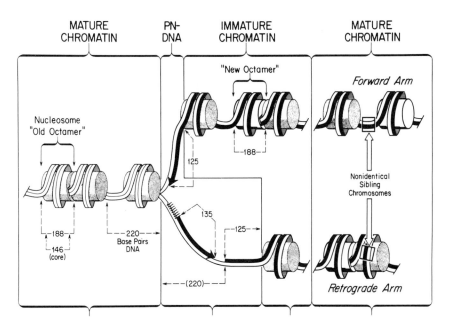

Fig. 9-8. Model for replication forks in SV-40 chromatin. Numbers give the average distances in nucleotides. Newly synthesized DNA is represented by black ribbons, with direction of synthesis indicated by arrowheads. PN-DNA is the prenucleosomal DNA (see text). The specification of "nonidentical sibling chromosomes" is based on evidence indicating a random distribution of nucleosomes on mature SV-40. From Cusick et al., 1981. (Reprinted with permission from Biochemistry *20*, p. 6648, Cusick et al. Copyright 1981 American Chemical Society.)

report evidence for nucleosomes very close to the fork. Even concerning the reorganization of chromatin after replication, there remain unexplored and fascinating problems. It is clear from several studies (Weintraub, 1979; Groudine and Weintraub, 1982; Lowenhaupt et al., 1983) that specific features of transcriptionally competent chromatin are reproduced on both daughter strands. How is this accomplished? The problem becomes particularly perplexing if we take the now-prevalent position that new and old nucleosomes are distributed to both new arms of the fork. The information cannot lie solely in the DNA itself, for the same DNA, in different tissues, is complexed into different chromatin structures. It would seem essential that *some* features of the chromatin structure peculiar to a given somatic cell type be retained undisturbed through the replication process and that this could somehow guide the organization of the daughter duplexes. The most fascinating work may still be ahead.

References

Absolom, D. & van Regenmortel, M.H. (1977) FEBS Lett. *81*, 419–422.

Achwal, C.W., Ganguly, P. & Chandra, H.S. (1984) EMBO J. *3*, 263–266.

Adamietz, P. & Rudolph, A. (1984) J. Biol. Chem. *259*, 6841–6846.

Adamietz, P., Bredehorst, R. & Hilz, H. (1978) Eur. J. Biochem. *91*, 317–326.

Adamson, E.D. & Woodland, H.R. (1974) J. Mol. Biol. *88*, 263–285.

Adler, A.J., Ross, D.G., Chen, K., Stafford, P.A., Woiszwillo, M.J. & Fasman, G.D. (1974) Biochemistry *13*, 616–623.

Adler, A.J., Moran, E.C. & Fasman, G.D. (1975) Biochemistry *14*, 4179–4185.

Adolph, K.W. (1981) Eur. J. Cell Biol. *24*, 146–153.

Adolph, K.W. (1984) FEBS Lett. *165*, 211–215.

Adolph, K.W., Cheng, S.M. & Laemmli, U.K. (1977) Cell *12*, 805–816.

Agard, D.A. & Sedat, J.W. (1983) Nature (London) *302*, 676–681.

Agell, N., Chiva, M. & Mezquita, C. (1983) FEBS Lett. *155*, 209–212.

Agutter, P.S. & Richardson, J.C.W. (1980) J. Cell Sci. *44*, 395–425.

Ahmed, C.M.I. & Padayatty, J.D. (1982) Indian J. Biochem. Biophys. *19*, 155–159.

Aitken, A. & Rouviere-Yaniv, J. (1979) Biochem. Biophys. Res. Comm. *91*, 461–467.

Ajiro, K. & Nishimoto, T. (1985) J. Biol. Chem. *260*, 15379–15381.

Ajiro, K., Borun, T.W. & Cohen, L.H. (1981a) Biochemistry *20*, 1445–1454.

Ajiro, K., Borun, T.W., Shulman, S.D., McFadden, G.M. & Cohen, L.H. (1981b) Biochemistry *20*, 1454–1464.

Albright, S.C., Nelson, P.P. & Garrard, W.T. (1979) J. Biol. Chem. *254*, 1065–1073.

Albright, S.C., Wiseman, J.M., Lange, R.A. & Garrard, W.T. (1980) J. Biol. Chem. *255*, 3673–3684.

Allan, J., Hartman, P.G., Crane-Robinson, C. & Aviles, F.X. (1980) Nature (London) *288*, 675–679.

Allan, J., Cowling, G.J., Harborne, N., Cattani, P., Cragie, R. & Gould, H. (1981) J. Cell Biol. *90*, 279–288.

Allan, J., Harborne, N., Rau, D.C. & Gould, H. (1982a) J. Cell Biol. *93*, 285–297.

Allan, J., Smith, B.J., Dunn, B. & Bustin, M. (1982b) J. Biol. Chem. *257*, 10533–10535.

Allan, J., Rau, D.C., Harborne, N. & Gould, H. (1983) J. Cell Biol. *98*, 1320–1327.

Allan, J., Mitchell, T., Harborne, N., Böhm, L. & Crane-Robinson, C. (1986) J. Mol. Biol. *187*, 591–601.

Allen, J.R., Roberts, T.H., Loeblich, A.R. & Klotz, L.C. (1975) Cell *6*, 161–169.

Allfrey, V. (1977) in *Chromatin and Chromosome Structure* (Li, H.J. & Eckhardt, R.A., eds.), pp. 167–191, Academic Press, N.Y.

Allfrey, V. (1980) in *Gene Expression: The Production of RNAs* (Goldstein, L. & Prescott, D.M., eds.), vol. 3, pp. 347–437, Academic Press, N.Y.

Allfrey, V. (1982) in *The HMG Chromosomal Proteins* (Johns, E.W., ed.), pp. 123–148, Academic Press, N.Y.

Allfrey, V. & Mirsky, A.E. (1964) in *The Nucleohistones* (Bonner, J. & Ts'o, P., eds.), pp. 267–288, Holden-Day, San Francisco.

Allfrey, V., Faulkner, R.M. & Mirsky, A.E. (1964) Proc. Natl. Acad. Sci. USA *51*, 786–794.

Allis, C.D. & Gorovsky, M.A. (1979) Proc. Natl. Acad. Sci. USA *76*, 4857–4861.

Allis, C.D. & Gorovsky, M.A. (1981) Biochemistry *20*, 3828–3833.

Allis, C.D. & Wiggins, J.C. (1984a) Exp. Cell Res. *153*, 287–298.

Allis, C.D. & Wiggins, J.C. (1984b) Devel. Biol. *101*, 282–294.

Allis, C.D., Bowen, J.K., Abraham, G.N., Glover, C.V.C. & Gorovsky, M.A. (1980a) Cell *20*, 55–64.

Allis, C.D., Glover, C.V.C., Bowen, J.K. & Gorovsky, M.A. (1980b) Cell *20*, 609–617.

Allis, C.D., Allen, R.L., Wiggins, J.C., Chicoine, L.G. & Richman, R. (1984) J. Cell Biol. *99*, 1669–1677.

Allis, C.D., Chicoine, L.G., Richman, R. & Schulman, I.G. (1985) Proc. Natl. Acad. Sci. USA *82*, 8048–8052.

Allis, C.D., Richman, R., Gorovsky, M.A., Ziegler, Y.S., Touchstone, B., Bradley, W.A. & Cook, R.G. (1986) J. Biol. Chem. *261*, 1941–1948.

Altman, R. (1889) Arch. fur Anatomie und Physiol. *1889*, 524–536.

Anachkova, B. & Russev, G. (1983) Biochim. Biophys. Acta *740*, 369–372.

Anachkova, B., Russev, G. & Tsanev, R. (1977) Int. J. Biochem. *8*, 619–621.

Andersen, M.W., Ballal, N.R., Goldknopf, I.L. & Busch, H. (1981a) Biochemistry *20*, 1100–1104.

Andersen, M.W., Goldknopf, I.L. & Busch, H. (1981b) FEBS Lett. *132*, 210–214.

Anderson, J.N., Vanderbilt, J.N., Lawson, G.M., Tsai, M-J. & O'Malley, B.W. (1983) Biochemistry *22*, 21–30.

Anderson, S. & DePamphilis, M.L. (1979) J. Biol. Chem. *254*, 11495–11504.

Angelier, N. & Lacroix, J.C. (1975) Chromosoma *51*, 323–335.

Annunziato, A.T. & Seale, R.L. (1982) Biochemistry *21*, 5431–5438.

Annunziato, A.T. & Seale, R.L. (1983a) Mol. C. Biochem. *55*, 99–112.

Annunziato, A.T. & Seale, R.L. (1983b) J. Biol. Chem. *258*, 12675–12684.

Annunziato, A.T. & Seale, R.L. (1984) Nucleic Acids Res. *12*, 6179–6196.

Annunziato, A.T., Schindler, R.K., Riggs, M.G. & Seale, R.L. (1982) J. Biol. Chem. *257*, 8507–8515.

Apriletti, J.W., David-Inouye, Y., Eberhardt, N.L. & Baxter, J.D. (1984) J. Biol. Chem. *259*, 10941–10948.

Arceci, R.J. & Gross, P.R. (1977) Proc. Natl. Acad. Sci. USA *74*, 5016–5020.

Arceci, R.J. & Gross, P.R. (1980) Devel. Biol. *80*, 186–209.

Arnott, S. & Chandrasekaran, R. (1981) in *Proceedings of the Second SUNYA Conversation in the Discipline Biomolecular Stereodynamics* (Sarma, R., ed.), vol. 1, pp. 99–122, Adenine Press, N.Y.

Arnott, S. & Hukins, D.W.L. (1972) Biochem. J. *130*, 453–465.

Arnott, S. & Selsing, E. (1974) J. Mol. Biol. *88*, 509–521.

Arnott, S., Hukins, D.W.L., Dover, S.D., Fuller, W. & Hodgson, A.R. (1973) J. Mol. Biol. *81*, 107–122.

Arnott, S., Chandrasekaran, R., Hukins, D.W.L., Smith, P.J.C., Watts, L. (1974) J. Mol. Biol. *88*, 523–533.

Arnott, S., Chandrasekaran, R., Birdsall, D.L., Leslie, A.G.W. & Ratliff, R.L. (1980) Nature (London) *283*, 743–745.

Arrigo, A.P. (1983) Nucleic Acids Res. *11*, 1389–1404.

Ashburner, M. & Bonner, J.J. (1979) Cell *17*, 241–254.

Ashikawa, I., Nishimura, Y., Tsuboi, M. & Watanabe, K. (1982a) J. Biochem. *91*, 2047–2055.

Ashikawa, I., Nishimura, Y., Tsuboi, M. & Zama, M. (1982b) J. Biochem. *92*, 1425–1430.

Ashikawa, I., Kinosita, K., Ikegami, A., Nishimura, Y., Tsuboi, M., Watanabe, K. & Iso, K. (1983a) J. Biochem. *93*, 665–668.

Ashikawa, I., Kinosita, K., Ikegami, A. Nishimura, Y., Tsuboi, M., Watanabe, K. & Iso, K. (1983b) Biochemistry *22*, 6018–6026.

Astbury, W. (1947) Symposia Soc. Exp. Biol. *1*, 66–76.

Aström, S. & von der Decken, A. (1980) Life Sciences *26*, 797–804.

Aubin, R.J., Frechetti, A., DeMurcia, G., Mandel, P., Lord, A., Grondin, G. & Poirier, G.G. (1983) EMBO J. *2*, 1685–1693.

Ausio, J. & van Holde, K.E. (1986) Biochemistry *25*, 1421–1428.

Ausio, J., Seger, D. & Eisenberg, H. (1984a) J. Mol. Biol. *176*, 77–104.

Ausio, J., Borochov, N., Seger, D. & Eisenberg, H. (1984b) J. Mol. Biol. *177*, 373–398.

Ausio, J., Zhou, G., & van Holde, K.E. (1987) Biochemistry *26*, 5595–5599.

Avery, O.T., MacLeod, C.M. & McCarty, M. (1944) J. Exp. Med. *79*, 137–158.

Aviles, F.J., Chapman, G.E., Kneale, G.C., Crane-Robinson, C. & Bradbury, E.M. (1978) Eur. J. Biochem. *88*, 363–371.

Avramova, Z., Dessev, G. & Tsanev, R. (1980) FEBS Lett. *118*, 58–62.

Axel, R., Cedar, H. & Felsenfeld, G. (1973) Cold Spring Harbor Symp. Quant. Biol. *38*, 773–783.

Azorin, F., Perez-Grau, L. & Subirana, J. (1982) Chromosoma *85*, 251–260.

Baase, W.D. & Johnson, W.C. (1979) Nucleic Acids Res. *6*, 797–814.

Baer, B.W. & Kornberg, R.D. (1979) J. Biol. Chem. *254*, 9678–9681.

Baer, B.W. & Rhodes, D. (1983) Nature (London) *301*, 482–488.

Bafus, N.L., Albright, S.C., Todd, R.D. & Garrard, W.T. (1978) J. Biol. Chem. *253*, 2568–2574.

Bailey, G., Nixon, J.E., Hendricks, J.D., Sinnhuber, R. & van Holde, K.E. (1980) Biochemistry *19*, 5836–5842.

Bakayev, V.V., Bakayeva, T.G. & Varshavsky, A.J. (1977) Cell *11*, 619–629.

Bakayev, V.V., Bakayeva, T.G., Schmatchenko, V.V. & Georgiev, G.P. (1978) Eur. J. Biochem. *91*, 291–301.

Bakayev, V.V., Schmatchenko, V.V. & Georgiev, G.P. (1979) Nucleic Acids Res. *7*, 1525–1540.

Bakayev, V.V., Domansky, N.N. & Bakayeva, T.G. (1981) FEBS Lett. *133*, 75–78.

Baker, C.C. & Isenberg, I. (1976) Biochemistry *15*, 629–634.

Bakke, A.C., Wu, J.R. & Bonner, J. (1978) Proc. Natl. Acad. Sci. USA *75*, 705–709.

Ballal, N.R., Kang, Y.-J., Olson, M.O.J. & Bank, H. (1975) J. Biol. Chem. *250*, 5921–5925.

Bannon, G.A. & Gorovsky, M.A. (1984) in *Histone Genes: Structure, Organization, and Regulation* (Stein, G., Stein, J. & Marzluff, W., eds.), pp. 163–179, John Wiley & Sons, N.Y.

Bannon, G.A., Bowen, J.K., Yao, M.-C., & Gorovsky, M.S. (1984) Nucleic Acids Res. *12*, 1961–1975.

Barbero, J.L., Franco, L., Montero, F. & Moran, F. (1980) Biochemistry *19*, 4080–4087.

Barnes, K.L., Craigie, R.A., Cattini, P.A. & Cavalier-Smith, T. (1982) J. Cell Sci. *57*, 151–160.

Barrack, E.R. & Coffey, D.S. (1982) Recent Prog. in Hormone Res. *18*, 133–194.

Barsh, G.S., Roush, C.L. & Gelinas, R.E. (1984) J. Biol. Chem. *259*, 14906–14913.

Barsoum, J. & Varshavsky, A. (1985) J. Biol. Chem. *260*, 7688–7697.

Barsoum, J., Levinger, L. & Varshavsky, A. (1982) J. Biol. Chem. *257*, 5274–5282.

Bates, D.L. & Thomas, J.O. (1981) Nucleic Acids Res. *9*, 5883–5894.

Bates, D.L., Butler, P.J.G., Pearson, E.C. & Thomas, J.O. (1981) Eur. J. Biochem. *119*, 469–476.

Bauer, W.R. (1978) Ann. Rev. Biophys. Bioeng. *7*, 287–313.

Beard, P. (1978) Cell *15*, 955–967.

Beaudette, N.V., Fulmer, A.W., Okabayashi, H. & Fasman, G.D. (1981) Biochemistry *20*, 6526–6535.

Behe, M. & Felsenfeld, G. (1981) Proc. Natl. Acad. Sci. USA *78*, 1619–1623.

Behe, M., Zimmerman, S. & Felsenfeld, G. (1981) Nature (London) *293*, 233–235.

Belikoff, E., Wong, L.J. & Alberts, B.M. (1980) J. Biol. Chem. *255*, 11448–11453.

Bellard, M., Oudet, P., Germond, J.-E. & Chambon, P. (1976) Eur. J. Biochem. *70*, 543–553.

Bellard, M., Gannon, F. & Chambon, P. (1977) Cold Spring Harbor Symp. Quant. Biol. *42*, 779–791.

Bellard, M., Kuo, M.T., Dretzen, G. & Chambon, P. (1980) Nucleic Acids Res. *8*, 2737–2750.

Bellard, M., Dretzen, G., Bellard, F., Oudet, P. & Chambon, P. (1982) EMBO J. *1*, 223–230.

Belyavsky, A.V., Bavykin, S.G., Goguadze, E.G. & Mirzabekov, A.D. (1980) J. Mol. Biol. *139*, 519–536.

Bendel, P., Laub, O. & James, T.L. (1982) J. Am. Chem. Soc. *104*, 6748–6754.

Benedict, R.C., Moudrianakis, E.N. & Ackers, G.K. (1984) Biochemistry *23*, 1214–1218.

Benezra, R., Cantor, C.R. & Axel, R. (1986) Cell *44*, 697–704.

Benham, C.J. (1980) Nature (London) *286*, 637–638.

Benham, C.J. (1981) J. Mol. Biol. *150*, 43–68.

Benjamin, R.C. & Gill, D.M. (1980) J. Biol. Chem. *255*, 10502–10508.

Bentley, G.A., Finch, J.T. & Lewit-Bentley, A. (1981) J. Mol. Biol. *145*, 771–784.

Bentley, G.A., Lewit-Bentley, A., Finch, J.T., Podjarny, A.D. & Roth, M. (1984) J. Mol. Biol. *176*, 55–75.

Benvenisty, N., Mencher, D., Meyuhas, O., Razin, A. & Reshef, L. (1985) Proc. Natl. Acad. Sci. USA *82*, 267–271.

Benyajati, C. & Worcel, A. (1976) Cell *9*, 393–407.

Berent, S.L. & Sevall, J.S. (1984) Biochemistry *23*, 2977–2983.

Berezney, R. & Burchholtz, L.A. (1981) Exp. Cell Res. *132*, 1–13.

Berezney, R. & Coffey, D.S. (1975) in *Advances in Enzyme Regulation* (Weber, G., ed.), vol. 14, pp. 63–100, Pergamon Press, Oxford.

Bergman, L.W. (1986) Mol. Cell. Biol. *6*, 2298–2304.

Bergman, L.W. & Kramer, R.A. (1983) J. Biol. Chem. *258*, 7223–7227.

Bergman, L.W., Stranathan, M.C. & Preis, L.H. (1986) Mol. Cell. Biol. *6*, 38–46.

Berkowitz, E.M. & Doty, P. (1975) Proc. Natl. Acad. Sci. USA *72*, 3328–3332.

Berkowitz, E.M. & Riggs, E.A. (1981) Biochemistry *20*, 7284–7290.

Berkowitz, E.M., Sanborn, A.C. & Vaughan, D.W. (1983) J. Neurochem. *41*, 516–523.

Bernúes, J., Querol, E., Martinez, P., Barris, A., Espel, E. & Lloberas, J. (1983) J. Biol. Chem. *258*, 11020–11024.

Berrios, M., Osheroff, N. & Fisher, P. (1985) Proc. Natl. Acad. Sci. USA *82*, 4142–4146.

Bertrand, E., Erard, M., Gomez-Lira, M. & Bode, J. (1984) Arch. Biochem. Biophys. *229*, 395–398.

Bhorjee, J.S. & Pederson, T. (1973) Biochemistry *12*, 2766–2773.

Bhorjee, J.S. & Pederson, T. (1976) Biochim. Biophys. Acta *418*, 154–159.

Bhullar, B.S., Hewitt, J. & Candido, E.P.M. (1981) J. Biol. Chem. *256*, 8801–8806.

Bidney, D.L. & Reeck, G.R. (1977) Biochemistry *16*, 1844–1849.

Bidney, D.L. & Reeck, G.R. (1978) Biochem. Biophys. Res. Comm. *85*, 1211–1218.

Bieker, J.J., Martin, P.L. & Roeder, R.G. (1985) Cell *40*, 119–127.

Bina-Stein, M. (1978) J. Biol. Chem. *253*, 5213–5219.

Bird, A.P. (1980) Nucleic Acids Res. *8*, 1499–1504.

Bloom, K.S. & Anderson, J.N. (1978) J. Biol. Chem. *253*, 4446–4450.

Bloom, K.S. & Anderson, J.N. (1979) J. Biol. Chem. *254*, 10532–10539.

Bloom, K.S. & Anderson, J.N. (1982) J. Biol. Chem. *257*, 13018–13027.

Bloom, K.S. & Carbon, J. (1982) Cell *29*, 305–317.

Bloom, K.S., Amaya, E., Carbon, J., Clarke, L., Hill, A. & Yeb E. (1984) J. Cell Biol. *99*, 1559–1568.

Blumenthal, A.B., Kreigstein, H. & Hogness, D.W. (1973) Cold Spring Harbor Symp. Quant. Biol. *38*, 205–223.

Bock, H., Abler, S., Zhang, X.-Y., Fritton, H. & Igo-Kemenes, T. (1984) J. Mol. Biol. *176*, 131–154.

Bode, J., Henco, K. & Wingender, E. (1980) Eur. J. Biochem. *110*, 143–152.

Bode, J., Gomez-Lira, M.M. & Schroter, H. (1983) Eur. J. Biochem. *130*, 437–445.

Boffa, L.C., Sterner, R., Vidali, G. & Allfrey, V.G. (1979) Biochem. Biophys. Res. Comm. *89*, 1322–1327.

Boffa, L.C., Gruss, R.J. & Allfrey, V.G. (1981) J. Biol. Chem. *256*, 9612–9621.

Bogenhagen, D.F., Sakonju, S. & Brown, D.D. (1980) Cell *19*, 27–35.

Böhm, L., Hayashi, H., Cary, P.D., Moss, T., Crane-Robinson, C. & Bradbury, E.M. (1977) Eur. J. Biochem. *77*, 487–493.

Böhm, L., Crane-Robinson, C. & Sautiere, P. (1980a) Eur. J. Biochem. *106*, 525–530.

Böhm, L., Schlaeger, E.J. & Knippers, R. (1980b) Eur. J. Biochem. *112*, 353–362.

Böhm, L., Briand, G., Sautiere, P. & Crane-Robinson, C. (1981) Eur. J. Biochem. *119*, 67–74.

Böhm, L., Briand, G., Sautiere, P. & Crane-Robinson, C. (1982) Eur. J. Biochem. *123*, 299–303.

Bolton, P.H. & James, T.L. (1980) Biochemistry *19*, 1388–1392.

Bolund, L.A. & Johns, E.W. (1973) Eur. J. Biochem. *35*, 546–553.

Bonne-Andrea, C., Harper, F., Sobczak, J. & De Recondo, A.M. (1984) EMBO J. *3*, 1193–1199.

Bonner, J. & Ts'o, P. (1964) *The Nucleohistones*, Holden Day, Inc., San Francisco.

Bonner, J. & Huang, R.-C. (1964) in *The Nucleohistones* (Bonner, J. & Ts'o, P., eds.), pp. 251–261, Holden Day, San Francisco.

Bonner, J., Chalkley, G.R., Dahmus, M., Fambrough, D., Fujimura, F., Huang, R.-C., Huberman, J., Jensen, R., Marushige, K., Ohlenbusch, H., Olivera, B., & Widholm, J. (1968) in *Methods in Enzymology* (Grossman, L. & Moldave, K., eds.), vol. 12B, p. 3, Academic Press, N.Y.

Bonner, W.M. (1975) J. Cell Biol. *64*, 421–430.

Bonner, W.M. (1978) Nucleic Acids Res. *5*, 71–85.

Bonner, W.M. & Stedman, J.D. (1979) Proc. Natl. Acad. Sci. USA *76*, 2190–2194.

Boothby, M., Tatchell, K. & van Holde, K.E. (1977) Biol. Bull. *153*, 416.

Borchsenius, S., Bonven, B., Leer, J.C. & Westergaard, O. (1981) Eur. J. Biochem. *117*, 245–250.

Borkhardt, B. & Neilsen, O.F. (1981) Chromosoma *84*, 131–143.

Borun, T.W., Pearson, D. & Paik, W.K. (1972) J. Biol. Chem. *247*, 4288–4298.

Bostock, C.J. & Sumner, A.T. (1978) *The Eukaryotic Chromosome*, North Holland Pub. Co., Amsterdam.

Botchan, P. & Dayton, A. (1982) Nature (London) *299*, 453–456.

Boulikas, T., Wiseman, J.M. & Garrard, W.T. (1980) Proc. Natl. Acad. Sci. USA *77*, 127–131.

Bouteille, M., Bouvier, D. & Seve, A.P. (1983) Int. Rev. Cytol. *83*, 135–182.

Bouvier, D. (1977) Cytobiology *15*, 420–437.

Bowen, B.C. (1981) Nucleic Acids Res. *9*, 5093–5108.

Bradbury, E.M. (1982) in *The HMG Chromosomal Proteins* (Johns, E.W., ed.), pp. 89–110, Academic Press, N.Y.

Bradbury, E.M. & Rattle, H.W.E. (1972) Eur. J. Biochem. *27*, 270–281.

Bradbury, E.M., Crane-Robinson, C., Phillips, D.M.P., Johns, E.W. & Murray, K. (1965) Nature (London) *205*, 1315–1316.

Bradbury, E.M., Cary, P.D., Crane-Robinson, C., Riches, P.L. & Johns, E.W. (1972) Eur. J. Biochem. *26*, 482–489.

Bradbury, E.M., Cary, P.D., Crane-Robinson, C. & Rattle, H.W.E. (1973a) Ann. N.Y. Acad. Sci. *222*, 266–288.

Bradbury, E.M., Inglis, R.J., Matthews, H.R. & Sarner, N. (1973b) Eur. J. Biochem. *33*, 131–139.

Bradbury, E.M., Inglis, R.J. & Matthews, H.R. (1974a) Nature (London) *247*, 257–261.

Bradbury, E.M., Inglis, R.J., Matthews, H.R. & Langan, T.A. (1974b) Nature (London) *249*, 553–556.

Bradbury, E.M., Cary, P.D., Chapman, G.E., Crane-Robinson, C., Danby, S.E. & Rattle, H.W.E. (1975a) Eur. J. Biochem. *52*, 605–613.

Bradbury, E.M., Cary, P.D., Crane-Robinson, C., Rattle, H.W.E., Boublik, M. & Sautiere, P. (1975b) Biochemistry *14*, 1876–1885.

Braddock, G.W., Baldwin, J.P. & Bradbury, E.M. (1981) Biopolymers *20*, 327–343.

Brahms, J. & Mommaerts, W.F.H.M. (1964) J. Mol. Biol. *10*, 73–88.

Brahms, J.G., Dargouge, O., Brahms, S., Ohara, Y. & Vagner, V. (1985) J. Mol. Biol. *181*, 455–465.

Brahms, S., Brahmachari, S.K., Angelier, N. & Brahms, J.G. (1981) Nucleic Acids Res. *9*, 4879–4893.

Brahms, S., Vergne, J., Brahms, J.G., Capua, E.D., Bucher, P. & Koller, T. (1982) J. Mol. Biol. *162*, 473–493.

Bram, S. (1971) J. Mol. Biol. *58*, 277–288.

Bram, S. & Ris, H. (1971) J. Mol. Biol. *55*, 325–336.

Brand, S.H., Kumar, N.M. & Walker, I.O. (1981) FEBS Lett. *133*, 63–66.

Brandt, W.F. & von Holt, C. (1974) Eur. J. Biochem. *46*, 407–417.

Brandt, W.F. & von Holt, C. (1978) Biochim. Biophys. Acta *537*, 177–181.

Brandt, W.F. & von Holt, C. (1982) Eur. J. Biochem. *121*, 501–510.

Brandt, W.F. & von Holt, C. (1986) FEBS Lett. *194*, 278–281.

Brandt, W.F., Strickland, W.N. & von Holt, C. (1974) FEBS Lett. *40*, 349–352.

Brandt, W.F., Strickland, W.N., Strickland, M., Carlisle, L., Woods, D. & von Holt, C. (1979) Eur. J. Biochem. *94*, 1–10.

Branno, M., De Franciscis, V. & Tosi, L. (1983) Biochim. Biophys. Acta *741*, 136–142.

Bremer, J.W., Busch, H. & Yoeman, L.C. (1981) Biochemistry *20*, 2013–2017.

Briand, G., Kmiecik, D., Sautiere, P., Wouters, D., Borlie-Loy, O., Biserte, G., Mazen, A. & Champagne, M. (1980) FEBS Lett. *112*, 147–151.

Briat, J.F., Letoffe, S., Mache, R. & Rouviere-Yaniv, J. (1984) FEBS Lett. *172*, 75–79.

Bricteux-Gregoire, S. & Verly, W.G. (1983) FEBS Lett. *157*, 115–118.

Brotherton, T.W. & Ginder, G. (1982) Clin. Res. *30*, A729.

Brotherton, T.W., Covault, J., Shires, A. & Chalkley, R. (1981) Nucleic Acids Res. *9*, 5061–5073.

Brown, A.P. (1983) J. Theo. Biol. *104*, 401–416.

Brown, D.D. & Gurdon, J. (1977) Proc. Natl. Acad. Sci. USA *74*, 2064–2068.

Brown, D.D. & Gurdon, J. (1978) Proc. Natl. Acad. Sci. USA *75*, 2849–2853.

Brown, E., Goodwin, G.H., Mayes, E.L.V., Hastings, J.R.B. & Johns, E.W. (1980) Biochem. J. *191*, 661–664.

Brown, I.R. (1978) Biochem. Biophys. Res. Comm. *84*, 285–292.

Broyles, S.S. & Pettijohn, D.E. (1986) J. Mol. Biol. *187*, 47–60.

Bruegger, B.B., DeLange, R.J., Smith, R.A. & Lin, Y.C. (1979) J. Chin. Chem. *26*, 5–10.

Brust, R. & Harbers, E. (1981) Eur. J. Biochem. *117*, 609–615.

Bryan, P.N., Wright, E.B. & Olins, D.E. (1979) Nucleic Acids Res. *6*, 1449–1465.

Bryan, P.N., Hofstetter, H. & Birnstiel, M.L. (1981) Cell *27*, 459–466.

Bryan, P.N., Hofstetter, H. & Birnstiel, M.L. (1983) Cell *33*, 843–848.

Bucci, L.R., Brock, W.A. & Meistrich, M.L. (1982) Exp. Cell Res. *140*, 111–118.

Burch, J.B.E. & Martinson, H.G. (1980) Nucleic Acids Res. *8*, 4969–4987.

Burch, J.B.E. & Weintraub, H. (1983) Cell *33*, 65–76.

Burgoyne, L.A. (1985) Cytobios *43*, 141–147.

Burgoyne, L.A. & Skinner, J.D. (1981) Biochem. Biophys. Res. Comm. *99*, 893–899.

Burgoyne, L.A. & Skinner, J.D. (1982) Nucleic Acids Res. *10*, 665–673.

Burgoyne, L.A., Hewish, D.R. & Mobbs, J. (1974) Biochem. J. *143*, 67–72.

Burlingame, R.W., Love, W.E. & Moudrianakis, E.N. (1984) Science *223*, 413–414.

Burlingame, R., Love, W.E., Wang, B.-C., Hamlin, R., Yuong, N.-H. & Moudrianakis, E.N. (1985) Science *228*, 546–553.

Burton, D.R., Butler, M.J., Hyde, J.E., Philips, D., Skidmore, C.J. & Walker, I.O. (1978) Nucleic Acids Res. *5*, 3643–3663.

Burzio, L.O., Riquelme, P.T. & Koide, S.C. (1979) J. Biol. Chem. *254*, 3029–3037.

Busch, H. & Goldknopf, I.L. (1981) Mol. Cell. Biochem. *40*, 173–187.

Busch, H. & Rothblum, L. (eds.) (1982) *The Cell Nucleus*, vol. 11, Academic Press, N.Y.

Busslinger, M., Portmann, R., Irminger, J.C. & Birnstiel, M.L. (1980) Nucleic Acids Res. *8*, 957–977.

Butler, A.P. & Olins, D.E. (1982) Biochim. Biophys. Acta *698*, 199–203.

Butler, A.P., Mardian, J.K.W. & Olins, D.E. (1985) J. Biol. Chem. *260*, 10613–10620.

Butler, M.J., Davies, K.E. & Walker, I.O. (1978) Nucleic Acids Res. *5*, 667–678.

Butler, P.J.G. (1983) CRC Crit. Reviews Biochem. *15*, 57–91.

Butler, P.J.G. (1984) EMBO J. *3*, 2599–2604.

Butler, P.J.G. & Thomas, J.O. (1980) J. Mol. Biol. *140*, 505–529.

Butt, T.R. & Smulson, M.E. (1980) Biochemistry *19*, 5235–5243.

Butt, T.R., Jump, D.B. & Smulson, M.E. (1979) Proc. Natl. Acad. Sci. USA *76*, 1628–1632.

Byvoet, P. (1971) Biochim. Biophys. Acta *238*, 375a–376b.

Byvoet, P., Sheperd, G.R., Hardin, J.M. & Noland, B.J. (1972) Arch. Biochem. Biophys. *148*, 558–567.

Caffarelli, E., DeSantis, P., Leoni, L., Savino, M. & Trotta, E. (1983) Biochim. Biophys. Acta *739*, 235–243.

Caffarelli, E., Leoni, L. & Savino, M. (1985) FEBS Lett. *181*, 69–73.

Calladine, C.R. (1982) J. Mol. Biol. *161*, 343–352.

Callaway, J.E., DeLange, R.J. & Martinson, H.G. (1985) Biochemistry *24*, 2686–2692.

Camato, R. & Tanguay, R.M. (1982) EMBO J. *1*, 1529–1532.

Camerini-Otero, R.D. & Felsenfeld, G. (1977a) Nucleic Acids Res. *4*, 1159–1181.
Camerini-Otero, R.D. & Felsenfeld, G. (1977b) Proc. Natl. Acad. Sci. USA *74*, 5519–5523.
Camerini-Otero, R.D. & Zasloff, M.A. (1980) Proc. Natl. Acad. Sci. USA *77*, 5079–5083.
Camerini-Otero, R.D., Sollner-Webb, B. & Felsenfeld, G. (1976) Cell *8*, 333–347.
Camerini-Otero, R.D., Sollner-Webb, B., Simon, R.H., Williamson, P., Zasloff, M. & Felsenfeld, G. (1977) Cold Spring Harbor Symp. Quant. Biol. *42*, 57–76.
Campbell, A.M., Cotter, R.I. & Pardon, J.F. (1978) Nucleic Acids Res. *5*, 1571–1580.
Candido, E.P.M. & Dixon, G.H. (1971) J. Biol. Chem. *246*, 3182–3188.
Candido, E.P.M. & Dixon, G.H. (1972a) J. Biol. Chem. *247*, 3868–3873.
Candido, E.P.M. & Dixon, G.H. (1972b) Proc. Natl. Acad. Sci. USA *69*, 2015–2019.
Candido, E.P.M., Reeves, R. & Davie, J.R. (1978) Cell *14*, 105–113.
Cantor & Schimmel, (1980) *Biophysical Chemistry* Part I, W.H. Freeman and Company, San Francisco.
Caplan, A., Kimara, T., Gould, H. & Allan, J. (1987) J. Mol. Biol. *193*, 57–70.
Carballo, M., Puigdomenech, P. & Palau, J. (1983) EMBO J. *2*, 1759–1764.
Carlson, R.D. (1984) Basic Life Sci. *27*, 47–72.
Caron, F. & Thomas, J.O. (1981) J. Mol. Biol. *146*, 513–537.
Caron, F., Jacq, C. & Rouviere-Yaniv, J. (1979) Proc. Natl. Acad. Sci. USA *76*, 4265–4269.
Carroll, A.G. & Ozaki, H. (1979) Exp. Cell Res. *119*, 307–315.
Carter, C.W. (1981) in *Structural Aspects of Recognition and Assembly in Biological Macromolecules* (Balaban, M., ed.), pp. 698–704, Balaban Int. Sci. Service, Rehovot, Israel.
Carter, C.W., Levinger, L.F. & Birinyi, F. (1980) J. Biol. Chem. *255*, 748–755.
Carter, D.B., Efird, P.H., Chae, C.-B. (1976) Biochemistry *15*, 2603–2607.
Cartwright, I.L. & Elgin, S.C.R. (1982a) Nucleic Acids Res. *10*, 5835–5852.
Cartwright, I.L. & Elgin, S.C.R. (1982b) UCLA Symp. Molec. Cell. Biol. *26*, 137–146.
Cartwright, I.L. & Elgin, S.C.R. (1984) EMBO J. *3*, 3101–3108.
Cartwright, I.L., Abmayr, S.M., Fleischman, G., Lowenhaupt, K., Elgin, S.C.R., Keene, M.A. & Howard, G.C. (1982) CRC Crit. Rev. Biochem. *13*, 1–86.
Cartwright, I.L., Herzberg, R.P., Dervan, P.B. & Elgin, S.C.R. (1983) Proc. Natl. Acad. Sci. USA *80*, 3213–3217.
Cary, P.D., Moss, T. & Bradbury, E.M. (1978) Eur. J. Biochem. *89*, 475–482.
Cary, P.D., Crane-Robinson, C., Bradbury, E.M. & Dixon, G.H. (1981a) Eur. J. Biochem. *119*, 545–551.
Cary, P.D., Hines, M.L., Bradbury, E.M., Smith, B.J. & Johns, E.W. (1981b) Eur. J. Biochem. *120*, 371–377.
Cary, P.D., Turner, C.H., Mayes, E. & Crane-Robinson, C. (1983) Eur. J. Biochem. *131*, 367–374.
Cary, P.D., Turner, C.H., Leung, I., Mayes, E., Crane-Robinson, C. (1984) Eur. J. Biochem. *143*, 323–330.
Caspersson, T. (1936) Skandinavisches Archiv fur Physiologie *73* (Supp. 8), 1–151.
Cavalier-Smith, T. (1976) Nature (London) *262*, 255–256.

Certa, U. & von Ehrenstein, G. (1981) Analyt. Biochem. *118*, 147–154.

Certa, U., Colavito-Shepanski, M. & Grunstein, M. (1984) Nucleic Acids Res. *12*, 7975–7985.

Chalkley, R. & Hunter, C. (1975) Proc. Natl. Acad. Sci. USA *72*, 1304–1308.

Chambers, S.A. & Shaw, B.R. (1984) J. Biol. Chem. *259*, 13458–13463.

Chambers, T.C., Langan, T.A., Matthews, H.R. & Bradbury, E.M. (1983) Biochemistry *22*, 30–37.

Chambon, P. (1975) Ann. Rev. Biochem. *44*, 613–638.

Chambon, P., Weill, J.D. & Mandel, P. (1963) Biochem. Biophys. Res. Comm. *11*, 39–43.

Chambon, P., Weill, J.D., Doly, J., Strosser, M.T. & Mandel, P. (1966) Biochem. Biophys. Res. Comm. *25*, 638–643.

Chambon, P., Gaub, M.P., Lepennec, J.P., Dierich, A. & Astinotti, D. (1984) in *Endocrinology: Proceedings of 7th International Congress of Endocrinology* (Labrie, F. & Prouix, L., eds.), pp. 3–10, Excerpts Medica, Amsterdam.

Chan, C.S.M. & Tye, B.-K. (1983) J. Mol. Biol. *168*, 505–527.

Chan, D.C.F., Biard-Roche, J., Gorka, C., Girardet, J.L., Lawrence, J.J. & Piette, L.H. (1984) J. Biomolec. Struct. Dynam. *2*, 319–332.

Chan, J.Y.H., Rodriguez, L.V. & Becker, F.F. (1977) Nucleic Acids Res. *4*, 2683–2695.

Chan, P.K. & Liew, C.C. (1979) Can. J. Biochem. *57*, 666–672.

Chandrasekaran, R., Annott, S., Bannerjee A., Campbell-Smith, S., Leslie A.G.W., & Puigjaner, L. (1980) in *Fiber Diffraction Methods* (French, A.D. & Gardner, K.H., eds.), ACS Publications, N.Y. pp. 483–502.

Chao, M.V., Gralla, J. & Martinson, H.G. (1979) Biochemistry *18*, 1068–1074.

Chao, M.V., Gralla, J.D. & Martinson, H.G. (1980a) Biochemistry *19*, 3254–3260.

Chao, M.V., Martinson, H.G. & Gralla, J.D. (1980b) Biochemistry *19*, 3260–3269.

Chapman, G.E., Hartman, P.G., Cary, P.D., Bradbury, E.M. & Lee, D.R. (1978) Eur. J. Biochem. *86*, 35–44.

Chargaff, E. (1950) Experimentia *6*, 201–209.

Cheah, K.S. & Osborne, D.J. (1977) Biochem. J. *163*, 141–144.

Chen, C.C., Smith, D.L., Bruegger, B.B., Halpern, R.M. & Smith, R.A. (1974) Biochemistry *13*, 3785–3789.

Chen, C.C., Bruegger, B.B., Kern, C.W., Lin, Y.C., Halpern, R.M. & Smith, R.A. (1977) Biochemistry *16*, 4852–4855.

Chestier, A. & Yaniv, M. (1979) Proc. Natl. Acad. Sci. USA *76*, 46–50.

Chicoine, L.G., Schulman, I.G., Richman, R., Cook, R.G. & Allis, C.D. (1986) J. Biol. Chem. *261*, 1071–1076.

Childs, G., Maxson, R. & Kedes, L.H. (1979) Dev. Biol. *73*, 153–173.

Childs, G., Nocente-McGrath, C., Lieber, T., Holt, C. & Knowles, J.A. (1982) Cell *31*, 383–393.

Chiu, M.L. (1982) Biochim. Biophys. Acta *699*, 110–120.

Chiu, M.L. & Irvin, J.L. (1983) Biochim. Biophys. Acta *740*, 342–345.

Chiva, M. & Mezquita, C. (1983) FEBS Lett. *162*, 324–328.

Choder, M., Bratosin, S. & Aloni, Y. (1984) EMBO J. *3*, 2929–2936.

Choe, J., Kolodrubetz, D. & Grunstein, M. (1982) Proc. Natl. Acad. Sci. USA *79*, 1484–1487.

Choie, D.D., Friedberg, E.C., Van den Berg, S.R. & Herman, M.M. (1977) J. Neurochem. *29*, 811–817.

Chong, M.T., Garrard, W.T. & Bonner, J. (1974) Biochemistry *13*, 5128–5134.

Chou, P.Y. & Fasman, G.D. (1974) Biochemistry *13*, 222–245.

Christensen, M.E. & Dixon, G.H. (1982) Develop. Biol. *93*, 404–415.

Christensen, M.E., Rattner, J.B. & Dixon, G.H. (1984) Nucleic Acids Res. *12*, 4575–4592.

Chung, D.G. & Lewis, P.N. (1986) Biochemistry *25*, 2048–2054.

Chung, S.Y., Folsom, V. & Wooley, J. (1983) Proc. Natl. Acad. Sci. USA *80*, 2427–2431.

Ciechanover, A., Heller, H., Elias, S., Haas, A.L. & Hersko, A. (1980) Proc. Natl. Acad. Sci. USA *77*, 1365–1368.

Ciechanover, A., Elias, S., Heller, H. & Hershko, A. (1982) J. Biol. Chem. *257*, 2537–2542.

Ciechanover, A., Finley, D. & Varshavsky, A. (1984) J. Cell Biochem. *24*, 27–53.

Ciejek, E.M., Tsai, M. & O'Malley, B.W. (1983) Nature (London) *306*, 607–609.

Clark, R.J. & Felsenfeld, G. (1971) Nature New Biology *229*, 101–106.

Clark, T.G. & Merriam, R.W. (1977) Cell *12*, 883–891.

Cockell, M., Rhodes, D. & Klug, A. (1983) J. Mol. Biol. *170*, 423–446.

Cockerill, P.N. & Garrard, W.T. (1986) Cell *44*, 272–282.

Cockerill, P.N. & Goodwin, G.H. (1983) Biochem. Biophys. Res. Comm. *112*, 547–554.

Cognetti, G., Platz, R.D., Meistrich, M.L. & DiLiegro, I. (1977) Cell Diff. *5*, 283–291.

Cohen, L.H., Newrock, K.M. & Zweidler, A. (1975) Science *190*, 994–997.

Cohen, S.S. (1945) J. Biol. Chem. *158*, 255–264.

Colavito-Shepanski, M. & Gorovsky, M. (1983) J. Biol. Chem. *258*, 5944–5954.

Colbert, D.A., Knoll, B.J., Woo, S.L.C., Mace, M.L., Tsai, M.-J. & O'Malley, B.W. (1980) Biochemistry *19*, 5586–5592.

Cole, K.D., York, R.G. & Kistler, S. (1984) J. Biol. Chem. *259*, 13695–13702.

Cole, R.D. (1977) in *The Molecular Biology of the Mammalian Genetic Apparatus* (Ts'o, P.O.P., ed.), vol. 1, pp. 93–104, North Holland Pub. Co., Amsterdam.

Cole, R.D. (1984) Anal. Biochem. *136*, 24–30.

Comings, D.E. (1968) Am. J. Hum. Genet. *20*, 440–460.

Comings, D.E. (1978) in *The Cell Nucleus* (Busch, H., ed.), vol. 4, part A, pp. 345–371, Academic Press, N.Y.

Comings, D.E. & Harris, D.C. (1976) J. Cell Biol. *70*, 440–452.

Compton, J.L. & Bonner, J.J. (1977) Cold Spring Harbor Symp. Quant. Biol. *42*, 835–838.

Compton, J.L., Bellard, M. & Chambon, P. (1976) Proc. Natl. Acad. Sci. USA *73*, 4382–4386.

Conner, B.N., Yoon, C., Dickerson, J. & Dickerson, R.E. (1984) J. Mol. Biol. *174*, 663–695.

Connor, W., States, J.C., Mezquita, J. & Dixon, G.H. (1984) J. Mol. Evol. *20*, 236–250.

Cook, P.R. & Brazell, I.A. (1976) J. Cell Sci. *22*, 287–302.

Cook, P.R. & Brazell, I.A. (1978) Eur. J. Biochem. *84*, 465–477.

Cook, P.R. & Brazell, I.A. (1980) Nucleic Acids Res. *8*, 2895–2906.

Cook, P.R. & Lasky, R.A. (eds.) (1984) *Journal of Cell Science, Suppl. 1*, Company of Cell Biologists Ltd., Cambridge.

Cook, P.R., Brazell, I.A. & Jost, E. (1976) J. Cell Sci. *22*, 303–324.

Cordingly, M.G., Riegel, A.T. & Hager, G.L. (1987) Cell *48*, 261–270.

Cornudella, L. & Rocha, E. (1979) Biochemistry *18*, 3724–3732.

Costlow, N.A., Simon, J.A. & Lis, J.T. (1985) Nature (London) *313*, 147–149.

Cotter, R.I. & Lilley, D.M.J. (1977) FEBS Lett. *82*, 63–68.

Cotton, R.W. & Hamkalo, B.A. (1981) Nucleic Acids Res. *9*, 445–467.

Cousens, L.S. & Alberts, B.M. (1982) J. Biol. Chem. *257*, 3945–3949.

Cousens, L.S., Gallwitz, D. & Alberts, B.M. (1979) J. Biol. Chem. *254*, 1716–1723.

Covault, J. & Chalkley, R. (1980) J. Biol. Chem. *255*, 9110–9116.

Cowman, M.K. & Fasman, G.D. (1980) Biochemistry *19*, 532–541.

Cox, R.F. (1973) Eur. J. Biochem. *39*, 49–61.

Craddock, V.M. & Henderson, A.R. (1980) Carcinogenesis *1*, 445–450.

Craine, B.L. & Kornberg, T. (1981) Cell *25*, 671–681.

Crampton, C.F., Stein, W.H. & Moore, S. (1957) J. Biol. Chem. *225*, 363–386.

Crane-Robinson, C. & Privalov, P.L. (1983) Biopolymers *22*, 113–118.

Crane-Robinson, C., Briand, G., Sautiere, P. & Champagne, M. (1977a) Biochim. Biophys. Acta *493*, 283–292.

Crane-Robinson, C., Hayashi, H., Cary, P.D., Briand, G., Sautiere, P., Krieger, D., Vidali, G., Lewis, P.N. & Tomkun, J. (1977b) Eur. J. Biochem. *79*, 535–548.

Crane-Robinson, C., Staynov, D.Z. & Baldwin, J.P. (1984) Comments Mol. Cell. Biophys. *2*, 219–265.

Cremisi, C. & Yaniv, M. (1980) Biochem. Biophys. Res. Comm. *92*, 1117–1123.

Cremisi, C., Pignatti, P.F. & Yaniv, M. (1976) Biochem. Biophys. Res. Comm. *73*, 548–554.

Cremisi, C., Chestier, A. & Yaniv, M. (1978) Cell *12*, 947–951.

Crick, F.H.C. (1976) Proc. Natl. Acad. Sci. USA *73*, 2639–2643.

Crick, F.H.C. & Klug, A. (1975) Nature (London) *255*, 530–533.

Crick, F.H.C. & Watson, J.D. (1954) Proc. Roy. Soc. (London) *A223*, 80–96.

Cross, M.E. & Ord, M.G. (1970) Biochem. J. *118*, 191–193.

Cruft, H.J., Mauritzen, C.M. & Stedman, E. (1954) Nature (London) *174*, 580.

Cusick, M.E., Herman, T.M., DePamphilis, M.L. & Wassarman, P.M. (1981) Biochemistry *20*, 6648–6658.

Cusick, M.E., Lee, K.S., DePamphilis, M.L. & Wassarman, P.M. (1983) Biochemistry *22*, 3873–3884.

Cusick, M.E., DePamphilis, M.L. & Wassarman, P.M. (1984) J. Mol. Biol. *178*, 249–271.

Czupryn, M. & Toczko, K. (1985) Eur. J. Biochem. *147*, 575–580.

D'Anna, J.A. & Isenberg, I. (1972) Biochemistry *11*, 4017–4024.

D'Anna, J.A. & Isenberg, I. (1973) Biochemistry *12*, 1035–1043.

D'Anna, J.A. & Isenberg, I. (1974a) Biochemistry *13*, 2098–2104.

D'Anna, J.A. & Isenberg, I. (1974b) Biochem. Biophys. Res. Comm. *61*, 343–347.

D'Anna, J.A. & Isenberg, I. (1974c) Biochemistry *13*, 4987–4992.

D'Anna, J.A. & Isenberg, I. (1974d) Biochemistry *13*, 4992–4997.

D'Anna, J.A. & Prentice, D.A. (1983) Biochemistry *22*, 5631–5640.

D'Anna, J.A., Gurley, L.R. & Deaven, L.L. (1978) Nucleic Acids Res. *5*, 3195–3207.

D'Anna, J.A., Tobey, R.A. & Gurley, L.R. (1980) Biochemistry *19*, 2656–2671.

D'Anna, J.A., Gurley, L.R., Walters, R.A., Tobey, R.A., Becker, R.R. & Barham, S.S. (1981) in *Protein Phosphorylation* (Rosen, O.M. & Krebs, E.G., eds.), Book B, pp. 1053–1072, Cold Spring Harbor Lab. Press.

Darlington, C.D. (1966) in *Chromosomes Today* (Darlington, C.D. & Lewis, K.R., eds.), pp. 1–6, Oliver and Boyd, Edinburgh and London.

Darzynkiewicz, Z., Traganos, F., Xue, S.B. & Melamed, M.R. (1981) Exp. Cell Res. *136*, 279–293.

Davidson, E. (1976) *Gene Activation in Early Development*, 2nd ed., Academic Press, N.Y.

Davie, J.R. & Candido, E.P.M. (1978) Proc. Natl. Acad. Sci. USA *75*, 3574–3577.

Davie, J.R. & Candido, E.P.M. (1980) FEBS Lett. *110*, 164–168.

Davie, J.R. & Saunders, C.A. (1981) J. Biol. Chem. *256*, 12574–12580.

Davies, H.G. (1968) J. Cell Sci. *3*, 129–150.

Davies, H.G., Murray, A.B. & Walmsley, M.E. (1974) J. Cell Sci. *16*, 261–299.

Davis, A.H., Reudelhuber, T.L. & Garrard, W.T. (1983) J. Mol. Biol. *167*, 133–155.

Davison, P.F., Conway, B.E. & Butler, J.A.V. (1954) Prog. Biophys. *4*, 148–194.

Dean, S.W., Tew, K.D., Clark, A.E. & Schein, P.S. (1985) Br. J. Canc. *52*, 377–382.

DeGroot, P., Strickland, W.N., Brandt, W.F. & von Holt, C. (1983) Biochim. Biophys. Acta *747*, 276–283.

Delabar, J.M. (1985) J. Biol. Chem. *260*,12622–12628.

DeLange, R.J. (1978) in *Methods in Cell Biology* (Stein, G., Stein, J., & Kleinsmith, L.J., eds.), vol. 18, pp. 169–188, Academic Press, N.Y.

DeLange, R.J., Fambrough, D.M., Smith, E.L. & Bonner, J. (1969a) J. Biol. Chem. *244*, 319–334.

DeLange, R.J., Fambrough, D.M., Smith, E.L. & Bonner, J. (1969b) J. Biol. Chem. *244*, 5669–5679.

DeLange, R.J., Hooper, J.A. & Smith, E.L. (1972) Proc. Natl. Acad. Sci. USA *69*, 882–884.

DeLange, R.J., Hooper, J.A. & Smith, E. (1973) J. Biol. Chem. *248*, 3261–3274.

DeLange, R.J., William, L.C. & Martinson, H.G. (1979) Biochemistry *18*, 1942–1946.

DeLange, R.J., Green, G.R. & Searcy, D.G. (1981a) J. Biol. Chem. *256*, 900–904.

DeLange, R.J., Williams, L.C. & Searcy, D.G. (1981b) J. Biol. Chem. *256*, 905–911.

Delotto, R. & Schedl, P. (1984) J. Mol. Biol. *179*, 607–628.

De Murcia, G. & Koller, T. (1981) Biologie Cellulaire *40*, 165–174.

De Murcia, G., Das, G.C., Erard, M. & Daune, M. (1978) Nucleic Acids Res. *5*, 523–535.

DePamphilis, M.L. & Wassarman, P.M. (1980) Ann. Rev. Biochem. *49*, 627–666.

DePamphilis, M.L., Chalifour, L.E., Charette, M.F., Cusick, M.E., Hay, R.T., Hendrickson, E.A., Pritchard, C.G., Tack, L.C., Wassarman, P.M., Weaver, D.T. & Wirak, D.O. (1983) in *Mechanisms of DNA Replication and Recombination* (Cozzarelli, N., ed.), pp. 423–447, Alan R. Liss, N.Y.

DePetrocellis, B., DePetrocellis, L., Lancieri, M. & Geraci, G. (1980) Cell Differ. *9*, 195–202.

Depew, R.F. & Wang, J.C. (1975) Proc. Natl. Acad. Sci. USA 72, 4275–4279.

Derenzini, M., Viron, A. & Puvion, E. (1984) Eur. J. Cell Biol. 33, 148–156.

Derenzini, M., Pession, A., Licastro, F. & Novello, F. (1985) Exp. Cell Res. 157, 50–62.

Dick, C. & Johns, E.W. (1969) Comp. Biochem. Physiol. 31, 529–533.

Dickerson, R.E. (1983a) J. Mol. Biol. 166, 419–441.

Dickerson, R.E. (1983b) Sci. Am. (December, 1983), 94–111.

Dickerson, R.E. & Drew, H.R. (1981) Proc. Natl. Acad. Sci. USA 78, 7318–7322.

Dickerson, R.E., Kopka, M.L. & Pjura, P. (1983) Proc. Natl. Acad. Sci. USA 80, 7099–7103.

Dierks, P., van Ooyen, A., Mantei, N. & Weissman, C. (1978) Proc. Natl. Acad. Sci. USA 75, 1411–1415.

Dieterich, A.E. & Cantor, C.R. (1981) Biopolymers 20, 111–127.

Dieterich, A.E., Axel, R. & Cantor, C.R. (1979) J. Mol. Biol. 129, 587–602.

Dieterich, A.E., Eshaghpour, H., Crothers, D.M. & Cantor, C.R. (1980) Nucleic Acids Res. 8, 2475–2487.

Diggle, J.H., McVittie, J.D. & Peacocke, A.R. (1975) Eur. J. Biochem. 56, 173–182.

Dijk, J., White, S.W., Wilson K.S., & Appelt, K. (1983) J. Biol. Chem. 268, 4003–4006.

Dimitriadis, G.J. & Tata, J.R. (1980) Biochem. J. 187, 467–477.

Dingwall, C. & Allan, J. (1984) EMBO J. 3, 1933–1937.

Dingwall, C., Lomonossoff, G.P. & Laskey, R.A. (1981) Nucleic Acids Res. 9, 2659–2673.

Dingwall, C., Sharnick, S.V. & Laskey, R.A. (1982) Cell 30, 449–458.

Djondjurov, L.P., Yancheva, N.Y. & Ivanova, E.C. (1983) Biochemistry 22, 4095–4102.

Dodge, J.D. (1966) in The Chromosomes of the Algae (Godward, M.B., ed.), St. Martins Press, N.Y., pp. 96–115.

Doenecke, D. & Gallwitz, D. (1982) Mol. Cell Biochem. 44, 113–128.

Doerfler, W. (1983) Ann. Rev. Biochem. 52, 93–124.

Domanskii, N.N., Bakaeva, T.G. & Bakaev, V.V. (1982) Mol. Biol. 16, 427–434.

Dougherty, A.M., Causley, G.C. & Johnson, W.C. Jr. (1983) Proc. Natl. Acad. Sci. USA 80, 2193–2195.

Douvas, A.S. & Bonner, J. (1977) in Mechanisms and Control of Cell Division (Rost, T.L. & Gifford, E.M., eds.), pp. 44–59, Dowden, Hutchinson & Ross, Stroudsburg, PA.

Drew, H. & Travers, A.A. (1986) J. Mol. Biol. 186, 773–790.

Drew, H., Takano, T., Tanaka, S., Itakura, H. & Dickerson, R.E. (1980) Nature (London) 286, 567–573.

Drew, H., Wing, R.M., Takano, T., Broka, C., Tanaka, S., Itakura, K. & Dickerson, R.E. (1981) Proc. Natl. Acad. Sci. USA 78, 2179–2183.

Dubochet, J. & Noll, M. (1978) Science 202, 280–286.

Duerre, J.A., Wallwork, J.C., Quick, D.P. & Ford, K.M. (1977) J. Biol. Chem. 252, 5981–5985.

Duerre, J.A., Quick, D.P., Traynor, M.D. & Onisk, D.V. (1982) Biochim. Biophys. Acta 719, 18–23.

Duncan, M.R., Robinson, M.J. & Dellorco, R.T. (1983) Biochim. Biophys. Acta 762, 221–226.

Dunn, K. & Griffith, J.D. (1980) Nucleic Acids Res. *8*, 555–566.

Dupressoir, T. & Sautiere, P. (1984) Biochem. Biophys. Res. Comm. *122*, 1136–1145.

Durban, E., Lee, H.W., Kim, S. & Paik, W.K. (1978) in *Methods in Cell Biol.* (Stein, G., Stein, J. & Kleinsmith, L.J., eds.), vol. 19, pp. 59–67, Academic Press, N.Y.

Dusenbery, D.B. & Uretz, R.B. (1972) J. Cell Biol. *52*, 639–647.

Dynan, W.S. & Tijan, R. (1983a) Cell *32*, 669–680.

Dynan, W.S. & Tijan, R. (1983b) Cell *35*, 79–87.

Dynan, W.S. & Tijan, R. (1985) Nature (London) *316*, 774–776.

Dyson, P.J. & Rabbitts, T.H. (1985) Proc. Natl. Acad. Sci. USA *82*, 1984–1988.

Dyson, P.J., Littlewood, T.D., Forster, A. & Rabbitts, T.H. (1985) EMBO J. *4*, 2885–2891.

Early, T.A. & Kearns, D.R. (1979) Proc. Natl. Acad. Sci. USA *76*, 4165–4169.

Earnshaw, W.C. & Heck, M.M.S. (1985) J. Cell Biol. *100*, 1716–1725.

Earnshaw, W.C. & Laemmli, U.K. (1983) J. Cell Biol. *96*, 84–93.

Earnshaw, W.C. & Laemmli, U.K. (1984) Chromosoma *89*, 186–192.

Earnshaw, W.C., Rekvig, O.P. & Haunestad, K. (1982) J. Cell Biol. *92*, 871–876.

Earnshaw, W.C., Halligan, B., Cooke, C.A., Heck, M.M.S. & Liu, L.F. (1985) J. Cell Biol. *100*, 1706–1715.

Ebralidse, K.K. & Mirzabekov, A.D. (1986) FEBS Lett. *194*, 69–72.

Eckols, T.K., Thompson, R.E. & Masaracchia, R.A. (1983) Eur. J. Biochem. *134*, 249–254.

Edwards, C.A. & Firtel, R.A. (1984) J. Mol. Biol. *180*,73–90.

Edwards, P.A. & Shooter, K.V. (1969) Biochem. J. *114*, 227–235.

Eichler, D.C. & Eales, S.J. (1982) J. Biol. Chem. *257*, 14384–14389.

Eichler, D.C. & Tatar, T.F. (1980) Biochemistry *19*, 3016–3022.

Eickbusch, T.H. & Moudrianakis, E.N. (1978a) Cell *13*, 295–306.

Eickbusch, T.H. & Moudrianakis, E.N. (1978b) Biochemistry *17*, 4955–4964.

Eickbusch, T.H., Watson, D.K. & Moudrianakis, E.N. (1976) Cell *9*, 785–792.

Eisenberg, H. & Felsenfeld, G. (1981) J. Mol. Biol. *150*, 537–555.

Eissenberg, J.C., Kimbrell, D.A., Fristrom, J.W. & Elgin, S.C.R. (1984) Nucleic Acids Res. *12*, 9025–9038.

Eissenberg, J.C., Cartwright, I.L., Thomas, G.H. & Elgin, S.C.R. (1985) Ann. Rev. Genetics *19*, 485–536.

Elgin, S.R.C. & Bonner, J. (1972) Biochemistry *11*, 772–781.

Elgin, S.R.C. & Hood, L.E. (1973) Biochemistry *12*, 4984–4991.

Elgin, S.R.C., Boyd, J.B., Hood, L.E., Wray, W. & Wu, F.C. (1973) Cold Spring Harbor Symp. Quant. Biol. *38*, 821–833.

Elgin, S.R.C., Schilling, J. & Hood, L.E. (1979) Biochemistry *18*, 5679–5685.

Ellison, M.J. & Pulleyblank, D.E. (1983a) J. Biol. Chem. *258*, 13307–13313.

Ellison, M.J. & Pulleyblank, D.E. (1983b) J. Biol. Chem. *258*, 13314–13320.

Ellison, M.J. & Pulleyblank, D.E. (1983c) J. Biol. Chem. *258*, 13321–13327.

Emerson, B.M. & Felsenfeld, G. (1984) Proc. Natl. Acad. Sci. USA *81*, 95–99.

Emerson, B.M., Lewis, C.D. & Felsenfeld, G. (1985) Cell *41*, 21–30.

Engelke, D.R., Ng, S.-Y., Shastry, B.S. & Roeder, R.G. (1980) Cell *19*, 717–728.

Erard, M., Das, G.C., De Murcia, G., Mazen, A., Pouyet, J., Champagne, M. & Daune, M. (1979) Nucleic Acids Res. *6*, 3231–3253.

Ermini, M. & Kuenzle, C.C. (1978) FEBS Lett. *90*, 167–172.

Eshaghpour, H., Dieterich, A.E., Cantor, C.R. & Crothers, D.M. (1980) Biochemistry *19*, 1797–1805.

Espel, E., Bernúes, J., Querol, E., Martinez, P., Barris, A. & Lloberas, J. (1983) Biochem. Biophys. Res. Comm. *117*, 817–822.

Espel, E., Bernúes, J., Perez-Pons, J.A. & Querol, E. (1985) Biochem. Biophys. Res. Comm. *132*, 1031–1037.

Estepa, I. & Pestana, A. (1981) Eur. J. Biochem. *119*, 431–436.

Estepa, I. & Pestana, A. (1983) Eur. J. Biochem. *132*, 249–254.

Faber, A.J., Cook, A. & Hancock, R. (1981) Eur. J. Biochem. *120*, 357–361.

Farr, R.M., Luckow, V. & Sundharadas, G. (1979) Exp. Cell Res. *121*, 428–432.

Fasman, G. (ed.) (1976) *Handbook of Molecular Biology*, 3rd ed., vol. II, CRC Pub. Co., Cleveland.

Fasman, G.D. (1979) in *Chromatin Structure and Function* (Nicolini, C., ed.), pp. 67–107, Plenum Press, N.Y.

Fasman, G.D., Chou, P.Y. & Adler, A.J. (1976) in *The Molecular Biology of the Mammalian Genetic Apparatus* (Ts'o, P.O.P., ed.), pp. 1–52, North Holland Pub. Co., Amsterdam.

Faulhaber, I. & Bernardi, G. (1967) Biochim. Biophys. Acta *140*, 561–564.

Feigon, J. & Kearns, D.R. (1979) Nucleic Acids Res. *6*, 2327–2337.

Felden, R.A., Saunders, M.M. & Morris, N.R. (1976) J. Cell Biol. *68*, 430–439.

Felsenfeld, G. & McGhee, J. (1982) Nature (London) *296*, 602–603.

Felsenfeld, G. & McGhee, J.D. (1986) Cell *44*, 375–377.

Ferl, R.J. (1985) Mol. Gen. Genet. *200*, 207–210.

Ferrer, P., Caizergu, M., Amalric, F. & Zalta, J.P. (1980) Biol. Cell *39*, 155–157.

Ferro, A.M. & Kun, E. (1976) Fed. Proc. *35*, 1541.

Feughelman, M., Langridge, R., Seeds, W.E., Stober, A.R., Wilson, H.R., Hooper, C.W., Wilbur, M.H.F., Barclay, R.K. & Hamilton, L.D. (1955) Nature (London) *175*, 834–836.

Finch, J.T. & Klug, A. (1976) Proc. Natl. Acad. Sci. USA *73*, 1897–1901.

Finch, J.T., Lutter, L.C., Rhodes, D., Brown, A.S., Rushton, B., Levitt, M. & Klug, A. (1977) Nature (London) *269*, 29–36.

Finch, J.T., Lewit-Bentley, A., Bentley, G.A., Roth, M. & Timmins, P.A. (1980) Phil. Trans. Roy. Soc. (London) *290B*, 635–638.

Finch, J.T., Brown, R.S., Rhodes, D., Richmond, T., Rushton, B., Lutter, L.C. & Klug, A. (1981) J. Mol. Biol. *145*, 757–769.

Finley, D., Ciechanover, A. & Varshavsky, A. (1984) Cell *37*, 43–55.

Fischman, G.J., Lambert, M.W. & Studzinski, G.P. (1979) Biochim. Biophys. Acta *567*, 464–471.

Fittler, F. & Zachau, H.G. (1979) Nucleic Acids Res. *7*, 1–13.

Fitzgerald, P.C. & Simpson, R.T. (1985) J. Biol. Chem. *260*, 15318–15324.

Fitzimmons, D.W. & Wolstenholme, G.E.W. (Eds.) (1975) *Ciba Foundation Symposium 28*, Elsevier-North Holland, Amsterdam.

Fleischer-Lambropoulos, H. & Pollow, K. (1978) Biochem. Biophys. Res. Comm. *80*, 773–780.

Fleischmann, G., Pflugfelder, G., Sterner, E.K., Javaherian, K., Howard, G.C., Wang, J.C. & Elgin, S.C.R. (1984) Proc. Natl. Acad. Sci. USA *81*, 6958–6962.

Flemming, W. (1880) Arkiv fur Microskopisch. Anat. *18*, 151–259.

Flenniken, A.M. (1984) Ph.D. Thesis, McGill University.

Flenniken, A.M. & Newrock, K.M. (1987) Devel. Biol. *124*, 457–468.

Foe, V.E. (1977) Cold Spring Harbor Symp. Quant. Biol. *42*, 723–740.

Foe, V.E., Wilkinson, L.E. & Laird, C.D. (1976) Cell *9*, 131–146.

Folger, K., Anderson, J.N., Hayward, M.A. & Shapiro, D.J. (1983) J. Biol. Chem. *258*, 8908–8914.

Forrester, W.C., Thompson, C., Elder, J.T. & Groudine, M. (1986) Proc. Natl. Acad. Sci. USA *83*, 1359–1363.

Fowler, E., Farb, R. & Elsaidy, S. (1982) Nucleic Acids Res. *10*, 735–748.

Franke, W.W., Scheer, U., Trendelenburg, M.F., Spring, H. & Zentgraf, H. (1976) Cytobiologie *13*, 401–434.

Franke, W.W., Scheer, U., Trendelenburg, M., Zentgraf, H. & Spring, H. (1977) Cold Spring Harbor Symp. Quant. Biol. *42*, 755–772.

Franke, W.W., Scheer, U., Zentgraf, H., Trendelenburg, M.F., Muller, U., Krohne, G. & Spring, H. (1980) in *Results and Problems in Cell Differentiation, Vol. 11, Differentiation and Neoplasia* (McKinnel, R.G., et al., eds.), Springer-Verlag, Berlin.

Franke, W.W., Scheer, U. & Zentgraf, H. (1984) Ber. Deutsche Bot. *97*, 315–326.

Franklin, S.G. & Zweidler, A. (1977) Nature (London) *266*, 273–275.

Franks, R.R. & Davis, F.C. (1983) Develop. Biol. *98*, 101–109.

Fratini, A.V., Kopka, M.L., Drew, H.R. & Dickerson, R.E. (1982) J. Biol. Chem. *257*, 14686–14707.

Frederick, C.A., Grable, G., Melia, M., Samudzi, C., Jen-Jacobson, L., Wang, B.-C., Greene, P., Boyer, H.W. & Rosenberg, J.M. (1984) Nature (London) *309*, 327–331.

Fritton, H.P., Sippel, A.E. & Igo-Kemenes, T. (1983) Nucleic Acids Res. *11*, 3467–3485.

Fujii, S., Wang, A.H.-J., van der Marel, G., van Boom, J.H. & Rich, A. (1982) Nucleic Acids Res. *10*, 7879–7892.

Fujimoto, D. & Segawa, K. (1973) FEBS Lett. *32*, 59–61.

Fujitaki, J.M., Fang, G., Oh, E.Y. & Smith, R.A. (1981) Biochemistry *20*, 3658–3664.

Fuller, F.B. (1971) Proc. Natl. Acad. Sci. USA *68*, 815–819.

Fulmer, A.W. & Bloomfield, V.A. (1982) Biochemistry *21*, 985–992.

Furberg, S. (1950) Acta Cryst. *3*, 325–333.

Fusauchi, Y. & Iwai, K. (1983) J. Biochem. *93*, 1487–1497.

Fusauchi, Y. & Iwai, K. (1984) J. Biochem. *95*, 147–154.

Futai, M. & Mizuno, D. (1967) J. Biol. Chem. *242*, 5301–5307.

Futai, M., Mizuno, D. & Sugimura, T. (1968) J. Biol. Chem. *243*, 6325–6329.

Gabrielli, F. & Baglioni, C. (1975) Develop. Biol. *43*, 254–263.

Gabrielli, F. & Baglioni, C. (1977) Nature (London) *269*, 529–531.

Gabrielli, F., Hancock, R. & Faber, A.J. (1981) Eur. J. Biochem. *120*, 363–369.

Gallwitz, D. (1973) in *Regulation of Transcription and Translation* (Bautz, E.K.F., ed.), pp. 39–57, Springer-Verlag, Berlin.

Gallwitz, D. & Sures, I. (1972) Biochim. Biophys. Acta *263*, 315–328.

Ganguly, A. & Bagchi, B. (1984) Biochim. Biophys. Acta *782*, 415–421.

Ganguly, A., Das, A., Mondal, H., Mandal, R.K. & Biswas, B.B. (1973) FEBS Lett. *34*, 27–30.

Ganguly, A., Bagchi, B., Bera, M., Ghosh, A.N. & Sen, A. (1983) Biochim. Biophys. Acta *739*, 286–290.

Garcea, R.L. & Alberts, B.M. (1980) J. Biol. Chem. *255*, 11454–11463.

Garel, A. & Axel, R. (1977) Cold Spring Harbor Symp. Quant. Biol. *42*, 701–708.

Garel, A., Kovacs, M.M., Champagne, M. & Daune, M. (1975) Biochim. Biophys. Acta *395*, 16–27.

Garel, A., Zolan, M. & Axel, R. (1977) Proc. Natl. Acad. Sci. USA *74*, 4867–4871.

Gargiulo, G. & Worcel, A. (1983) J. Mol. Biol. *170*, 699–722.

Gargiulo, G., Razvi, F. & Worcel, A. (1984) Cell *38*, 511–521.

Gargiulo, G., Razvi, F., Ruberti, I., Mohr, I. & Worcel, A. (1985) J. Mol. Biol. *181*, 339–349.

Garrard, W.T. & Bonner, J. (1974) J. Biol. Chem. *249*, 5570–5579.

Garrard, W.T., Todd, R.D., Boatwright, D.T. & Albright, S.C. (1976) in *10th Int. Congress of Biochemistry, Hamburg, Germany.* Abst. 02-3-178.

Garrells, J.I., Elgin, S.C.R., & Bonner, J. (1972) Biochem Biophys. Res. Commun. *46*, 545–551.

Gasser, S.M. & Laemmli, U.K. (1986) EMBO J. *5*, 511–518.

Gasser, S.M., Laroche, T., Falquet, J., Boy de la Tour, E. & Laemmli, U.K. (1986) J. Mol. Biol. *188*, 613–629.

Gates, D.M. & Bekhor, I. (1979a) Nucleic Acids Res. *6*, 1617–1630.

Gates, D.M. & Bekhor, I. (1979b) Nucleic Acids Res. *6*, 3411–3426.

Gaubatz, J., Ellis, M. & Chalkley, R. (1979) Fed. Proc. *38*, 1973–1978.

Gaviez, A.I. & Kuzin, A.M. (1973) Eur. J. Biochem. *37*, 7–11.

Gazit, B., Panet, A. & Cedar, H. (1980) Proc. Natl. Acad. Sci. USA *77*, 1787–1790.

Gellert, M. (1981) Ann. Rev. Biochem. *50*, 879–910.

Genest, D. & Wahl, P. (1981) Biochimie *63*, 561–564.

Geoghegan, T.E., Keller, G.H. & Roark, D.E. (1974) Fed. Proc. *33*, 1598.

Gerace, L. & Blobel, G. (1980) Cell *19*, 277–287.

Gerace, L., Comeau, C. & Benson, M. (1984) J. Cell Sci. *Suppl. 1*, 137–160.

Germond, J.E., Hirt, P., Oudet, P., Gross-Bellard, M. & Chambon, P. (1975) Proc. Natl. Acad. Sci. USA *72*, 1843–1847.

Gershey, E.L., Haslett, G.W., Vidali, G. & Allfrey, V.G. (1969) J. Biol. Chem. *244*, 4871–4877.

Gidoni, D., Kadonaga, J.T., Berrera-Saldana, H., Takahashi, K., Chambon, P. & Tijan, R. (1985) Science *230*, 511–517.

Gierer, A. (1966) Nature (London) *212*, 1480–1481.

Gilmour, D.S., Pflugfelder, G., Wang, J.C. & Lis, J.T. (1986) Cell *44*, 401–407.

Gilmour, R.S. & Paul, J. (1970) FEBS Lett. *9*, 242–244.

Giri, C.P., West, M.H.P. & Smulson, M. (1978a) Biochemistry *17*, 3495–3500.

Giri, C.P., West, M.H.P., Ramirez, M.L. & Smulson, M. (1978b) Biochemistry *17*, 3501–3504.

Gjerset, R.A. & Martin, D.W. (1982) J. Biol. Chem. *257*, 8581–8583.

Glass, D.B. & Krebs, E.G. (1979) J. Biol. Chem. *254*, 9728–9738.

Glikin, G.C., Ruberti, I. & Worcel, A. (1984) Cell *37*, 33–41.

Glotov, B.O., Itkes, A.V., Nikolaev, L.G. & Severin, E.S. (1978) FEBS Lett. *91*, 149–152.

Glotov, B.O., Trakht, I.N., Grozdova, I.D., Nikolaev, L.G., Itkes, A.V. & Severin, E.S. (1980) Adv. Enzyme Reg. *18*, 261–273.

Glotov, B.O., Rudin, A.V. & Severin, E.S. (1982) Biochim. Biophys. Acta *696*, 275–284.

Glover, C.V.C. & Gorovsky, M.A. (1978) Biochemistry *17*, 5705–5713.

Godfrey, J.E., Eickbush, T.H., Moudrianakis, E.N. (1980) Biochemistry *19*, 1339–1346.

Goff, C.G. (1976) J. Biol. Chem. *251*, 4131–4138.

Goffin, C. & Verly, W.G. (1984) Biochem. J. *220*, 133–137.

Goldknopf, I.A. & Busch, H. (1978) in *The Cell Nucleus* (Busch, H., ed.), vol. VI, Part C, pp. 149–180, Academic Press, N.Y.

Goldknopf, I.A. & Busch, H. (1980) Biochem. Biophys. Res. Comm. *96*, 1724–1731.

Goldknopf, I.A., Taylor, C.W., Baum, R.M., Yoeman, L.C., Olson, M.O., Prestayk, A.W. & Busch, H. (1975) J. Biol. Chem. *250*, 7182–7187.

Goldknopf, I.L., French, M.F., Musso, R. & Busch, H. (1977) Proc. Natl. Acad. Sci. USA *74*, 5492–5495.

Goldknopf, I.L., French, M.F., Daskal, Y. & Busch, H. (1978) Biochem. Biophys. Res. Comm. *84*, 786–793.

Goldknopf, I.L., Sudhakar, S., Rosenbaum, F. & Busch, H. (1980a) Biochem. Biophys. Res. Comm. *95*, 1253–1260.

Goldknopf, I.L., Wilson, G., Ballal, N.R. & Busch, H. (1980b) J. Biol. Chem. *255*, 10555–10558.

Goldstein, G., Scheid, M., Hammerling, V., Boyse, E.A., Schlessinger, D.H. & Nail, H.D. (1975) Proc. Natl. Acad. Sci. USA *72*, 11–15.

Goldstein, L., Rubin, R.W. & Ko, C. (1977) Cell *12*, 601–608.

Goodwin, D.C. & Brahms, J. (1978) Nucleic Acids Res. *5*, 835–850.

Goodwin, D.C., Vergne, J., Brahms, J., Defer, N. & Krush, J. (1979) Biochemistry *18*, 2057–2064.

Goodwin, G.H. & Mathew, C.G.P. (1982) in *The HMG Chromosomal Proteins* (Johns, E.W., ed.), pp. 193–221, Academic Press, N.Y.

Goodwin, G.H., Sanders, C. & Johns, E.W. (1973) Eur. J. Biochem. *38*, 14–19.

Goodwin, G.H., Nicolas, R.H. & Johns, E.W. (1975) Biochim. Biophys. Acta *405*, 280–291.

Goodwin, G.H., Woodhead, L. & Johns, E.W. (1977a) FEBS Lett. *73*, 85–88.

Goodwin, G.H., Nicolas, R.H. & Johns, E.W. (1977b) Biochem. J. *167*, 485–488.

Goodwin, G.H., Walker, J.M. & Johns, E.W. (1978a) in *The Cell Nucleus* (Busch, H., ed.), vol. VI, pp. 182–219, Academic Press, N.Y.

Goodwin, G.H., Walker, J.M. & Johns, E.W. (1978b) Biochim. Biophys. Acta *519*, 233–242.

Goodwin, G.H., Brown, E., Walker, J.M. & Johns, E.W. (1980) Biochim. Biophys. Acta *623*, 329–338.

Goodwin, G.H., Wright, C.A. & Johns, E.W. (1981) Nucleic Acids Res. *9*, 2761–2775.

Goodwin, G.H., Nicolas, R.H., Cockerill, P.N., Zavou, S. & Wright, C.A. (1985) Nucleic Acids Res. *13*, 3561–3579.

Gordon, J.S., Rosenfeld, B.I., Kaufman, R. & Williams, D.L. (1980) Biochemistry *19*, 4395–4402.

Gordon, V.C., Knobler, C.M., Olins, D.E. & Schumaker, V.N. (1978) Proc. Natl. Acad. Sci. USA *75*, 660–663.

Gordon, V.C., Schumaker, V.N., Olins, D.E., Knobler, C.M. & Horowitz, J. (1979) Nucleic Acids Res. *6*, 3845–3858.

Gorovsky, M.A., Glover, C., Johmann, C.A., Keevert, J.B., Mathis, D.J. & Samuelson, M. (1977) Cold Spring Harbor Symp. Quant. Biol. *42*, 493–503.

Gorski, J., Toft, D., Shyamala, G., Smith, D. & Notides, A. (1968) Recent Progr. Hormone Res. *24*, 45–72.

Gorski, J., Welshons, W.V., Sakai, D., Hansen, J., Walent, J., Kassis, J., Shull, J., Stack, G. & Campen, C. (1986) Recent Prog. Hormone Res. *42*, 297–322.

Gottesfeld, J.M. (1980) Nucleic Acids Res. *8*, 905–922.

Gottesfeld, J.M. & Bloomer, L.S. (1980) Cell *21*, 751–760.

Gottesfeld, J.M. & Butler, P.J.G. (1977) Nucleic Acids Res. *4*, 3155–3173.

Gottesfeld, J.M. & Melton, D.A. (1978) Nature (London) *273*, 317–319.

Gottesfeld, J.M. & Partington, G.A. (1977) Cell *12*, 953–962.

Gottesfeld, J.M., Murphy, R.F. & Bonner, J. (1975) Proc. Natl. Acad. Sci. USA *72*, 4404–4408.

Gottschling, D.E. & Cech, T.R. (1984) Cell *38*, 501–510.

Gottschling, D.E., Palen, T.E. & Cech, T.R. (1983) Nucleic Acids Res. *11*, 2093–2109.

Grainger, R.M. & Ogle, R.C. (1977) Chromosoma *65*, 115–126.

Grandy, D.K., Engel, J.D. & Dodgson, J.D. (1982) J. Biol. Chem. *257*, 8577–8580.

Graves, R.A. & Marzluff, W.F. (1984) Mol. Cell Biol. *4*, 351–357.

Gray, D.M., Edmondson, S.P., Lang, D. & Vaughan, M. (1979) Nucleic Acids Res. *6*, 2089–2107.

Grebamier, A.E. & Pogo, A.O. (1981) Biochemistry *20*, 1094–1099.

Green, G.R., Searcy, D.G. & DeLange, R.J. (1983) Biochim. Biophys. Acta *741*, 251–257.

Grellet, F., Penon, P. & Cooke, R. (1980) Planta *148*, 346–353.

Greyling, H.J., Schwager, S., Sewell, B.T. & Von Holt, C. (1983) Eur. J. Biochem. *137*, 221–226.

Griffith, J.D. (1976) Proc. Natl. Acad. Sci. USA *73*, 563–567.

Griffith, J.D. (1978) Science *201*, 525–527.

Grigera, P.R. & Tisminetzky, S.G. (1984) Virology *136*, 10–19.

Grigoryev, S.A. & Krasheninnikov, I.A. (1982) Eur. J. Biochem. *129*, 119–125.

Grimes, S.R. & Henderson, N. (1983) Arch. Biochem. Biophys. *221*, 108–116.

Grimes, S.R. & Henderson, N. (1984a) Exp. Cell Res. *152*, 91–97.

Grimes, S.R. & Henderson, N. (1984b) Dev. Biol. *101*, 516–521.

Groppi, V.E. Jr. & Coffino, P. (1980) Cell *21*, 195–204.

Groudine, M. & Weintraub, H. (1982) Cell *30*, 131–139.

Groudine, M., Kohwi-Shigematsu, T., Gelinas, R., Stamatoyannopoulos, G. & Papayannopdulou, T. (1983) Proc. Natl. Acad. Sci. USA *80*, 7551–7555.

Grove, G.W. & Zweidler, A. (1984) Biochemistry *23*, 4436–4443.

Gruenbaum, Y., Cedar, H. & Razin, A. (1982) Nature (London) *295*, 620–622.

Gruenbaum, Y., Hochstrasser, M., Mathog, D., Saumweber, H., Agard, D.A. & Sedat, J.W. (1984) J. Cell Sci. *Suppl. 1*, 223–234.

Grunstein, M., Diamond, K.E., Knoppel, E. & Grunstein, J.E. (1981) Biochemistry *20*, 1216–1223.

Grunstein, M., Rykowski, M., Kolodrubetz, D., Choe, J. & Wallis, J. (1984) in *Histone Genes* (Stein, G. & Stein, J., eds.), pp. 35–63, John Wiley & Sons, N.Y.

Gruol, D.J. (1980) Endrocrinol. *107*, 994–999.

Gupta, A., Jensen, D. & Kim, S. (1982) J. Biol. Chem. *257*, 9677–9683.

Gurley, L.R., Walters, R.A. & Tobey, R.A. (1972) Arch. Biochem. Biophys. *148*, 633–641.

Gurley, L.R., Walters, R.A. & Tobey, R.A. (1973) Biochem. Biophys. Res. Comm. *50*, 744–750.

Gurley, L.R., Walters, R.A. & Tobey, R.A. (1975) J. Biol. Chem. *250*, 3936–3944.

Gurley, L.R., D'Anna, J.A., Barham, S.S., Deavan, L.L. & Tobey, R.A. (1978a) Eur. J. Biochem. *84*, 1–15.

Gurley, L.R., Tobey, R.A., Walters, R.A., Hildebrand, C.E., Hohmann, P.G., D'Anna, J.A., Barham, S.S. & Deavan, L.L. (1978b) in *Cell Cycle Regulation* (Jeter, J., Cameron, I.L., Padilla, G.M. & Zimmerman, A.M., eds.), pp. 37–60, Academic Press, N.Y.

Gurley, L.R., Valdez, J.G., Prentice, D.A. & Spall, W.D. (1983a) Anal. Biochem. *129*, 132–144.

Gurley, L.R., Prentice, D.A., Valdez, J.G. & Spall, W.D. (1983b) Anal. Biochem. *131*, 465–477.

Gurley, L.R., Prentice, D.A., Valdez, J.G. & Spall, W.D. (1983c) J. Chromatog. *266*, 609–627.

Gurley, L.R., D'Anna, J.A., Blumenfeld, M., Valdez, J.G., Sebring, R.J., Donahue, P.R., Prentice, D.A. & Spall, W.D. (1984) J. Chromatog. *297*, 147–165.

Haas, A.L., Warms, J.V.B., Hershko, A. & Rose, I.A. (1982) J. Biol. Chem. *257*, 2543–2548.

Hadlaczky, G., Sumner, A.T. & Ross, A. (1981a) Chromosoma *81*, 537–555.

Hadlaczky, G., Sumner, A.T. & Ross, A. (1981b) Chromosoma *81*, 557–567.

Hagerman, P.J. (1984) Proc. Natl. Acad. Sci. USA *81*, 4632–4636.

Haggis, G.H. & Bond, E.F. (1979) J. Microscopy *115*, 225–234.

Hagiwara, H., Miyazaki, K., Matuo, Y., Yamashita, J. & Horio, H. (1980) Biochem. Biophys. Res. Comm. *94*, 988–995.

Hamana, K. & Iwai, K. (1979) J. Biochem. *86*, 789–794.

Hamkalo, B.A. & Rattner, J.B. (1977) Chromosoma *60*, 39–47.

Hammarsten, E. (1924) Biochem. Z. *144*, 383–466.

Han, S., Udvardy, A. & Schedl, P. (1985a) J. Mol. Biol. *183*, 13–29.

Han, S., Udvardy, A. & Schedl, P. (1985b) J. Mol. Biol. *184*, 657–665.

Hancock, R. (1982) Biol. Cell *46*, 105–122.

Hancock, R. & Boulikas, T. (1982) Int. Rev. Cytol. *79*, 165–213.

Hancock, R. & Hughes, M.E. (1982) Biol. Cell *44*, 201–212.

Hansen, J.C. & Gorski, J. (1985) Biochemistry *24*, 6078–6085.

Harborne, N. & Allan, J. (1983) FEBS Lett. *155*, 88–92.

Hardison, R.C., Zeitler, D.P., Murphy, J.M. & Chalkley, R. (1977) Cell *12*, 417–427.

Harland, R.M., Weintraub, H. & McKnight, S.L. (1983) Nature (London) *302*, 38–43.

Harrington, R.E. (1982) Biochemistry *21*, 1177–1186.

Harris, M.R. & Smith, B.J. (1983) Biochem. J. *211*, 763–766.

Harris, M.R., Harborne, N., Smith, B.J. & Allan, J. (1982) Biochem. Biophys. Res. Comm. *109*, 78–82.

Harrison, M.F. & Wilt, F.H. (1982) J. Exp. Zool. *223*, 245–256.

Hartman, P.G., Chapman, G.E., Moss, T. & Bradbury, E.M. (1977) Eur. J. Biochem. *77*, 45–51.

Harvey, R.P., Robins, A.J. & Wells, J.R.E. (1982) Nucleic Acids Res. *10*, 7851–7863.

Harvey, R.P., Whiting, J.A., Coles, L.S., Krieg, P.A. & Wells, J.R.E. (1983) Proc. Natl. Acad. Sci. USA *80*, 2819–2823.

Haselkorn, R. & Rouviere-Yaniv, J. (1976) Proc. Natl. Acad. Sci. USA *73*, 1917–1920.

Hashimoto, E., Takeda, M., Nishizuka, Y., Hamana, K. & Iwai, K. (1976) J. Biol. Chem. *251*, 6287–6293.

Hatayama, T., Inaba, M. & Yukioka, M. (1982) J. Biochem. *92*, 749–755.

Hatayama, T., Nakamura, T. & Yukioka, M. (1984) Biochem. Int. *9*, 251–258.

Hawkins, A.R. & Wootton, J.C. (1981) FEBS Lett. *130*, 275–278.

Hay, C.W. & Candido, E.P.M. (1983a) J. Biol. Chem. *258*, 3726–3741.

Hay, C.W. & Candido, E.P.M. (1983b) Biochemistry *22*, 6175–6180.

Hayaishi, O. & Ueda, K. (1977) Ann. Rev. Biochem. *46*, 95–116.

Hayashi, H., Nomoto, M. & Iwai, K. (1980) Proc. Jpn. Acad. Sci. *56B*, 579–584.

Hayashi, H., Nomoto, M. & Iwai, K. (1984) J. Biochem. *96*, 1449–1456.

Hayashi, K., Hofstaetler, T. & Yakua, N. (1978) Biochemistry *17*, 1880–1883.

Hayashi, T., Ohe, Y., Hayashi, H. & Iwai, K. (1980) J. Biochem. *88*, 27–34.

Hayashi, T., Ohe, Y., Hayashi, H. & Iwai, K. (1982) J. Biochem. *92*, 1995–2000.

Hayashi, T., Hayashi, H., Fusauchi, Y. & Iwai, K. (1984a) J. Biochem. *95*, 1741–1749.

Hempel, K., Thomas, G., Roos, G., Stocker, W., Lange, H.W. (1979) Hoppe-Seyler's Z. Physiol. Chem. *360*, 869–876.

Hendrick, D., Tolstoshev, P. & Randlett, D. (1977) Gene *2*, 147–158.

Hentschel, C.C. & Birnstiel, M.L. (1981) Cell *25*, 301–313.

Hereford, L.M., Fahrner, K., Woolford, J., Jr., Rosbach, M. & Kaback, D.B. (1979) Cell *18*, 1261–1271.

Hereford, L.M., Osley, M.A., Ludwig, J.R. & McLaughlin, C.S. (1981) Cell *24*, 367–375.

Hereford, L.M., Bromley, S. & Osley, M.A. (1982) Cell *30*, 305–310.

Herlands, L., Allfrey, V.G. & Poccia, D. (1982) J. Cell Biol. *94*, 219–223.

Herman, T.M., DePamphilis, M.L. & Wassarman, P.M. (1981) Biochemistry *20*, 621–630.

Hershko, A., Heller, H., Elias, S. & Ciechanover, A. (1983) J. Biol. Chem. *258*, 8206–8214.

Hertwig, O. (1885) Jenaische Z. Naturwiss. *18*, 276–318.

Herzog, M. & Maroteaux, L. (1986) Proc. Natl. Acad. Sci. USA *83*, 8644–8648.

Herzog, M. & Soyer, M.O. (1981) Eur. J. Cell *23*, 295–302.

Hewish, D.R. & Burgoyne, L.A. (1973) Biochem. Biophys. Res. Comm. *52*, 504–510.

Hieter, P.A., Hendricks, M.B., Hemminki, K. & Weinberg, E.S. (1979) Biochemistry *18*, 2707–2716.

Higashi, Y. (1985) Nucleic Acids Res. *13*, 5157–5172.

Hill, R.J., Mott, M.R., Burnett, E.J., Abmayr, S.M., Lowenhaupt, K. & Elgin, S.C.R. (1982) J. Cell Biol. *95*, 262–266.

Hilliard, P.R., Smith, R.M. & Rill, R.L. (1986) J. Biol. Chem. *261*, 5992–5998.

Hilz, H. & Stone, P. (1976) Rev. Physiol. Biochem. Pharmacol. *76*, 1–58.

Hinnebusch, A.G., Klotz, L.C., Immergut, E. & Loeblich, A.R. (1980) Biochemistry *19*, 1744–1754.

Hinnebusch, A.G., Klotz, L.C., Blankar, R.L. & Loeblich, A.R. (1981) J. Mol. Evol. *17*, 334–347.

Hirasawa, E., Takahashi, E. & Matsumoto, H. (1978) Plant Cell Physiol. *19*, 1095–1098.

Hiremath, S.T., Loor, R.M. & Wang, T.Y. (1980) Biochem. Biophys. Res. Comm. *97*, 981–986.

Hofer, E., Hofer-Warbinck, R. & Darnell, J.E. (1982) Cell *29*, 887–893.

Hofstetter, H., Kressman, A. & Birnstiel, M.L. (1981) Cell *24*, 573–585.

Hogan, M.E. & Jardetzky, O. (1980) Biochemistry *19*, 3460–3468.

Hogan, M., Dattagupta, N. & Crothers, D.M. (1978) Proc. Natl. Acad. Sci. USA *75*, 195–199.

Hohmann, P. (1983) Mol. Cell. Biochem. *57*, 81–92.

Hohmann, P., Tobey, R.A. & Gurley, L.R. (1976) J. Biol. Chem. *251*, 3685–3692.

Holliday, R. & Pugh, J.E. (1975) Science *187*, 226–232.

Holmgren, P., Rasmuson, B., Johansson, T. & Sundquist, G. (1976) Chromosoma *54*, 99–116.

Honda, B.M., Dixon, G.H. & Candido, E.P.M. (1975a) J. Biol. Chem. *250*, 8681–8685.

Honda, B.M., Candido, E.P.M. & Dixon, G.H. (1975b) J. Biol. Chem. *250*, 8686–8689.

Hooper, J.A., Smith, E.L., Sommer, K.R. & Chalkley, R. (1973) J. Biol. Chem. *248*, 3275–3279.

Hörz, W. & Zachau, H.G. (1980) J. Mol. Biol. *144*, 305–328.

Hörz, W., Fittler, F. & Zachau, H.G. (1983) Nucleic Acids Res. *11*, 4275–4285.

Hovatter, K.R. & Martinson, H.G. (1987) Proc. Natl. Acad. Sci. USA *84*, 1162–1166.

Howard, G.C., Javaherian, K., Abmayr, S., Fleischmann, G. & Elgin, S.R.C. (1980) J. Cell Biol. *87*, 42a.

Howell, W.M., & Hsu, T.C. (1979) Chromosoma *73*, 61–66.

Hozier, J., Renz, M. & Nehls, P. (1977) Chromosoma *62*, 301–317.

Hsiang, M.W. & Cole, R.D. (1978) *Methods in Cell Biol.* (Stein G., Stein J., & Kleinsmith, L.J., eds.), vol. 18, pp. 189–228.

Hsiao, C.-L. & Carbon, J.A. (1979) Proc. Natl. Acad. Sci. USA, *76*, 3829–3833.

Huang, H.C. & Cole, R.D. (1984) J. Biol. Chem. *259*, 14237–14242.

Hübscher, U., Lutz, H. & Kornberg, A. (1980) Proc. Natl. Acad. Sci. USA *77*, 5097–5101.

Huiskamp, W. (1901a) Z. Physiol. Chem. *32*, 145–196.

Huiskamp, W. (1901b) Hoppe-Seyler's Z. Physiol. Chem. *34*, 32–54.

Huiskamp, W. (1903) Hoppe-Seyler's Z. Physiol. Chem. *39*, 55–72.

Humphries, S.E., Young, D. & Carroll, D. (1979) Biochemistry *18*, 3223–3231.

Hunt, L.T. & Dayhoff, M.O. (1977) Biochem. Biophys. Res. Comm. *74*, 650–655.

Hutcheon, T., Dixon, G.H. & Levy-Wilson, B. (1980) J. Biol. Chem. *256*, 681–685.

Hyde, J.E. (1982) Exp. Cell Res. *140*, 63–70.

Hyde, J.E., Igo-Kemenes, T. & Zachau, H.G. (1979) Nucleic Acids Res. *7*, 31–48.

Ichimura, S., Mita, K. & Zama, M. (1982) Biochemistry *21*, 5329–5334.

Igo-Kemenes, T. & Zachau, H.G. (1977) Cold Spring Harbor Symp. Quant. Biol. *42*, 109–118.

Igo-Kemenes, T., Omori, A. & Zachau, H.G. (1980) Nucleic Acids Res. *8*, 5377–5390.

Ilyin, Y.V., Varshavsky, A., Michelsaar, U.N. & Georgiev, G.P. (1971) Eur. J. Biochem. *22*, 235–245.

Imai, B.S., Yau, P., Baldwin, J.P., Ibel, K., May, R.P. & Bradbury, E.M. (1986) J. Biol. Chem. *261*, 8784–8792.

Imber, R., Bachinger, H. & Bickle, T.A. (1982) Eur. J. Biochem. *122*, 627–632.

Innis, J.W. & Scott, W.A. (1984) Mol. Cell. Biol. *4*, 1499–1507.

Isackson, P.J., Fishback, J.L., Bidney, D.L. & Reeck, G.R. (1979) J. Biol. Chem. *254*, 5569–5572.

Isackson, P.J., Debold, W.A. & Reeck, G.R. (1980a) FEBS Lett. *119*, 337–342.

Isackson, P.J., Bidney, D.I., Reeck, G.R., Neihart, N.K. & Bustin, M. (1980b) Biochemistry *19*, 4466–4471.

Isenberg, I. (1979) Ann. Rev. Biochem. *48*, 159–191.

Ishida, R., Akiyoshi, H. & Takahashi, T. (1976) Biochem. Biophys. Res. Comm. *56*, 703–710.

Ishimi, Y., Yasuda, H., Ohba, Y. & Yamada, M.A. (1981) J. Biol. Chem. *256*, 8249–8251.

Ishimi, Y., Yasuda, H., Hirosumi, J., Hanaoka, F. & Yamada, M. (1983) J. Biochem. *94*, 735–744.

Ishimi, Y., Hirosumi, J., Sato, W., Sugasawa, K., Yokota, S., Hanaoka, F. & Yamada, M.A. (1984) Eur. J. Biochem. *142*, 431–439.

Ito, S., Shizuta, Y. & Hayaishi, O. (1979) J. Biol. Chem. *254*, 3647–3651.

Itzhaki, R.E. (1971) Biochem. J. *125*, 221–224.

Iwai, K., Hayashi, H. & Ishikawa, K. (1972) J. Biochem. *72*, 357–367.

Iwasa, Y., Takai, Y., Kikkawa, U. & Nishizuka, Y. (1980) Biochem. Biophys. Res. Comm. *96*, 180–187.

Jack, R.S., Gehring, W.J. & Brack, C. (1981) Cell *24*, 321–331.

Jackson, D.A., McCready, S.J. & Cook, P.R. (1984) J. Cell Sci. *Suppl. 1*, 59–79.

Jackson, J.B. & Rill, R.L. (1981) Biochemistry *20*, 1042–1046.

Jackson, J.B., Pollock, J.M. & Rill, R.L. (1979) Biochemistry *18*, 3739–3748.

Jackson, P.D. & Felsenfeld, G. (1985) Proc. Natl. Acad. Sci. USA *82*, 2296–2300.

Jackson, P.S. & Gurley, L.R. (1985) J. Chromatog. *326*, 199–216.

Jackson, V. (1987a) Biochemistry *26*, 2315–2325.

Jackson, V. (1987b) Biochemistry *26*, 2325–2334.

Jackson, V. & Chalkley, R. (1981a) Cell *23*, 121–134.

Jackson, V. & Chalkley, R. (1981b) J. Biol. Chem. *256*, 5095–5103.

Jackson, V. & Chalkley, R. (1985a) Biochemistry *24*, 6921–6930.

Jackson, V. & Chalkley, R. (1985b) Biochemistry *24*, 6930–6938.

Jackson, V., Shires, A., Tanphaichitr, N. & Chalkley, R. (1976) J. Mol. Biol. *104*, 471–483.

Jackson, V., Marshall, S. & Chalkley, R. (1981) Nucleic Acids Res. *9*, 4563–4581.

Jacquet, M., Cukier-Kahn, K., Pla, J., and Gros, F. (1971) Biochem. Biophys. Res. Commun. *45*, 1597–1607.

Jaeger, A.W. & Kuenzle, C.C. (1982) EMBO J. *1*, 811–816.

Jakob, K.M., Ben-Yosef, S. & Tal, I. (1984) Nucleic Acids Res. *12*, 5015–5024.

Jamaluddin, M. & Philip, M. (1982) FEBS Lett. *150*, 429–433.

Jamrich, M., Greenleaf, A.L. & Bautz, E.F.K. (1977) Proc. Natl. Acad. Sci. USA *74*, 2079–2083.

Jaurez-Salinas, H., Levi, V., Jacobson, E.L. & Jacobson, M.K. (1982) J. Biol. Chem. *257*, 607–642.

Javaherian, K. (1977) in *The Organization and Expression of the Eukaryotic Genome* (Bradbury, E.M. & Javaherian, K., eds.), pp. 51–65, Academic Press, N.Y.

Javaherian, K. & Amini, S. (1977) Biochim. Biophys. Acta *478*, 295–304.

Javaherian, K. & Amini, S. (1978) Biochem. Biophys. Res. Comm. *85*, 1385–1391.

Javaherian, K. & Liu, L.F. (1983) Nucleic Acids Res. *11*, 461–472.

Javaherian, K., Liu, L.F. & Wang, J.C. (1978) Science *199*, 1345–1346.

Javaherian, K., Sadeghi, M. & Liu, L.F. (1979) Nucleic Acids Res. *6*, 3569–3579.

Jensen, E.V., Suzuki, T., Kawashima, T., Stumpf, W.E., Jungblut, P.W. & De Sombre, E.R. (1968) Proc. Natl. Acad. Sci. USA *59*, 632–636.

Jensen, E.V., Greene, G.L., Closs, L.E., DeSombre, E.R. & Nadji, M. (1982) Recent Prog. in Hormone Res. *38*, 1–34.

Jenson, J.C., Chin-Lin, P., Gerber-Jenson, B. & Litman, G.W. (1980) Proc. Natl. Acad. Sci. USA *77*, 1389–1393.

Jeppesen, P.G. & Bankier, A.T. (1979) Nucleic Acids Res. *7*, 49–67.

Jeppesen, P.G., Bankier, A.T. & Sanders, L. (1978) Exp. Cell Res. *115*, 293–302.

Jerzmanowski, A. & Maleszewski, M. (1985) Biochemistry *24*, 2360–2367.

Jessee, B., Gargiulo, G., Razvi, F. & Worcel, A. (1982) Nucleic. Acids. Res. *10*, 5823–5834.

Jirgensons, B. & Hnilica, L.S. (1965) Biochim. Biophys. Acta *109*, 241–249.

Joffe, J., Keene, M. & Weintraub, H. (1977) Biochemistry *16*, 1236–1238.

Johns, E.W. (1964) Biochem. J. *92*, 55–59.

Johns, E.W. (1967) Biochem. J. *104*, 78–82.

Johns, E.W. (1977) in *Methods in Cell Biology* (Stein, G., Stein, J. & Kleinsmith, L.J., eds.), vol. 16, pp. 183–203, Academic Press, N.Y.

Johns, E.W. (ed.) (1982a) *The HMG Chromosomal Proteins*, Academic Press, N.Y.

Johns, E.W. (1982b) in *The HMG Chromosomal Proteins*, (Johns, E.W., ed.), pp. 1–7, Academic Press, N.Y.

Johns, E.W. & Butler, J.A.V. (1962) Biochem. J. *82*, 15–18.

Johns, E.W. & Forrester, S. (1969) Eur. J. Biochem. *8*, 547–551.

Johns, E.W., Phillips, D.M.P., Simson, P. & Butler, J.A.V. (1960) Biochem. J. *77*, 631–636.

Johnson, E.M. & Allfrey, V.G. (1978) in *Biochemical Actions of Hormones* (Litwack, G., ed.), vol. 5, pp. 1–51, Academic Press, N.Y.

Johnson, E.M., Littau, V.C., Allfrey, V.G., Bradbury, E.M. & Matthews, H.R. (1976) Nucleic Acids Res. *3*, 3313–3329.

Johnson, E.M., Matthews, H.R., Littau, V.C., Lothstein, L., Bradbury, E.M. & Allfrey, V.G. (1978) Arch. Biochem. *191*, 537–550.

Johnson, E.M., Campbell, G.R. & Allfrey, V.G. (1979) Science *206*, 1192–1194.

Johnson, R.M. & Albert, S. (1953) J. Biol. Chem. *200*, 335–344.

Jolles, B., Laigle, A., Chinsky, L. & Turpin, P.Y. (1985) Nucleic Acids Res. *13*, 2075–2085.

Jones, K.A., Kadonaga, J.T., Rosenfeld, P.J., Kelly, T.J. & Tijan, R. (1987) Cell *48*, 79–89.

Jones, R.W. (1978) Biochem. J. *173*, 155–164.

Jordano, J., Montero, F. & Palacian, E. (1984a) Biochemistry *23*, 4280–4284.

Jordano, J., Montero, F. & Palacian, E. (1984b) Biochemistry *23*, 4285–4289.

Jordano, J., Montero, F. & Palacian, E. (1984c) FEBS Lett. *172*, 70–74.

Joseph, G., Caizergues-Ferrer, M. & Amalric, F. (1983) Eur. J. Biochem. *135*, 143–149.

Jost, E. & Johnson, R.T. (1981) J. Cell Sci. *47*, 25–53.

Jovin, T.M., McIntosh, L.P., Arndt-Jovin, D.J., Zarling, D.A., Robert-Nicoud, M., van de Sande, J.H., Jorgenson, K. & Eckstein, G.F. (1983) J. Biomol. Struct. Dynam. *1*, 2–57.

Jump, D.B. & Oppenheimer, J.H. (1980) Science *209*, 811–813.

Jump, D.B. & Oppenheimer, J.H. (1983) Mol. Cell Biochem. *55*, 159–176.

Jump, D.B. & Smulson, M.E. (1980) Biochemistry *19*, 1024–1031.

Jump, D.B., Butt, T.R. & Smulson, M.E. (1979) Biochemistry *18*, 983–990.

Jump, D.B., Butt, T.R. & Smulson, M.E. (1980) Biochemistry *19*, 1031–1037.

Jump, D.B., Seelig, S., Schwartz, H.L. & Oppenheimer, J.H. (1981) Biochemistry *20*, 6781–6789.

Kanai, Y., Miwa, M., Matsushina, T. & Sugimura, T. (1981) Proc. Natl. Acad. Sci. USA *78*, 2801–2804.

Kaneta, H. & Fujimoto, D. (1974) J. Biochem. *76*, 905–907.

Karpov, V.L., Bavykin, S.G., Preobrazhenskaya, O.V., Belyavsky, A.V. & Mirzabekov, A.D. (1982) Nucleic Acids Res. *10*, 4321–4337.

Karpov, V.L., Preobrazhenskaya, O.V. & Mirzabekov, A.D. (1984) Cell *36*, 423–431.

Katula, K.S. (1983) Develop. Biol. *98*, 15–27.

Kaufman, G. (1981) J. Mol. Biol. *147*, 25–39.

Kaufman, G., Bar-Shavit, R. & DePamphilis, M.L. (1978) Nucleic Acids Res. *5*, 2535–2545.

Kavenoff, R. & Zimm, B.H. (1973) Chromosoma *41*, 1–27.

Kawashima, S. & Imahori, K. (1982) J. Biochem. *91*, 959–966.

Kaye, A.M. & Sheratzky, D. (1969) Biochim. Biophys. Acta *190*, 527–538.

Kaye, J.S., Pratt-Kaye, S., Bellard, M., Dretzen, G., Bellard, F. & Chambon, P. (1986) EMBO J. *5*, 277–285.

Kaye, P.L. & Wales, R.G. (1981) J. Exp. Zool. *216*, 453–459.

Kearns, D.R. (1984) CRC Critical Reviews in Biochem. *15*, 237–290.

Kedes, L.H. (1979) Ann. Rev. Biochem. *48*, 837–870.

Keene, M.A., Corces, V., Lowenhaupt, K. & Elgin, S.C.R. (1981) Proc. Natl. Acad. Sci. USA *78*, 143–146.

Keepers, J.W., Kollman, P.A., Weiner, P.K. & James, T.L. (1982) Proc. Natl. Acad. Sci. USA *79*, 5537–5541.

Keichline, L.D. & Wassarman, P.M. (1977) Biochim. Biophys. Acta *475*, 139–151.

Keichline, L.D. & Wassarman, P.M. (1979) Biochemistry *18*, 214–219.

Kelley, R.I. (1973) Biochem. Biophys. Res. Comm. *54*, 1588–1593.

Kelly, P.M., Schofield, P.N. & Walker. I.O. (1983) FEBS Lett. *161*, 79–83.

Kennedy, B.P. & Davies, P.L. (1980) J. Biol. Chem. *255*, 2533–2539.

Keppel, F., Allet, B., & Eisen, H. (1977) Proc. Natl. Acad. Sci. USA *74*, 653–656.

Keshet, I., Lieman-Hurwitz, J. & Cedar, H. (1986) Cell *44*, 535–543.

Khrapunov, S.N., Dragan, A.I., Protas, A.F. & Berdyshev, G.D. (1984) Biochim. Biophys. Acta *787*, 97–104.

Kidwell, W.R. & Mage, M.G. (1976) Biochemistry *15*, 1213–1217.

Kim, S. & Paik, W.K. (1978) in *Methods in Cell Biology* (Stein, G., Stein, J. & Kleinsmith, L.J., eds.), vol. 19, pp. 79–88, Academic Press, N.Y.

Kimura, M. & Wilson, K.S. (1983) J. Biol. Chem. *258*, 4007–4011.

Kimura, T., Mills, F.C., Allan, J. & Gould, H. (1983) Nature (London) *306*, 709–712.

Kinkade, J.M. & Cole, R.D. (1966) J. Biol. Chem. *241*, 5798–5805.

Kiryanov, G.I., Manamshjan, T.A., Polyakov, V.Y., Fais, D. & Chentsov, J.S. (1976) FEBS Lett. *67*, 323–327.

Kitzis, A., Tichonicky, L., Defer, N. & Kruh, J. (1980) Eur. J. Biochem. *111*, 237–244.

Kitzis, A., Leibovitch, S.A., Leibovitch, M.P., Tichonicky, L., Harel, J. & Kruh, J. (1982) Biochim. Biophys. Res. Comm. *697*, 60–70.

Kleinschmidt, A.M. & Martinson, H.G. (1981) Nucleic Acids Res. *9*, 2423–2431.

Kleinschmidt, A.M. & Martinson, H.G. (1984) J. Biol. Chem. *259*, 497–503.

Kleinschmidt, J.A. & Franke, W.W. (1982) Cell *29*, 799–809.

Kleinsmith, L.J., Allfrey, V.G., & Mirsky, A.E. (1966) Proc. Natl. Acad. Sci. USA *55*, 1182–1189.

Klempnauer, K.-H., Fanning, E., Otto, B. & Knippen, R. (1980) J. Mol. Biol. *136*, 359–374.

Klevan, L., Dattagupta, N., Hogan, M. & Crothers, D.M. (1978) Biochemistry *17*, 4533–4540.

Klevan, L., Armitage, I.M. & Crothers, D.M. (1979) Nucleic Acids Res. *6*, 1607–1616.

Klug, A. & Lutter, L.C. (1981) Nucleic Acids Res. *9*, 4267–4283.

Klug, A., Rhodes, D., Smith, J., Finch, J.T. & Thomas, J.O. (1980) Nature (London) *287*, 509–516.

Klug, A., Finch, J.T. & Richmond, T.J. (1985) Science *229*, 1109–1110.

Klump, H.H. & Falk, H. (1984) Hoppe-Seyler's Z. Physiol. Chem. *365*, 661–665.

Kmiec, E.B., Ryoji, M. & Worcel, A. (1986a) Proc. Natl. Acad. Sci. USA *83*, 1305–1309.

Kmiec, E., Razvl, F. & Worcel, A. (1986b) Cell *45*, 209–218.

Kmiecik, D., Couppez, M., Belaiche, D. & Sautiere, P. (1983) Eur. J. Biochem. *135*, 113–121.

Kmiecik, D., Sellos, D., Belaiche, D., Sautiere, P. (1985) Eur. J. Biochem. *150*, 359–370.

Knippers, R. & Bohme, R. (1978) FEBS Lett. *89*, 253–256.

Knowles, J.A. & Childs, G.J. (1984) Proc. Natl. Acad. Sci. USA *81*, 2411–2415.

Kohlstaedt, L.A., King, D.S. & Cole, R.D. (1986) Biochemistry *25*, 4562–4565.

Kok, K., Snippe, L., Ab, G. & Gruber, M. (1985) Nucleic Acids Res. *13*, 5189–5202.

Kolodrubetz, D., Rykowski, M.C. & Grunstein, M. (1982) Proc. Natl. Acad. Sci. USA *79*, 7814–7818.

Komaiko, W. & Felsenfeld, G. (1985) Biochemistry *24*, 1186–1193.

Komura, H., Iwashita, T., Naoki, H., Nakanishi, K., Oka, J., Ueda, K. & Hayaishi, O. (1983) J. Am. Chem. Soc. *105*, 5164–5165.

Kootstra, A. & Bailey, G.S. (1978) Biochemistry *17*, 2504–2510.

Kopka, M.L., Fratini, A.V., Drew, H.R. & Dickerson, R.E. (1983) J. Mol. Biol. *163*, 129–146.

Kornberg, R. (1974) Science *184*, 868–871.

Kornberg, R. (1975) in *Proceedings of the Tenth FEBS Meeting* (Bernardi, G. & Gros, F., eds.), pp. 73–79, North Holland/American Elsevier, Amsterdam.

Kornberg, R. (1981) Nature (London) *292*, 579–580.

Kornberg, R. & Klug, A. (1981) Scientific American *244*, 52–64.

Kornberg, R. & Thomas, J.O. (1974) Science *184*, 865–868.

Kornberg, T., Lockwood, A. & Worcel, A. (1974) Proc. Natl. Acad. Sci. USA *71*, 3189–3193.

Kossel, A. (1884) Z. Physiol. Chem. *8*, 511–515.

Kossel, A. (1892) Ver. Physiol. Ges. Berlin *21* (October).

Kossel, A. & Schenck, E.G. (1928) Hoppe-Seyler's Z. Physiol. Chem. *173*, 278–308.

Kowalski, J. & Denhardt, D.T. (1979) Nature (London) *281*, 704–706.

Krajewska, W. & Klyszerjko-Stefanowicz, L. (1980) Biochim. Biophys. Acta *624*, 522–530.

Kreimeyer, A., Wielckens, K., Adamietz, P. & Hilz, H. (1984) J. Biol. Chem. *259*, 890–896.

Krohne, G. & Franke, W.W. (1980) Proc. Natl. Acad. Sci. USA *77*, 1034–1038.

Krust, A., Green, S., Argos, P., Kumar, V., Walter, P., Bornert, J.-M. & Chambon, P. (1986) EMBO J. *5*, 891–897.

Kubai, D.F. & Ris, H. (1969) J. Cell Biol. *40*, 508–528.

Kuehl, L., Salmond, B. & Tran, L.E. (1984) J. Cell Biol. *99*, 648–654.

Kumar, S. & Leffak, M. (1986) Biochemistry *25*, 2055–2060.

Kumar, V., Green, S., Staub, A. & Chambon, P. (1986) EMBO J. *5*, 2231–2236.

Kunkel, G.R. & Martinson, H.G. (1981) Nucleic Acids Res. *9*, 6869–6887.

Kunkel, N.S. & Weinberg, E.S. (1978) Cell *14*, 313–326.

Kuo, M.T. (1982) J. Cell Biol. *93*, 278–284.

Kurochkin, S.N., Trakht, I.N. & Severin, E.S. (1977) FEBS Lett. *84*, 163–166.

Labhart, P. & Koller, T. (1982a) Cell *28*, 279–292.

Labhart, P. & Koller, T. (1982b) Prog. Clin. Biol. Res. *85*, *Pt.A*, 173–186.

Labhart, P., Koller, T. & Wunderli, H. (1982) Cell *30*, 115–121.

Lacy, E. & Axel, R. (1975) Proc. Natl. Acad. Sci. USA *72*, 3978–3982.

Laemmli, U.K., Cheng, S.M., Adolph, K.W., Paulson, J.R., Brown, J.A. & Baumbach, W.R. (1977) Cold Spring Harbor Symp. Quant. Biol. *42*, 109–118.

LaFond, R.E., Goguen, J., Einck, L. & Woodcock, C.L.F. (1981) Biochemistry *20*, 2127–2132.

Laine, B., Sautiere, P. & Biserte, G. (1976) Biochemistry *15*, 1640–1645.

Laine, B., Kmiecik, D., Sautiere, P. & Biserte, G. (1978) Biochimie *60*, 147–150.

Laine, B., Kmiecik, D., Sautiere, P., Biserte, G. & Cohen-Solal, M. (1980) Eur. J. Biochem. *103*, 447–461.

Laine, B., Belaiche, D., Sautiere, P. & Biserte, G. (1982) Biochem. Biophys. Res. Comm. *106*, 101–107.

Laine, B., Belaiche, D., Khanaka, H. & Sautiere, P. (1983) Eur. J. Biochem. *131*, 325–331.

Laine, B., Sautiere, P., Spassky, A. & Rimsky, S. (1984) Biochem. Biophys. Res. Comm. *119*, 1147–1153.

Laird, C.D., Wilkinson, L.E., Foe, V.E. & Chooi, W.Y. (1976) Chromosoma *58*, 169–192.

Lake, R.S. (1973) J. Cell Biol. *58*, 317–331.

Lake, R.S., Goidl, J.A. & Salzman, N.P. (1972) Exp. Cell Res. *73*, 113–121.
Langan, T.A. (1971) Ann. N.Y. Acad. Sci. *185*, 166–180.
Langan, T.A. (1978) in *Methods in Cell Biology* (Stein, G., Stein, J. & Kleinsmith, G.J., eds.), vol. 19, pp. 127–142, Academic Press, N.Y.
Langan, T.A. (1982) J. Biol. Chem. *257*, 14835–14846.
Langan, T.A., Rall, S.C. & Cole, R.D. (1971) J. Biol. Chem. *246*, 1942–1944.
Langan, T.A., Zellig, C. & Leichtling, B. (1981) in *Protein Phosphorylation* (Rosen, O.M. & Krebs, E.G., eds.), Book B, pp. 1039–1052, Cold Spring Harbor Laboratories.
Langmore, J.P. & Paulson, J.R. (1983) J. Cell Biol. *96*, 1120–1131.
Langmore, J.P. & Wooley, J.C. (1975) Proc. Natl. Acad. Sci. USA *72*, 2691–2695.
Larsen, A. & Weintraub, H. (1982) Cell *29*, 609–622.
Larue, H., Bissonnette, E. & Belanger, L. (1983) Can. J. Bioch. Cell Biol. *61*, 1197–1200.
Lasater, L.S. & Eichler, D.C. (1984) Biochemistry *23*, 4367–4373.
Laskey, R.A. & Earnshaw, W.C. (1980) Nature (London) *286*, 763–767.
Laskey, R.A. & Harland, R.M. (1981) Cell *24*, 283–284.
Laskey, R.A., Honda, B.M., Mills, A.D., Morris, N.R., Wyllie, A.H., Mertz, J.E., DeRobertis, E.M. & Gurdon, J.B. (1977) Cold Spring Harbor Symp. Quant. Biol. *42*, 171–178.
Laskey, R.A., Honda, B.M., Mills, A.D. & Finch, J.T. (1978) Nature (London) *275*, 416–420.
Latham, K.R., Apriletti, J.W., Eberhardt, W.L. & Baxter, J.D. (1981) J. Biol. Chem. *256*, 12094–12101.
Lathe, R., Buc, H., Lecoco, J.P. & Bautz, E.K.F. (1980) Proc. Natl. Acad. Sci. USA *77*, 3548–3552.
Lawn, R.M., Heumann, J.M., Herrick, G.A. & Prescott, D.M. (1977) Cold Spring Harbor Symp. Quant. Biol. *42*, 483–492.
Lawrence, J.-J. & Goeltz, P. (1981) Nucleic Acids Res. *9*, 859–866.
Lawson, G.M. & Cole, R.D. (1979) Biochemistry *18*, 2160–2166.
Lawson, G.M., Tsai, M.-J. & O'Malley, B.W. (1980) Biochemistry *19*, 4403–4411.
Lawson, G.M., Knoll, B.J., March, C.J., Woo, S.L., Tsai, M.J. & O'Malley, B.W. (1982) J. Biol. Chem. *257*, 1501–1507.
Lazarus, H.M. & Sporn, M.B. (1967) Proc. Natl. Acad. Sci. USA *57*, 1386–1393.
Leber, B. & Hemleben, V. (1979a) Nucleic Acids Res. *7*, 1263–1281.
Leber, B. & Hemleben, V. (1979b) Pl. Syst. Evol. *Suppl. 2*, 187–199.
Lebkowski, J.S. & Laemmli, U.K. (1982a) J. Mol. Biol. *156*, 309–324.
Lebkowski, J.S. & Laemmli, U.K. (1982b) J. Mol. Biol. *156*, 325–344.
Lee, C.-H. & Charney, E. (1982) J. Mol. Biol. *161*, 289–303.
Lee, C.-H., Mizusawa, H. & Kakefuda, T. (1981) Proc. Natl. Acad. Sci. USA *78*, 2838–2842.
Lee, K.S., Mandelkern, M. & Crothers, D. (1981) Biochemistry *20*, 1438–1445.
Leffak, I.M. (1983) Nucleic Acids Res. *11*, 2717–2732.
Leffak, I.M. (1984) Nature (London) *307*, 82–85.
Leffak, I.M., Grainger, R. & Weintraub, H. (1977) Cell *12*, 837–845.
Leiter, J.M.E., Helliger, W. & Puschendorf, B. (1984) Exp. Cell Res. *155*, 222–231.
Lennox, R.W. (1984) J. Biol. Chem. *259*, 669–672.
Lennox, R.W. & Cohen, L.H. (1983) J. Biol. Chem. *258*, 262–268.

Lennox, R.W. & Cohen, L.H. (1984a) Develop. Biol. *103*, 80–84.

Lennox, R.W. & Cohen, L.H. (1984b) in *Histone Genes: Structure, Organization, and Regulation* (Stein, G., Stein, J. & Marzluff, W., eds.), pp. 373–395, John Wiley & Sons, N.Y.

Leslie, A.G.W., Arnett, S., Chandrasekaran, R., & Ratliff R.L. (1980) J. Mol. Biol. *143*, 49–72.

LeStourgeon, W.M. (1978) in *The Cell Nucleus* (Busch, H., ed.), vol. 6, pp. 305–326, Academic Press, N.Y.

LeStourgeon, W.M., Forer, A., Yang, Y., Bertram, J.S. & Rusch, H.P. (1975) Biochim. Biophys. Acta *379*, 529–552.

Levene, P.A. & Jacobs, W.A. (1909) Ber. Deut. Chem. Ges. *42*, 335–338.

Levene, P.A. & London, E.S. (1929) J. Biol. Chem. *83*, 793–802.

Levina, E.S. & Mirzabekov, A.D. (1975) Dokl. Akad. Nauk. SSSR *221*, 1222–1225.

Levinger, L. & Varshavsky, A. (1980) Proc. Natl. Acad. Sci. USA *77*, 3244–3248.

Levinger, L. & Varshavsky, A. (1982) Cell *28*, 375–385.

Levitt, M. (1978) Proc. Natl. Acad. Sci. USA *75*, 640–644.

Levy, A. & Jakob, K.M. (1978) Cell *14*, 259–267.

Levy, A. & Noll, M. (1980) Nucleic Acids Res. *8*, 6059–6067.

Levy, A. & Noll, M. (1981) Nature (London) *289*, 198–203.

Levy, G.C., Hilliard, P.R., Levy, L.F., Rill, R.L. & Inners, R. (1981) J. Biol. Chem. *256*, 9986–9989.

Levy, S., Sures, I. & Kedes, L. (1982) J. Biol. Chem. *257*, 9438–9443.

Levy-Wilson, B. (1981a) Arch. Biochem. Biophys. *208*, 528–534.

Levy-Wilson, B. (1981b) Proc. Natl. Acad. Sci. USA *78*, 2189–2193.

Levy-Wilson, B. (1983a) Biochemistry *22*, 484–489.

Levy-Wilson, B. (1983b) DNA *2*, 9–13.

Levy-Wilson, B. & Dixon, G.H. (1978) Can. J. Biochem. *56*, 480–491.

Levy-Wilson, B. & Dixon, G.H. (1979) Proc. Natl. Acad. Sci. USA *76*, 1682–1686.

Levy-Wilson, B., Wong, N.C.W. & Dixon, G.H. (1977) Proc. Natl. Acad. Sci. USA *74*, 2810–2814.

Levy-Wilson, B., Connor, W. & Dixon, G.H. (1979) J. Biol. Chem. *254*, 609–620.

Levy-Wilson, B., Denker, M.S. & Ito, E. (1983) Biochemistry *22*, 1715–1721.

Lewis, C.D. & Laemmli, U.K. (1982) Cell *29*, 171–181.

Lewis, C.D., Lebkowski, J.S., Daly, A. & Laemmli, U.K. (1984) J. Cell Sci. *Suppl. 1*, 103–122.

Lewis, E.A. & Reams, R.R. (1983) Arch. Biochem. Biophys. *223*, 185–192.

Lewis, M.K. & Burgess, R.R. (1982) in *The Enzymes* (Boyer, P., ed.), vol. 15, Part B, pp. 109–153, Academic Press, N.Y.

Lewis, P.N. (1976) Biochem. Biophys. Res. Comm. *68*, 329–335.

Lewis, P.N. & Chiu, S.S. (1980) Eur. J. Biochem. *109*, 369–376.

Li, H.J. (1977) in *Chromatin and Chromosome Structure* (Li, H.J. & Eckhardt, R.A., eds.), pp. 1–36, Academic Press, N.Y.

Li, H.J., Wickett, R., Craig, A.M. & Isenberg, I. (1972) Biopolymers *11*, 375–397.

Liao, L.W. & Cole, R.D. (1981a) J. Biol. Chem. *256*, 10124–10128.

Liao, L.W. & Cole, R.D. (1981b) J. Biol. Chem. *256*, 11145–11150.

Libby, P.R. (1978) J. Biol. Chem. *253*, 233–236.

Libby, P.R. (1980) Arch. Biochem. Biophys. *203*, 384–389.

Libertini, L.J. & Small, E.W. (1980) Nucleic Acids Res. *8*, 3517–3534.

Libertini, L.J. & Small, E.W. (1982) Biochemistry *21*, 3327–3334.

Libertini, L. & Small, E.W. (1984) Nucleic Acids Res. *12*, 4351–4359.

Lichtenwald, D.M. & Suhadolnik, R.J. (1979) Biochemistry *18*, 3749–3755.

Lichtler, A.C., Detke, S., Phillips, I.R., Stein, G.S. & Stein, J.L. (1980) Proc. Natl. Acad. Sci. USA *77*, 1942–1946.

Lichtler, A.C., Sierra, F., Stein, J., and Stein, G. (1982) Nature (London) *298*, 195–198.

Liew, C.C. & Chan, P.K. (1976) Proc. Natl. Acad. Sci. USA *73*, 3458–3462.

Liew, C.C., Haslett, G.W. & Allfrey, V. (1970) Nature *226*, 414–417.

Lilienfeld, L. (1894) Z. Physiol. Chem. *18*, 473–486.

Lilley, D.M.J. (1980) Proc. Natl. Acad. Sci. USA *77*, 6468–6472.

Lilley, D.M.J. (1981a) Nature *292*, 380–382.

Lilley, D.M.J. (1981b) Nucleic Acids Res. *9*, 1271–1289.

Lilley, D.M.J. (1984) in *Twentieth Colworth Memorial Lecture* Biochemical Soc. Trans. *12*, 127–140.

Lilley, D.M.J. & Tatchell, K. (1977) Nucleic Acids Res. *4*, 2039–2065.

Lilley, D.M.J., Jacobs, M.F. & Houghton, M. (1979) Nucleic Acids Res. *7*, 377–399.

Lin, P.P.C. & Key, J.L. (1980) Plant Physiol. *66*, 360–367.

Lindsey, G.G., Thompson, P., Purves, L.R. & von Holt, C. (1982) FEBS Lett. *145*, 131–136.

Lipps, H.J. & Morris, N.R. (1977) Biochem. Biophys. Res. Comm. *74*, 230–234.

Littlefield, B.A., Cidlowski, N.B. & Cidlowski, J.A. (1982) Exp. Cell Res. *141*, 283–291.

Liu, L.F. & Wang, J.C. (1978) Cell *15*, 979–984.

Loeb, L.A. (1970) Nature *226*, 448–449.

Loeblich, A.R. (1976) J. Protozoology *23*, 13–28.

Lohr, D. (1983a) Biochemistry *22*, 927–934.

Lohr, D. (1983b) Nucleic Acids Res. *11*, 6755–6773.

Lohr, D. (1984a) Cell Biophys. *6*, 87–102.

Lohr, D. (1984b) Nucleic Acids Res. *12*, 8457–8474.

Lohr, D. & Ide, G. (1979) Nucleic Acids Res. *6*, 1909–1927.

Lohr, D. & van Holde, K.E. (1975) Science *188*, 165–166.

Lohr, D. & van Holde, K.E. (1979) Proc. Natl. Acad. Sci. USA *76*, 6326–6330.

Lohr, D., Corden, J., Tatchell, K., Kovacic, R.T. & van Holde, K.E. (1977a) Proc. Natl. Acad. Sci. USA *74*, 79–83.

Lohr, D., Tatchell, K., van Holde, K.E. (1977b) Cell *12*, 829–836.

Loidl, P., Loidl, A., Puschendorf, B. & Grobner, P. (1983) Nature (London) *305*, 446–448.

Long, B.H., Huang, C.Y. & Pogo, A.U. (1979) Cell *18*, 1079–1090.

Lorimer, W.S., Stone, P.R. & Kidwell, W.R. (1977) Exp. Cell Res. *106*, 261–266.

Lossius, I., Sjastad, K., Haarr, L. & Kleppe, K. (1984) J. Gen. Micro. *130*, 3153–3157.

Louis, C., Schedl, P., Samal, B. & Worcel, A. (1980) Cell *22*, 387–392.

Louters, L. & Chalkley, R. (1985) Biochemistry *24*, 3080–3085.

Lowenhaupt, K., Cartwright, I.L., Keene, M.A., Zimmerman, J.L. & Elgin, S.C.R. (1983) Develop. Biol. *99*, 194–201.

Luchnik, A.N., Bakayev, V.V., Zbarsky, I.B. & Georgiev, G.P. (1982) EMBO J. *1*, 1353–1358.

Luck, J.M., Cook, H.A., Eldredge, N.T., Haley, M.I., Kupke, P.W. & Rasmussen, P.S. (1956) Arch. Biochem. Biophys. *65*, 449–467.

Luck, J.M., Rasmussen, P.S., Satake, K. & Tsvetikov, A.N. (1958) J. Biol. Chem. *233*, 1407–1414.

Lund, T., Holtland, J. & Lalaud, S.G. (1985) FEBS Lett. *180*, 275–279.

Lutter, L.C. (1978) J. Mol. Biol. *124*, 391–420.

Lutter, L.C. (1979) Nucleic Acids Res. *6*, 41–56.

Lutter, L.C. (1981) Nucleic Acids Res. *9*, 4251–4265.

Luzzati, V. & Nicolaieff (1963) J. Mol. Biol. *7*, 142–163.

Mace, H.A.F., Pelham, H.R.B. & Travers, A.A. (1983) Nature (London) *304*, 555–557.

Mace, M.L., Darkal, Y., Wray, V.P., Wray, W. (1977) Cytobios *19*, 27–40.

Machray, G.C. & Bonner, J. (1981) Biochemistry *20*, 5466–5470.

MacKay, S. & Newrock, K.M. (1982) Develop. Biol. *93*, 430–437.

MacLeod, A.R., Wong, N.C.W. & Dixon, G.H. (1977) Eur. J. Biochem. *78*, 281–291.

Magnaval, R., Valencia, R. & Paoletti, J. (1980) Biochem. Biophys. Res. Comm. *92*, 1415–1421.

Makarov, V.L., Dimitrov, S.I., Tsaneva, I.R., Pashev, I.G. (1984) Biochem. Biophys. Res. Comm. *122*, 1021–1027.

Makiguchi, K., Chida, Y., Yoshida, M. & Shimura, K. (1984) J. Biochem. *95*, 423–429.

Malik, N. & Smulson, M. (1984) Biochemistry *23*, 3721–3725.

Malik, N., Miwa, M., Sugimura, T., Thraves, P. & Smulson, M. (1983) Proc. Natl. Acad. Sci. USA *80*, 2554–2558.

Malik, N., Smulson, M. & Bustin, M. (1984) J. Biol. Chem. *259*, 699–702.

Man, N.T. & Shall, S. (1982) Eur. J. Biochem. *126*, 83–88.

Mandelkern, M., Dattagupta, N. & Crothers, D.M. (1981) Proc. Natl. Acad. Sci USA *78*, 4294–4298.

Mansy, S., Engstrom, S.K. & Peticolas, W.L. (1976) Biochem. Biophys. Res. Comm. *68*, 1242–1243.

Marcus-Sekura, C.J. & Carter, B.J. (1983) J. Virology *48*, 79–87.

Mardian, J.K.W. & Isenberg, I. (1978) Biochemistry *17*, 3825–3833.

Mardian, J.K.W., Paton, A.E., Bunick, G.J. & Olins, D.E. (1980) Science *209*, 1534–1536.

Marian, B. & Wintersberger, V. (1982) FEBS Lett. *139*, 72–76.

Marini, J.C., Levene, S.D., Crothers, D.M. & Englund, P.T. (1982) Proc. Natl. Acad. Sci. USA *79*, 7664–7668.

Marion, C., Bezot, P., Hesse-Bezot, C., Roux, B. & Bernengo, J.C. (1981) Eur. J. Biochem. *120*, 169–176.

Marion, C., Roux, B. & Coulet, P.R. (1983) FEBS Lett. *157*, 317–321.

Marks, D.B., Paik, W.K. & Borun, T.W. (1973) J. Biol. Chem. *248*, 5668–5677.

Marks, D.B., Kanefsky, T., Keller, B.J. & Marks, A.D. (1975) Cancer Res. *35*, 886–889.

Marquez, G., Moran, F., Franco, L. & Montero, F. (1982) Eur. J. Biochem. *123*, 165–170.

Marsden, M.P.F. & Laemmli, U.K. (1979) Cell *17*, 849–858.

Marsh, W.H. & Fitzgerald, P.S. (1973) Fed. Proc. *32*, 2119.

Martinage, A., Mangeat, P., Sautiere, P., Marchis-Mouren, G. & Biserte, G. (1979) Biochimie *61*, 61–69.

Martinage, A., Mangeat, P., Laine, B., Couppez, M., Sautiere, P., Marchis-Mouren, G. & Biserte, G. (1980) FEBS Lett. *118*, 323–329.

Martinage, A., Quirin-Stricker, C., Champagne, M. & Sautiere, P. (1981a) FEBS Lett. *134*, 103–106.

Martinage, A., Mangeat, P., Sautiere, P., Couppez, M., Marchis-Mouren, G. & Biserte, G. (1981b) FEBS Lett. *134*, 107–121.

Martinage, A., Belaiche, D., Dupressoir, T. & Sautiere, P. (1983) Eur. J. Biochem. *130*, 465–472.

Martinson, H.G. & True, R.J. (1979a) Biochemistry *18*, 1089–1094.

Martinson, H.G. & True, R.J. (1979b) Biochemistry *18*, 1947–1951.

Martinson, H.G., True, R., Burch, J.B.E. & Kunkel, G. (1979a) Proc. Natl. Acad. Sci. USA *76*, 1030–1034.

Martinson, H.G., True, R., Lau, C.K. & Mehrabian, M. (1979b) Biochemistry *18*, 1075–1082.

Martinson, H.G., True, R.J. & Burch, J.B.E. (1979c) Biochemistry *18*, 1082–1089.

Masaracchia, R.A., Kemp, B.E. & Walsh, D.A. (1977) J. Biol. Chem. *252*, 7109–7117.

Mathew, C.G.P., Goodwin, G.H., Gooderham, K., Walker, J.M. & Johns, E.W. (1979) Biochem. Biophys. Res. Comm. *87*, 1243–1251.

Mathew, C.G.P., Goodwin, G.H. & Johns, E.W. (1980) J. Chromatog. *198*, 80–83.

Mathis, D.J. & Gorovsky, M.A. (1977) Cold Spring Harbor Symp. Quant. Biol. *42*, 773–778.

Mathis, D., Oudet, P. & Chambon, P. (1980) Prog. Nucleic Acids *24*, 1–55.

Mathog, D., Hockstrasser, M., Gruenbaum, Y., Saumweber, H. & Sedat, J.W. (1984) Nature (London) *308*, 414–421.

Matsui, S., Seon, B.K. & Sandberg, A.A. (1979) Proc. Natl. Acad. Sci. USA *76*, 6386–6390.

Matsui, S., Sandberg, A.A., Negoro, S., Seon, B.K. & Goldstein, G.G. (1982) Proc. Natl. Acad. Sci. USA *79*, 1535–1539.

Mattern, M.R. & Painter, R.B. (1979) Biochim. Biophys. Acta *563*, 306–312.

Mauron, A., Kedes, L., Hough-Evans, B.R. & Davidson, E.H. (1982) Develop. Biol. *94*, 425–434.

Maxson, R.E. & Wilt, F.H. (1982) Develop. Biol. *94*, 435–440.

Mayes, E.L.V. (1982) in *The HMG Chromosomal Proteins* (Johns, E.W., ed.), pp. 9–40, Academic Press, N.Y.

Mayes, E.L.V. & Walker, J.M. (1984) Int. J. Pept. & Prot. Res. *23*, 516–520.

Mazen, A., Champagne, M., Wilhelm, M. & Wilhelm, F.X. (1978) Exp. Cell Res. *117*, 431–438.

Mazen, A., DeMurcia, G., Bernard, S., Pouyet, J. & Champagne, M. (1982) Eur. J. Biochem. *127*, 169–176.

McCarroll, R., Olsen, G.J., Stahl, Y.D., Woese, C.R. & Sogin, M.L. (1983) Biochemistry *22*, 5858–5868.

McCarty, M. (1946) J. Gen. Physiol. *29*, 123–139.

McCready, S.J., Akrigg, A. & Cook, P.R. (1979) J. Cell Sci. *39*, 53–62.

McGhee, J.D. & Felsenfeld, G. (1979) Proc. Natl. Acad. Sci. USA *76*, 2133–2137.

McGhee, J.D. & Felsenfeld, G. (1980a) Ann. Rev. Biochem. *49* 1115–1156.

McGhee, J.D. & Felsenfeld, G. (1980b) Nucleic Acids Res. *8*, 2751–2769.

McGhee, J.D. & Felsenfeld, G. (1982) J. Mol. Biol. *158*, 685–698.

McGhee, J.D. & Felsenfeld, G. (1983) Cell *32*, 1205–1215.

McGhee, J.D., Rau, D.C., Charney, E. & Felsenfeld, G. (1980) Cell *22*, 87–96.

McGhee, J.D., Wood, W.I., Dolan, M., Engel, J.D. & Felsenfeld, G. (1981) Cell *27*, 45–55.

McGhee, J.D., Rau, D.C. & Felsenfeld, G. (1982) Nucleic Acids Res. *10*, 2007–2016.

McGhee, J.D., Nickol, J.M., Felsenfeld, G. & Rau, D.C. (1983a) Cell *33*, 831–841.

McGhee, J.D., Rau, D.C. & Felsenfeld, G. (1983b) Proc. Clin. Biol. Res. *134*, 143–157.

McGuire, M.S., Center, M.S. & Consigli, R.A. (1976) J. Biol. Chem. *251*, 7746–7752.

McKeon, F., Tuffanelli, F., Kobayashi, S. & Kirschner, M. (1984) Cell *36*, 83–92.

McKnight, S.L. & Miller, O.S. (1977) Cell *12*, 795–804.

McKnight, S.L., Sullivan, N.L. & Miller, O.L. (1976) in *Progress in Nucleic Acid Research and Molecular Biology* (Cohn, W.E. & Volkin, E., eds.), vol. 19, pp. 313–318, Academic Press, N.Y.

McMurray, C. (1987) Ph.D. Thesis, Oregon State University.

McMurray, C. & van Holde, K.E. (1986) Proc. Natl. Acad. Sci. USA *83*, 8472–8476.

McMurray, C., van Holde, K.E., Jones, R.L. & Wilson, W.D. (1985) Biochemistry *24*, 7037–7044.

Meistrich, M.L., Bucci, L.R., Trrostle-Weige, P.K. & Broch, W.A. (1985) Develop. Biol. *112*, 230–240.

Mende, L., Timm, B. & Subramanian, A.R. (1978) FEBS Lett. *96*, 395–398.

Mengeritsky, G. & Trifonov, E.N. (1984) Cell Biophys. *6*, 1–8.

Menko, A.S. & Tan, K.B. (1980) Biochim. Biophys. Acta *629*, 359–370.

Mezquita, J., Connor, W., Winkfein, R.J. & Dixon, G.H. (1985) J. Mol. Evol. *21*, 209–219.

Michalski-Scrive, C., Aubert, J.P., Couppez, M., Biserte, G. & Loucheux-Lefebvre, M.H. (1982) Biochimie *64*, 347–355.

Miescher, F. (1871) Med-Chem Untersuch. (Hoppe-Seyler) pt. 4, pp. 441–460.

Miescher, F. (1874) Ver. Naturforsch. Ges. Basel *6*, 138–208.

Miller, F.D., Dixon, G.H., Rattner, J.B. & van de Sande, J.H. (1985) Biochemistry *24*, 102–109.

Miller, J., McLachlan, A.D. & Klug, A. (1985) EMBO J. *4*, 1609–1614.

Miller, O.L., Jr. & Bakken, A.H. (1972) Acta Endocrinol. *Suppl. 168*, 155–177.

Miller, O.L., Jr. & Beatty, B.R. (1969) Science *164*, 955–957.

Mills, F.C., Fisher, L.M., Kuroda, R., Ford, A.M. & Gould, H.J. (1983) Nature (London) *306*, 809–812.

Mirault, M.-E., Goldschmidt-Clermont, M., Artavanis-Tsakonas, S. & Schedl, P. (1979) Proc. Natl. Acad. Sci. USA *76*, 5254–5258.

Mirkovitch, J., Mirault, M.-E. & Laemmli, U.K. (1984) Cell *39*, 223–232.

Mirsky, A.E. (1947) Cold Spring Harbor Symp. Quant. Biol. *12*, 143–146.

Mirsky, A.E. (1968) Sci. Am. *218*, 78–88.

Mirsky, A.E. & Pollister, A.W. (1942) Proc. Natl. Acad. Sci. USA *28*, 344–352.

Mirsky, A.E. & Pollister, A.W. (1946) J. Gen. Physiol. *30*, 117–148.

Mirsky, A.E. & Ris, H. (1947) J. Gen. Physiol. *31*, 7–18.

Mirzabekov, A.D., Shick, V.V., Belyavsky, A.V., Karpov, V.L. & Bavykin, S.G. (1977) Cold Spring Harbor Symp. Quant. Biol. *42*, 149–155.

Mirzabekov, A.D., Shick, V.V., Belyavsky, A.V. & Bavykin, S.G. (1978) Proc. Natl. Acad. Sci. USA *75*, 4184–4188.

Mirzabekov, A.D., Bavykin, S.G., Karpov, V.L., Preobrazhenskaya, O.V., Ebralidze, K.K., Tuneev, V.M., Melinkova, A.F., Goguadze, E.G., Chenchick, A.A. & Beabealashvili, R.S. (1982) Cold Spring Harbor Symp. Quant. Biol. *47*, 503–509.

Mita, K., Zama, M., Ichimura, S. & Hamana, K. (1981) Biochem. Biophys. Res. Comm. *98*, 330–336.

Mita, K., Zama, M., Ichimura, S., Niimura, N., Kaji, K., Hirai, M. & Ishikawa, Y. (1983) Physica *120B*, 436–439.

Mitchelson, K., Chambers, T., Bradbury, E.M. & Matthews, H.R. (1978) FEBS Lett. *92*, 339–342.

Mitra, S., Sen, D. & Crothers, D.M. (1984) Nature (London) *308*, 247–250.

Mitsui, Y., Langridge, R., Shortle, B., Cantor, C., Grant, R., Kodama, M. & Wells, R.D. (1970) Nature (London) *228*, 1166–1169.

Miwa, M., Tanaka, M., Matsushima, T. & Sugimura, T. (1974) J. Biol. Chem. *249*, 3475–3482.

Miwa, M., Saikawa, N., Yamaizumi, Z., Nishimura, S. & Sugimura, T. (1979) Proc. Natl. Acad. Sci. USA *76*, 595–598.

Miwa, M., Ishihara, M., Takishima, S., Takasuka, N., Maeda, M., Yamaizumi, Z., Sugimara, T., Yokoyama, S. & Miyazawa, T. (1981) J. Biol. Chem. *256*, 2916–2921.

Miyakawa, N., Veda, K. & Hayaishi, O. (1972) Biochem. Biophys. Res. Comm. *49*, 239–245.

Miyazaki, K., Hagiwara, H., Nagao, Y., Matuo, Y. & Horio, T. (1978) J. Biochem. *84*, 135–143.

Mizuuchi, K., Mizuuchi, M. & Gellert, M. (1982) J. Mol. Biol. *156*, 229–242.

Modak, S.P., Lawrence, J.J. & Gorka, C. (1980) Mol. Biol. Rep. *6*, 235–243.

Moore, M., Jackson, V., Sealy, L. & Chalkley, R. (1979) Biochim. Biophys. Acta *561*, 248–260.

Moorman, A.F.M., Deboer, P.A.J., Delaaf, R.T.M., van Dongen, W.M. & Destree, O.H.J. (1981) FEBS Lett. *136*, 45–52.

Moorman, A.F.M., Deboer, P.A.J., Delaaf, R.T.M. & Destree, O.H.J. (1982) FEBS Lett. *144*, 235–241.

Moorman, A.F.M., Deboer, P.A.J., Linders, M.T. & Charles, R. (1984) Cell Differ. *14*, 113–123.

Morris, G. & Lewis, P.N. (1977) Eur. J. Biochem. *77*, 471–477.

Morris, N.R. (1976a) Cell *8*, 357–363.

Morris, N.R. (1976b) Cell *9*, 627–632.

Moss, T., Cary, P.D., Crane-Robinson, C. & Bradbury, E.M. (1976a) Biochemistry *15*, 2261–2276.

Moss, T., Cary, P.D., Abercrombie, B.D., Crane-Robinson, C. & Bradbury, E.M. (1976b) Eur. J. Biochem. *71*, 337–350.

Moudrianakis, E.N., Love, W.E., Wang, B.C., Xuong, N.G. & Burlingame, R.W. (1985a) Science *229*, 1110–1112.

Moudrianakis, E.N., Love, W.E. & Burlingame, R.W. (1985b) Science 229, 1113.

Muller, S., Himmelspach, K. & van Regenmortel, M.H. (1982) EMBO J. 1, 421–425.

Mullins, D.W., Jr., Giri, C.P. & Smulson, M. (1977) Biochemistry 16, 506–513.

Multhaup, I., Csordas, A., Grunicke, H., Pfister, R. & Puschendorf, B. (1983) Arch. Biochem. Biophys. 222, 497–503.

Murphy, R.F., Wallace, R.B. & Bonner, J. (1978) Proc. Natl. Acad. Sci. USA 75, 5903–5907.

Murphy, R.F., Wallace, R.B. & Bonner, J. (1980) Proc. Natl. Acad. Sci. USA 77, 3336–3340.

Murray, K. (1964) Biochemistry 3, 10–15.

Murray, K. (1969) J. Mol. Biol. 39, 125–144.

Murray, M.G. & Kennard, W.C. (1984) Biochemistry 23, 4225–4232.

Murray, M.G., Guilfoyle, T.J. & Key, J.L. (1978) Plant Physiol. 61, 1023–1030.

Musich, P.R., Brown, F.L. & Maio, J.J. (1977a) Proc. Natl. Acad. Sci. USA 74, 3297–3301.

Musich, P.R., Maio, J.J. & Brown, F.L. (1977b) J. Mol. Biol. 117, 657–677.

Musich, P.R., Brown, F.L., Maio, J.J. (1982) Proc. Natl. Acad. Sci. USA 79, 118–122.

Muyldermans, S., Lasters, I. & Wyns, L. (1980a) Nucleic Acids Res. 8, 731–739.

Muyldermans, S., Lasters, I., Wyns, L. & Hamers, R. (1980b) Nucleic Acids Res. 8, 2165–2172.

Nadeau, P., Oliver, D.R. & Chalkley, R. (1978) Biochemistry 17, 4885–4893.

Nahon, J.L., Gal, A., Erdos, T. & Sala-Trepat, J.M. (1984) Proc. Natl. Acad. Sci. USA 81, 5031–5035.

Nakane, M., Ide, T., Anzai, K., Ohara, S. & Andoh, T. (1978) J. Biochem. 84, 145–157.

Nakayama, T. (1980) Biochem. Biophys. Res. Comm. 97, 318–324.

Nathan, D.G. (1983) in Globin Gene Expression and Hematopoietic Differention (Stamatoyannopoulos, G. & Nienhuis, A.W., eds.) pp. 399–410, Alan R. Liss, N.Y.

Naveh-Many, T. & Cedar, H. (1981) Proc. Natl. Acad. Sci. USA 78, 4246–4250.

Nedospasov, S.A. & Georgiev, G.P. (1980) Biochem. Biophys. Res. Comm. 92, 532–539.

Nedospasov, S., Shakhov, A. & Georgiev, G. (1981) FEBS Lett. 125, 35–38.

Nelson, D.A., Oosterhof, D.K. & Rill, R.L. (1977) Nucleic Acids Res. 4, 4223–4234.

Nelson, D.A., Perry, M., Sealy, L. & Chalkley, R. (1978) Biochem. Biophys. Res. Comm. 82, 1346–1353.

Nelson, D.A., Mencke, A.J., Chambers, S.A., Oosterhof, D.K. & Rill, R.L. (1982) Biochemistry 21, 4350–4362.

Nelson, P.P., Albright, S.C., Wiseman, J.M. & Garrard, W.T. (1979) J. Biol. Chem. 254, 11751–11760.

Nelson, R.G. & Fangman, W.L. (1979) Proc. Natl. Acad. Sci. USA 76, 6515–6519.

Ness, P.J., Labhart, P., Banz, E., Koller, T. & Parish, R.W. (1983) J. Mol. Biol. 166, 361–381.

Newrock, K.M., Alfageme, C.R., Nardi, R.V. & Cohen, L.H. (1977) Cold Spring Harbor Symp. Quant. Biol. 42, 421–431.

Newrock, K.M., Cohen, L.H., Hendricks, M.B., Donnelly, R.J. & Weinberg, E.S. (1978) Cell *14*, 327–336.

Newrock, K.M., Freedman, N., Alfageme, C.R. & Cohen, L.H. (1982) Develop. Biol. *89*, 248–253.

Nickol, J. & Felsenfeld, G. (1983) Cell *35*, 467–477.

Nickol, J. & Martin, R.G. (1983) Proc. Natl. Acad. Sci. USA *80*, 4669–4673.

Nickol, J., Behe, M. & Felsenfeld, G. (1982) Proc. Natl. Acad. Sci. USA *79*, 1771–1775.

Nielsen, P.E. (1981) FEBS Lett. *135*, 173–176.

Nishizuka, Y., Ueda, K., Honjo, T. & Hayaishi, O. (1968) J. Biol. Chem. *243*, 3765–3767.

Nochumson, S., Kim, S. & Paik, W.K. (1978) Meth. Cell Biol. *19*, 69–77.

Nolan, N.L., Butt, T.R., Wong, M., Lambrianidou, A. & Smulson, M.E. (1980) Eur. J. Biochem. *113*, 15–25.

Noll, M. (1974a) Nature (London) *251*, 249–251.

Noll, M. (1974b) Nucleic Acids Res. *1*, 1573–1578.

Noll, M. (1976) Cell *8*, 349–355.

Noll, M. (1977) J. Mol. Biol. *116*, 49–71.

Noll, M. & Kornberg, R. (1977) J. Mol. Biol. *109*, 393–404.

Noll, M., Thomas, J.O., Kornberg, R. (1975) Science *187*, 1203–1206.

Nomoto, M., Hayashi, H. & Iwai, K. (1982) J. Biochem. *91*, 897–904.

North, G. (1985) Nature (London) *316*, 394–395.

Nose, K., Tanaka, A. & Okamoto, H. (1981) J. Biochem. *89*, 1711–1719.

Notbohm, H. (1982) Biochim. Biophys. Acta *696*, 223–225.

Nurse, P. (1983) Nature (London) *302*, 378.

O'Connor, P.J. (1969) Biochem. Biophys. Res. Comm. *35*, 805–810.

Ogata, N., Ueda, K. & Hayaishi, O. (1980a) J. Biol. Chem. *255*, 7610–7615.

Ogata, N., Ueda, K., Kagamiyama, H. & Hayaishi, O. (1980b) J. Biol. Chem. *255*, 7616–7620.

Ogawa, Y., Quagliarotti, G., Jordan, J., Taylor, C.W., Starbuck, W.C. & Busch, H. (1969) J. Biol. Chem. *244*, 4387–4392.

Ohashi, Y., Ueda, K., Kawaichi, M. & Hayaishi, O. (1983) Proc. Natl. Acad. Sci. USA *80*, 3604–3607.

Ohba, M. & Oshima, T. (1980) Stud. Biophys. *81*, 141–142.

Ohe, Y. & Iwai, K. (1981) J. Biochem. *90*, 1205–1211.

Ohe, Y., Hanashi, H., and Iwai, K. (1979) J. Biochem. *85*, 615–624.

Ohghushi, H., Yoshihara, K. & Kamiya, T. (1980) J. Biol. Chem. *255*, 6205–6211.

Ohlenbusch, H.H., Olivera, B.M., Tuan, D. & Davidson, N. (1967) J. Mol. Biol. *25*, 299–315.

Okada, T.A. & Comings, D.E. (1980) Am. J. Hum. Genet. *32*, 814–832.

Okai, Y. (1984) Mol. Biol. Rep. *10*, 19–22.

Okayama, H., Edson, C.M., Fukushima, M., Ueda, K. & Hayaishi, O. (1977) J. Biol. Chem. *252*, 7000–7005.

Okayama, H., Honda, M. & Hayaishi, O. (1978) Proc. Natl. Acad. Sci. USA *75*, 2254–2257.

Olby, R. (1974) *The Path to the Double Helix*, Univ. of Washington Press, Seattle.

Old, R.W. & Woodland, H.R. (1984) Cell *38*, 624–626.

Olins, A.L. (1979) in *Chromatin Structure and Function* (Nicolini, C., ed.), part A, pp. 31–40, Plenum Press, N.Y.

Olins, A.L. & Olins, D.E. (1973) J. Cell Biol. *59*, 252a.

Olins, A.L. & Olins, D.E. (1974) Science *183*, 330–332.

Olins, A.L. & Olins, D.E. (1983) Current Contents (Life Sciences) *26*, No. 10, 25.

Olins, A.L., Carlson, R.D., Wright, E.B. & Olins, D.E. (1976) Nucleic Acids Res. *3*, 3271–3291.

Olins, A.L., Breillatt, J.P., Carlson, R.D., Senior, M.B., Wright, E.B. & Olins, D.E. (1977a) in *The Molecular Biology of the Mammalian Genetic Apparatus* (Ts'o, P.O.P., ed.), pp. 211–237, Elsevier, North Holland.

Olins, A.L., Olins, D.E., Zentgraf, H. & Franke, W.W. (1980) J. Cell Biol. *87*, 833–836.

Olins, A.L., Olins, D.E., Levy, H.A., Durfee, R.C., Margle, S.M., Tinnel, E.P., Hingerty, B.E., Dover, S.D. & Fuchs, H. (1984) Eur. J. Cell Biol. *35*, 129–142.

Olins, A.L., Olins, D.E., Levy, H.A., Durfee, R.C., Margle, S.M. & Tinnel, E.P. (1986) Eur. J. Cell Biol. *40*, 105–110.

Olins, D.E., Bryan, P.N., Harrington, R.E., Hill, W.E. & Olins, A.L. (1977b) Nucleic Acids Res. *4*, 1911–1931.

Oliva, R. & Mezquita, C. (1982) Nucleic Acids Res. *10*, 8049–8059.

Oliver, D., Balhorn, R., Granner, D. & Chalkley, R. (1972) Biochemistry *11*, 3921–3925.

Olson, M.O.J., Goldknopf, I.L., Guetzow, K.A., James, G.T., Hawkins, T.C., Mays-Rothberg, C.J. & Busch, H. (1976) J. Biol. Chem. *251*, 5901–5903.

Omori, A., Igo-Kemenes, T. & Zachau, H.G. (1980) Nucleic Acids Res. *8*, 5363–5375.

Ord, M.G. & Stocken, L.A. (1966) Biochem. J. *98*, 888–897.

Orrick, L.R., Olson, M.O. & Bush, H. (1973) Proc. Natl. Acad. Sci. USA *70*, 1316–1320.

Osborne, H.B. & Chabanas, A. (1984) Exp. Cell Res. *152*, 449–458.

Osborne, T.B. & Harris, I.F. (1902) Hoppe-Seyler's Z. Physiol. Chem. *36*, 85–133.

Oshima, R., Curiel, D. & Linney, E. (1980) J. Supramol. Struct. *14*, 85–96.

Osipova, T.N. (1980) Mol. Biol. *14*, 374–381.

Osipova, T.N., Pospelov, V.A., Svetlikova, S.B. & Vorobev, V.I. (1980) Eur. J. Biochem. *113*, 183–188.

Osley, M.A. & Hereford, L. (1981) Cell *24*, 377–384.

Otto, B., Böhm, J. & Knippers, R. (1980) Eur. J. Biochem. *112*, 363–366.

Oudet, P., Gross-Bellard, M. & Chambon, P. (1975) Cell *4*, 281–300.

Oudet, P., Germond, J.E., Sures, M., Gallwitz, D., Bellard, M. & Chambon, P. (1977a) Cold Spring Harbor Symp. Quant. Biol. *42*, 287–300.

Oudet, P., Spadafora, C. & Chambon, P. (1977b) Cold Spring Harbor Symp. Quant. Biol. *42*, 301–312.

Ozaki, H. (1971) Develop. Biol. *26*, 209–219.

Paik, W.K. & Kim, S. (1973) Biochem. Biophys. Res. Comm. *51*, 781–788.

Paik, W.K. & Kim, S. (1974) Arch. Biochem. Biophys. *165*, 369–378.

Paik, W.K. & Kim, S. (eds.) (1980) *Protein Methylation*, John Wiley & Sons, N.Y.

Palau, J. & Padros, E. (1975) Eur. J. Biochem. *52*, 555–560.

Palen, T.E. & Cech, T.R. (1984) Cell *36*, 933–942.

Palter, K.B., Foe, V.E., & Alberts, B.M. (1979) Cell *18*, 451–467.

Pantazis, P. & Bonner, W.M. (1981) J. Biol. Chem. *256*, 4669–4675.

Panyim, S. & Chalkley, R. (1969a) Biochem. Biophys. Res. Comm. *37*, 1042–1049.

Panyim, S. & Chalkley, R. (1969b) Biochemistry *8*, 3972–3979.

Pardoll, D.M., Vogelstein, B. & Coffey, D.S. (1980) Cell *19*, 527–536.

Pardon, J.F. & Wilkins, M.H.F. (1972) J. Mol. Biol. *68*, 115–124.

Pardon, J.F., Wilkins, M.H.F. & Richards, B.M. (1967) Nature (London) *215*, 508–509.

Pardon, J.F., Richards, B.M. & Cotter, R.J. (1973) Cold Spring Harbor Symp. Quant. Biol. *38*, 75–81.

Pardon, J.F., Worcester, D.L., Wooley, J.C., Tatchell, K., van Holde, K.E. & Richards, B.M. (1975) Nucleic Acids Res. *2*, 2163–2175.

Pardon, J.F., Worcester, D.L., Wooley, J.C., Cotter, R.I., Lilley, D.M.J. & Richards, B.M. (1977a) Nucleic Acids Res. *4*, 3199–3214.

Pardon, J.F., Cotter, R.I., Lilley, D.M.J., Worcester, D.L., Campbell, A.M., Wooley, J.C. & Richards, B.M. (1977b) Cold Spring Harbor Symp. Quant. Biol. *42*, 11–22.

Parish, R.W. & Schmidlin, S. (1985) Nucleic Acids Res. *13*, 15–30.

Parish, R.W., Banz, E. & Ness, P.J. (1986) Nucleic Acids Res. *14*, 2089–2107.

Parker, C.S. & Roeder, R.G. (1977) Proc. Natl. Acad. Sci. USA *74*, 44–48.

Parker, C.S. & Topol, J. (1984a) Cell *36*, 273–283.

Parker, C.S. & Topol, J. (1984b) Cell *37*, 357–369.

Parslow, T.G. & Granner, D.K. (1982) Nature (London) *299*, 449–451.

Parslow, T.G. & Granner, D.K. (1983) Nucleic Acids Res. *11*, 4775–4792.

Pashev, I.G., Nencheva, M.M. & Markov, G.G. (1980) Biochim. Biophys. Acta *607*, 269–276.

Pastink, A., Berkhout, T.A., Mager, W.H. & Planta, R.J. (1979) Biochem. J. *177*, 917–923.

Paton, A.E., Wilkinson-Singley, E. & Olins, D.E. (1983) J. Biol. Chem. *258*, 13221–13229.

Patthy, L. & Smith, E.L. (1975) J. Biol. Chem. *250*, 1919–1920.

Patthy, L., Smith, E.L. & Johnson, J. (1973) J. Biol. Chem. *248*, 6834–6840.

Paul, I.J. & Duerksen, J.D. (1977) Can. J. Biochem. *55*, 1140–1144.

Paulson, J.R. & Laemmli, U.K. (1977) Cell *12*, 817–828.

Paulson, J.R. & Taylor, S.S. (1982) J. Biol. Chem. *257*, 6064–6072.

Payvar, F., deFranco, D., Firestone, G.L., Edgar, B., Wrange, O., Okret, S., Gustafsson, J.-A. & Yamamoto, K. (1983) Cell *35*, 381–392.

Pearson, E.C., Butler, P.J.G. & Thomas, J.O. (1983) EMBO J. *2*, 1367–1372.

Pearson, E.C., Bates, D.L., Prospero, T.D. & Thomas, J.O. (1984) Eur. J. Biochem. *144*, 353–360.

Peck, L.J. & Wang, J.L. (1983) Proc. Natl. Acad. Sci. USA *80*, 6206–6210.

Peck, L.J., Nordheim, A., Rich, A. & Wang, J.C. (1982) Proc. Natl. Acad. Sci. USA *79*, 4560–4564.

Pedersen, T. & Bhorjee, J.S. (1975) Biochemistry *14*, 3238–3242.

Pehrson, J.R. & Cohen, L.H. (1984) Biochemistry *23*, 6761–6764.

Pehrson, J.R. & Cole, R.D. (1980) Nature (London) *285*, 43–44.

Pehrson, J.R. & Cole, R.D. (1981) Biochemistry *20*, 2298–2301.

Pekary, A.E., Li, H.J., Chan, S.I., Hsu, C.J. & Wagner, T.E. (1975) Biochemistry *14*, 1177–1184.

Pelham, H.R.B. (1982) Cell *30*, 517–528.

Pelham, H.R.B. (1985) Trends Genet. *1*, 31–35.

Pentecost, B.T., Wright, J.M. & Dixon, G.H. (1985) Nucleic Acids Res. *13*, 4871–4888.

Perdone, F., Filetici, P. & Ballario, P. (1982) Nucleic Acids Res. *10*, 5197–5208.

Perlman, D. & Huberman, J.A. (1977) Cell *12*, 1029–1043.

Perry, M. & Chalkley, R. (1981) J. Biol. Chem. *256*, 3313–3318.

Perry, M. & Chalkley, R. (1982) J. Biol. Chem. *257*, 7336–7347.

Peterman, M.L. & Lamb, C.M. (1948) J. Biol. Chem. *176*, 685–693.

Peters, E.H., Levy-Wilson, B. & Dixon, G.H. (1979) J. Biol. Chem. *254*, 3358–3361.

Petersen, L.G.H. & Sheridan, W.F. (1978) Carlsberg Res. Comm. *43*, 415–422.

Peterson, J.L. & McConkey, E.H. (1976) J. Biol. Chem. *251*, 548–554.

Pezolet, M., Savoie, R., Guillot, J.G. & Pigeon-Gosselium, M. (1980) Can. J. Biochem. *58*, 633–640.

Philip, M., Jamaluddin, M., Sastry, R.V.R. & Chandra, H.S. (1979) Proc. Natl. Acad. Sci. USA *76*, 5178–5182.

Phillips, D.M.P. (1963) Biochem. J. *87*, 258–263.

Phillips, D.M.P. (1968) Biochem. J. *107*, 135–138.

Phillips, D.M.P. & Johns, E.W. (1965) Biochem. J. *94*, 127–130.

Philipps, G. & Gigot, C. (1977) Nucleic Acids Res. *4*, 3617–3626.

Pieler, C., Adolf, G.R. & Swetly, P. (1981) Eur. J. Biochem. *115*, 329–333.

Pinon, R. & Salts, Y. (1977) Proc. Natl. Acad. Sci. USA *74*, 2850–2854.

Plumb, M., Stein, J. & Stein, G. (1983) Nucleic Acids Res. *11*, 2391–2410.

Plumb, M.A., Nicolas, R.H., Wright, C.A. & Goodwin, G.H. (1985) Nucleic Acids Res. *13*, 4047–4065.

Plumb, M.A., Lobanenkov, V.V., Nicolas, R.H., Wright, C.A., Zavon, S. & Goodwin, G.H. (1986) Nucleic Acids Res. *14*, 7675–7693.

Poccia, D., Salik, J. & Krystal, G. (1981) Develop. Biol. *82*, 287–296.

Pohl, F.W. (1976) Nature (London) *260*, 365–366.

Pohl, F.W. & Jovin, T.M. (1972) J. Mol. Biol. *67*, 375–396.

Poirier, G.G. & Savard, P. (1980) Can. J. Biochem. *58*, 509–515.

Poirier, G.G., De Murcia, G., Jongstra-Bilen, J., Niedergang, C. & Mandel, P. (1982) Proc. Natl. Acad. Sci. USA *79*, 3423–3427.

Ponder, B.A.J., Crew, F. & Crawford, L.V. (1978) J. Virology *25*, 175–186.

Poon, N.H. & Seligy, V.L. (1980) Exp. Cell Res. *128*, 333–341.

Pospelov, V.A. & Svetlikova, S.B. (1982) FEBS Lett. *146*, 157–160.

Pospelov, V.A., Svetlikova, S.B. & Vorobev, V.I. (1977) Mol. Biol. *11*, 605–611.

Pospelov, V.A., Russev, G., Varsilev, L. & Tsanev, R. (1982a) J. Mol. Biol. *156*, 79–91.

Pospelov, V.A., Anachkova, B. & Russev, G. (1982b) Biochim. Biophys. Acta *699*, 241–246.

Prentice, D.A., Loechel, S.C. & Kitos, P.A. (1982) Biochemistry *21*, 2412–2420.

Prescott, D. (1966) J. Cell Biol. *31*, 1–9.

Prevelige, P.E. & Fasman, G.D. (1983) Biochim. Biophys. Acta *739*, 85–96.

Prince, D.J., Cummings, D.J. & Seale, R.L. (1977) Biochem. Biophys. Res. Comm. *79*, 190–197.

Prior, C.P., Cantor, C.R., Johnson, E.M. & Allfrey, V.G. (1980) Cell *20*, 597–608.

Prior, C.P., Cantor, C.R., Johnson, E.M., Littau, V.C. & Allfrey, V.G. (1983) Cell *34*, 1033–1042.

Proffit, J.H., Davie, J.R., Swinton, D. & Hattman, S. (1984) Mol. Cell. Biol. *4*, 985–988.

Pruitt, S.C. & Grainger, R.M. (1980) Chromosoma *78*, 257–274.

Prunell, A. (1982) EMBO J. *1*, 173–179.

Prunell, A. (1983) Biochemistry *22*, 4887–4894.

Prunell, A. & Kornberg, R.D. (1982) J. Mol. Biol. *154*, 515–523.

Prunell, A., Kornberg, R.D., Lutter, L.C., Klug, A., Levitt, M. & Crick, F.H.C. (1979) Science *204*, 855–858.

Prunell, A., Goulet, I., Jacob, Y. & Goutorbe, F. (1984) Eur. J. Biochem. *138*, 253–257.

Puigdomenech, P., Jose, M., Ruiz-Carrillo, A. & Crane-Robinson, C. (1983) FEBS Lett. *154*, 151–155.

Pulleyblank, D.E., Shure, M., Tang, D., Vinograd, J. & Vosbey, H.-P. (1975) Proc. Natl. Acad. Sci. USA *72*, 4280–4284.

Rabbani, A., Goodwin, G.H. & Johns, E.W. (1978) Biochem. Biophys. Res. Comm. *81*, 351–358.

Rabbani, A., Goodwin, G.H., Walker, J.M., Brown, E. & Johns, E.W. (1980) FEBS Lett. *109*, 294–298.

Rall, S.C., Okinaka, R.T. & Strniste, G.F. (1977) Biochemistry *16*, 4940–4944.

Ralph-Edwards, A. & Silver, J.C. (1983) Exp. Cell Res. *148*, 363–376.

Ramponi, G., Nassai, P., Liguri, G., Cappugi, G. & Grisolia, S. (1978) FEBS Lett. *90*, 228–232.

Ramsay, N. (1986) J. Mol. Biol. *189*, 179–188.

Ramsay, N., Felsenfeld, G., Rushton, B.M. & McGhee, J.D. (1984) EMBO J. *3*, 2605–2611.

Razin, A. & Riggs, A.D. (1980) Science *210*, 604–610.

Razin, A., Cedar, H. & Riggs, A.D. (eds.) (1984) *DNA Methylation: Biochemistry and Biological Significance*, Springer-Verlag, Berlin.

Razin, S.V., Mantieva, V.L. & Georgiev, G.P. (1979) Nucleic Acids Res. *7*, 1713–1735.

Rechsteiner, M. & Kuehl, L. (1979) Cell *16*, 901–908.

Record, M.T., Anderson, C.F. & Lohman, T.M. (1978) Quart. Rev. Biophys. *11*, 103–178.

Reddy, G.P.V. & Pardee, A.B. (1980) Proc. Natl. Acad. Sci. USA *77*, 3312–3316.

Reeck, G.R., Swanson, E. & Teller, D.C. (1978) J. Mol. Evol. *10*, 309–317.

Reeck, G.R., Isackson, P.J. & Teller, D.C. (1982) Nature (London) *300*, 76–78.

Reeves, R. (1977) Cold Spring Harbor Symp. Quant. Biol. *42*, 709–722.

Reeves, R. (1978) Biochemistry *17*, 4908–4916.

Reeves, R. (1984) Biochim. Biophys. Acta *782*, 343–393.

Reeves, R. & Candido, E.P.M. (1980) Nucleic Acids Res. *8*, 1947–1963.

Reeves, R. & Chang, D. (1983) J. Biol. Chem. *258*, 679–687.

Reeves, R., Chang, D. & Chung, S.-C. (1981) Proc. Natl. Acad. Sci. USA *78*, 6704–6708.

Renard, A. & Verly, W.G. (1983) Eur. J. Biochem. *136*, 453–460.

Renaud, J. & Ruiz-Carrillo, A. (1986) J. Mol. Biol. *189*, 217–226.

Renz, M. (1979) Nucleic Acids Res. *6*, 2761–2767.

Renz, M., Nehls, P. & Hozier, J. (1977) Proc. Natl. Acad. Sci. USA *74*, 1879–1883.

Reynolds, W., Smith, R.D., Bloomer, L.S. & Gottesfeld, J.M. (1982) in *The Cell Nucleus* (Busch, H. & Rothblum, L., eds.), vol. 11, pp. 63–87, Academic Press, N.Y.

Rhodes, D. (1979) Nucleic Acids Res. *6*, 1805–1816.

Rhodes, D. (1985) EMBO J. *4*, 3473–3482.

Rhodes, D. & Klug, A. (1980) Nature (London) *286*, 573–578.

Riazance, J.H., Baase, W.A., Johnson, W.C., Hall, K., Cruz, P. & Tinoco, I. (1985) Nucleic Acids Res. *13*, 4983–4989.

Richards, B.M. & Pardon, J.F. (1970) Exp. Cell Res. *62*, 184–196.

Richards, R.G. & Shaw, B.R. (1984) Biochemistry *23*, 2095–2102.

Richmond, T.J., Finch, J.T., Rushton, B., Rhodes, D. & Klug, A. (1984) Nature (London) *311*, 532–537.

Riggs, A.D. (1975) Cytogenet. Cell Genet. *14*, 9–25.

Riggs, M.G., Whittaker, R.G., Neumann, J.R. & Ingram, V.M. (1977) Nature (London) *268*, 462–464.

Riley, D.E. (1980) Biochemistry *19*, 2977–2992.

Riley, D. & Weintraub, H. (1978) Cell *13*, 281–293.

Riley, D. & Weintraub, H. (1979) Proc. Natl. Acad. Sci. USA *76*, 328–332.

Rill, R. & Nelson, D.A. (1977) Cold Spring Harbor Symp. Quant. Biol. *42*, 475–482.

Rill, R.L. & Oosterhof, D.K. (1982) J. Biol. Chem. *257*, 14875–14880.

Rill, R. & van Holde, K.E. (1973) J. Biol. Chem. *248*, 1080–1083.

Rill, R. & van Holde, K.E. (1974) J. Mol. Biol. *83*, 459–471.

Rill, R.L., Oosterhof, D.K., Hozier, J.C. & Nelson, D.A. (1975) Nucleic Acids Res. *2*, 1525–1538.

Rill, R.L., Oosterhof, D.K., Andrean, B. & Nelson, D.A. (1980) Fed. Proc. *39*, 1886.

Rill, R.L., Hilliard, P.R., Levy, G.C. (1983) J. Biol. Chem. *258*, 250–256.

Ring, D. & Cole, R.D. (1979) J. Biol. Chem. *254*, 11688–11695.

Ring, D. & Cole, R.D. (1983) J. Biol. Chem. *258*, 15361–15364.

Ringold, G.M. (1985) Ann. Rev. Pharmacol. Toxicol. *25*, 529–566.

Ringold, G.M., Dobson, D.E., Grove, R.J., Hall, C.V., Lee, F. & Vannice, J. (1983) Recent Prog. Hormone Res. *39*, 387–421.

Riquelme, P.T., Barzio, L.O. & Koide, S.S. (1979) J. Biol. Chem. *254*, 3018–3028.

Ris, H. & Kubai, D.F. (1970) Ann. Rev. Genetics *4*, 263–294.

Rizzo, P.J. (1982) J. Protozool. *29*, 98–103.

Rizzo, P.J. (1985) Biosystems *18*, 249–262.

Rizzo, P.J. & Burghardt, R.C. (1980) Chromosoma *76*, 91–99.

Rizzo, P.J. & Cox, E.R. (1977) Science *198*, 1258–1260.

Rizzo, P.J. & Morris, R.L. (1983) Biosystems *16*, 211–216.

Rizzo, P.J. & Nooden, L.D. (1974) Biochim. Biophys. Acta *349*, 415–427.

Rizzo, P.J., Choi, J. & Morris, R.L. (1984) J. Phycology *20*, 95–100.

Rizzo, P.J., Bradley, W. & Morris, R.L. (1985) Biochemistry *24*, 1727–1732.

Roark, D.E. (1978) in *Methods in Cell Biology* (Stein, G., Stein, J. & Kleinschmidt, L.J., eds.), vol. 18, pp. 417–428, Academic Press, N.Y.

Roark, D.E., Geoghegan, T.E. & Keller, G.H. (1974) Biochem. Biophys. Res. Comm. *59*, 542–547.

Roark, E.E., Goeghegan, T.E., Keller, G.H., Matter, K.V. & Engle, R.C. (1976) Biochemistry *15*, 3019–3025.

Robbins, E. & Borun, T.W. (1967) Proc. Natl. Acad. Sci. USA *57*, 409–416.

Robinson, S.J., Nelkin, B.D. & Vogelstein, B. (1982) Cell *28*, 99–106.

Rocha, E., Davie, J., van Holde, K.E. & Weintraub, H. (1984) J. Biol. Chem. *259*, 4212–4222.

Rodrigues, J.D.A., Brandt, W.T. & von Holt, C. (1979) Biochim. Biophys. Acta *578*, 196–206.

Rodrigues, J.D.A., Brandt, W.F., von Holt, C. (1985) Eur. J. Biochem. *150*, 499–506.

Rodriguez-Alfagame, C., Rudka, G.T. & Cohen, L.H. (1980) Chromosoma *78*, 1–31.

Roeder, R. (1976) in *RNA Polymerase* (Losick, R. & Chamberlin, M., eds.), pp. 285–329, Cold Spring Harbor Lab., Cold Spring Harbor, N.Y.

Romhanyi, T., Seprodi, J., Antoni, F., Nikolics, K., Meszaros, G. & Farago, A. (1982) Biochim. Biophys. Acta *70*, 57–62.

Rosenberg, N.L., Smith, R.M. & Rill, R.L. (1986) J. Biol. Chem. *261*, 12375–12383.

Ross, W. & Landy, A. (1982) J. Mol. Biol. *156*, 523–529.

Roufa, D.J. & Marchionni, M.A. (1982) Proc. Natl. Acad. Sci. USA *79*, 1810–1814.

Rouviere-Yaniv, J. (1977) Cold Spring Harbor Symp. Quant. Biol. *42*, 439–447.

Rouviere-Yaniv, J. & Gros, F. (1975) Proc. Natl. Acad. Sci. USA *72*, 3428–3432.

Rouviere-Yaniv, J., Yaniv, M. & Germond, J.E. (1979) Cell *17*, 265–274.

Rubio, J., Rosado, Y. & Castaneda, M. (1980) Can. J. Biochem. *58*, 1247–1251.

Ruiz-Carrillo, A. (1984) Nucleic Acids Res. *12*, 6473–6492.

Ruiz-Carrillo, A., Wangh, L.J. & Allfrey, V.G. (1975) Science *190*, 117–128.

Ruiz-Carrillo, A., Puigdomenech, P., Eder, G. & Lurz, R. (1980) Biochemistry *19*, 2544–2554.

Ruiz-Carrillo, A., Affolter, M. & Renaud, J. (1983) J. Mol. Biol. *170*, 843–859.

Ruderman, J.V. & Gross, P.R. (1974) Develop. Biol. *36*, 286–298.

Ruggieri, S. & Magni, G. (1982) Physiol. Chem. Phys. *14*, 315–322.

Russev, G. & Hancock, R. (1981) Nucleic Acids Res. *9*, 4129–4137.

Russo, E., Giancotti, V., Crane-Robinson, C. & Geraci, G. (1983) Int. J. Biochem. *15*, 487–493.

Rykowski, M.C., Wallis, J.W., Choe, J. & Grunstein, M. (1981) Cell *25*, 477–487.

Ryoji, M. & Worcel, A. (1984) Cell *37*, 21–32.

Ryoji, M. & Worcel, A. (1985) Cell *40*, 923–932.

Saffer, J.D. & Coleman, J.E. (1980) Biochemistry *19*, 5874–5883.

Saffer, J.D. & Glazer, R.I. (1980) Biochem. Biophys. Res. Comm. *93*, 1280–1285.

Saffer, J.D. & Glazer, R.I. (1982) J. Biol. Chem. *257*, 4655–4660.

Sahasrabuddhe, C.G. & van Holde, K.E. (1974) J. Biol. Chem. *249*, 152–156.

Sahyoun, N., Levine, H., McConnell, R., Bronson, D. & Cuatrecasas, P. (1983) Proc. Natl. Acad. Sci. USA *80*, 6760–6764.

Saito, H., Kameyama, M., Kodama, M. & Nagata, C. (1982) J. Biochem. *92*, 233–241.

Sakonju, S. & Brown, D.D. (1982) Cell *31*, 395–405.

Sakonju, S., Bogenhagen, D.F. & Brown, D.D. (1980) Cell *19*, 13–25.

Sakuma, K., Matsumura, Y. & Senshu, T. (1984) Nucleic Acids Res. *12*, 1415–1426.

Salditt-Georgieff, M., Sheffery, M., Krauter, K., Darnell, J.E., Rifkind, R. & Marks, P.A. (1984) J. Mol. Biol. *172*, 437–450.

Salik, J., Herlands, L., Hoffman, H.P. & Poccia, D. (1981) J. Cell Biol. *90*, 385–395.

Samal, B., Worcel, A., Louis, C. & Schedl, P. (1981) Cell *23*, 401–409.

Samuels, H.H., Perlman, A.J., Raaka, B.M. & Stanley, F. (1982) Recent Prog. in Hormone Res. *38*, 557–592.

Sandeen, G., Wood, W.I. & Felsenfeld, G. (1980) Nucleic Acids Res. *8*, 3757–3778.

Sanders, M.M. (1978) J. Cell Biol. *79*, 97–109.

Sanders, M.M. (1981) J. Cell Biol. *91*, 579–583.

Sargan, D.R. & Butterworth, P.H. (1985) Nucleic Acids Res. *13*, 3805–3822.

Sarkander, H.I. & Dulce, H.-J. (1979) Exp. Brain Res. *35*, 109–125.

Sarnow, P., Rasched, I. & Knippers, R. (1981) Biochim. Biophys. Acta *655*, 349–358.

Sasaki, K. & Sugita, M. (1982a) Physiol. Plant *56*, 148–154.

Sasaki, K. & Sugita, M. (1982b) Plant Physiol. *69*, 543–545.

Sasaki, K. & Tazawa, T. (1973) Biochem. Biophys. Res. Comm. *52*, 1440–1449.

Sassone-Corsi, P., Wildeman, A. & Chambon, P. (1985) Nature (London) *313*, 458–463.

Sautiere, P., Tyrou, D., Moschetto, Y. & Biserte, G. (1971a) Biochimie *53*, 479–483.

Sautiere, P., Lamberlin-Breynaert, M.P., Moschetto, Y. & Biserte, G. (1971b) Biochimie *53*, 711–715.

Savic, A., Richman, P., Williamson, P. & Poccia, D. (1981) Proc. Natl. Acad. Sci. USA *78*, 3706–3710.

Schafer, R. & Zillig, W. (1973) Eur. J. Biochem. *33*, 201–206.

Schaffner, W., Kunz, H.O., Daetwyler, H., Telford, J., Smith, H.O. & Birnstien, M.L. (1978) Cell *14*, 655–671.

Scheer, U. (1980) Eur. J. Cell Biol. *23*, 189–196.

Scheer, U., Sommerville, J. & Muller, U. (1980) Exp. Cell Res. *129*, 115–126.

Scheer, U., Zentgraf, H. & Sauer, H.W. (1981) Chromosoma *84*, 279–290.

Scheer, U., Hinssen, H., Franke, W.W. & Jockusch, B.M. (1984) Cell *39*, 111–122.

Schlaeger, E.-J., van Telgen, H.-J., Klempnauer, K.-H., & Knippers, R. (1978) Eur. J. Biochem. *84*, 95–102.

Schlaeger, E.-J., Pulm, W. & Knippers, R. (1983) FEBS Lett. *156*, 281–286.

Schlegel, R.A., Haye, K.R., Litwack, A.H. & Phelps, B.M. (1980a) Biochim. Biophys. Acta *606*, 316–330.

Schlegel, R.A., Litwack, A.H. & Phelps, B.M. (1980b) Mol. Biol. Rep. *6*, 115–118.

Schlepper, J. & Knippers, R. (1975) Eur. J. Biochem. *60*, 209–220.

Schlessinger, D.H., Goldstein, G. & Nial, H.D. (1975) Biochemistry *14*, 2214–2218.

Schlessinger, F.B., Dattagupta, N. & Crothers, D.M. (1982) Biochemistry *21*, 664–669.

Schlissel, M.S. & Brown, D.D. (1984) Cell *37*, 903–913.

Schmidt, G. & Levine, P.A. (1938) Science *88*, 172–173.

Schmitz, K.S. (1982) J. Theor. Bio. *98*, 29–43.

Schröter, H. & Bode, J. (1982) Eur. J. Biochem. *127*, 429–436.

Schröter, H., Hasse, E. & Arfman, H. (1980) Eur. J. Cell Biol. *22*, Abs. G241.

Schröter, H., Maier, G., Ponstingl, H. & Nordheim, A. (1985) EMBO J. *4*, 3867–3872.

Schurr, J.M. & Schurr, R.L. (1985) Biopolymers *24*, 1931–1940.

Schwager, S., Brandt, W.F. & von Holt, C. (1983) Biochim. Biophys. Acta *741*, 315–321.

Schwager, S., Retief, J.D., de Groot, P. & von Holt, C. (1985) FEBS Lett. *189*, 305–309.

Seale, R.L. (1975) Nature (London) *255*, 247–249.

Seale, R.L. (1976) Cell *9*, 423–429.

Seale, R.L. (1981) Biochemistry *20*, 6432–6437.

Seale, R.L. & Simpson, R.T. (1975) J. Mol. Biol. *94*, 479–501.

Seale, R.L., Annunziato, A.T. & Smith, R.D. (1983) Biochemistry *22*, 5008–5015.

Sealy, L. & Chalkley, R. (1978a) Cell *14*, 115–121.

Sealy, L. & Chalkley, R. (1978b) Nucleic Acids Res. *5*, 1863–1876.

Searcy, D.G. (1975) Biochim. Biophys. Acta *395*, 535–547.

Searcy, D.G. & Stein, D.B. (1980) Biochim. Biophys. Acta *609*, 180–195.

Sedat, J. & Manelidis, L. (1977) Cold Spring Harbor Symp. Quant. Biol. *42*, 331–350.

Segawa, K. & Oda, K. (1978) Biochim. Biophys. Acta *521*, 374–386.

Seidl, A. & Hinz, H.-J. (1984) Proc. Natl. Acad. Sci. USA *81*, 1312–1316.

Seidman, M., Levine, A. & Weintraub, H. (1979) Cell *18*, 439–449.

Seiler-Tuyns, A. & Birnstiel, M.L. (1981) J. Mol. Biol. *151*, 607–625.

Sellos, D. & van Wormhaudt, A. (1979) Biochimie *61*, 393–404.

Sen, D. & Crothers, D.M. (1986) Biochemistry *25*, 1495–1503.

Sen, D., Mitra, S. & Crothers, D.M. (1986) Biochemistry *25*, 3441–3447.

Sen, S., Siciliano, M.J., Johnston, D.A., Schwartz, R.J. & Kuo, T. (1985) J. Biol. Chem. *260*, 3071–3078.

Senger, D.R., Arceci, R.J. & Gross, P.R. (1978) Develop. Biol. *65*, 416–425.

Senshu, T., Fukuda, M. & Ohashi, M. (1978) J. Biochem. *84*, 985–988.

Seyedin, S.M. & Kistler, W.S. (1979a) Biochemistry *18*, 1376–1379.

Seyedin, S.M. & Kistler, W.S. (1979b) J. Biol. Chem. *254*, 11264–11272.

Seyedin, S.M. & Kistler, W.S. (1980) J. Biol. Chem. *255*, 5949–5954.

Seyedin, S.M. & Kistler, W.S. (1983) Exp. Cell Res. *143*, 451–454.

Seyedin, S.M., Cole, R.D. & Kistler, W.S. (1981) Exp. Cell Res. *136*, 399–405.

Shaw, B.R., Corden, J.L., Sahasrabuddhe, C. & van Holde, K.E. (1974) Biochem. Biophys. Res. Comm. *61*, 1193–1198.

Shaw, B.R., Cognetti, G., Sholes, W.M. & Richards, R.G. (1981) Biochemistry *20*, 4971–4978.

Shaw, J.E., Levinger, L. & Carter, C.W. (1979) J. Virol. *29*, 657–665.

Sheffery, M., Rifkind, R.A. & Marks, P.A. (1982) Proc. Natl. Acad. Sci. USA *79*, 1180–1184.

Sheffery, M., Marks, P.A. & Rifkind, R.A. (1984) J. Mol. Biol. *172*, 417–436.

Sheinin, R. & Humbert, J. (1978) Ann. Rev. Biochem. *47*, 277–316.

Shelton, E.R., Wassarman, P.M. & DePamphilis, M.L. (1978) J. Mol. Biol. *125*, 491–514.

Shen, C.K.J. (1983) Nucleic Acids Res. *11*, 7899–7910.

Shepherd, G.R., Hardin, J.M. & Noland, B.J. (1971) Arch. Biochem. Biophys. *143*, 1–5.

Shermoen, A.W. & Beckendorf, S.K. (1982) Cell *29*, 601–607.

Shestopalov, B.V. & Chirgadze, Y.N. (1976) Eur. J. Biochem. *67*, 123–128.

Shick, V.V., Belyavsky, A.V., Bavykin, S.G. & Mirzabekov, A.D. (1980) J. Mol. Biol. *139*, 491–517.

Shick, V.V., Belyavsky, A.V. & Mirzabekov, A.D. (1985) J. Mol. Biol. *185*, 329–339.

Shih, T.Y. & Lake, R.S. (1972) Biochemistry *11*, 4811–4817.

Shimada, T. & Nienhuis, A.W. (1984) J. Biol. Chem. *260*, 2468–2474.

Shindo, H., McGhee, J.D. & Cohen, J.S. (1980) Biopolymers *19*, 523–537.

Shlyapnikov, S.V., Arutyunyan, A.A., Kurochin, S.N., Menelova, L.V., Nesterova, M.H., Sashchenko, L.P. & Severin, E.S. (1975) FEBS Lett. *53*, 316–319.

Shoemaker, C.B. & Chalkley, R. (1978) J. Biol. Chem. *253*, 5802–5807.

Shoemaker, C.B. & Chalkley, R. (1980) J. Biol. Chem. *255*, 11048–11055.

Shooter, K.V., Goodwin, G.H. & Johns, E.W. (1974) Eur. J. Biochem. *47*, 263–270.

Shupe, K. & Rizzo, P.J. (1983) J. Protozool. *30*, 599–606.

Shupe, K., Rizzo, P.J. & Johnson, J.R. (1980) FEBS Lett. *115*, 221–224.

Sibbet, G.J. & Carpenter, B.G. (1983) Biochim. Biophys. Acta *740*, 331–338.

Sibbet, G.J., Carpenter, B.G., Ibel, K., May, R.P., Kneale, G.G., Bradbury, E.M. & Baldwin, J.P. (1983) Eur. J. Biochem. *133*, 393–398.

Sierra, F., Stein, G. & Stein, J. (1983) Nucleic Acids Res. *11*, 7069–7086.

Sigee, D.D. (1984) Biosystems *16*, 203–210.

Signer, R., Caspersson, T. & Hammarsten, E. (1938) Nature (London) *141*, 122.

Silva, J.E. (1983) Endocrinology *113*, 699–705.

Silver, J.C. (1979) Biochim. Biophys. Acta *561*, 261–264.

Simon, R.H. & Felsenfeld, G. (1979) Nucleic Acids Res. *6*, 689–696.

Simpson, R.T. (1976) Proc. Natl. Acad. Sci. USA *73*, 4400–4404.

Simpson, R.T. (1978a) Cell *13*, 691–699.

Simpson, R.T. (1978b) Biochemistry *17*, 5524–5531.

Simpson, R.T. (1978c) Nucleic Acids Res. *5*, 1109–1119.

Simpson, R.T. (1979) J. Biol. Chem. *254*, 10123–10127.

Simpson, R.T. (1981) Proc. Natl. Acad. Sci. USA *78*, 6803–6807.

Simpson, R.T. & Bergman, L.W. (1980) J. Biol. Chem. *255*, 10702–10709.

Simpson, R.T. & Kunzler, P. (1979) Nucleic Acids Res. *6*, 1387–1415.

Simpson, R.T. & Stafford, D.W. (1983) Proc. Natl. Acad. Sci. USA *80*, 51–55.

Simpson, R.T. & Whitlock, J.P. (1976) Cell *9*, 347–353.

Simpson, R.T., Thoma, F. & Brubaker, J.M. (1985) Cell *42*, 799–808.

Sinden, R.R., Carlson, J.O. & Pettijohn, D.E. (1980) Cell *21*, 773–783.

Singer, D.S. (1979) J. Biol. Chem. *254*, 5506–5514.

Singer, D.S. & Singer, M.F. (1976) Nucleic Acids Res. *3*, 2531–2547.

Singer, D.S. & Singer, M.F. (1978) Biochemistry *17*, 2086–2095.

Singleton, C.K., Kilpatrick, M.W. & Wells, R.D. (1984) J. Biol. Chem. *259*, 1963–1967.

Sittman, D.B., Chiu, I.M., Pan, C.I., Cohn, R.H., Kedes, L.H. & Marzluff, W.F. (1981) Proc. Natl. Acad. Sci. USA *78*, 4078–4082.

Sittman, D.B., Graven, R.A. & Marzluff, W.F. (1983) Nucleic Acids Res. *11*, 6679–6697.

Skandrani, E., Mizon, J., Sautiere, P. & Bizerte, G. (1972) Biochimie *54*, 1267–1272.

Sledziewski, A. & Young, E.T. (1982) Proc. Natl. Acad. Sci. USA *79*, 253–256.

Small, D., Chou, P.Y., Fasman, G.D. (1977) Biochem. Biophys. Res. Comm. *79*, 341–346.

Smerdon, M.J. & Isenberg, I. (1976a) Biochemistry *15*, 4046–4049.

Smerdon, M.J. & Isenberg, I. (1976b) Biochemistry *15*, 4233–4242.

Smerdon, M.J. & Isenberg, I. (1976c) Biochemistry *15*, 4242–4247.

Smerdon, M.J. & Lieberman, M.W. (1981) J. Biol. Chem. *256*, 2480–2483.

Smith, A.J. & Billett, M.A. (1982a) Biochim. Biophys. Acta *697*, 121–133.

Smith, A.J. & Billett, M.A. (1982b) Biochim. Biophys. Acta *697*, 134–147.

Smith, B.J., Toogood, C.I.A. & Johns, E.W. (1980) J. Chromatog. *200*, 200–205.

Smith, D.L., Bruegger, B.B., Halpern, R.M. & Smith, R.A. (1973) Nature (London) *246*, 103–104.

Smith, D.L., Chen, C.-C., Bruegger, B.B., Holtz, S.L., Halpern, R.M. & Smith, R.A. (1974) Biochemistry *13*, 3780–3785.

Smith, D.R., Jackson, I.J. & Brown, D.D. (1984) Cell *37*, 645–652.

Smith, G.R. (1981) Cell *24*, 599–600.

Smith, J.S. & Stocken, L.A. (1973) Biochem. Biophys. Res. Comm. *54*, 297–300.

Smith, M.M. (1984) in *Histone Genes; Structure, Organization, and Regulation* (Stein, G., Stein, J. & Marzluff, W., eds.), pp. 1–33, John Wiley & Sons, N.Y.

Smith, M.M. & Andresson, O.S. (1983) J. Mol. Biol. *169*, 663–690.

Smith, M.M. & Murray, K. (1983) J. Mol. Biol. *169*, 641–661.

Smith, M.R. & Lieberman, M.W. (1984) Nucleic Acids Res. *12*, 6493–6510.

Smith, P.A., Jackson, V. & Chalkley, R. (1984) Biochemistry *23*, 1576–1581.

Smith, R.A. (1982) *Experiences in Biochemical Perception*, Academic Press, N.Y.

Smith, R.A., Halpern, R.M., Bruegger, B.B., Dunlap, A.K. & Fricke, O. (1978) in *Methods in Cell Biology* (Stein, G., Stein, J. & Kleinsmith, L.J., eds.), vol. 19, pp. 153–159, Academic Press, N.Y.

Smith, R.D. & Yu, J. (1984) J. Biol. Chem. *259*, 4609–4615.

Smith, R.D., Seale, R.L. & Yu, J. (1983) Proc. Natl. Acad. Sci. USA *80*, 5505–5509.

Smith, R.D., Yu, J. & Seale, R.L. (1984a) Biochemistry *23*, 785–790.

Smith, R.D., Yu, J., Annunziato, A. & Seale, R.L. (1984b) Biochemistry *23*, 2970–2976.

Smulson, M. (1984) Meth. Enzym. *106*, 512–522.

Smulson, M. & Shall, S. (1976) Nature (London) *263*, 14.

Smulson, M. and Sugimura T. (eds.) (1980) *Novel Post-Translational ADP Ribosylation of Regulatory Enzymes and Proteins*, Elsevier-North Holland, N.Y.

Smulson, M. & Sugimura, T. (1984) Adv. Enzym. *106*, 438–440.

Soeda, E., Arrand, J.R., Smolar, N. & Griffin, B.E. (1979) Cell *17*, 357–370.

Sogin, M.L., Elwood, H.J. & Gunderson, J.H. (1986) Proc. Natl. Acad. Sci. USA *83*, 1383–1387.

Sogo, J.M., Ness, P.J., Widmer, R.M., Parish, R.W. & Koller, T. (1984) J. Mol. Biol. *178*, 897–928.

Sogo, J.M., Stahl, H., Koller, T., & Knippers, R. (1986) J. Mol. Biol. *189*, 189–204.

Sollner-Webb, B. & Felsenfeld, G. (1975) Biochemistry *14*, 2915–2920.

Solomon, M.J., Strauss, F. & Varshavsky, A. (1986) Proc. Natl. Acad. Sci. USA *83*, 1276–1280.

Sommer, A. (1978) Mol. Gen. Genet. *161*, 323–331.

Southern, E.M. (1975) J. Mol. Biol. *98*, 503–517.

Spadafora, C., Bellard, M., Compton, J.L. & Chambon, P. (1976) FEBS Lett. *69*, 281–285.

Spadafora, C., Oudet, P. & Chambon, P. (1979) Eur. J. Biochem. *100*, 225–235.

Spalding, J., Kajiwara, K. & Mueller, G.C. (1966) Proc. Natl. Acad. Sci. USA *56*, 1535–1542.

Spassky, A., Rimsky, S., Garreau, H. & Buc, H. (1984) Nucleic Acids Res. *12*, 5321–5340.

Spelsberg, T.C., Littlefield, B.A., Seelke, R., Dani, G.M., Toyoda, H., Boyd-Leinen, P., Thrall, C. & Kon, O.L. (1983) Recent Prog. in Hormone Res. *39*, 463–513.

Spelsberg, T.C., Gosse, B.J., Littlefield, B.A., Toyoda, H. & Seelke, R. (1984) Biochemistry *23*, 5103–5113.

Spencer, M., Fuller, M., Wilkins, M.H.F. & Brown, G.L. (1962) Nature (London) *194*, 1014–1020.

Sperling, L. & Klug, A. (1977) J. Mol. Biol. *112*, 253–263.

Sperling, L. & Weiss, M.C. (1980) Proc. Natl. Acad. Sci. USA *77*, 3412–3416.

Sperling, L., Tardieu, A. & Weiss, M.C. (1980) Proc. Natl. Acad. Sci. USA *77*, 2716–2720.

Sperling, R. & Amos, L.A. (1977) Proc. Natl. Acad. Sci. USA *74*, 3772–3776.

Sperling, R. & Bustin, M. (1974) Proc. Natl. Acad. Sci. USA *71*, 4625–4629.

Sperling, R. & Bustin, M. (1975) Biochemistry *14*, 3322–3331.

Spiker, S. & Isenberg, I. (1977a) Biochemistry *16*, 1819–1826.

Spiker, S. & Isenberg, I. (1977b) Cold Spring Harbor Symp. Quant. Biol. *42*, 157–163.

Spiker, S., Mardian, J.K.W. & Isenberg, I. (1978) Biochem. Biophys. Res. Comm. *82*, 129–135.

Spinelli, G., Albanese, I., Anello, L., Ciaccio, M. & Diliegro, I. (1982) Nucleic Acids Res. *10*, 7977–7991.

Spring, T.J. & Cole, R.D. (1977) in *Methods in Cell Biology* (Stein, G., Stein, J. & Kleinsmith, L.J., eds.), vol. 16, pp. 227–240, Academic Press, N.Y.

Srebreva, L.N., Andreeva, N.B., Gasaryan, K.G., Tsanev, R.G. & Zlatanova, J.S. (1983) Differentiation *25*, 113–120.

Srivastava, B.I.S. (1972) Biochem. Biophys. Res. Comm. *48*, 270–273.

Stacks, P.C. & Schumaker, V.N. (1979) Nucleic Acids Res. *7*, 2457–2467.

Stahl, H. & Knippers, R. (1980) Biochim. Biophys. Acta *614*, 71–80.

Stalder, J. & Braun, R. (1978) FEBS Lett. *90*, 223–227.

Stalder, J., Groudine, M., Dodgson, J., Engel, J.D. & Weintraub, H. (1980a) Cell *19*, 973–980.

Stalder, J., Larsen, A., Engel, J.D., Dolan, M., Groudine, M. & Weintraub, H. (1980b) Cell *20*, 451–460.

Stamatoyannopoulos, G. & Nienhuis, A.W. (eds.) (1983) *Progress in Clinical and Biological Research*, Vol. 134, Alan R. Liss, N.Y.

Staynov, D.Z. (1983) Int. J. Biol. Med. *5*, 3–9.

Staynov, D.Z., Dunn, S., Baldwin, J.P. & Crane-Robinson, C. (1983) FEBS Lett. *157*, 311–315.

Stedman, E. & Stedman, Ellen (1951) Phil. Trans. Roy. Soc. London. *235*, 565–595.

Stein, A. (1979) J. Mol. Biol. *130*, 103–130.

Stein, A. & Bina, M. (1984) J. Mol. Biol. *178*, 341–363.

Stein, A. & Page, D. (1980) J. Biol. Chem. *255*, 3629–3637.

Stein, A., Bina-Stein, M. & Simpson, R.T. (1977) Proc. Natl. Acad. Sci. USA *74*, 2780–2784.

Stein, G., Stein, J., & Marzluff, W. (eds.) (1984) *Histone Genes: Structure, Organization, and Regulation*, John Wiley & Sons, N.Y.

Stein, R., Gruenbaum, Y., Pollack, Y., Razin, A. & Cedar, H. (1982) Proc. Natl. Acad. Sci. USA *79*, 61–65.

Steiner, R.F. (1952) Trans. Far. Soc. *48*, 1185–1196.

Steinmetz, M., Streek, R.E., & Zachau, H.G. (1975) Nature (London) *258*, 447–449.

Steinmetz, M., Streek, R.E. & Zachau, H.G. (1978) Eur. J. Biochem. *83*, 615–628.

Stern, K.G. (1949) Exp. Cell Res. *1, Suppl. 1*, 97–99.

Sterner, R. & Allfrey, V.G. (1982) J. Biol. Chem. *257*, 13872–13876.

Sterner, R., Boffa, L.C. & Vidali, G. (1978) J. Biol. Chem. *253*, 3830–3836.

Sterner, B., Vidali, G. & Allfrey, V.G. (1979) J. Biol. Chem. *254*, 11577–11583.

Sterner, R., Vidali, G. & Allfrey, V.G. (1981) J. Biol. Chem. *256*, 8892–8895.

Stockley, P.G. & Thomas, J.O. (1979) FEBS Lett. *99*, 129–135.

Stoeckert, C.J., Beer, M., Wiggins, J.W. & Wierman, J.C. (1984) J. Mol. Biol. *177*, 483–505.

Stone, G.R., Baldwin, J.P. & Carpenter, B.G. (1985) Biochem. Biophys. Res. Comm. *131*, 230–238.

Stone, P.R., Lorimer, W.S. & Kidwell, W.R. (1977) Eur. J. Biochem. *81*, 9–18.

Stone, P.R., Lorimer, W.S., Ranchalis, J., Danley, M. & Kidwell, W.R. (1978) Nucleic Acids Res. *5*, 173–184.

Stonington, G.O. & Pettijohn, D.E. (1971) Proc. Natl. Acad. Sci. USA *68*, 6–9.

Stoute, J.A. & Marzluff, W.F. (1982) Biochem. Biophys. Res. Comm. *107*, 1279–1284.

Strasburger, E. (1909) in *Darwin and Modern Science* (Seward, A.C., ed.), pp. 102–111, Cambridge Press, Cambridge.

Strätling, W.H. (1979) Biochemistry *18*, 596–603.

Strätling, W.H., Muller, O. & Zentgraf, H. (1978) Exp. Cell Res. *117*, 303–311.

Strauss, F. & Prunell, A. (1983) EMBO J. *2*, 51–56.

Strauss, F. & Varshavsky, A. (1984) Cell *37*, 889–901.

Strauss, F., Gaillard, C., Prunell, A. (1981) Eur. J. Biochem. *118*, 215–222.

Strickland, M., Strickland, W.N., Brandt, W.F. & von Holt, C. (1977a) Eur. J. Biochem. *77*, 263–275.

Strickland, W.N., Strickland, M., Brandt, W.F., von Holt, C. (1977b) Eur. J. Biochem. *77*, 277–286.

Strickland, M., Strickland, W.N., Brandt, W.F., von Holt, C., Wittmann-Liebold, B. & Lehmann, A. (1978) Eur. J. Biochem. *89*, 443–452.

Strickland, W.N., Strickland, M., Brandt, W.F., von Holt, C., Lehmann, A. & Wittmann-Liebold, B. (1980a) Eur. J. Biochem. *104*, 567–578.

Strickland, W.N., Strickland, M., DeGroot, P.C. & von Holt, C. (1980b) Eur. J. Biochem. *109*, 151–158.

Strickland, M., Strickland, W.N. & von Holt, C. (1981) FEBS Lett. *135*, 86–88.

Strogatz, S. (1983) J. Theor. Biol. *103*, 601–607.

Struhl, K., Stinchcomb, D.T., Scherer, S. & Davis, R.W. (1979) Proc. Natl. Acad. Sci. USA *76*, 1035–1039.

Stubblefield, E. & Wray, W. (1971) Chromosoma *32*, 262–294.

Stumph, W.E., Baez, M., Lawson, G.M., Tsai, M.J. & O'Malley, B.W. (1982) in *UCLA Symp. Mol. & Cell Biol.* (O'Malley, B.W., ed.), vol. 26, pp. 87–104.

Suau, P., Kneale, G.G., Braddock, G.W., Baldwin, J.P. & Bradbury, E.M. (1977) Nucleic Acids Res. *4*, 3769–3786.

Suau, P., Bradbury, E.M. & Baldwin, J.P. (1979) Eur. J. Biochem. *97*, 593–602.

Subirana, J.A., Munoz-Guerra, S., Aymami, J., Radermacher, M. & Frank, J. (1985) Chromosoma *91*, 377–390.

Subramanian, K.N. & Shenk, T. (1978) Nucleic Acids Res. *5*, 3635–3642.

Suchilene, S.P. & Gineitis, A.A. (1978a) Exp. Cell Res. *114*, 454–458.

Suchilene, S.P. & Gineitis, A.A. (1978b) Mol. Biol. *12*, 657–660.

Suda, M. & Iwai, K. (1979) J. Biochem. *86*, 1659–1670.

Sugarman, B.J., Dodgson, J.B. & Engel, J.D. (1983) J. Biol. Chem. *258*, 9005–9016.

Sun, Y.L., Yuan, Z.X., Bellard, M. & Chambon, P. (1986) EMBO J. *5*, 293–300.

Sung, M.T. (1977) Biochemistry *16*, 286–290.

Sung, M.T. & Dixon, G.H. (1970) Proc. Natl. Acad. Sci. USA *67*, 1616–1623.

Sung, M.T. & Freedlender, E. (1978) Biochemistry *17*, 1884–1890.

Sung, M.T., Harford, J., Bundman, M. & Vidalakis, G. (1977) Biochemistry *16*, 279–285.

Sures, I. & Gallwitz, D. (1980) Biochemistry *19*, 943–951.

Sures, I., Lowry, J. & Kedes, L.H. (1978) Cell *15*, 1033–1044.

Suzuki, Y. & Murachi, T. (1978) J. Biochem. *84*, 977–984.

Sweet, R.W., Chao, M.V. & Axel, R. (1982) Cell *31*, 347–353.

Swerdlow, P.S. & Varshavsky, A. (1983) Nucleic Acids Res. *11*, 387–401.

Szent-Gyorgyi, C., Finkelstein, D.B. & Garrard, W.T. (1987) J. Mol. Biol. *193*, 71–80.

Szopa, J., Jacob, G. & Arfmann, H.A. (1980) Biochemistry *19*, 987–990.

Tabata, T. & Iwabuchi, M. (1984) Gene *31*, 285–289.

Tabata, T., Sasaki, K. & Iwabuchi, M. (1983) Nucleic Acids Res. *17*, 5865–5875.

Tabata, T., Fukasawa, M. & Iwabuchi, M. (1984) Mol. Gen. Genet. *196*, 397–400.

Tack, L.C., Wassarman, P.M. & DePamphilis, M.L. (1981) J. Biol. Chem. *256*, 8821–8828.

Taillandier, E., Fort, L., Liquier, J., Couppez, M. & Sautiere, P. (1984) Biochemistry *23*, 2644–2650.

Tamura, S. & Tsuiki, S. (1980) Eur. J. Biochem. *111*, 217–224.

Tamura, S., Kikuchi, K., Hiraga, A., Kikuchi, H., Hosokawa, M. & Tsuiki, S. (1978) Biochim. Biophys. Acta *524*, 349–356.

Tanaka, I., Appelt, K., Dijk, J., White, S.W. & Wilson, K.S. (1984) Nature (London) *310*, 376–381.

Tanaka, Y., Hashida, T., Yoshihara, H. & Yoshihara, K. (1979) J. Biol. Chem. *254*, 12433–12488.

Tanigawa, Y., Nomura, H., Imai, Y. & Shimoyama, M. (1981) Biochem. Int. *2*, 319–325.

Tanigawa, Y., Tsuchiya, M., Imai, Y. & Shimoyama, M. (1983) FEBS Lett. *160*, 217–220.

Tata, J. & Baker, B. (1974) Exp. Cell Res. *83*, 125–138.

Tatchell, K. (1978) Ph.D. Thesis, Oregon State University.

Tatchell, K. & van Holde, K.E. (1977) Biochemistry *16*, 5295–5303.

Tatchell, K. & van Holde, K.E. (1979) Biochemistry *18*, 2870–2880.

Tate, V.E. & Philipson, L. (1979) Nucleic Acids Res. *6*, 2769–2785.

Taylor, F.J.R. (1980) Biosystems *13*, 65–108.

Taylor, J.H. (1984) *DNA Methylation and Cellular Differentiation*, Springer-Verlag, Berlin.

Taylor, S.S. (1982) J. Biol. Chem. *257*, 6056–6063.

Teng, C.S., Andrews, K. & Teng, C.T. (1979) Biochem. J. *181*, 585–591.

Thoma, F. (1986) J. Mol. Biol. *190*, 177–190.

Thoma, F. & Koller, T. (1977) Cell *12*, 101–107.

Thoma, F. & Koller, T. (1981) J. Mol. Biol. *149*, 709–733.

Thoma, F. & Simpson, R.T. (1985) Nature (London) *315*, 250–252.

Thoma, F., Koller, T. & Klug, A. (1979) J. Cell Biol. *83*, 403–427.

Thoma, F., Bergman, L.W. & Simpson, R.T. (1984) J. Mol. Biol. *177*, 715–733.

Thomas, G., Lange, H.W. & Hempel, K. (1975) Eur. J. Biochem. *51*, 609–615.

Thomas, G.J., Prescott, B. & Olins, D.E. (1977) Science *197*, 385–388.

Thomas, J.O. (1977) in *The Molecular Biology of the Mammalian Genetic Apparatus* (T'so, P.O.P., ed.), pp. 199–209, North Holland Pub. Co., Amsterdam.

Thomas, J.O. (1984) J. Cell Sci. *Suppl. 1*, 1–20.

Thomas, J.O. & Butler, P.J.G. (1977) J. Mol. Biol. *116*, 769–781.

Thomas, J.O. & Butler, P.J.G. (1980) J. Mol. Biol. *144*, 89–93.

Thomas, J.O. & Furber, V. (1976) FEBS Lett. *66*, 274–280.

Thomas, J.O. & Khabaza, A.J.A. (1980) Eur. J. Biochem. *112*, 501–511.

Thomas, J.O. & Rees, C. (1983) Eur. J. Biochem. *134*, 109–115.

Thomas, J.O. & Thompson, R.J. (1977) Cell *10*, 633–640.

Thornburg, W., O'Malley, A. & Lindell, T.J. (1978) J. Biol. Chem. *253*, 4638–4641.

Thraves, P.J. & Smulson, M.E. (1982) Carcinogenesis *3*, 1143–1148.

Tiktopulo, E.I., Privalov, P.L., Odintsova, T.I., Ermokhina, T.M., Krasheninnikov, I.A., Aviles, F.X., Cary, P.D. & Crane-Robinson, C. (1982) Eur. J. Biochem. *122*, 327–331.

Till, J.E. (1982) J. Cell. Physiol. *Suppl. 1*, 3–11.

Todd, R.D. & Garrard, W.T. (1977) J. Biol. Chem. *252*, 4729–4738.

Todd, R.D. & Garrard, W.T. (1979) J. Biol. Chem. *254*, 3074–3083.

Tomasz, M., Barton, J.K., Magliozzo, C.C., Tucker, D., Lafer, E.M. & Stollar, B.D. (1983) Proc. Natl. Acad. Sci. USA *80*, 2874–2878.

Tonjes, R. & Doenecke, D. (1984) Hoppe-Seyler's Z. Physiol. Chem. *365*, 1071.

Trautner, T.A., ed. (1984) *Methylation of DNA*, Springer-Verlag, Berlin.

Travis, G.H., Colavito-Shepanski, M. & Grunstein, M. (1984) J. Biol. Chem. *259*, 14406–14412.

Trendelenburg, M.F., Mathis, D. & Oudet, P. (1980) Proc. Natl. Acad. Sci. USA *77*, 5984–5998.

Triebel, H., Osipova, T.N., Bar, H., Zalenskaya, I.A. & Hartmann, M. (1984) Stud. Biophys. *101*, 63–64.

Trifonov, E.N. (1980) Nucleic Acids Res. *8*, 4041–4043.

Trifonov, E.N. (1985) CRC Critical Reviews in Biochem. *19*, 89–106.

Trifonov, E.N. & Bettecken, T. (1979) Biochemistry *18*, 454–456.

Trifonov, E.N. & Sussman, J.L. (1980) Proc. Natl. Acad. Sci. USA *77*, 3816–3620.

Tsanev, R. & Tsaneva, I. (1986) in *Methods and Achievements in Experimental Pathology* (Jasmin, G. & Simard, R., eds.), vol. 12, pp. 63–104.

Tseng, B.Y., Erickson, J.M. & Goulian, M. (1979) J. Mol. Biol. *129*, 531–545.

Tsuruo, T. & Ukita, T. (1974) Biochem. Biophys. Acta *353*, 146–159.

Tuan, D.Y.H. & Bonner, J. (1969) J. Mol. Biol. *45*, 59–76.

Tunis-Schneider, M.J.B. & Maestre, M.F. (1970) J. Mol. Biol. *52*, 521–541.

Turner, G. & Hancock, R. (1974) Biochem. Biophys. Res. Comm. *58*, 437–445.

Turner, P.C. & Woodland, H.R. (1982) Nucleic Acids Res. *10*, 3769–3780.

Turner, P.C., Aldridge, T.C., Woodland, H.R. & Old, R.W. (1983) Nucleic Acids Res. *11*, 4093–4107.

Uberbacher, E.C. & Bunick, G.J. (1985a) J. Biomol. Struct. Dynam. *2*, 1033–1055.

Uberbacher, E.C. & Bunick, G.J. (1985b) Science *229*, 1112–1113.

Uberbacher, E.C., Mardian, J.K.W., Rossi, R.M., Olins, D.E. & Bunick, G.J. (1982) Proc. Natl. Acad. Sci. USA *79*, 5258–5262.

Uberbacher, E.C., Ramakrishnan, V., Olins, D. & Bunick, G. (1983) Biochemistry *22*, 4916–4923.

Uberbacher, E.C., Harp, J., Wilkinson-Singley, E. & Bunick, G.J. (1986) Science *232*, 1247–1249.

Udvardy, A. & Schedl, P. (1984) J. Mol. Biol. *172*, 385–403.

Udvardy, A., Louis, C., Han, S. & Schedl, P. (1984) J. Mol. Biol. *175*, 113–130.

Udvardy, A., Maine, E. & Schedl, P. (1985) J. Mol. Biol. *185*, 341–358.

Ueda, K., Okamaya, H., Fukushima, M. & Hayaishi, O. (1975a) J. Biochem. *77*, 1p.

Ueda, K., Omachi, A., Kawaichi, M. & Hayaishi, O. (1975b) Proc. Natl. Acad. Sci. USA *72*, 205–209.

Ueda, K., Kawaichi, M., Okayama, H. & Hayaishi, O. (1979) J. Biol. Chem. *254*, 679–687.

Uemura, T. & Yanagida, M. (1984) EMBO J. *3*, 1737–1744.

Uhr, M.L. & Smulson, M. (1982) Eur. J. Biochem. *128*, 435–443.

Ulanovsky, L., Bodner, M., Trifonov, E. & Choder, M. (1986) Proc. Natl. Acad. Sci. USA *83*, 862–866.

Urban, M.K., Franklin, S.C. & Zweidler, A. (1979) Biochemistry *18*, 3952–3960.

Urban, M.K., Neelin, J.M. & Betz, T.W. (1980) Can. J. Biochem. *58*, 726–731.

Urbanczyk, J. & Studzinski, G.P. (1974) Biochem. Biophys. Res. Comm. *59*, 616–622.

Urieli-Shoval, S., Gruenbaum, Y., Sedat, J. & Razin, A. (1982) FEBS Lett. *146*, 148–152.

Uschewa, A., Avramova, Z. & Tsanev, R. (1982) FEBS Lett. *138*, 50–54.

van der Westhuyzen, D.R. & von Holt, C. (1971) FEBS Lett. *14*, 333–337.

van der Westhuyzen, D.R., Boyd, M.C.D., Fitschen, W. & von Holt, C. (1973) FEBS Lett. *30*, 195–198.

van Helden, P.D. (1982) J. Theor. Biol. *96*, 327–336.

van Helden, P.D., Strickland, W.N., Brandt, W.F. & von Holt, C. (1979) Eur. J. Biochem. *93*, 71–78.

van Helden, P.D., Strickland, W.N., Strickland, M. & von Holt, C. (1982) Biochim. Biophys. Acta *703*, 17–20.

van Holde, K.E. (1975) in *The Proteins* (R. Hill & H. Neurath, eds.) 3rd. Ed., Academic Press, N.Y., pp. 225–291.

van Holde, K.E. & Yager, T.D. (1985) in *Structure and Function of the Genetic Apparatus* (Nicolini, C. & Ts'o, P.O.P., eds.), Plenum Pub. Co., N.Y.

van Holde, K.E., Sahasrabuddhe, C.G., Shaw, B.R., van Bruggen, E.F.J. & Arnberg, A.C. (1974a) Biochem. Biophys. Res. Comm. *60*, 1365–1370.

van Holde, K.E., Sahasrabuddhe, C.G. & Shaw, B.R. (1974b) Nucleic Acids Res. *1*, 1579–1586.

van Holde, K.E., Shaw, B.R., Lohr, D., Herman, T.M. & Kovacic, R.T. (1975) in *Proc. Tenth FEBS Meeting* (Bernardi, G. & Gros, F., eds.), vol. 38, pp. 57–72, North Holland/American Elsevier, Amsterdam.

van Holde, K.E., Allen, J.R., Cordon, J., Lohr, D., Tatchell, K. & Weischet, W.O. (1979) in *Chromatin Structure and Function* (Nicolini, C., ed.), Part B, pp. 389–411, Plenum Pub. Corp., N.Y.

van Holde, K.E., Allen, J.R., Tatchell, K., Weischet, W.O. & Lohr, D. (1980) Biophys. J. *32*, 271–282.

Varshavsky, A.J., Bakayev, V.V. & Georgiev, G.P. (1976) Nucleic Acids Res. *3*, 477–492.

Varshavsky, A.J., Nedospasov, S.A., Bakayev, V.V., Bakayeva, T.G. & Georgiev, G.P. (1977a) Nucleic Acids Res. *4*, 2725–2745.

Varshavsky, A.J., Bakayev, V.V., Nedospasov, S.A. & Georgiev, G.P. (1977b) Cold Spring Harbor Symp. Quant. Biol. *42*, 457–473.

Varshavsky, A., Levinger, L., Sundin, O., Barsoum, J., Ozkaynak, E., Swerdlow, P. & Finley, D. (1982) Cold Spring Harbor Symp. Quant. Biol. *47*, 511–528.

Vaury, C., Gilly, C., Alix, D. & Lawrence, J.J. (1983) Biochem. Biophys. Res. Comm. *110*, 811–818.

Vavra, K.J., Allis, C.D. & Gorovsky, M.A. (1982) J. Biol. Chem. *257*, 2591–2598.

Vidali, G., Boffa, L.C. & Allfrey, V.G. (1972) J. Biol. Chem. *247*, 7365–7373.

Vidali, G., Boffa, L.C. & Allfrey, V.G. (1977) Cell *12*, 409–415.

Vidali, G., Boffa, L.C., Bradbury, E.M. & Allfrey, V.G. (1978) Proc. Natl. Acad. Sci. USA *75*, 2239–2243.

Villasante, A., Corces, V.G., Manso-Martinez, R. & Avila, J. (1981) Nucleic Acids Res. *9*, 895–908.

Villeponteau, B., Lundell, M. & Martinson, H. (1984) Cell *39*, 469–478.

Villeponteau, B., Pribyl, T.M., Grant, M.H. & Martinson, H.G. (1986) J. Biol. Chem. *261*, 10359–10365.

Vinograd, J. & Lebowitz, J. (1966) J. Gen. Physiol. *49*, 103–125.

Vinograd, J., Lebowitz, J., Radloff, R., Watson, R. & Laipis, P. (1965) Proc. Natl. Acad. Sci. USA *53*, 1104–1111.

Vinograd, J., Lebowitz, J. & Watson, R. (1968) J. Mol. Biol. *33*, 173–197.

Voeller, B.R. (1968) *The Chromosomal Theory of Inheritance*, Appleton-Century-Crofts, N.Y.

Vogel, T. & Singer, M. (1975a) J. Biol. Chem. *250*, 796–798.

Vogel, T. & Singer, M. (1975b) Proc. Natl. Acad. Sci. USA *72*, 2579–2583.

Vogelstein, B., Pardoll, D.M. & Coffey, D.S. (1980) Cell *22*, 79–85.

Vollenweider, H.J., James, A. & Szybalski, W. (1978) Proc. Natl. Acad. Sci. USA *75*, 710–714.

Vologodskii, A.V. & Frank-Kamenetskii, M.D. (1982) FEBS Lett. *143*, 257–260.

von Beroldingen, C.H., Reynolds, W.F., Milstein, L., Bazett-Jones, D.P. & Gottesfeld, J.M. (1984) Mol. Cell. Biochem. *62*, 97–108.

von Holt, C. & Brandt, W.F. (1977) in *Methods in Cell Biology* (Stein, G., Stein, J. & Kleinsmith, L.G., eds.), vol. 16, pp. 205–225, Academic Press, N.Y.

von Holt, C., Strickland, W.N., Brandt, W.F. & Strickland, M.S. (1979) FEBS Lett. *100*, 201–218.

von Holt, C., de Groot, P., Schwager, S. & Brandt, W.F. (1984) in *Histone Genes: Structure, Organization, and Regulation* (Stein, G., Stein, J. & Mazluff, W., eds.), pp. 65–105, John Wiley & Sons, N.Y.

Wachtel, E.J. & Sperling, R. (1979) Nucleic Acids Res. *6*, 139–152.

Wada, R.K. & Spear, B.B. (1980) Cell Different. *9*, 261–268.

Waldeck, W. & Saver, G. (1981) Biochemistry *20*, 4203–4209.

Walker, J.M. (1982) in *The HMG Chromosomal Proteins* (Johns, E.W., ed.), pp. 69–87, Academic Press, N.Y.

Walker, J.M., Goodwin, G.H. & Johns, E.W. (1978) FEBS Lett. *90*, 327–330.

Walker, J.M., Brown, E., Goodwin, G.H., Stearn, C. & Johns, E.W. (1980) FEBS Lett. *113*, 253–257.

Wallace, R.B., Dube, S.K. & Bonner, J. (1977) Science *198*, 1166–1168.

Wallis, J.W., Hereford, L. & Grunstein, M. (1980) Cell *22*, 799–805.

Wallis, J.W., Rykowski, M. & Grunstein, M. (1983) Cell *35*, 711–719.

Walmsley, M.E. & Davies, H.G. (1975) J. Cell Sci. *17*, 113–139.

Walton, G.M. & Gill, G.N. (1981) Biochim. Biophys. Acta *656*, 155–159.

Wang, A.H.-J., Quigley, G.J., Kolpack, F.J., Crawford, J.L., van Boom, J.H., van der Marel, G. & Rich, A. (1979) Nature (London) *282*, 680–686.

Wang, J.C. (1974) J. Mol. Biol. *87*, 797–816.

Wang, J.C. (1979) Proc. Natl. Acad. Sci. USA *76*, 200–203.

Wang, J.N.C. & Hogan, M. (1985) J. Biol. Chem. *260*, 8194–8202.

Wang, J.N.C., Hogan, M. & Austin, R.H. (1982) Proc. Natl. Acad. Sci. USA *79*, 5896–5900.

Wang, T.Y. (1968) Arch. Biochem. Biophys. *127*, 235–240.

Waqar, M.A. & Huberman, J.A. (1975) Biochim. Biophys. Acta *383*, 410–420.

Wassarman, P.M. & Mrozak, S.C. (1981) Devel. Biol. *84*, 364–371.

Wasylyk, B. & Chambon, P. (1979) Eur. J. Biochem. *98*, 317–327.

Watanabe, F. (1984) FEBS Lett. *170*, 19–22.

Waterborg, J.H. & Matthews, H.R. (1982) Exp. Cell Res. *138*, 462–466.

Waterborg, J.H. & Matthews, H.R. (1983a) FEBS Lett. *162*, 416–419.

Waterborg, J.H. & Matthews, H.R. (1983b) Cell Biophys. *5*, 265–279.

Waterborg, J.H. & Matthews, H.R. (1984) Eur. J. Biochem. *142*, 329–335.

Waterborg, J.H., Fried, S.R. & Matthews, H.R. (1983) Eur. J. Biochem. *136*, 245–252.

Watson, D.C., Peters, E.H. & Dixon, G.H. (1977) Eur. J. Biochem. *74*, 53–60.

Watson, D.C., Wong, N.C.W., Dixon, G.H. (1979) Eur. J. Biochem. *95*, 193–202.

Watson, D.K. & Moudrianakis, E.N. (1982) Biochemistry *21*, 248–256.

Watson, J.D. & Crick, F.H.C. (1953) Nature (London) *171*, 737–738.

Webb-Walker, B., Lothstein, L., Baker, C.L. & LeStourgeon, W.M. (1980) Nucleic Acids Res. *8*, 3639–3657.

Weber, J.L. & Cole, R.D. (1982) J. Biol. Chem. *257*, 11784–11790.

Weber, S. & Isenberg, I. (1980) Biochemistry *19*, 2236–2240.

Weigand, R.C. & Brutlag, D.L. (1981) J. Biol. Chem. *256*, 4578–4583.

Weinberg, E.S., Hendricks, M.B., Hemminki, K., Kuwabara, P.E. & Farrelly, L.A. (1983) Develop. Biol. *98*, 117–129.

Weintraub, H. (1973) Cold Spring Harbor Symp. Quant. Biol. *38*, 247–256.

Weintraub, H. (1975) in *Results and Problems in Cell Differentiation* (Reinert, J. & Holzer, H., eds.), vol. 7, pp. 27–42, Springer-Verlag, N.Y.

Weintraub, H. (1976) Cell *9*, 419–422.

Weintraub, H. (1978) Nucleic Acids Res. *5*, 1179–1188.

Weintraub, H. (1979) Nucleic Acids Res. *7*, 781–792.

Weintraub, H. (1984) Cell *38*, 17–27.

Weintraub, H. (1985) Cell *42*, 705–711.

Weintraub, H. & Groudine, M. (1976) Science *193*, 848–856.

Weintraub, H. & van Lente, F. (1974) Proc. Natl. Acad. Sci. USA *71*, 4249–4253.

Weintraub, H., Larson, A. & Groudine, M. (1981a) Cell *24*, 333–344.

Weintraub, H., Weisbrod, S., Larsen, A. & Groudine, M. (1981b) in *Organization and Expression of Globin Genes* (Stamatoyannopoulos, G., Nienhuis, A.W., eds.), pp. 175–190, A.R. Liss, N.Y.

Weiss, E., Ruhlmann, C. & Oudet, P. (1986) Nucleic Acids Res. *14*, 2045–2058.

Weisbrod, S. (1982) Nucleic Acids Res. *10*, 2017–2042.

Weisbrod, S. & Weintraub, H. (1979) Proc. Natl. Acad. Sci. USA *76*, 630–634.

Weisbrod, S. & Weintraub, H. (1981) Cell *23*, 391–400.

Weisbrod, S., Groudine, M. & Weintraub, H. (1980) Cell *19*, 289–301.

Weischet, W.O. (1979) Nucleic Acids Res. *7*, 291–304.

Weischet, W.O. & van Holde, K.E. (1980) Nucleic Acids Res. *8*, 3743–3755.

Weischet, W.O., Tatchell, K., van Holde, K.E. & Klump, H. (1978) Nucleic Acids Res. *5*, 139–160.

Weischet, W.O., Allen, J.R., Riedel, G. & van Holde, K.E. (1979) Nucleic Acids Res. *6*, 1843–1862.

Weischet, W.O., Glotov, B.O., Schnell, H. & Zachau, H.G. (1982) Nucleic Acids Res. *10*, 3627–3645.

Weischet, W.O., Glotov, B.O. & Zachau, H.G. (1983a) Nucleic Acids Res. *11*, 3593–3612.

Weischet, W.O., Glotov, B.O. & Zachau, H.G. (1983b) Nucleic Acids Res. *11*, 3613–3630.

Wenkert, D. & Allis, C.D. (1984) J. Cell Biol. *98*, 2107–2117.

West, M.H.P. & Bonner, W.M. (1980) Nucleic Acids Res. *8*, 4671–4680.

West, M.H.P. & Bonner, W.M. (1984) Comp. Biochem. Physiol. *76B*, 455–464.

Whatley, S.A., Hall, C. & Lim, L. (1981) Biochem. J. *196*, 115–119.

Whitlock, J.P. (1979) J. Biol. Chem. *254*, 5684–5689.

Whitlock, J.P. & Simpson, R.T. (1976a) Nucleic Acids Res. *3*, 2255–2266.

Whitlock, J.P. & Simpson, R.T. (1976b) Biochemistry *15*, 3307–3314.

Whitlock, J.P. & Simpson, R.T. (1977) J. Biol. Chem. *252*, 6516–6520.

Whitlock, J.P. & Stein, A. (1978) J. Biol. Chem. *253*, 3857–3861.

Whitlock, J.P., Rushizky, G.W. & Simpson, R.T. (1977) J. Biol. Chem. *252*, 3003–3006.

Whitlock, J.P., Augustine, R. & Schulman, H. (1980) Nature (London) *287*, 74–76.

Whitlock, J.P., Galeazzi, D. & Schulman, H. (1983) J. Biol. Chem. *258*, 1299–1304.

Wickett, R.R., Li, H.J. & Isenberg, I. (1972) Biochemistry *11*, 2952–2957.

Widmer, R.M., Lucchini, R., Lezzi, M., Meyer, B., Sogo, J.M., Edstrom, J.E. & Koller, T. (1984) EMBO J. *3*, 1635–1641.

Widom, J. & Klug, A. (1985) Cell *43*, 207–213.

Widom, J., Finch, J.T. & Thomas, J.O. (1985) EMBO J. *4*, 3189–3194.

Wijns, L., Nieuwenhuysen, P., Muyldermans, S. & Clauwaert, J. (1982) Arch. Int. Phys. *90*, BP10–BP11.

Wildeman, A.G., Zenke, M., Schatz, C., Wintzerith, M., Grundstrom, T., Matthes, H., Takahachi, K. & Chambon, P. (1986) Mol. Cell. Biol. *6*, 2098–2105.

Wilhelm, M.L. & Wilhelm, F.X. (1980) Biochemistry *19*, 4327–4331.

Wilhelm, M.L. & Wilhelm, F.X. (1984) FEBS Lett. *168*, 249–254.

Wilhelm, M.L., Mazen, A. & Wilhelm, F.X. (1977) FEBS Lett. *79*, 404–408.

Wilhelm, M.L., Wilhelm, F.X., Toublan, B. & Jalouzot, R. (1982) FEBS Lett. *150*, 439–444.

Wilkins, M.H.F. (1956) Cold Spring Harbor Symp. Quant. Biol. *21*, 75–88.

Wilkins, M.H.F., Zubay, G. & Wilson, H.R. (1959) J. Mol. Biol. *1*, 179–185.

Wilkinson, D.J., Shinde, B.G. & Hohmann, P. (1982) J. Biol. Chem. *257*, 1247–1252.

Wilkinson, K.D., Urban, M.K. & Haas, A.L. (1980) J. Biol. Chem. *255*, 7529–7532.

Williams, R.E. (1976) Science *192*, 473–474.

Williams, S.P., Athey, B.D., Muglia, L.J., Schappe, R.S., Gough, A.H. & Langmore, J.P. (1986) Biophys. J. *49*, 233–248.

Williamson, R. (1970) J. Mol. Biol. *51*, 157–168.

Wilson, A.C., Carlson, S.S. & White, T.J. (1977) Ann. Rev. Biochem. *46*, 573–639.

Wilson, E.B. (1896) *The Cell in Development and Inheritance*, MacMillan, N.Y.

Wing, R., Drew, H., Tanako, T., Broka, C., Tanaka, S., Itakura, K. & Dickerson, R.E. (1980) Nature (London) *287*, 755–758.

Winkfein, R.J., Connor, W., Mezquita, J. & Dixon, G.H. (1985) J. Mol. Evol. *22*, 1–19.

Wittig, B. & Wittig, S. (1979) Cell *18*, 1173–1183.

Wittig, S. & Wittig, B. (1982) Nature (London) *297*, 31–38.

Wong, M. & Smulson, M. (1984) Biochemistry *23*, 3726–3730.

Wong, M., Kanai, Y., Miwa, M., Bustin, M. & Smulson, M. (1983a) Proc. Natl. Acad. Sci. USA *80*, 205–209.

Wong, M., Miwa, M., Sugimura, T. & Smulson, M. (1983b) Biochemistry *22*, 2384–2389.

Wong, M., Allan, J. & Smulson, M. (1984) J. Biol. Chem. *259*, 7963–7969.

Wong, Y.C., O'Connell, P., Rosbash, M. & Elgin, S.C.R. (1981) Nucleic Acids Res. *9*, 6749–6762.

Wood, W.I. & Felsenfeld, G. (1982) J. Biol. Chem. *257*, 7730–7736.

Woodcock, C.L.F. (1973) J. Cell. Biol. *59*, 368a.

Woodcock, C.L.F. & Frado, L.-L.Y. (1977) Cold Spring Harbor Symp. Quant. Biol. *42*, 43–55.

Woodcock, C.L.F., Maguire, D.I. & Stanchfield, J.E. (1974) J. Cell. Biol. *63*, 377a.

Woodcock, C.L.F., Frado, L.-L.Y., Hatch, C.L. & Ricciardiello, L. (1976) Chromosoma *58*, 33–39.

Woodcock, C.L.F., Frado, L.-L.Y., Rattner, J.B. (1984) J. Cell. Biol. *99*, 42–52.

Woodland, H.R. (1979) Develop. Biol. *68*, 360–370.

Woodland, H.R. & Adamson, E.D. (1977) Develop. Biol. *57*, 118–135.

Wooley, J.C. & Langmore, J.P. (1977) in *Molecular Human Cytogenetics* (Sparks, R.S., Comings, D.E. & Fox, C.F., eds.), pp. 41–51, Academic Press, N.Y.

Worcel, A. & Burgi, E. (1972) J. Mol. Biol. *71*, 127–147.

Worcel, A., Han, S. & Wong, M.L. (1978) Cell *15*, 969–977.

Worcel, A., Strogatz, S. & Riley, D. (1981) Proc. Natl. Acad. Sci. USA *78*, 1461–1465.

Worcel, A., Gargiulo, G., Jessee, B., Udvardy, A., Louis, C. & Schedl, P. (1983) Nucleic Acids Res. *11*, 421–439.

Workman, J.L. & Langmore, J.P. (1985a) Biochemistry *24*, 4731–4738.

Workman, J.L. & Langmore, J.P. (1985b) Biochemistry *24*, 7486–7497.

Woudt, L.P., Pastink, A., Kempers-Veenstra, A.E., Jansen, A.E.M., Mager, W.H. & Planta, R.J. (1983) Nucleic Acids Res. *11*, 5347–5360.

Wouters, D., Sautiere, P. & Biserte, G. (1978) Eur. J. Biochem. *90*, 231–239.

Wouters-Tyrou, D., Sautiere, P. & Biserte, G. (1976) FEBS Lett. *65*, 225–228.

Wouters-Tyrou, D., Martin-Ponthieu, A., Briand, G., Sautiere, P. & Biserte, G. (1982) Eur. J. Biochem. *124*, 489–498.

Wray, V.P., Elgin, S.C.R. & Wray, W. (1980) Nucleic Acids Res. *8*, 4155–4163.

Wray, W. & Stubblefield, T.E. (1970) Exp. Cell Res. *59*, 469–478.

Wray, W., Mace, M., Daskel, Y. & Stubblefield, E. (1977) Cold Spring Harbor Symp. Quant. Biol. *42*, 361–365.

Wrinch, D.M. (1936) Protoplasma *25*, 550–569.

Wu, C. (1980) Nature (London) *286*, 854–860.

Wu, C. (1982) UCLA Symp. Mol. Cell. Biol. *26*, 147–156.

Wu, C. (1984a) Nature (London) *309*, 229–234.

Wu, C. (1984b) Nature (London) *311*, 81–84.

Wu, C. & Gilbert, W. (1981) Proc. Natl. Acad. Sci. USA *78*, 1577–1580.

Wu, C., Bingham, P.M., Livak, K.J., Holmgren, R. & Elgin, S.C.R. (1979a) Cell *16*, 797–806.

Wu, C., Wong, Y.C. & Elgin, S.C.R. (1979b) Cell *16*, 807–814.

Wu, H.-M., Dattagupta, N., Hogan, M. & Crothers, D.M. (1979) Biochemistry *18*, 3960–3965.

Wu, H.-M., Dattagupta, N., Hogan, M. & Crothers, D.M. (1980) Biochemistry *19*, 626–634.

Wu, H.M., Dattagupta, N. & Crothers, D.M. (1981) Proc. Natl. Acad. Sci. USA *78*, 6808–6811.

Wu, K.C., Strauss, F. & Varshavsky, A. (1983) J. Mol. Biol. *170*, 93–117.

Wu, R.S. & Bonner, W.M. (1981) Cell *27*, 321–330.

Wu, R.S., Kohn, K.W. & Bonner, W.M. (1981) J. Biol. Chem. *256*, 5916–5920.

Wu, R.S., Nishioka, D. & Bonner, W.M. (1982) J. Cell Biol. *93*, 426–431.

Wuilmart, C. & Wyns, L. (1977) J. Theo. Biol. *65*, 231–252.

Wurtz, T. & Fakan, S. (1983) Biol. Cell *48*, 109–120.

Yabuki, H., Dattagupta, N. & Crothers, D.M. (1982) Biochemistry *21*, 5015–5020.

Yager, T. & van Holde, K.E. (1984) J. Biol. Chem. *259*, 4212–4222.

Yaguchi, M., Roy, C., Seligy, V.L. (1979) Biochem. Biophys. Res. Comm. *90*, 1400–1405.

Yakura, K. & Tanifuji, S. (1980) Biochim. Biophys. Acta *609*, 448–455.

Yamamoto, K. (1985) Ann. Rev. Genet. *19*, 209–252.

Yang, L., Rowe, T.C., Nelson, E.M. & Liu, L.F. (1985) Cell *41*, 127–132.

Yasuda, H., Logan, K.A. & Bradbury, E.M. (1984) FEBS Lett. *166*, 263–266.

Yoeman, L.C., Olson, W.O., Sugano, N., Jordan, J.J., Taylor, C.W., Starbuck, W.C. & Bush, H. (1972) J. Biol. Chem. *247*, 6018–6023.

Yoshida, M. & Shimura, K. (1984) J. Biochem. *95*, 117–124.

Yoshihara, K., Hashida, T., Yoshihara, H., Tanaka, Y. & Ohgushi, H. (1977) Biochem. Biophys. Res. Comm. *78*, 1281–1288.

Yoshihara, K., Hashida, T., Tanaka, Y., Ohgushi, H., Yoshihara, H. & Kamiya, T. (1978) J. Biol. Chem. *253*, 6459–6466.

Yoshihara, K., Hashida, T., Tanaka, Y., Matsunami, N., Yamaguchi, A. & Kamiya, T. (1981) J. Biol. Chem. *256*, 3471–3478.

Young, D. & Carroll, D. (1983) Mol. Cell. Biol. *3*, 720–730.

Yu, J. & Smith, R.D. (1985) J. Biol. Chem. *260*, 3035–3040.

Yu, S.H. & Spring, T.G. (1977) Biochim. Biophys. Acta *492*, 20–28.

Yu, S.S., Li, H.J., Goodwin, G.H. & Johns, E.W. (1977) Eur. J. Biochem. *78*, 497–502.

Yukioka, M., Ukai, Y., Hasuma, T. & Inoue, A. (1978) FEBS Lett. *86*, 85–88.

Yukioka, M., Omori, K., Okai, Y. & Inoue, A. (1979) FEBS Lett. *104*, 169–172.

Yukioka, M., Sasaki, S., Henmi, S., Matsuo, M., Hatayama, T. & Inoue, A. (1983) FEBS Lett. *158*, 281–284.

Zacharias, E. (1881) Bot. Zeitung *39*, 169–176.

Zacharias, E. (1893a) Ber. d. Bot. Ges. *11*, 188–195.

Zacharias, E. (1893b) Ber. d. Bot. Ges. *11*, 293–307.

Zalenskaya, I.A., Pospelov, V.A., Zalensky, A.O. & Vorobev, V.I. (1981) Nucleic Acids Res. *9*, 473–487.

Zama, M., Olins, D.E., Prescott, B. & Thomas, G.J. (1978a) Nucleic Acids Res. *5*, 3881–3897.

Zama, M., Olins, D.E., Wilkinson, E. & Olins, A.L. (1978b) Biochem. Biophys. Res. Comm. *85*, 1446–1452.

Zama, M., Mita, K. & Ichimura, S. (1984) Biochim. Biophys. Acta *783*, 100–104.

Zaret, K.S. & Yamamoto, K.R. (1984) Cell *38*, 29–38.

Zayetz, V.M., Bavykin, S.G., Karpos, V.L. & Mirzabekov, A.D. (1981) Nucleic Acids Res. *9*, 1053–1068.

Zeilig, C.E., Langan, T.A. & Glass, D.B. (1981) J. Biol. Chem. *256*, 994–1001.

Zentgraf, H. & Franke, W.W. (1984) J. Cell. Biol. *99*, 272–286.

Zentgraf, H., Muller, U. & Franke, W. (1980a) Eur. J. Cell Biol. *20*, 254–264.

Zentgraf, H., Muller, U. & Franke, W. (1980b) Eur. J. Cell Biol. *23*, 171–188.

Zhang, X.-Y. & Hörz, W. (1984) J. Mol. Biol. *176*, 105–129.

Zhang, X.-Y., Fittler, F. & Hörz, W. (1983) Nucleic Acids Res. *11*, 4287–4306.

Zheng, H.-Z. & Burkholder, G.D. (1982) Exp. Cell Res. *141*, 117–125.

Zhong, R., Roeder, R.G. & Heintz, N. (1983) Nucleic Acids Res. *11*, 7409–7425.

Zhu, J.D., Allan, M. & Paul, J. (1984) Nucleic Acids Res. *12*, 9191–9204.

Zhurkin, V.B. (1981) Nucleic Acids Res. *9*, 1963–1971.

Zimmerman, S.B. & Pheiffer, B.H. (1979a) Proc. Natl. Acad. Sci. USA *76*, 2703–2707.

Zimmerman, S.B. & Pheiffer, B.H. (1979b) J. Mol. Biol. *135*, 1023–1027.

Zubay, G. & Doty, P. (1959) J. Mol. Biol. *1*, 1–20.

Zweidler, A. (1980) Develop. Biochem. *15*, 47–56.

Zweidler, A. (1984) in *Histone Genes: Structure, Organization and Regulation* (Stein, G.S., Stein, J.L. & Marzluff, W.F., eds.), pp. 339–371, J. Wiley & Sons, N.Y.

Zweidler, A., Urban, M.K. & Goldman, P. (1978) Miami Winter Symposium *15*, 531.

Index